CHALLENGES IN PANCHROMATIC MODELLING
WITH NEXT GENERATION FACILITIES

IAU SYMPOSIUM 341

COVER ILLUSTRATION:

The cover illustration is an art work by H. Chihara who was inspired by the autumn leaves of forest in Minoh quasi-national park near the venue, Osaka University. The image of NGC3034 (M82) taken with the Subaru Telescope operated by the National Astronomical Observatory of Japan and its SED plot provided by F. Galliano are superimposed.

IAU SYMPOSIUM PROCEEDINGS SERIES

INTERNATIONAL ASTRONOMICAL UNION

UNION ASTRONOMIQUE INTERNATIONALE

International Astronomical Union

IAU

CHALLENGES IN PANCHROMATIC MODELLING WITH NEXT GENERATION FACILITIES

PROCEEDINGS OF THE 341st SYMPOSIUM OF THE INTERNATIONAL ASTRONOMICAL UNION HELD IN OSAKA, JAPAN 12–16 NOVEMBER, 2018

Edited by

MÉDÉRIC BOQUIEN
Universidad de Antofagasta, Chile

ELISABETA LUSSO
Durham University, United Kingdom

CARLOTTA GRUPPIONI
Astrophysics and Space Science Observatory - INAF, Italy

and

PATRICIA TISSERA
Universidad Andrés Bello, Chile

CAMBRIDGE
UNIVERSITY PRESS

CAMBRIDGE UNIVERSITY PRESS
University Printing House, Cambridge CB2 8BS, United Kingdom
1 Liberty Plaza, Floor 20, New York, NY 10006, USA
10 Stamford Road, Oakleigh, Melbourne 3166, Australia

First published 2020

Printed in the UK by Bell & Bain, Glasgow, UK

Typeset in System LaTeX 2ε

A catalogue record for this book is available from the British Library Library of Congress Cataloguing in Publication data

ISBN 9781108471473 hardback
ISSN 1743-9213

Table of Contents

Preface xi

Editors xiii

Conference Photograph xiv

Participants xv

A historical overview of galaxy surveys 1
Sadanori Okamura

Panchromatic study of the first galaxies with large ALMA programs 12
A. Faisst, M. Béthermin, P. Capak, P. Cassata, O. Le Fèvre, D. Schaerer, J. Silverman, L. Yan and the ALPINE *team*

The average FIR SED of proto-clusters at $z = 4$ 17
Mariko Kubo, Jun Toshikawa, Nobunari Kashikawa, Hisakazu Uchiyama and Kei Ito

The spectral energy distributions of active galactic nuclei 21
M. J. I. Brown, K. J. Duncan, H. Landt, M. Kirk, C. Ricci and N. Kamraj

Panchromatic SED fitting codes and modelling techniques 26
Maarten Baes

Evolutionary Population Synthesis model with binary stars – Yunnan-II model . . . 35
F. Zhang, Z. Han and L. Li

HELP project - a dreamed-of multiwavelength dataset for SED fitting: The influence of used models for the main physical properties of galaxies 39
Katarzyna Małek, Veronique Buat, Denis Burgarella, Yannick Roehlly, Raphael Shirley and the HELP team

Modelling the emission of passive galaxies at $z \sim 3$ 44
C. D'Eugenio, E. Daddi, R. Gobat, V. Strazzullo and S. Jin

Evolution histories of massive galaxies at $z \sim 2$ over the past 3 Gyr 50
T. Morishita, L. E. Abramson, T. Treu, G. B. Brammer, T. Jones, P. Kelly, M. Stiavelli, M. Trenti, B. Vulcani and X. Wang

Spatially resolved stellar mass buildup and quenching in massive disk galaxies over the last 10 Gyr revealed with spatially resolved SED fitting 55
Abdurro'uf and Masayuki Akiyama

Exploring the star formation histories of galaxies in different environments from MaNGA spectra 60
Maria Argudo-Fernández, Médéric Boquien, Shiyin Shen, Fangting Yuan, Jun Yin, Ruixiang Chang and Lei Hao

High-resolution radiation transfer modelling of barred galaxies 65
A. Nersesian, S. Verstocken, S. Viaene and M. Baes

Which attenuation curves for star-forming galaxies? 70
Véronique Buat, David Corre, Médéric Boquien and Katarzyna Małek

Dust attenuation on and off the galaxy Main Sequence at $z \geqslant 1$ 74
Annagrazia Puglisi

Spatially resolved dust-to-gas mass ratios in nearby galaxies 78
Basilio Solís-Castillo and Marcus Albrecht

A MUSE inquiry into the physical processes taking place within the Abell 2667 Brightest Cluster Galaxy 83
E. Iani, G. Rodighiero, J. Fritz, G. Cresci, C. Mancini, P. Tozzi, L. Rodríguez-Muñoz, P. Rosati, G. B. Caminha, S. Berta, P. Cassata, A. Concas, A. Enia, D. Fadda, A. Franceschini, A. Liu, A. Mercurio, L. Morselli, P. G. Pérez-González, P. Popesso, G. Sabatini and A. Zanella

Pushing the technical frontier: From overwhelmingly large data sets to machine learning 88
Viviana Acquaviva

Modeling with the crowd: Optimizing the human-machine partnership with Zooniverse 99
Hugh Dickinson, Lucy Fortson, Claudia Scarlata, Melanie Beck and Mike Walmsley

Deep learning for galaxy mergers in the galaxy main sequence 104
William J. Pearson, Lingyu Wang, James Trayford, Carlo E. Petrillo and Floris F. S. van der Tak

Automatic classification of sources in large astronomical catalogs 109
Agnieszka Pollo, Aleksandra Solarz, Małgorzata Siudek, Katarzyna Małek, Maciej Bilicki, Tomasz Krakowski, Tsutomu Takeuchi and the VIPERS team

Predicting the global far-infrared emission of galaxies 114
Wouter Dobbels and Maarten Baes

`Mentari`: A pipeline to model the galaxy SED using semi analytic models 119
Dian Triani, Darren Croton and Manodeep Sinha

Variations of the stellar Initial Mass Function in semi-analytic models: Implications for the mass assembly of galaxies in the GAEA model 124
Fabio Fontanot

Modeling the panchromatic emission of galaxies with `CIGALE` 129
M. Boquien, D. Burgarella, Y. Roehlly, V. Buat, L. Ciesla, D. Corre, A. K. Inoue and H. Salas

Going beyond galaxy ages with dense basis star formation history reconstruction 134
Kartheik G. Iyer and Eric Gawiser

A hierarchical bayesian dust SED model and its application to the nearby universe 138
Frédéric Galliano

Bayesian discrimination of the panchromatic spectral energy distribution modelings of galaxies 143
Yunkun Han, Zhanwen Han and Lulu Fan

Stellar population synthesis of galaxies with chemical evolution model 147
Shiyin Shen and Jun Yin

A new galaxy Spectral Energy Distribution model with the evolution of dust consistent with chemical evolution 152
Kazuki Y. Nishida, Tsutomu T. Takeuchi, Takuma Nagata, Ryosuke S. Asano and Akio K. Inoue

The Spitzer Extragalactic Representative Volume Survey - measuring photometric redshifts for ~4 million galaxies - challenges and ways forward 157
Janine Pforr

Probing the building blocks of galaxies: Sub-galactic scaling relations between X-ray luminosity, SFR and stellar mass 162
K. Kouroumpatzakis, A. Zezas, P. H. Sell, P. Bonfini, M. L. N. Ashby and S. P. Willner

Tips learned from panchromatic modeling of AGNs 167
Y. Sophia Dai (戴昱)

X-ray - Infrared relation of AGNs and search for highly obscured accretion in the *AKARI* NEP Field 172
Takamitsu Miyaji and AKARI *NEP Survey Team*

The Diagnostic power of radio spectra from star-forming galaxies 177
Eric J. Murphy

Challenges in modelling the rest-frame ultraviolet/optical spectra of galaxies at the high-redshift frontier 187
Erik Zackrisson and Anton Vikaeus

Systematic errors in dust mass fits: The role of dust opacity 196
Lapo Fanciullo, Francisca Kemper, Sundar Srinivasan, Peter Scicluna and James M. Simpson

Properties of LBGs with [OIII] detection at z > 3: The importance of including nebular emission data in SED fitting 201
Fang-Ting Yuan, Denis Burgarella, David Corre, Veronique Buat, Médéric Boquien and Shiyin Shen

Census of Lyα, [OIII]5007, Hα, and [CII]158μm line emission with ~1000 galaxies at $z = 4.9 - 7.0$ revealed with Subaru/HSC, Spitzer, and ALMA 206
Yuichi Harikane

Detections of far-infrared [OIII] and dust emission in a galaxy at $z = 8.312$: Early metal enrichment in the heart of the reionization era 211
Y. Tamura, K. Mawatari, T. Hashimoto, A. K. Inoue, E. Zackrissonm, L. Christensen, C. Binggeli, Y. Matsuda, H. Matsuo, T. T. Takeuchi, R. S. Asano, K. Sunaga, I. Shimizu, T. Okamoto, N. Yoshida, M. Lee, T. Shibuya, Y. Taniguchi, H. Umehata, B. Hatsukade, K. Kohno and K. Ota

Dust mass and dust production efficiencies on the redshift frontier 216
Hiroyuki Hirashita, Denis Burgarella and Rychard J. Bouwens

The onset of star formation 250 million years after the Big Bang 221
Takuya Hashimoto

Extreme variations in star formation activity in the first galaxies 226
Christian Binggeli, Erik Zackrisson, Xiangcheng Ma, Akio K. Inoue, Anton Vikaeus, Takuya Hashimoto, Ken Mawatari, Ikkoh Shimizu and Philip F. Hopkins

Subaru/HSC identifications of protocluster candidates at $z \sim 6-7$: Implications for cosmic reionization 231
Ryo Higuchi, Masami Ouchi, Yoshiaki Ono, Takatoshi Shibuya, Jun Toshikawa, Yuichi Harikane, Takashi Kojima, Yi-Kuan Chiang, Eiichi Egami, Nobunari Kashikawa, Roderik Overzier, Akira Konno, Akio K. Inoue, Kenji Hasegawa, Seiji Fujimoto, Tomotsugu Goto, Shogo Ishikawa, Kei Ito, Yutaka Komiyama and Masayuki Tanaka

Probing the ISM of HeII λ1640 emitters at $z = 2-4$ via MUSE 235
Themiya Nanayakkara, Jarle Brinchmann and The MUSE Collaboration

Panchromatic study of the first galaxies in cosmological simulations 240
Hidenobu Yajima, Shohei Arata, Makito Abe and Kentaro Nagamine

FIRSTLIGHT: Cosmological simulations of first galaxies at cosmic dawn 245
Daniel Ceverino

The distribution and physical properties of high-redshift [OIII] emitters in a cosmological hydrodynamics simulation 249
Kana Moriwaki

Pop III supernova feedback on the formation of the first galaxies 253
Li-Hsin Chen and Ke-Jung Chen

Chemical enrichment of Pop III supernovae in the first galaxies 257
Ke-Jung Chen

Radiative properties of the first galaxies: Rapid transition from blue to red 261
Shohei Arata, Hidenobu Yajima, Kentaro Nagamine, Yuexing Li and Sadegh Khochfar

Helium rich stars produce the UV upturn 264
R. De Propris, S. Ali, M. N. Bremer and S. Phillipps

Stellar and gas mass distributions for understanding the nature of spiral arms 266
Fumi Egusa, Erin Mentuch Cooper, Jin Koda and Junichi Baba

Panchromatic modeling of the extremely luminous dust-obscured quasars at the cosmic noon . . . 268
Lulu Fan, Yunkun Han and Kirsten K. Knudsen

The new fundamental plane dictating galaxy cluster evolution . . . 271
Yutaka Fujita, Keiichi Umetsu, Elena Rasia, Massimo Meneghetti, Megan Donahue, Elinor Medezinski, Nobuhiro Okabe, Marc Postman and Stefano Ettori

Radial resolved galaxy disks models based on semi-analytic models of galaxy formation . . . 273
Jian Fu

A comparison of star formation history between NGC 300 and M33 . . . 275
Xiaoyu Kang, Fenghui Zhang and Ruixiang Chang

The environment of large equivalent width Lyα emitters at $z \sim 3$ . . . 277
Arisa Kida, Satoshi Kikuta, Yuichi Matsuda, et al.

DEIMOS and MOSFIRE spectroscopy of star-forming galaxies in the AKARI NEP-Deep field . . . 279
Helen K. Kim, Matt Malkan, Nagisa Oi, Toshinobu Takagi, Denis Burgarella, Véronique Buat, Samir Salim, Chris Pearson and Hideo Matsuhara

Mid-infrared PAH emission from star-forming galaxies selected at 250 μm . . . 281
Seong Jin Kim and AKARI NEP Collaboration

First [N II] 122 μm line detection in a starburst pair at $z = 4.7$ . . . 283
Minju M. Lee, Tohru Nagao, Carlos De Breuck, et al.

Panchromatic Analysis for Nature of HIgh-z galaxies Tool (PANHIT) . . . 285
Ken Mawatari, Akio K. Inoue, Satoshi Yamanaka, Takuya Hashimoto and Yoichi Tamura

First galaxy SED: Contribution from pre-main-sequence stars . . . 287
Hiroto Mitani, Naoki Yoshida, Kazuyuki Omukai and Takashi Hosokawa

Direct collapse to SMBH seeds in cosmological halos with radiation transfer . . . 289
Kentaro Nagamine, Isaac Shlosman and Yang Luo

Optical properties of infrared-bright dust-obscured galaxies viewed with Subaru Hyper Suprime-Cam . . . 292
A. Noboriguchi, T. Nagao, Y. Toba, M. Niida, M. Kajisawa, M. Onoue, Y. Matsuoka, T. Yamashita, Y. Chang, T. Kawaguchi, Y. Komiyama, K. Nobuhara, Y. Terashima and Y. Ueda

Investigating the early phase of the galaxy evolution through high-z damped Lyα absorption systems . . . 294
Kazuyuki Ogura, Tohru Nagao, Masatoshi Imanishi, Nobunari Kashikawa, Yoshiaki Taniguchi, Masaru Kajisawa, Masakazu A. R. Kobayashi and Yoshiki Toba

Strong FeII emission in NLS1s: An unsolved mystery . . . 297
Swayamtrupta Panda, Katarzyna Małek, Marzena Śniegowska and Bożena Czerny

Modeling galaxy evolution at high-redshift in highly overdense and normal regions . . . 299
Raphael Sadoun, Emilio Romano-Díaz, Isaac Shlosman and Zheng Zheng

Spectral study of scattered light by interstellar dust grains . . . 302
Kei Sano

Caught in the web: A tale of filament galaxies . . . 304
Ankit Singh, Smriti Mahajan and Devika Shobhana

Galactic outflows in star-forming galaxies at $z \sim 6$ studied with deep UV spectra and ALMA emission line . . . 307
Yuma Sugahara, Masami Ouchi, Yuichi Harikane, Nicolas Bouché, Peter D. Mitchell and Jérémy Blaizot

ALMA detection of the [OIII] 88 μm line in a highly-magnified Lyman break galaxy at z = 6.1 . . . 309
Kaho Sunaga, Yoichi Tamura, Minju Lee, Ken Mawatari, Akio K. Inoue, Takuya Hashimoto, Hiroshi Matsuo and Akio Taniguchi

Dust evolution in galaxies at $z > 7$ . . . 312
Tsutomu T. Takeuchi, Ryosuke S. Asano, Sayaka Nagasaki, Takaya Nozawa, Yoichi Tamura, Ken Mawatari and Akio K. Inoue

X-Ray spectral model from clumpy torus and its application to circinus galaxy . . . 314
Atsushi Tanimoto

Propeties of H alpha emitters at $z \sim 2.3$: Derivation of H alpha luminosity from multi-band photometry . . . 316
Yasunori Terao, Lee Spitler and Kentaro Motohara

Probing the luminous and dark matter profiles in the inner regions of a group-scale lens at z = 0.6 . . . 318
Mônica Tergolina, Cristina Furlanetto and Marina Trevisan

Magneto-Hydrodynamic simulations on galaxy modeling . . . 320
Wei-Chen Wang and Ke-Jung Chen

Active and dust obscured star-forming galaxies at $z \sim 4$ probed with UV spectral slope beta . . . 323
Satoshi Yamanaka and Toru Yamada

Preface

How galaxies form and evolve across cosmic times is one of the fundamental questions in modern astronomy. Over the past decade, modeling the panchromatic emission of galaxies has become one of the key tools in measuring their properties. As new and next-generation facilities progressively open a new era in astronomy, we face new and specific challenges in this endeavor: LSST and Euclid will provide us with an avalanche of data; the advent of e-ROSITA and the preparation for Athena make it ever more pressing to include X-ray emission into the standard UV-to-IR panchromatic models while EVLA and LOFAR expand our view of galaxies far into the radio domain; JWST will observe the first galaxies with extreme stellar populations and in the meantime ALMA is already starting to provide us with remarkable dust and metal observations at high redshift. As we are pivoting into this new era, IAU symposium 341 took place at Ōsaka University Hall from November 12 to November 16, only the third IAU symposium being held in Japan over the past 21 years after the general assembly in Kyoto in 1997. Besides the support from IAU, it also benefited of the sponsorship of the *Science Council of Japan*, *Ōsaka Sangyo University*, the *Department of Earth and Space Science of Ōsaka University*, the *National Astronomical Observatory of Japan*, and the *Society for Promotion of Space Science*. The meeting was attended by 127 people (90 men and 37 women), 36 of whom were supported by travel grants obtained with the support of IAU and NAOJ.

The programme consisted of 8 invited talks, 65 regular talks and 44 posters divided over four themes.

1. "State-of-the-art panchromatic surveys and studies". Three major reviews were given. Sadanori Okamura (University of Tokyo, Japan) gave a historical overview of galaxy surveys. Stéphane Charlot (Institut d'Astrophysique de Paris, France) presented the physical ingredients for panchromatic modeling, with a particular emphasis on stellar populations. Finally Maarten Baes gave a presentation on panchromatic codes and modeling techniques.
2. "Pushing the technical frontier: from overwhelmingly large datasets to machine learning". An introductory review on the new technique of machine learning was given by Viviana Acquaviva (New York City College of Technology, USA)
3. "Pushing the wavelength frontier: extending models towards X-rays and radio". Two reviews focused on each end of the electromagnetic spectrum. First, Andrea Merloni (Max-Planck-Institut für Extraterrestrische Physik, Germany) extensively presented the different sources of X-ray emission in galaxies before presenting in detail the eROSITA plans to map the universe in the X-rays. Eric Murphy (National Radio Astronomy Observatory, USA) then focused his review on the modeling of the energetic processes powering radio continuum emission from galaxies.
4. "Pushing the redshift frontier: modeling the first galaxies". Two reviews were given. Erik Zackrisson (Uppsala University, Sweden) presented the modeling challenges in the rest-frame UV/optical at the high-redshift frontier, focusing in particular on population III stars. Toru Yamada (JAXA/ISAS, Japan) presented future missions that will be critical to shape the field of panchromatic modeling in the next decades.

In order to avoid possible unconscious biases based on career stage, gender, country of origin, affiliation, etc., abstracts were anonymised and graded independently by 15 of the 16 SOC members. Talk allocations were granted purely based on the final ranking until the scheduled was filled. This resulted in a selection rate of abstracts submitted by women almost identical to the proportion of women among applicants (29%) and while giving the opportunity to numerous more junior members of the community to present their investigations.

To encourage the interaction between attendants and poster presenters, posters were installed in the room reserved for refreshments during breaks. Poster presenters gave flash presentations during three poster sessions over the first two days of the meeting. In parallel, a small jury made of a few SOC/LOC members and invited speakers evaluated posters. This led to an award to 8 poster presenters (3 women and 5 men), with the top-ranked person, Dr. Minju Lee (Nagoya University/National Astronomical Observatory of Japan), being offered the opportunity to give a talk.

Denis Burgarella (Laboratory of Astrophysics of Marseille, France) and Michael Brown (Monash University, Australia) also presented the activities of IAU commission J1 to the attendants in an effort to raise awareness of the existence and gather what the community expects from this commission.

Editors

Médéric Boquien
Universidad de Antofagasta, Chile

Elisabeta Lusso
Durham University, United Kingdom

Carlotta Gruppioni
Astrophysics and Space Science Observatory - INAF, Italy

Patricia Tissera
Universidad Andrés Bello, Chile

Organising Committee

Scientific Organising Committee

SOC Chairs

Médéric Boquien	Universidad de Antofagasta, Chile

SOC Members

Véronique Buat	Aix-Marseille University, France
Laure Ciesla	Laboratory of Astrophysics of Marseille, France
Andrew Connolly	University of Washington USA
Daniel Dale	University of Wyoming, USA
Ilse De Looze	University College London, UK
Maud Galametz	CEA Saclay, France
Carlotta Gruppioni	Astrophysics and Space Science Observatory - INAF, Italy
Hiroyuki Hirashita	ASIAA, Taiwan
Akio Inoue	Osaka Sangyo University, Japan
Elisabeta Lusso	Durham University, UK
Kentaro Nagamine	Osaka University, Japan
Giulia Rodighiero	University of Padova, Italy
Daniel Schaerer	University of Geneva, Switzerland
Renske Smit	University of Cambridge, UK
Patricia Tissera	Universidad Andrés Bello, Chile

Local Organising Committee

LOC Chair

Akio Inoue	Osaka Sangyo University, Japan

LOC Members

Denis Burgarella	Laboratory of Astrophysics of Marseille, France
Takuya Hashimoto	Osaka Sangyo University, Japan
Ken Mawatari	Osaka Sangyo University, Japan
Tohru Nagao	Ehime University, Japan
Hiroshi Shibai	Osaka University, Japan

CONFERENCE PHOTOGRAPH

Participants

Name & Institution	Email
Abdurrouf Abdurrouf, Academia Sinica Institute of Astronomy and Astrophysics	abdurrouf@astr.tohoku.ac.jp
Viviana Acquaviva, Physics Department, CUNY NYC College of Technology	VAcquaviva@citytech.cuny.edu
Shohei Aoyama, Academia Sinica (IAA)	saoyama@asiaa.sinica.edu.tw
Shohei Arata, Osaka University	arata@astro-osaka.jp
Maria Argudo Fernández, Universidad de Antofagasta	maria.argudo@uantof.cl
Maarten Baes, Universiteit Gent	maarten.baes@ugent.be
Laia Barrufet, ESA	lbarrufet@sciops.esa.int
Andrew John Battisti, Australian National University	andrew.battisti@anu.edu.au
Christian Binggeli, Uppsala University	christian.binggeli@physics.uu.se
Mederic Boquien, Universidad de Antofagasta	mederic.boquien@uantof.cl
Fabio Bresolin, Institute for Astronomy, University of Hawaii	bresolin@ifa.hawaii.edu
Michael Brown, Monash University	Michael.Brown@monash.edu
Véronique Buat, Laboratoire d'Astrophysique de Marseille	veronique.buat@lam.fr
Denis Burgarella, Laboratoire d'Astrophysique de Marseille, Aix-Marseille Univ.	denis.burgarella@lam.fr
Gabriela Calistro-Rivera, Leiden Observatory	gcalistrorivera@gmail.com
Adam Christopher Carnall, Observatory Edinburgh	adamc@roe.ac.uk Royal
Daniel Ceverino, Heidelberg University	ceverino@uni-heidelberg.de
Stephane Charlot, CNRS / Institut d'Astrophysique de Paris	charlot@iap.fr
Li-Hsin Chen, National Taiwan University/Institut of Astronomy and Astrophysics, Academia Sinica	r06244001@ntu.edu.tw
Ke-Jung Chen, ASIAA	chenken1229@gmail.com
Benjamin A. Cook, Harvard-Smithsonian Center for Astrophysics	bacook17@gmail.com
Asantha Cooray, UC Irvine	acooray@uci.edu
Chiara D'Eugenio, CEA/Irfu, AIM Service d'Astrophysique, Université Paris 7	chiara.deugenio@cea.fr
Yu Sophia Dai, NAOC	daysophia@gmail.com
Roberto De Propris, FINCA, University of Turku	rodepr@utu.fi
Hugh John Dickinson, University of Minnesota	hdickins@umn.edu
Wouter Dobbels, Ghent University	wouter.dobbels@ugent.be
Fumi Egusa, Institute of Astronomy, University of Tokyo	fegusa@ioa.s.u-tokyo.ac.jp
Andreas Lukas, Caltech/IPAC	Faisst afaisst@ipac.caltech.edu
Lulu Fan, Shandong University, Weihai	llfan@sdu.edu.cn
Lapo Fanciullo, ASIAA	lfanciullo@asiaa.sinica.edu.tw
Fabio Fontanot, INAF - Trieste Observatory	fabio.fontanot@inaf.it
Jian Fu, Shanghai Astronomical Observatory	fujian@shao.ac.cn
Yutaka Fujita, Osaka University	fujita@astro-osaka.jp
Maud Galametz, CEA Astrophysics Department	maud.galametz@cea.fr
Frédéric Galliano, AIM	frederic.galliano@cea.fr
Xiaobo Gong, Yunnan Observatories, Chinese Academy of Sciences	gxbo@ynao.ac.cn
Tomotsugu Goto, National Tsing Hua University	tomo@gapp.nthu.edu.tw
Yanxin Guo, National Astronomical Observatories, Chinese Academy of Sciences	yxguo@bao.ac.cn
Yunkun Han, Yunnan Observatories, Chinese Academy of Sciences	hanyk@ynao.ac.cn
Yuichi Harikane, The University of Tokyo	hari@icrr.u-tokyo.ac.jp
Takuya Hashimoto, Osaka Sangyo University, NAOJ	thashimoto@est.osaka-sandai.ac.jp
Ryo Higuchi, ICRR (Institute for Cosmic Ray Research), the University of Tokyo	rhiguchi@icrr.u-tokyo.ac.jp
Hiroyuki Hirashita, Academia Sinica Institute of Astronomy and Astrophysics	hirashita@asiaa.sinica.edu.tw
Edoardo Iani, Università degli Studi di Padova	edoardo.iani@phd.unipd.it
Akio Inoue, Osaka Sangyo University,	akinoue@est.osaka-sandai.ac.jp
Kartheik G Iyer, Rutgers, the State University of New Jersey	kgi1@physics.rutgers.edu
Yipeng Jing, Shanghai Jiao Tong University, Department of Astronomy	ypjing@sjtu.edu.cn
Xiaoyu Kang, Yunnan Observatories, Chinese Academy of Sciences	kxyysl@ynao.ac.cn
Arisa Kida, Kwansei Gakuin University	arisa.kd@gmail.com
Seong-Jin Kim, National Tsing Hua University	seongini@gmail.com
Helen Kyung Kim, University of California, Los Angeles	hkim@astro.ucla.edu
Eunbin Kim, Astronomy & Space Science Institute	ebkim@kasi.re.kr Korea

Name & Institution	Email
Kotaro Kohno, The University of Tokyo	kkohno@ioa.s.u-tokyo.ac.jp
Konstantinos Kouroumpatzakis, FORTH	kkouroub@physics.uoc.gr
Mariko Kubo, NAOJ	mariko.kubo@nao.ac.jp
Minju Lee, Nagoya University / NAOJ	minju.lee@nao.ac.jp
Kai-yang Lin, Institute of Astronomy and Astrophysics, Academia Sinica (ASIAA)	kylin@asiaa.sinica.edu.tw
Lihwai Lin, ASIAA	lihwailin@asiaa.sinica.edu.tw
Elisabeta Lusso, Durham University, Physics Department	elisabeta.lusso@durham.ac.uk
Katarzyna Ewa Malek, National Centre for Nuclear Research	katarzyna.malek@ncbj.gov.pl
Yuichi Matsuda, NAOJ	yuichi.matsuda@nao.ac.jp
Ken Mawatari, University of Tokyo	mawatari@icrr.u-tokyo.ac.jp
Andrea Merloni, Max-Planck Institute fuer Extraterrestrische Physik (MPE)	am@mpe.mpg.de
Hiroto Mitani, Graduate School of Science, The University of Tokyo	mitani@utap.phys.s.u-tokyo.ac.jp
Takamitsu Miyaji, Unversidad Nacional Autónoma de México, Instituto de Astronomía sede Ensenada	miyaji@astro.unam.mx
Amanda Moffett, Vanderbilt University	amanda.moffett@vanderbilt.edu
Takahiro Morishita Space Telescope Science Institute	tmorishita@stsci.edu
Kana Moriwaki, The University of Tokyo	kana.moriwaki@utap.phys.s.u-tokyo.ac.jp
Kentaro Motohara, University of Tokyo	kmotohara@ioa.s.u-tokyo.ac.jp
Eric Murphy, National Radio Astronomy Observatory	emurphy@nrao.edu
Kentaro Nagamine, Osaka University	kn@astro-osaka.jp
Tohru Nagao, Ehime University	tohru@cosmos.ehime-u.ac.jp
Themiya Nanayakkara, Leiden Observatory	themiyananayakkara@gmail.com
Kazuki Y Nishida, Division of Particle and Astrophysical Science, Nagoya University	nishida.kazuki@nagoya-u.jp
Akatoki Noboriguchi, Ehime University	noboriguchi@cosmos.phys.sci.ehime-u.ac.jp
Shoji Ogawa, Kyoto University	ogawa@kusastro.kyoto-u.ac.jp
Kazuyuki Ogura, Bunkyo University	kazu8astro@gmail.com
Sadanori Okamura, University of Tokyo	sadanori.okamura@emp.u-tokyo.ac.jp
Swayamtrupta Panda, Center for Theoretical Physics - Polish Academy of Sciences	panda@cft.edu.pl
William James Pearson, SRON Netherlands Institute for Space Research	w.j.pearson@sron.nl
Janine Pforr, ESA/ESTEC	janine.pforr@esa.int
Agnieszka Pollo, Jagiellonian University and National Center for Nuclear Research	agnieszka.pollo@gmail.com
Gergö Popping, MPIA	popping@mpia.de
Annagrazia Puglisi, IRFU CEA-Saclay (France)	annagrazia.puglisi@cea.fr
Joel R, Harvard University	joel.leja@cfa.harvard.edu
Sadoun Raphael, Department of Earth & Space Science, Graduate School of Science, Osaka University	sadoun@astro-osaka.jp
Chang Hee Ree, Korea Astronomy and Space Science Institute	chr@kasi.re.kr
David Rosario, Centre for Extragalactic Astronomy, Durham University	david.rosario@durham.ac.uk
Brett Salmon, Telescope Science Institute	bsalmon@stsci.edu Space
Kei Sano, Kwansei Gakuin University	sano0410@kwansei.ac.jp
Daniel Schaerer, Geneva Observatory, University of Geneva	daniel.schaerer@unige.ch
Shiyin Shen, Astronomical Observatory, Chinese Academy of Sciences	ssy@shao.ac.cn Shanghai
Hiroshi Shibai, Osaka University	shibai@ess.sci.osaka-u.ac.jp
Ankit Singh, Indian Institute of Science Education and Research	ansingh16@gmail.com
Marzena Sniegowska, Center for Theoretical Physics PAS	marzena.sniegowska@gmail.com
Basilio Solis, Argelander Institute for Astronomy, University of Bonn	bsolis1984@gmail.com
Mimi Song, NASA Goddard Space Flight Center	mmsong.rustedfaith@gmail.com
Yihan Song, National Astronomical Observatories, Chinese Academy of Sciences	yhsong@bao.ac.cn
Yuma Sugahara, The University of Tokyo	sugayu@icrr.u-tokyo.ac.jp

Name & Institution	Email
Kaho Sunaga, Nagoya University	sunaga@a.phys.nagoya-u.ac.jp
Koyo Suzuki, Nagoya University	koyo@a.phys.nagoya-u.ac.jp
Nao Suzuki, Kavli IPMU, Univ of Tokyo	nao.suzuki@kavli-ipmu.jp
Tsutomu T. Takeuchi, Division of Particle and Astrophysical Science, Nagoya University	takeuchi.tsutomu@g.mbox.nagoya-u.ac.jp
Yoichi Tamura, Nagoya University	ytamura@nagoya-u.jp
Yoshiaki Taniguchi, Open University of Japan	tani@cosmos.phys.sci.ehime-u.ac.jp
Atsushi Tanimoto, Kyoto University	tanimoto@kusastro.kyoto-u.ac.jp
Dan Taranu, Princeton University	dtaranu@astro.princeton.edu
Yasunori Terao, Institute of Astronomy, The University of Tokyo	y_terao@ioa.s.u-tokyo.ac.jp
MÔnica Tergolina, Federal University of Rio Grande do Sul	monica.tergolina@ufrgs.br
Akihiko Tomita, Wakayama University	atomita@center.wakayama-u.ac.jp
Dian Triani, Swinburne University	dtriani@swin.edu.au
Yoshihiro Ueda, Kyoto University	ueda@kusastro.kyoto-u.ac.jp
Etsuko UJIKE, The Open University of Japan	e.ujike@ouj.ac.jp
Alexa Villaume, University of California Santa Cruz	avillaum@ucsc.edu
Wei-Chen Wang, Academia Sinica Institute of Astronomy and Astrophysics	weichenwang.network@gmail.com
Mengxin Wang, NAOC	beawmx@163.com
Tao Wang, University of Tokyo	taowang@ioa.s.u-tokyo.ac.jp
Hidenobu Yajima, University of Tsukuba	yajima@ccs.tsukuba.ac.jp
Toru Yamada, ISAS, JAXA	yamada@ir.isas.jaxa.jp
Satoshi Yamada, Kyoto University (Department of Astronomy)	styamada@kusastro.kyoto-u.ac.jp
Satoshi Yamanaka, Osaka Sangyo University	s.yamanaka@est.osaka-sandai.ac.jp
Takuji Yamashita, Center for Space and Cosmic Evolution, Ehime University	takuji@cosmos.phys.sci.ehime-u.ac.jp
Tova Yoast-Hull, Canadian Institute for Theoretical Astrophysics	yoasthull@cita.utoronto.ca
Fangting Yuan, Shanghai Astronomical Observatory	yuanft@shao.ac.cn
Erik Zackrisson, Uppsala University	erik.zackrisson@physics.uu.se
Fenghui Zhang, Yunnan Observatories, Chinese Academy of Sciences	zhangfh@ynao.ac.cn

Challenges in Panchromatic Modelling with Next Generation Facilities
Proceedings IAU Symposium No. 341, 2018
M. Boquien, E. Lusso, C. Gruppioni & P. Tissera, eds.

doi:10.1017/S1743921319004800

A historical overview of galaxy surveys

Sadanori Okamura

Executive Management Program, The University of Tokyo,
7-3-1 Hongo, Bunkyo-ku, Tokyo 113-0033 Japan
email: sadanori.okamura@emp.u-tokyo.ac.jp

Abstract. Astronomy has changed from data-starving science to data-flooding science about 20 years ago due to advances of observational technology in all the wavelength regimes of electromagnetic waves. This paper gives a historical overview of galaxy surveys. We start from the impact of the technology development. Then, old imaging surveys and redshift surveys based on photography, especially using Schmidt telescopes, in the era of data-starving science are described in some detail. Several features of modern surveys are given and two highlights obtained from the exploitation of modern galaxy surveys are introduced.

Keywords. galaxy survey, history, Schmidt surveys, redshift surveys, modern surveys

1. Technological Impact: From Naked Eye to CCD

Let us begin with the impact of survey technology development in terms of both telescope aperture and detectors. Messier (1771) published practically the first catalog of non-stellar objects, *Messier Catalogue*, in the era of visual observation. It included about 100 objects; mixture of nebulae, star clusters, and galaxies. He used a refracting telescope of 10 cm aperture. About 100 year later, Dreyer (1888) compiled *New General Catalogue of Nebulae and Clusters of Stars* (NGC) and *Index Catalogues* (IC) by supplementing the data to the *General Catalogue* (GC) by John Herschel. NGC and the two ICs included more than 10,000 non-stellar objects. Currently used version is the error-corrected *Revised NGC Catalog* (RNGC) by Steinicke (2018), and an image atlas of all the NGC/IC objects by Numazawa & Wakiya (2009) is available.

In order to see the impact of survey technology development, I take the Virgo cluster survey. Binggeli *et al.* (1984) carried out the survey of the Virgo cluster based on 67 large photographic plates taken with the 2.5-m du Pont Telescope. This Las Campanas survey revealed a large number of dwarf galaxies (Fig. 1, left), and enabled the first study of dwarf galaxies beyond the Local Group. The resulting catalog of 2096 galaxies is called the *Virgo Cluster Catalog* (VCC). The Next Generation Virgo Survey (NGVS; Ferrarese *et al.* 2012) is the CCD version of the Las Campanas Survey carried out with the 3.8-m Canada France Hawaii Telescope. In the NGVS image, VCC dwarf galaxies look like giant galaxies in the photograph of Las Campanas Survey (Fig. 1).

Figure 2 shows the limiting magnitudes and limiting surface brightnesses of various surveys, visual survey, the Palomar Sky Survey, the Las Campanas Survey, the NGVS, and the recent Subaru HSC survey, overlaid on the sequence of Virgo galaxies. Near the faint end of the Virgo sequence, plotted are the Local Group dwarf galaxies placed at the distance of the Virgo Cluster. The improvement of limiting magnitude from visual to CCD was 15 magnitudes, and now it reaches 19 magnitudes ($\sim$12 magnitudes in recent 30 years). Limiting surface brightness improved by more than 6 magnitudes.

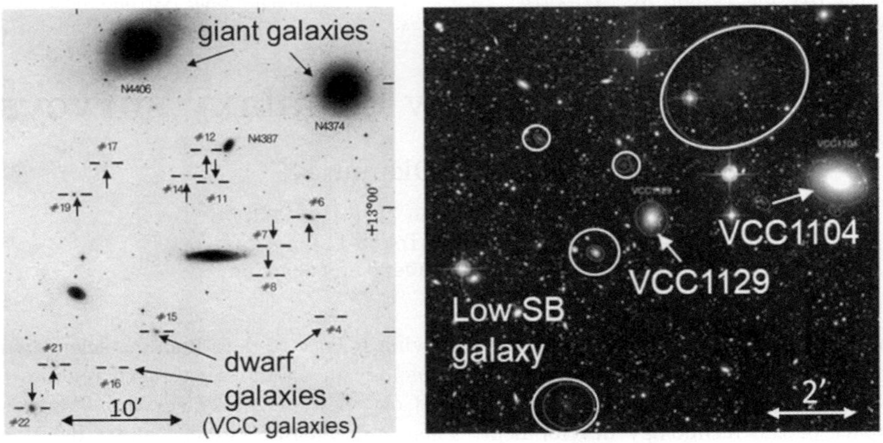

Figure 1. Images from the Las Campanas Survey (left, modified from PLATE 32 of Ichikawa *et al.* (1986) and the Next Generation Virgo Survey (right, modified from Fig. 15 of Ferrarese *et al.* 2012). Two dwarf galaxies in the right image look like giant galaxies in the left image.

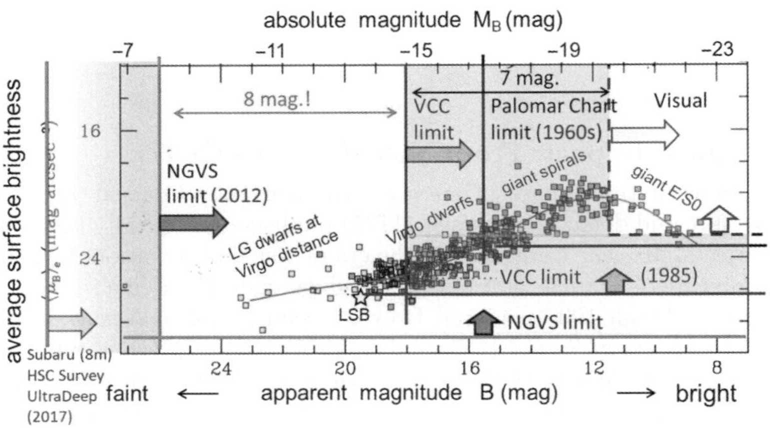

Figure 2. Improvement of limiting magnitude and limiting surface brightness. (modified from Fig. 16 of Ferrarese *et al.* 2012

2. Era of Photographic Imaging

2.1. *Early Surveys and Catalogs*

The Shapley-Ames Catalog is the first catalog dedicated to galaxies based on photographic observations. Objects are more than 1000 nearby galaxies in what we now call the Local Supercluster. Three critical factors of surveys, i.e., completeness, homogeneity, and wide sky coverage were already stated here. This is the first magnitude-limited catalog with the nominal limit of 13.2 magnitude. Large efforts were made to give calibrated magnitudes.

The Shapley-Ames catalog was succeeded by *A Revised Shapley-Ames Catalog* (Sandage & Tammann 1981). They made the original catalog complete by obtaining new redshifts, morphological types, and magnitudes. The RSA catalog is the first complete galaxy catalog in terms of redshifts (distances) and therefore it enabled the study of the structure of the Local Supercluster (Yahil *et al.* 1980) and the luminosity function of field galaxies (Binggeli *et al.* 1988).

Table 1. Major surveys with Schmidt telescopes.

Designation	Survey Region[1]	Emulsion+Filter	No. of plates	year
Northern Sky: Palomar Schmidt (Gal *et al.* 2004, AJ, 128, 3082)				
POSS-I(B)[2]	$-30° < \delta < +90°$	103a-O+none	936	1949-1956
POSS-I(R)	$-30° < \delta < +90°$	103a-E+plexiglass 2444	936	1949-1956
POSS-II(J)	$0° < \delta < +90°$	IIIa-J+GG395	897 (894[3])	1987-2000?
POSS-II(R)	$0° < \delta < +90°$	IIIa-F+RG610	897 (891[3])	1987-2000?
POSS-II(I)	$0° < \delta < +90°$	IV-N+RG9	897 (780[3])	1987-2000?
USNO(J)	$0° < \delta < +90°$	IIIa-J	894	1996-2000?
Southern Sky: UK Schmidt (http://www.roe.ac.uk/ifa/wfau/ukstu/)				
Designation	**Survey Region[1]**	**Emulsion+Filter**	**No. of plates**	**year**
SERC(J)[4]	$-90° < \delta < -20°$	IIIa-J+GG395	606	1974-1987
E(J)	$-15° < \delta < 0°$	IIIa-J+GG395	288	1979-1994
E(R)	$-15° < \delta < 0°$	IIIa-F+RG630	288	1984-1998
SERC(I)	$-90° < \delta < 0°$	IV-N+RG715	882	1978-2002
SERC-II(R)	$-90° < \delta < -20°$	IIIa-F+OG590	606	1985-1999
Hα	Galactic Plane	Tech-Pan4415+Hα659	233 (108[3])	1997-2001?
Hα	Magellanic Clouds	Tech-Pan4415+Hα659	40 (24[3])	1997-2001?
Southern Sky: ESO Schmidt				
ESO(B)	$-90° < \delta < -20°$	IIa-O+GG385	606	1973-1978
ESO(R)	$-90° < \delta < -20°$	IIIa-F+RG630	606	1973-1988?

Notes: [1]plate center, [2]POSS=Palomar Observatory Sky Survey, [3]as of April 1999, [4]SERC=Science and Engineering Research Council

Another important early survey is the *Shane-Wirtanen Counts* (Shane & Wirtanen 1954, 1967). They counted the number of galaxies brighter than 18.4 mag in the Northern sky on the photographic plates taken at the Lick Observatory. Counts were made in $10' \times 10'$ meshes but the data were published for $1° \times 1°$ meshes. The total number of galaxies amounted to 800,000. Their data were used for the analysis of two-point correlation function of galaxies, first by Totsuji & Kihara (1969), followed by Peebles (1973) and others.

With the Palomar Schmidt telescope built in 1949, we entered the era of Schmidt telescopes. It culminated in the 1970s when among others Palomar Schmidt in USA and its twins, UK Schmidt, in Australia, ESO Schmidt in Chile, Kiso Schmidt in Japan, and Tautenburg Schmidt in Germany were in operation. The limiting magnitude of imaging with $\sim$1-m Schmidt telescopes, which was about 16 magnitude for galaxies in visual inspection, was nearly the same as that of spectroscopy with 4-m class telescopes. This good match of limiting magnitude was the basis of ideal role sharing, i.e., Schmidt telescopes provide the source list and 4-m telescopes use it for follow-up spectroscopy.

A lot of surveys were carried out with Schmidt telescopes (Table 1). The Northern sky was covered with the Palomar Schmidt. The Southern sky was covered with the UK Schmidt and in part with the ESO Schmidt. Many galaxy catalogs were made, especially from the POSS-I survey, by the visual inspection of the survey plates or printed charts. Major catalogs include *Morphological Catalogue of Galaxies* (MCG) by Vorontsov-Velyaminov and Arkhipova (1962-68; Trudy Gosud. Astron. Sternberga), *Catalogue of Galaxies and Clusters of Galaxies* (CGCG) by Zwicky *et al.* (1961-1968; California Institute of Technology), *Uppsala General Catalogue of Galaxies* (UGC) by Nilson (1973; Uppsala Astronomiska Observatoriums Annaler), and *The ESO/Uppsala Survey of ESO(B) Atlas* by Lauberts (1982; European Southern Observatory). The total numbers of galaxies in these catalogs are 10-30 thousands. This may reflect the man-power limit of visual inspection.

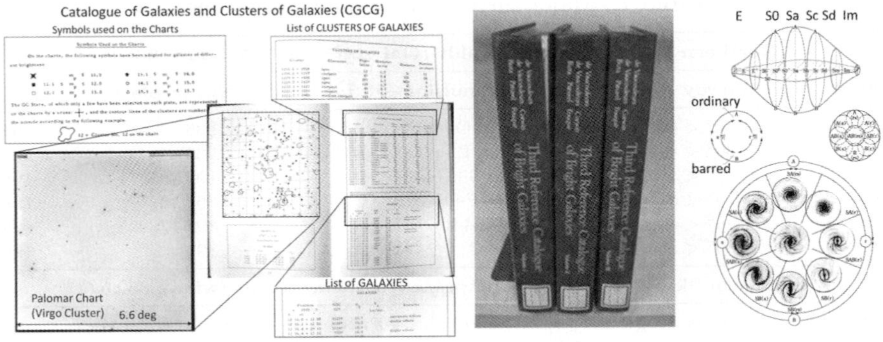

Figure 3. (left) A part of CGCG corresponding to the central part of the Virgo Cluster. (right) RC3 catalog together with the revised Hubble classification scheme shown in RC1.

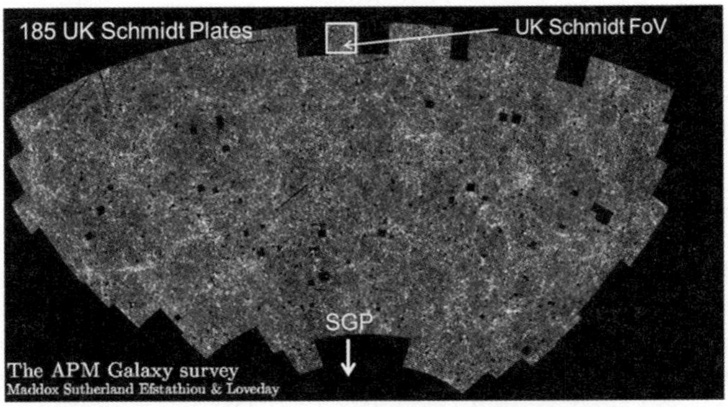

Figure 4. Sky distribution of galaxies in the APM survey (Maddox *et al.* 1990)

CGCG, consisting of 6 volumes, includes 28,000 galaxies complete down to Zwicky magnitude $m_z = 15.5$ mag and nearly 10,000 clusters of galaxies. UGC includes fewer galaxies (13,000) to a brighter magnitude limit ($m = 14.5$ mag), but has better homogeneity than CGCG. These catalogs had been an important and indispensable basis of observational cosmology until recently. Figure 3 (left) shows how CGCG looks like. For each of the POSS-I survey plates, a pictorial illustration and lists of objects are given.

There is another important catalog, the three editions of the *Reference Catalogue of Bright Galaxies* (RC). They were compiled by Gerard de Vaucouleurs *et al.* for over 30 years (RC1 in 1964 and RC2 in 1976; Univ Texas Press, and RC3 in 1991; Springer). They are not complete in any sense, but the most comprehensive compilation of then available data. Numbers of galaxies included in the catalog increased exponentially; 2599 in RC1, 4364 in RC2, and 23,024 in RC3. They remain one of the fundamental references, especially for the morphological type (Figure 3 right).

In the meantime, visual inspection was taken over by the computer analysis of digital data created by plate scanning machines. Several high performance machines include super COSMOS in Edinburgh, APM in Cambridge, APS in Minnesota, and PDS in many places. Digitized data of photographic plates resulted in several catalogs including *The APM galaxy survey* (Maddox *et al.* 1990) and *The Digitized Sky Survey* (DSS; see ref. 'Useful data centers'). Figure 4 shows the distribution of 2 million galaxies detected in the APM survey in 4300 square degrees covered with the UK Schmidt telescope. These APM galaxies are used as the input catalog of *2dF Galaxy Redshift Survey* described in Section 3.

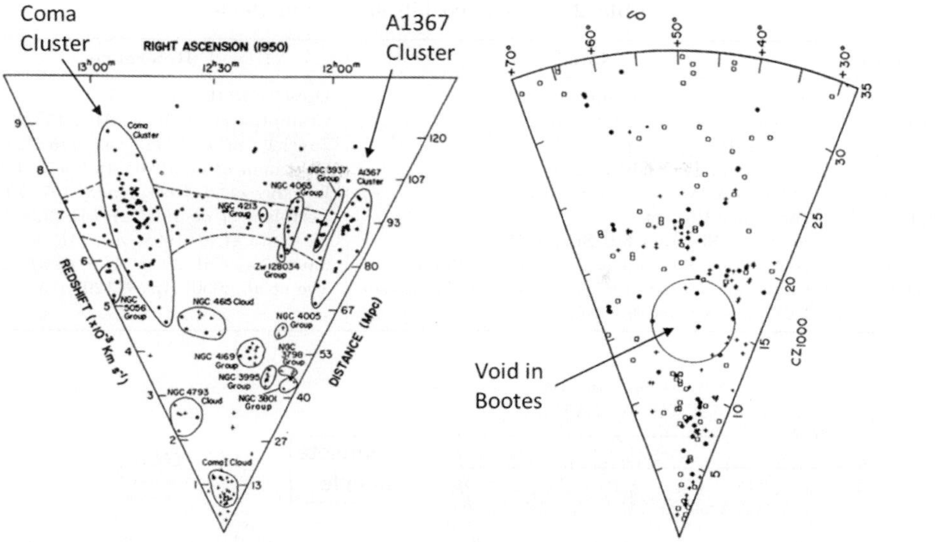

Figure 5. (left) First hint of supercluster (Gregory & Thompson 1978). (right) Void in Bootes (Kirshner *et al.* 1981).

3. Early Redshift Surveys

Gregory & Thompson (1978) carried out a redshift survey in the Coma/A1367 cluster region. Figure 5 (left) shows the wedge diagram of 238 CGCG galaxies brighter than 15 mag. The critical point is the completeness. They obtained 44 new redshifts to make this sample complete by combining their data with redshifts in the literature. The bridge-like structure connecting the two known clusters is visible. There is also an empty space, although the term *void* had not appeared yet. This is the first hint of the large scale structure. Kirshner *et al.* (1981) made a redshift survey of galaxies in the Bootes area and found a large volume devoid of galaxies. They called it void (Fig. 5 right).

A group of astronomers at the Center for Astrophysics, Cambridge, started an extensive wide-area galaxy redshift survey (*The CfA Survey*) in 1977 (Tonry & Davis 1979). The input source list came from the combination of CGCG and UGC. The first survey (CfA-I), which was completed in 1981, included 2401 galaxies brighter than 14.5 mag (Huchra *et al.* 1983). The second survey (CfA-II), which started in 1985 and completed in 1999, went 1 mag deeper (Huchra *et al.* 1999). The wedge diagram of CfA-I for a strip at $26.5° \leqslant \delta \leqslant 38.5°$ and $8^{\mathrm{h}} \leqslant \alpha \leqslant 17^{\mathrm{h}}$ showed only a hint of some structure. However, CfA-II revealed clearly the large scale structure for the same strip (Fig. 6).

CfA-II demonstrated the remarkable power of going to 1 mag deeper. Astonishingly, however, the structure was so large that its scale was not known due to the limited survey area. This motivated many wide or deep surveys as given in Table 2.

During the 1990s, some pencil-beam surveys reached 21-22 magnitude. However, the limiting magnitude of wide surveys did not change much because input source catalogs were based on Schmidt surveys.

4. Features of Modern Surveys

After around 2000, we entered the era of modern surveys, which are characterized by several features described below.

Table 2. Major redshift surveys in 1990s.

Abbrev.	Name	Reference
SPS	Supergalactic Plane Survey	Dressler 1991, *ApJS*, 75, 241
CFRS	Canada-France Redshift Survey	Crampton *et al.* 1995, *ApJ*, 455, 96
ORS	Optical Redshift Survey	Santiago *et al.* 1995, *ApJ*, 446, 457
LCRS	Las Campanas Redshift Survey	Shectman *et al.* 1996, *ApJ*, 470, 172
SAPM	Stromlo-APM Redshift Survey	Loveday *et al.* 1996, *ApJS*, 107, 201
ESP	ESO Slice Project	Vettolani *et al.* 1997, *A&Ap*, 325, 954
SSRS2	Southern Sky Redshift Survey II	da Costa *et al.* 1998, *AJ*, 116, 1
IRASz	IRAS Point Source Catalog Redshift Survey	Saunders *et al.* 2000, *ASP Conf.*, 201, 223
CNOC2	Canadian Network for Observational Cosmology Field Galaxy Redshift Survey	Yee *et al.* 2000, *ApJS*, 129, 475

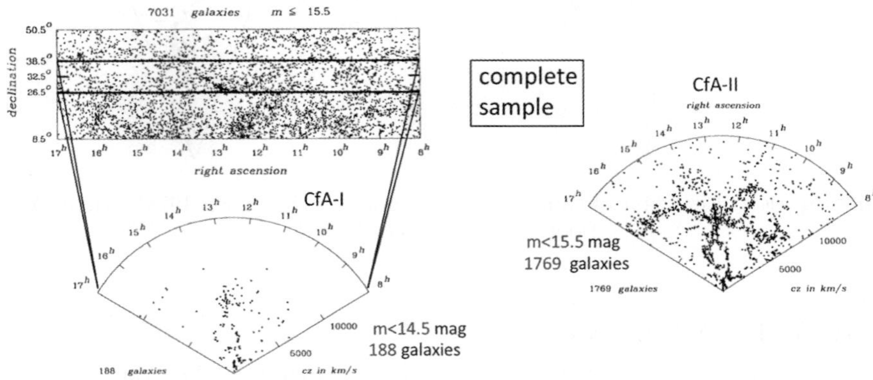

Figure 6. Wedge diagrams of CfA-I and CfA-II surveys. Going to 1 magnitude deeper resulted in a big difference in terms of the large scale structure. (Geller *et al.* 1987)

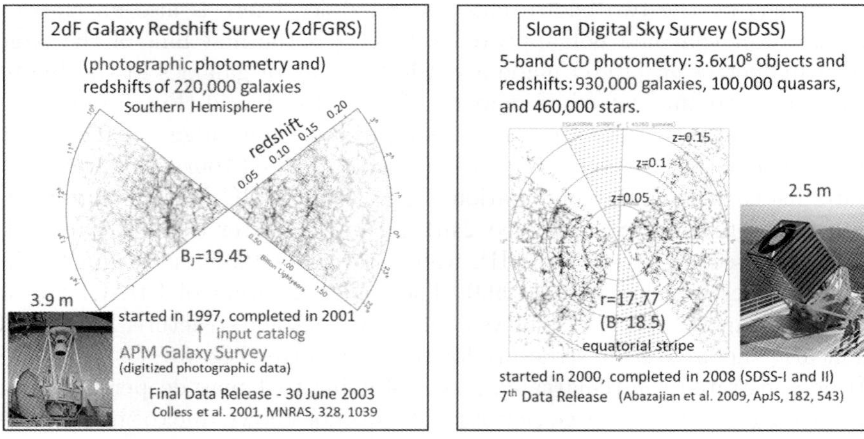

Figure 7. Forerunners of modern wide surveys. 2dFGRS (left) and SDSS (right).

4.1. *Parallel Survey and Multi-Object Spectrograph*

In order to go deeper in wide surveys, a deeper input source catalog was necessary. Such input catalog could only be made from imaging survey deeper than the Schmidt surveys. This means that new surveys should do both imaging and spectroscopy in parallel.

2dF Galaxy Redshift Survey (2dFGRS; Colless *et al.* 2001; Fig. 7 left), which was carried out with Anglo-Australian Telescope, may be regarded as the first modern wide survey, although the input catalog came from the *APM galaxy survey*, i.e., digitized photographic

Table 3. Major surveys of distant Universe.

Abbrev.	Name
	NOAO Deep Wide Field Survey (1999-)
GOODS	Great Observatories Origins Deep Survey (2004-)
SDF	Subaru Deep Field (2000-2003)
ECDFS	Extended Chandra Deep Field South (2000-)
COSMOS	Cosmic Evolution Survey (2003- ; 640 HST orbits)
UKIDSS	UK Infrared Deep Sky Survey-UDS (Ultradeep Survey) (2005-)
AEGIS	All-Wavelength Extended Groth Strip International Survey (1994-/2006)
CANDELS	Cosmic Assembly Near-Infrared Deep Extragalactic Legacy Survey (2010-2013; 902 HST orbits)
ALPINE	The ALMA Large Program to Investigate CII at Early Times (2017-2018)

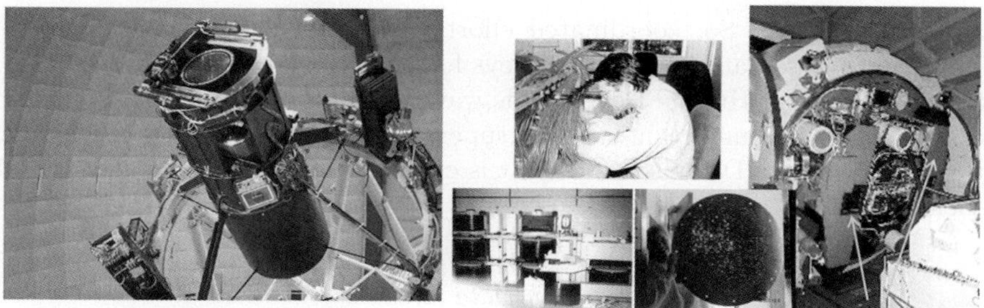

Figure 8. Multi-object spectrographs of 2dFGRS (left) and SDSS (right)

data. *Sloan Digital Sky Survey* (SDSS; York *et al.* 2000; Fig. 7 right) carried out 5-band CCD imaging and spectroscopy in parallel using the same telescope. SDSS is really the first modern survey.

The multi-object spectrograph has become a critical component of modern surveys. Figure 8 shows the multi-object spectrograph of 2dFGRS and SDSS. Both are fiber-fed spectrographs. 2dfGRS uses computer-controlled fiber positioner over 400 channels. On the other hand, SDSS uses a plug plate with 640 holes drilled at the locations of targets. Fibers are plugged into the holes manually.

4.2. *Wide Surveys over All the Wavelengths*

The next feature of modern surveys is the availability of wide surveys over almost all wavelengths. There is a useful review by Djorgovski *et al.* (2012). References to various surveys could also be found through search for available data of individual objects in large databases such as NED and SIMBAD (see ref. 'Useful data centers').

In the context of galaxy evolution studies described in Section 5.1, surveys in UV, Optical, and IR are particularly important to probe star formation, dust extinction, and stellar mass of galaxies at various redshifts.

4.3. *Coordinated Surveys of Distant Universe*

The Hubble Deep/Ultradeep Fields (HDF/HUDF) opened the window to distant universe. They are deep enough to reveal the galaxies in the early stage of evolution. However,

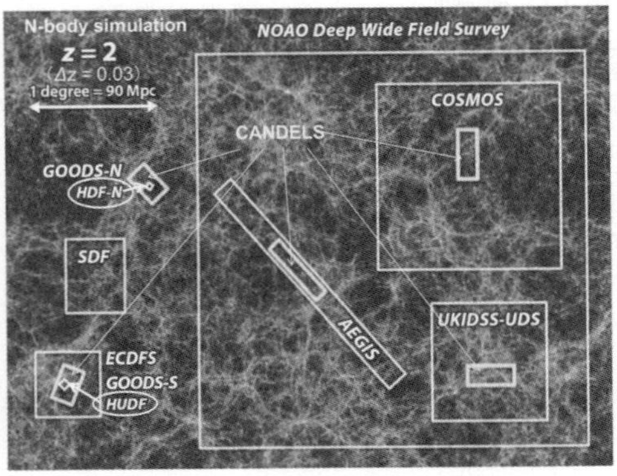

Figure 9. Relative sky coverage of surveys of distant Universe (Fig. 6 of Madau & Dickinson (2014), slightly modified)

they were very narrow. So, coordinated efforts began to cover wider area with deep multi-wavelength observations. Many surveys followed as shown in Table 3.

Wider coverage than Hubble Deep Fields was particularly important. Figure 9 shows relative sky coverage of important surveys, superposed on the simulated large-scale structure at $z = 2$ (Madau & Dickinson 2014). It is evident that small fields like HDF/HUDF could yield highly biased results due to cosmic variance.

4.4. *Big Data Archive*

Astronomy was a data-starving science until about 20 years ago. It has now changed to a data-flooding science due to the advances in observational technology. The Obama initiative in 2012 is known as a strong trigger for the big data movement. However, in 2010, two years earlier than this, *The Economist* carried a special report '*Data, data everywhere*' (ref. 'Economist'), where SDSS and Large Synoptic Survey Telescope (LSST) were quoted in the beginning.

In the astronomical community, there are a lot of useful data centers where data from surveys are archived. Such data centers have been making a large contribution to the nice research environment where astronomers all over the world can access and use various survey data easily (ref. 'Useful data centers').

5. Examples Exploiting Modern Galaxy Surveys

5.1. *Cosmic Star Formation History*

The left panel of Fig. 10, which is taken from Madau & Dickinson (2014), is a famous plot showing cosmic star formation history. The abscissa is the redshift and the ordinate is the star formation rate density (SFRD), i.e., the total star formation rate of galaxies divided by the comoving volume of the universe at that redshift. Star formation rate (SFR) is estimated from UV or NIR continuum depending upon redshifts of galaxies. For dusty galaxies, SFR is estimated from FIR luminosity. The data points here came from 13 papers based on various galaxy surveys described in Section 4.3.

The right panel is the stellar mass density of the Universe as a function of redshift. The ordinate is the total stellar mass in galaxies divided by the volume of the Universe. Stellar mass of each galaxy is estimated from light, i.e., luminosity and spectral energy distribution (SED), of galaxies measured from galaxy surveys. Conversion from light to stellar mass is done by SED fitting based on population synthesis models. The data points here came from 17 papers.

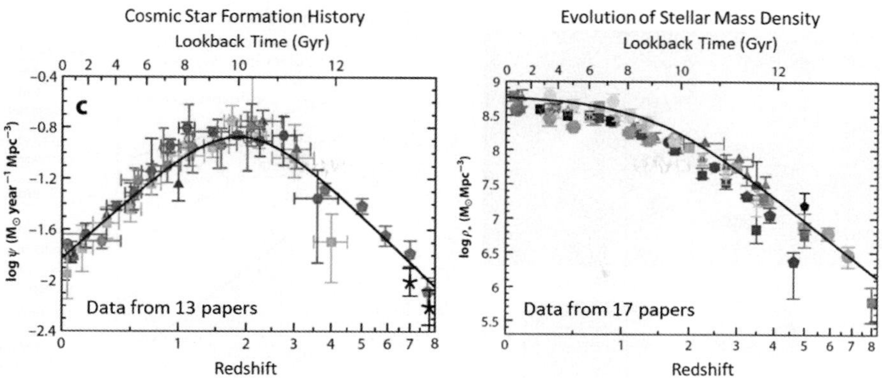

Figure 10. (left) Cosmic star formation history. (right) Evolution of stellar mass density. (Madau & Dickinson 2014)

The SFRD increased from $z \sim 8$, and peaked at $z \sim 1.9$, 3.5 Gyr after big bang, and declined exponentially toward the present. Theoretically, the integration of the SFRD from the beginning ($z \sim \infty$) to a particular redshift gives the stellar mass density at that redshift though the return fraction should be taken into account properly. The solid line in the left panel is the best-fit to the observed SFRD evolution expressed by equation (15) of Madau & Dickinson (2014). The solid line in the right panel is obtained by integrating the best-fit SFRD, which slightly overshoots the observational data at $0 < z < 3$. This ~0.2-dex discrepancy might be among the topics of this symposium, but it is beyond the scope of this review. Here, I would say that we now have a fairly consistent picture of cosmic star formation history due to the progress of deep multi-wavelength coordinated galaxy surveys.

5.2. *Subaru HSC Survey*

The Subaru telescope is unique among 8-m class telescopes in that it has a prime focus. The original prime-focus camera was the Suprime-Cam which worked for 17 years (2000-2017). The author was the PI of the Suprime-Cam. According to Arimoto (2015), the Suprime-Cam was most heavily used (24% of available nights during 2000-2013) and produced by far the largest number of papers (446 during 2000-2012, 48% of the total Subaru papers) among the Subaru instruments.

The successor of the Suprime-Cam, the Hyper Suprime-Cam (HSC), came in operation in 2013. The PI is Satoshi Miyazaki. Figure 11 (left) shows the comparison of field of views (FoVs) between the Suprime-Cam and the HSC. The Suprime-Cam, which carried out several important surveys such as the Subaru Deep Survey and the COSMOS survey, had a half degree FoV, while the FoV of the HSC is 10 times wider. With the much more powerful HSC we started a large survey in 2014. About 300 nights will be dedicated to this HSC survey (Hyper Suprime-Cam Subaru Strategic Program, https://hsc.mtk.nao.ac.jp/ssp/).

The diagram in Fig. 11 (right) was first presented by the HSC team in 2012, when the HSC survey had not been approved. It shows the limiting magnitude of existing, upcoming, and future surveys as a function of survey area. The four lines, data amounts and data rates are added by the author. Black symbols are existing surveys, magenta ones are upcoming surveys and blue ones are future surveys by LSST. It is clear that HSC survey surpasses other surveys by a wide margin, except for those by LSST.

The HSC survey consists of three layers, wide, deep and ultradeep in g, r, i, z, and y bands. Three NB filters centered at λ=818 nm, 921 nm, and 1009 nm are added to the

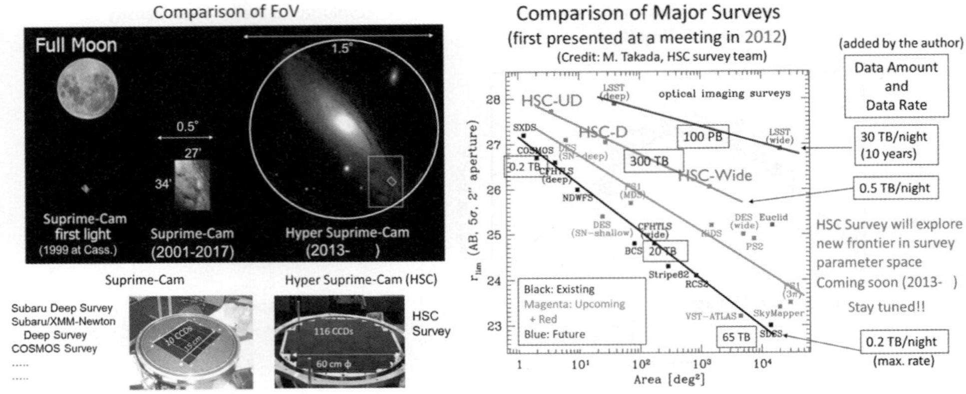

Figure 11. (left) Comparison of field of views of Suprime-Cam and HSC (credit NAOJ). (right) Comparison of major surveys (first presented in 2012).

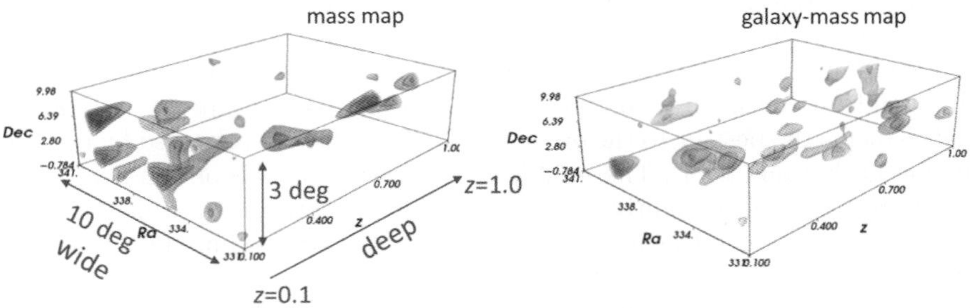

Figure 12. Three-dimensional mass map and galaxy-mass map of the VVDS region. The contours are drawn from 2σ to 6σ with the 1σ interval. (Oguri *et al.* 2018)

deep and the ultradeep surveys. The wide survey consists of 916 HSC pointings covering 1400 square degrees to the limiting magnitude of $i \sim 26$ mag. The deep/ultradeep survey consists of 15/2 pointings covering 26/3.5 square degrees to $i \sim 27/28$ magnitudes.

First Data Release was made in February 2018, and the Special Issue of Publications of the Astronomical Society of Japan (PASJ) published in January 2018 was dedicated to the HSC survey. It includes 40 papers, one of which presented a weak-lensing mass map for an unprecedentedly large volume (Fig. 12).

There is the *HSC viewer*, which has various functions to investigate and enjoy the HSC survey data. It is also used in citizen science programs.

References

Arimoto, N. 2015, *Subaru Telescope, Director's Report 2009-2013*, Presented at the External Review of National Astronomical Observatory, held in 24-26 Feb, 2015
Binggeli, B., Sandage, A., Tarenghi, M., *et al.* 1984, *AJ*, 89, 64
Binggeli, B., Sandage, A., Tammann, G. A., *et al.* 1988, *ARAA*, 26, 509
Colless, M. *et al.* 2001, *MNRAS*, 328, 1039
Djorgovski, S. G. *et al.* 2012, in *Astronomical Techniques, Software, and Data*, H. Bond (ed.) https://arxiv.org/ftp/arxiv/papers/1203/1203.5111.pdf
Dreyer, J. L. E. 1888, *Memoirs of the Royal Astronomical Society*, 49, 1
Economist, https://www.economist.com/special-report/2010/02/25/data-data-everywhere
Ferrarese, L. *et al.* 2012, *ApJS*, 200, 4
Geller, M. J., Huchra, J. P., de Lapparent, V., *et al.* 1987, *IAU Symp.*, 124, 301
Gregory, S. A. & Thompson, L. A. 1978, *ApJ*, 222, 784

Huchra, J., Davis, M., Latham, D., Tonry, J., *et al.* 1983, *ApJS*, 52, 89
Huchra, J. P., Vogeley, M. S., Geller, M., *et al.* 1999, *ApJS*, 121, 287
Ichikawa, S., Wakamatsu, K., Okamura, S., *et al.* 1986, *ApJS*, 60, 475
Kirshner, R. P., Oemler, A., Jr., Schechter, P. L., Shectman, S. A., *et al.* 1981, *ApJ*, 248, L5
Madau, P. & Dickinson, M. 2014, *ARAA*, 52, 415
Maddox, S. J. *et al.* 1990, *MNRAS*, 243, 692
Messier, C. 1771, *Memoirs of the French Academy of Sciences for 1771*, 435-461 http://www.messier.seds.org/
Numazawa, S. & Wakiya, N. 2009, http://skywatch.jp/ngc/
Oguri, M. *et al.* 2018, *PASJ*, 70, SP-1, S26-1
Peebles, P. J. E. 1973, *AJ*, 185, 413
Sandage, A. & Tammann, G. 1981, *A Revised Shapley-Ames Catalog* (RSA), Carnegie Institution of Washington
Shane, C. D. & Wirtanen, C. A. 1954, *AJ*, 59, 285; 1967, *Publ. Lick Obs.*, 22, Part I
Steinicke, W. 2018, http://www.klima-luft.de/steinicke/ngcic/rev2000/Explan.htm
Tonry, J. & Davis, M. 1979, *AJ*, 84, 1511
Totsuji, H. & Kihara, T. 1969, *PASJ*, 21, 221
Useful data centers
NED: https://ned.ipac.caltech.edu/, SIMBAD: http://simbad.u-strasbg.fr/simbad/
IVOA: http://www.ivoa.net/, HEASARC: https://heasarc.gsfc.nasa.gov/
MAST: https://archive.stsci.edu/, DSS: https://archive.stsci.edu/cgi-bin/dss_form
IPAC: https://www.ipac.caltech.edu/, SDSS: https://www.sdss.org/
SMOKA: https://smoka.nao.ac.jp/, DARTS: https://www.darts.isas.jaxa.jp/
Yahil, A., Sandage, A., Tammann, G. A., *et al.* 1980, *ApJ*, 242, 448
York, D. G. *et al.* 2000, *AJ*, 120, 1579

Discussion

TOMO GOTO: Why CGCG citation history has a valley in 1980s?

SADANORI OKAMURA: It may be better to say that the citation history shows a rapid rise since around 1990. This may be due to increased interests in the large scale structure revealed by the redshift surveys such as CfA-II survey and IRAS galaxy redshift survey.

DENIS BURGARELLA: What recommendations for future surveys?

SADANORI OKAMURA: One important area is the time-domain survey, especially, with time scales less than a minute. Wide and deeper surveys in mid to far infrared are also important to measure star formation from dust continuum for low-z galaxies and stellar mass for high-z galaxies.

YUTAKA FUJITA: What kind of objects will be observed in the future time-domain surveys?

SADANORI OKAMURA: GRB afterglows, AGNs, novae, supernovae, variable stars, etc., and of course, unknown objects and/or phenomena. Time-domain surveys with time scale less than 1 second will open a new frontier.

DIAN TRIANI: What is the redshift limit of Sloan?

SADANORI OKAMURA: The spectroscopic samples of SDSS are magnitude limited and have no fixed redshift limits. But roughly speaking, they are $z \sim 0.2$ (median $z \sim 0.1$) for galaxies in the main sample, $z \sim 0.4$ for luminous red galaxies, and $z \sim 6$ for quasars.

Challenges in Panchromatic Modelling with Next Generation Facilities
Proceedings IAU Symposium No. 341, 2018
M. Boquien, E. Lusso, C. Gruppioni & P. Tissera, eds.

doi:10.1017/S1743921319001881

Panchromatic study of the first galaxies with large ALMA programs

A. Faisst[1], M. Béthermin[2], P. Capak[1], P. Cassata[3], O. Le Fèvre[2], D. Schaerer[4], J. Silverman[5], L. Yan[1,6] and the *ALPINE* team

[1]IPAC, California Institute of Technology, 1200 East California Boulevard, Pasadena, CA 91125, USA
email: afaisst@ipac.caltech.edu

[2]Laboratoire d'Astrophysique de Marseille UMR 7326, 13388, Marseille, France

[3]University of Padova, Department of Physics and Astronomy Vicolo Osservatorio 3, 35122, Padova, Italy

[4]Observatoire de Genève, Université de Genève 51 Ch. des Maillettes, 1290 Versoix, Switzerland

[5]Kavli Institute for the Physics and Mathematics of the Universe, The University of Tokyo Kashiwa, Chiba 277-8583, Japan

[6]Caltech Optical Observatories, Cahill Center for Astronomy and Astrophysics 1200 East California Boulevard, Pasadena, CA 91125, USA

Abstract. Thanks to deep optical to near-IR imaging and spectroscopy, significant progress is made in characterizing the rest-frame UV to optical properties of galaxies in the early universe ($z > 4$). Surveys with Hubble, Spitzer, and ground-based facilities (Keck, Subaru, and VLT) provide spectroscopic and photometric redshifts, measurements of the spatial structure, stellar masses, and optical emission lines for large samples of galaxies. Recently, the Atacama Large (Sub) Millimeter Array (ALMA) has become a major player in pushing studies of high redshift galaxies to far-infrared wavelengths, hence making panchromatic surveys over many orders of frequencies possible. While past studies focused mostly on bright sub-millimeter galaxies, the sensitivity of ALMA now enables surveys like *ALPINE*, which focuses on measuring the gas and dust properties of a large sample of *normal* main-sequence galaxies at $z > 4$. Combining observations across different wavelengths into a single, panchromatic picture of galaxy formation and evolution is currently and in the future an important focus of the astronomical community.

Keywords. Galaxy: formation, Galaxy: evolution, galaxies: ISM, dust, extinction, ISM: evolution, ISM: kinematics and dynamics, ISM: evolution, infrared: galaxies, surveys

1. Overview

Galaxies evolve significantly over cosmic time in both internal as well as external properties. The cosmic star formation rate (SFR) density of the early universe increases rapidly to a redshift $z = 2$ (Figure 1) and galaxies in the early universe double their stellar mass at > 10 times faster rates compared to today (Faisst *et al.* 2016a). This steep mass growth goes in hand with a significant increase of the dust and metal content (Maiolino *et al.* 2008; Bouwens *et al.* 2009; Faisst *et al.* 2016b). In addition, star-forming galaxies evolve from lumpy and turbulent (dominated by gas inflow, mergers, and star-forming clumps) to smooth disks as cosmic time progresses (Förster Schreiber *et al.* 2011).

In only 500 Myrs between redshifts $4 < z < 6$, galaxies significantly increase their stellar, dust, and gas mass as well as metal content and in addition change their structure by building up the first stellar disks. Hence, this *early growth phase* is an important link

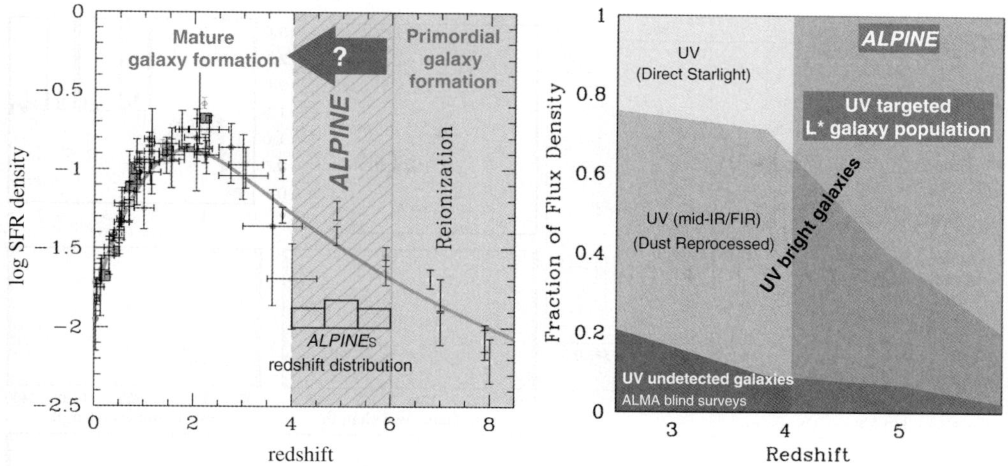

Figure 1. *Left:* SFR density across cosmic time (see Madau & Dickinson (2014)). *ALPINE* targets galaxies in the early growth phase, a transition phase between primordial and mature galaxy formation where galaxies substantially increase their stellar mass, dust, and metal content. *Right:* Fraction of flux density emitted by galaxies directly in the UV (top shading), by UV-selected galaxies in the far-IR (middle shading), and by galaxies only probed in blind far-IR surveys (bottom shading). Data from Bouwens *et al.* (2009).

between primordial (during the Epoch of Reionization at $z > 4$) and mature (at the peak of cosmic SFR density at $z \sim 2-4$) galaxy formation.

To date, more than 1000 galaxies have been identified at $4 < z < 6$ by photometric redshifts, color-color selections, and narrow-band observations targeting Lyα (see review by Dayal *et al.* 2017). Several hundred galaxies have been confirmed spectroscopically. Large surveys with HST specifically focus on the galaxies' UV properties including their luminosity function, UV dust obscuration, and morphology (e.g., Bouwens *et al.* 2015; Shibuya *et al.* 2016). Observations with Spitzer at $3-5\,\mu$m (rest-frame optical) are used to determine robust stellar masses and to estimate SFRs and stellar mass growth rates from optical emission lines such as Hα (Davidzon *et al.* 2017; Faisst *et al.* 2016a). In addition, deep rest-frame UV spectroscopy from 8-10 meter telescopes (e.g., Keck and VLT) characterize UV emission lines (such as Lyα or He II) and provide important constraints on metallicity from the measurement of various absorption features (Ando *et al.* 2007; Faisst *et al.* 2016b).

These large UV to optical surveys lay the foundation for a comprehensive panchromatic study over many orders of frequencies with the newest and next-generation facilities. The *ALPINE* survey with ALMA is a current example of this. The combination of ALMA's sensitivity and efficiency with carefully pre-selected samples from complete optical surveys opens doors for the first large-sample study of dust and gas properties of galaxies in the early growth phase at $z > 4$.

2. *ALPINE*: A Textbook Example of a Panchromatic Study

The *ALMA Large Program to Investigate [C II] at Early Times* is a 70-hour ALMA cycle 5 program (#2017.1.00428.L) to carry out targeted Band-7 observations of 122 main-sequence galaxies in the early growth phase at $4 < z < 6$ (Figure 1). *ALPINE* is a diverse, multi-national collaboration of ~ 45 scientists, led by the PI-team of O. Le Fèvre (PI), A. Faisst (North America science lead), M. Béthermin (ALMA data reduction lead), P. Cassata, P. Capak, D. Schaerer, J. Silverman, and L. Yan. As of today

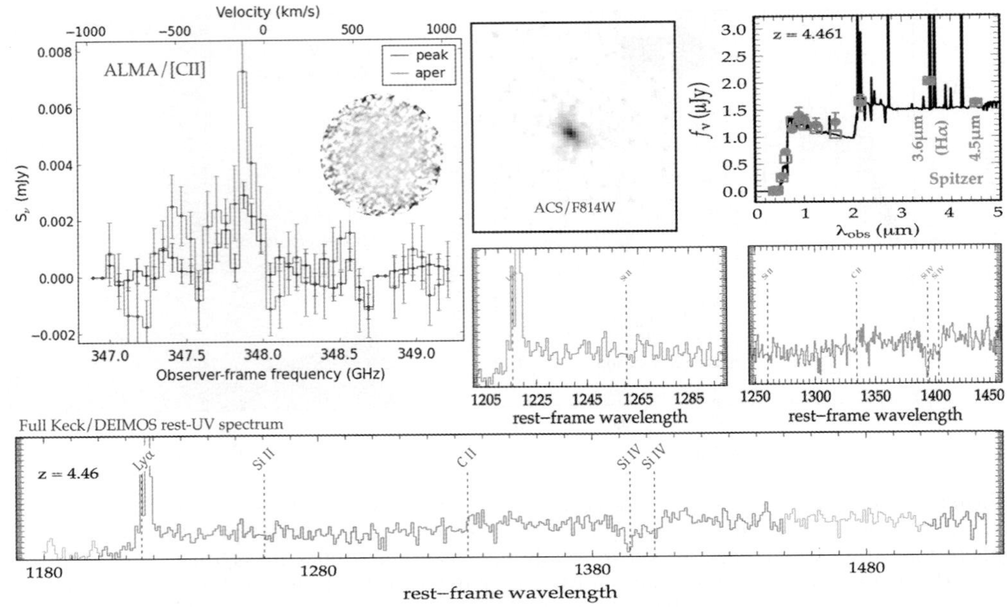

Figure 2. Demonstration of the power of panchromatic data on a galaxy from *ALPINE*. *Clockwise:* [C II] emission and dust continuum from ALMA; HST imaging; multi-band SED and Hα from Spitzer color; rest-frame UV spectrum from Keck/DEIMOS (Lyα and absorption lines).

(December 2018), 85% of the data has been acquired and reduced. For more information on *ALPINE*, we refer to future survey papers.

ALPINE maximizes its efficiency and impact by building upon a carefully pre-selected sample in the *Cosmic Evolution Survey* (COSMOS, Scoville *et al.* 2007) and the *Extended Chandra Deep Field South* (E-CDFS, Giacconi *et al.* 2002) fields, well studied fields at UV to near-IR wavelengths. The sample is drawn from large spectroscopic surveys undertaken at Keck (Hasinger *et al.* 2018) and VLT (Le Fèvre *et al.* 2015), including galaxies with diverse spectroscopic properties in rest-frame UV emission and absorption lines. This approach leads to a complete probe of a diverse galaxy population above SFR $\sim 10\,\mathrm{M}_\odot$/yr in a stellar mass range of $9 < \log(M/\mathrm{M}_\odot) < 11$. By following up UV-selected galaxies, *ALPINE* probes more than 80% of the total missing flux density at these redshifts, while blind surveys with ALMA cover the remaining less than 20% (Figure 1). In combination with its multi-wavelength ancillary data, *ALPINE* is a textbook example of one of the first large panchromatic surveys probing galaxies in the early Universe (Figure 2).

2.1. *ALPINE Science in a Nutshell*

ALPINE builds upon its successful small-scale precursor study based on 10 galaxies at $z \sim 5.5$ (Riechers *et al.* 2014; Capak *et al.* 2015) and reaches a similar sensitivities (continuum: 50 μJy or $\log(L_{\mathrm{FIR}}/\mathrm{L}_\odot) \sim 10.3$; line: $\mathrm{S}_{\mathrm{CII}}\Delta v \sim 0.05$ Jy km/s or $\log(L_{\mathrm{CII}}/\mathrm{L}_\odot) \sim 7.5$) and resolution ($0.5'' - 0.7''$). Its main goal is to quantify the evolution of obscured star-formation, the gas and dust content, and gas kinematics of galaxies in the early universe via [C II] emission and far-IR continuum measurements at rest-frame $\sim 150\,\mu$m. [C II] is one of the dominant coolants, hence one of the strongest far-IR lines. In addition the atmospheric transmission is high, making these observations favorable. *ALPINE* combines several science goals and is designed to answer the following questions.

How does the total star-formation evolve at $4 < z < 6$? *ALPINE* provides a comprehensive study of the total SFR density at these redshifts from UV and far-IR continuum measurements. Independent SFRs from far-IR, [C II], and Hα (derived from Spitzer colors, see Faisst *et al.* 2016a) provide the first measurement of the SFR versus stellar mass main-sequence on different timescales of star-formation at these redshifts.

Is [C II] a reliable SFR indicator at high-z? The prevalence of [C II] emission in high-redshift galaxies is a promising tool to estimate SFRs of far-IR continuum faint galaxies. *ALPINE* allows to calibrate this locally well established (De Looze *et al.* 2014) relation at $z > 4$ where galaxies are more metal poor.

How dusty are galaxies in the early universe? Small-sample studies of the IRX$-\beta$ relation, which relates the total dust extinction to the line-of-sight dust obscuration, suggest diverse dust properties at $z \sim 5$ (Faisst *et al.* 2017). *ALPINE*'s large sample constrains the diversity and studies its causes by a comprehensive analysis of its ancillary data.

How does the gas fraction evolve at $z > 4$? *ALPINE* constrains gas masses from far-IR continuum, [C II], and dynamical mass measurements and measures gas kinematics from [C II] profiles. It is the first study to trace the evolution of gas mass out to $z = 6$.

ALPINE unites current UV to far-IR facilities and will in the future build a benchmark sample for various simulations and set the basis for follow-up observations with the James Webb Space Telescope and other next-generation ground- and space-based facilities.

3. Conclusions

Large photometric and spectroscopic surveys at UV to near-IR wavelengths provide a solid basis to build a full panchromatic sample over the next couple of years. Importantly, these samples build on a variety of galaxies, which makes optimally planned targeted follow-up observations with ALMA and later JWST and TMT/E-ELT efficient in covering a significant amount of the parameter space of galaxy properties. A current example is *ALPINE*, which is specifically built to extend the wavelength coverage of a carefully chosen large spectroscopic sample with ground- and space-based multi-wavelength photometric ancillary data. This makes it to one of the currently largest panchromatic samples to study the formation and evolution of normal main-sequence galaxies in the early Universe. In the future, the *ALPINE* sample will provide a solid basis for follow-up observations with various next-generation facilities.

References

Ando, M., Ohta, K., Iwata, I., Akiyama, M., Aoki, K., Tamura, N., *et al.* 2007, *PASJ*, 59, 717

Bouwens, R. J., Illingworth, G. D., Franx, M., Chary, R.-R., Meurer, G. R., Conselice, C. J., Ford, H., Giavalisco, M., *et al.* 2009, *ApJ*, 705, 936

Bouwens, R. J., Illingworth, G. D., Oesch, P. A., Trenti, M., Labbé, I., Bradley, L., Carollo, M., van Dokkum, P. G., *et al.* 2015, *ApJ*, 803, 34

Capak, P. L., Carilli, C., Jones, G., Casey, C. M., Riechers, D., Sheth, K., Carollo, C. M., Ilbert, O., *et al.* 2015, *Nature*, 522, 455

Dayal, P. & Ferrara, A. 2017, *Physics Reports*, 780, 1

Davidzon, I., Ilbert, O., Laigle, C., Coupon, J., McCracken, H. J., Delvecchio, I., Masters, D., Capak, P., *et al.* 2017, *A&A*, 605, A70

De Looze, I., Cormier, D., Lebouteiller, V., Madden, S., Baes, M., Bendo, G. J., Boquien, M., Boselli, A., *et al.* 2014, *A&A*, 568, A62

Faisst, A. L., Capak, P., Hsieh, B. C., Laigle, C., Salvato, M., Tasca, L., Cassata, P., Davidzon, I., *et al.* 2016, *ApJ*, 821, 122

Faisst, A. L., Capak, P. L., Davidzon, I., Salvato, M., Laigle, C., Ilbert, O., Onodera, M., Hasinger, G., *et al.* 2016, *ApJ*, 822, 29

Faisst, A. L., Capak, P. L., Yan, L., Pavesi, R., Riechers, D. A., Barišić, I., Cooke, K., Kartaltepe, J. S., *et al.* 2017, *ApJ*, 847, 21

Förster Schreiber, N. M., Shapley, A. E., Erb, D. K., Genzel, R., Steidel, C. C., Bouché, N., Cresci, G., Davies, R., *et al.* 2011, *ApJ*, 731, 65

Giacconi, R., Zirm, A., Wang, J., Rosati, P., Nonino, M., Tozzi, P., Gilli, R., Mainieri, V., *et al.* 2002, *ApJS*, 139, 369

Hasinger, G., Capak, P., Salvato, M., Barger, A. J., Cowie, L. L., Faisst, A., Hemmati, S., Kakazu, Y., *et al.* 2018, *ApJ*, 858, 77

Le Fèvre, O., Tasca, L. A. M., Cassata, P., Garilli, B., Le Brun, V., Maccagni, D., Pentericci, L., Thomas, R., *et al.* 2015, *A&A*, 576, A79

Madau, P. & Dickinson, M. 2014, *ARAA*, 52, 415

Maiolino, R., Nagao, T., Grazian, A., Cocchia, F., Marconi, A., Mannucci, F., Cimatti, A., *et al.* 2008, *A&A*, 488, 463

Riechers, D. A., Carilli, C. L., Capak, P. L., Scoville, N. Z., Smolčić, V., Schinnerer, E., Yun, M., Cox, P., *et al.* 2014, *ApJ*, 796, 84

Scoville, N., Aussel, H., Brusa, M., Capak, P., Carollo, C. M., Elvis, M., Giavalisco, M., Guzzo, L., *et al.* 2007, *ApJS*, 172, 1

Shibuya, T., Ouchi, M., Kubo, M. & Harikane, Y. 2016, *ApJ*, 821, 72

Discussion

T. Goto: Is there a correlation between IRX$-\beta$ and geometry of the galaxies?

A. Faisst: We generally see that galaxies with a high $L_{\rm FIR}/L_{\rm UV}$ are more compact. Galaxies with a low ratio look more clumpy and irregular. However, the sample size is not yet big enough to draw any conclusions, but we will with *ALPINE*.

D. Burgarella: Why would you expect galaxies to move from the local galaxy region (i.e. Meurer) to below the SMC one in the IRX$-\beta$ diagram?

A. Faisst: We expect lower dust masses at high-z, so galaxies will likely evolve downwards. For β, that question is still open, spatial geometry effects might play a role.

F. Fontanot: I have a question about your initial sample of 10 galaxies. Do you have a mass estimate for these objects? And if so, do you see a significant difference in mass or redshift range between the dust-rich and the dust-poor population?

A. Faisst: The current sample (10 galaxies) is too small to see trends on the IRX$-\beta$ diagram. Also, the mass-range is pretty small. The masses are around $\log(M/{\rm M}_{\odot}) = 10 \pm 0.3$. Also, the redshift range in the current sample is small ($5.2 < z < 5.6$). We currently do not see any significant trends of dust attenuation with mass and redshift. However, this is something we will study with *ALPINE*. Specifically, it covers more than a magnitude in stellar mass ($9.5 < \log(M/{\rm M}_{\odot}) < 11.5$) and a large redshift range ($4.4 < z < 5.8$). If there are trends with stellar mass and redshift, we will likely see it in *ALPINE*.

D. Rosario: Does the data allow a comparison of Hα and [C II]?

A. Faisst: Hα can be estimated from the excess in the Spitzer/IRAC bands at 3.6 and 4.5 μm. A statistical approach is possible. A direct comparison can be made in individual cases. This is something the team wants to do, for studying gas physics and comparison of different star-formation tracers.

Challenges in Panchromatic Modelling with Next Generation Facilities
Proceedings IAU Symposium No. 341, 2018
M. Boquien, E. Lusso, C. Gruppioni & P. Tissera, eds.

doi:10.1017/S1743921319002904

The average FIR SED of proto-clusters at $z = 4$

Mariko Kubo[1], Jun Toshikawa[2], Nobunari Kashikawa[1,3], Hisakazu Uchiyama[1] and Kei Ito[1]

[1]National Astronomical Observatory of Japan, Mitaka, Tokyo 181-8588, Japan
[2]Institute for Cosmic Ray Research, The University of Tokyo, 5-1-5 Kashiwanoha, Kashiwa, Chiba 277-8582, Japan
[3]Department of Astronomy, School of Science, The University of Tokyo, 7-3-1 Hongo, Bunkyo-ku, Tokyo, JAPAN, 113-0033
email: mariko.kubo@nao.ac.jp

Abstract. One of the key questions of the observational cosmology is how the environmental dependence of galaxies today formed. Proto-clusters, galaxy overdense regions at high redshift are important laboratory to study the formation history of clusters of galaxies. We perform the first statistic study of far-infrared spectral energy distribution(SED)s of proto-clusters at $z \sim 4$ by the stacking analysis of *Planck*/*AKARI*/*IRAS* images of proto-clusters at $z \sim 4$ selected from the Hyper Suprime-Cam Subaru Strategic Program (HSC-SSP) survey. By stacking ~ 200 proto-clusters, we successfully constrain their average total SEDs in $60 - 850$ μm. Our results imply the excess of dusty starburst galaxies with star formation rate several 1000 $M_{\odot} yr^{-1}$ in total and obscured AGNs in proto-clusters at $z \sim 4$.

Keywords. galaxies: formation, large-scale structure of universe

1. Introduction

One of the key question of observational cosmology is when and how the environmental dependence of galaxies today formed. Proto-clusters, galaxy overdense regions at high redshift are thought to be progenitors of clusters of galaxies today. They are important targets to discuss how the environmental dependence of galaxies today evolved.

Proto-clusters have been studied a lot, however, there are several problems with observational method. One is the statistics. The number of the known proto-clusters is not so large. Since the number density of proto-clusters at a given epoch is not so large (~ 1 per 1 deg^2 at a given epoch according to the cosmological numerical simulations), they are generally searched around radio galaxies/QSOs at high redshift, which are thought to evolve into the brightest cluster galaxies today (e.g., Venemans *et al.* 2007), and sometimes found by chance (e.g., Steidel *et al.* 1998). In addition, the selection bias on them is not clear. According to the past few studies, there is a large variation in proto-clusters. Large statistical studies of proto-clusters are needed to show their typical properties and variation.

Another observational issue is their properties obscured by dust. The large fraction of the UV radiation from newly formed stars in starburst galaxies is absorbed by dust and re-emitted as thermal emission from dust at infrared (IR) wavelength (Casey *et al.* 2014). Dusty starburst galaxies at high redshift found at sub-mm referred as sub-mm galaxies (SMGs) are the most intense (SFR> several 100 $M_{\odot}/yr$) starburst galaxies. They comprise the greater part of the observed cosmic star formation history (Madau & Dickinson 2014). Such SMGs are of particular importance in proto-clusters. Indeed, significant density excess of SMGs in proto-clusters have been reported by the deep mm/sub-mm imaging and spectroscopic observations (e.g., Tamura *et al.* 2009; Wang *et al.* 2016; Umehata *et al.* 2018; Miller *et al.* 2018). However the capability of current telescopes is not enough to perform a statistical study of IR properties of galaxies in proto-clusters.

In this study, we perform the statistical study of the IR SEDs of proto-clusters by the stacking analysis of the archival IR images *Planck/IRAS/AKARI* of a large proto-cluster catalog from the Hyper Suprime-Cam Subaru Strategic Program (HSC-SSP) survey.

2. Data/Method

We use the catalog of protoclusters at $z \sim 3.8$ selected systematically based on the HSC-SSP survey (Aihara *et al.* 2018) by Toshikawa *et al.* (2018). HSC is the prime focus camera on Subaru Telescope. It has a large field-of-view (FoV) of 1.8 deg^2 and high sensitivity (Miyazaki *et al.* 2012). The HSC-SSP survey is an on-going multi-color ($griz$ + narrow band) survey in optical consisting of three layers, UltraDeep (UD; 3.5 deg^2, $i \sim 28$ mag), Deep (26 deg^2, $i \sim 27$ mag) and Wide (1400 deg^2, $i \sim 26$ mag) layers. In Toshikawa *et al.* (2018), they construct the catalog of Lyman break galaxies using gri-band images (here after g-drop galaxies) from an area of 121 deg^2 of the current HSC-SSP wide layer. This color cut is sensitive for galaxies at $3.3 \leqslant z \leqslant 4.2$. They measure the surface number density of g-drop galaxies with $i < 25.0$ mag within an aperture of $1'.8$ (0.75 Mpc in physical), and select the regions with overdensity significance larger than 4σ as proto-cluster candidates. This radius is the typical extent of the region which will collapse into a single massive halo of $> 10^{14}\ M_\odot$ by $z = 0$, according to the cosmological numerical simulations (e.g., Chiang *et al.* 2013). In total, they identified 179 unique proto-cluster candidates.

We cut out the archival IR all sky maps at the positions of the proto-clusters; the 353, 545 and 857 GHz images taken with *Planck* High Frequency Instrument (HFI); 60 and 100 μm taken with *IRAS*, and *N60*, *WIDE-S*, *WIDE-L* and *N160* band images taken with *AKARI* Far Infrared Surveyor (FIS; Kawada *et al.* 2007). The major contaminant from foreground sources on these IR images is diffuse emission from Galactic cirrus. In case of *Planck*, we use the 353, 545 and 857 GHz cosmic infrared background (CIB) products (Planck Collaboration *et al.* 2016 XLVIII) from the *Planck* legacy archive† which are subtracted of the Galactic cirrus component. *IRAS*/*AKARI* images are stacked after evaluating and subtracting the sky values and smoothed to have the FWHM PSF sizes similar to *Planck* images (≈ 5 arcmin). We measure the fluxes and errors by the bootstrap resampling.

Before the stacking analysis, we checked whether the HSC-SSP proto-clusters are detected individually on the *Planck* images. We found that none of them are detected above 5 σ significance. On the other hand, the overall flux distribution of them is slightly brighter than random points. We make a sub-sample, "brightest half", based on this tentative detections on *Planck*.

3. Result and Discussion

Figure 1 shows the stacked images of the brightest half sample of proto-clusters. There are clear detections with $3 - 10\sigma$ significance except for *AKARI/N60*-band. The red circles in Fig. 2 show the observed values. It is the first time to constrain the average total FIR SED of proto-clusters at $z \sim 4$.

The IR emission at rest-frame $> 70\mu$m is likely from the cold dust emission originated in starburst. This implies significant star formation activities hidden by dust. If the whole

† https://wiki.cosmos.esa.int/planckpla2015/index.php/CMB_and_astrophysical_component_maps

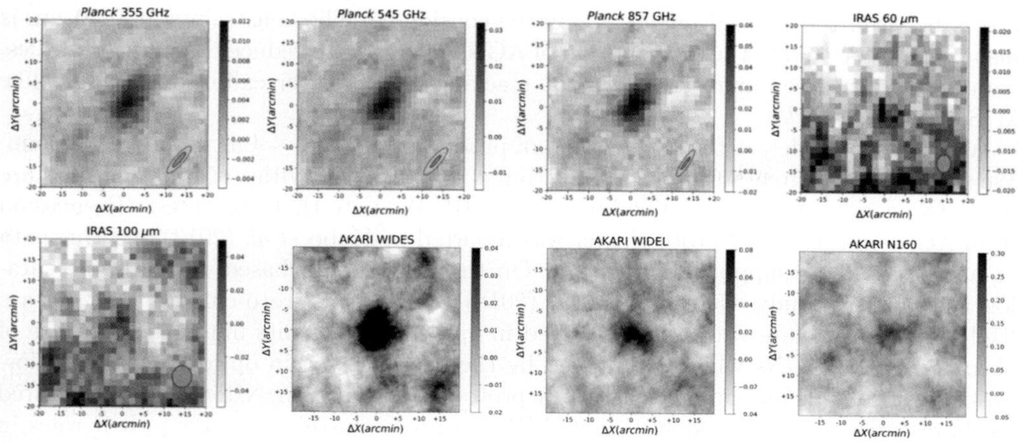

Figure 1. The stacked images of *Planck* 353, 545 and 857 GHz, *IRAS* 60 and 100 μm, and *AKARI WIDES*, *WIDEL* and *N160*-band images of "brightest half" g-drop proto-clusters from *top left* to *bottom right*. The size of each image is 40 arcmin by side. The ellipses on the *Planck* stacked images show the simulated beam profiles. 50 and 90% of flux is contained within red and yellow filled ellipses, respectively. The red ellipse at the bottom right of the *IRAS* image show FWHM of the beam.

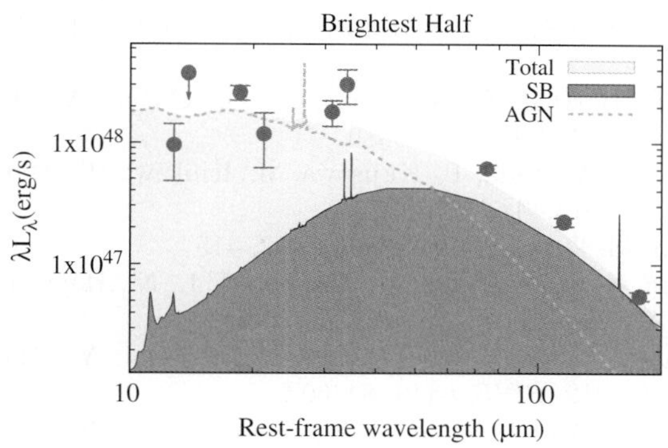

Figure 2. The best fit IR SED for the brightest half sample. Red filled circles show the observed points. The SED model is consisting of a starburst and an AGN components. The total SED is shown with the light blue filled region. The starburst component among the total SED is shown with the thick blue region. The SED of the AGN component is shown with the green dashed line.

IR emission is dominated by cold dust emission from starburst galaxies, the SED should have the shape characterized as a gray body with dust temperature $20 \sim 40K$, however, the observed SED shows significant excess from this at shorter wavelength. The excess at shorter wavelength is likely originated in a warm dust heated by AGNs.

We perform the SED fitting with the models composed from the starburst and AGN components using the templates in Dale *et al.* (2014). The light blue filled region in Fig. 2 shows the best fit SED. The best-fit models of the starburst and AGN components are shown with the thick blue region and the green dashed line, respectively. We found that, in case of the all proto-cluster sample, the total SFR is $\sim 1300\ M_{\odot}/yr$ while it comprises $\sim 8\%$ of the total IR emission from $8 - 1000\mu$ m. In case of the brightest half sample, the total SFR is $\sim 6700\ M_{\odot}/yr$ while it comprises $\sim 20\%$ of the total IR emission. This implies the wide variation in IR properties of proto-clusters while their properties in optical do not correlate significantly with IR properties. We note that the calibration errors between the stacked images should be considered. The constraints at $< 10\ \mu$m is also needed to evaluate the contribution of AGNs robustly. To reduce these uncertainties, we need to compare our results with the stacking of the better calibrated IR data like WISE and *Herschel*.

We successfully detect the IR emission of proto-clusters at $z \sim 4$ as their average sum. We show that the greater part of the starburst and AGN activities of proto-clusters are likely to be hidden in optical but luminous in IR. We note that the excess IR emission from AGNs in the SSA22 proto-cluster was reported by Kubo *et al.* (2013) and Umehata *et al.* (2017). This supports our results. On the other hand, based on the same catalog used in this study, Uchiyama *et al.* (2018) report that proto-clusters at $z \sim 4$ are likely to anti-correlate with QSOs selected in optical. Our results imply the presence of obscured QSOs/AGNs which are missed by the QSO catalog in optical. Further follow up observations of individual galaxies in proto-clusters at mm/NIR/X-ray are desired to characterize the obscured starburst/AGNs in proto-clusters and study their roles in galaxy evolution.

References

Aihara, H., Arimoto, N., Armstrong, R., Arnouts, S., Bahcall, N., Bickerton., Bosch, J., Bundy, K., *et al.* 2018, *PASJ*, 70, S4

Casey, C. M., Narayanan, D., & Cooray, A. 2014, *Physical Review*, 541, 45

Chiang, Y.-K., Overzier, R., & Gebhardt, K. 2013, *ApJ*, 779, 127

Dale, D. A., Helou, G., Magdis, G. E., Armus, L., Díaz-Santos, T., Shi, Y., *et al.* 2014 *ApJ*, 784, 83

Kawada, M., Baba, H., Barthel, P. D., Clements, D., Cohen, M., Doi, Y., Figueredo, E., Fujiwara, M., *et al.* 2007, *PASJ*, 59, S389

Kubo, M., Uchimoto, Y. K., Yamada, T., Kajisawa, M., Ichikawa, T., Matsuda, Y., Akiyama, M., Hayashino, T., *et al.* 2013, *ApJ*, 778, 170

Madau, P. & Dickinson, M. 2014, *Annual Review*, 52, 415

Miller, T. B., Chapman, S. C., Aravena, M., Ashby, M. L. N., Hayward, C. C., Vieira, J. D., Weiß, A., Babul, A., *et al.* 2018, *Nature*, 556, 469

Miyazaki, S., Komiyama, Y., Nakaya, H., Kamata, Y., Doi, Y., Hamana, T., Karoji, H., Furusawa, H., *et al.* 2012, *SPIE*, 8446, 84460Z

Planck Collaboration 2016 XLVIII, Aghanim, N., Ashdown, M., Baccigalupi, C., Ballardini, M., Banday, A. J., Barreiro, R. B., Bartolo, N., *et al.* 2016, *A&A*, 596, A109

Steidel, C. C., Adelberger, K. L., Dickinson, M., Giavalisco, M., Pettini, Max., Kellogg, M., *et al.* 1998, *ApJ*, 492, 428

Tamura, Y., Kohno, K., Nakanishi, K., Hatsukade, B., Iono, D., Wilson, G. W., Yun, M. S., Takata, T., *et al.* 2009, *Nature*, 459, 61

Toshikawa, J., Uchiyama, H., Kashikawa, N., Ouchi, M., Overzier, R., Ono, Y., Harikane, Y., Ishikawa, S., *et al.* 2018, *PASJ*, 70, S12

Uchiyama, H., Toshikawa, J., Kashikawa, N., Overzier, R., Chiang, Y., Marinello, M., Tanaka, M., Niino, Y., *et al.* 2018, *PASJ*, 70, S32

Umehata, H., Tamura, Y., Kohno, K., Ivison, R. J., Smail, I., Hatsukade, B., Nakanishi, K., Kato, Y. *et al.* 2017, *ApJ*, 835, 98

Umehata, H., Hatsukade, B., Smail, I., Alexander, D. M. Ivison, R. J. Matsuda, Y. Tamura, Y. Kohno, K., *et al.* 2018, *PASJ*, 70, 65

Venemans, B. P., Röttgering, H. J. A., Miley, G. K., van Breugel, W. J. M., de Breuck, C., Kurk, J. D., Pentericci, L., Stanford, S. A., *et al.* 2007, *A&A*, 461, 823

Wang, T., Elbaz, D., Daddi, E., *et al.* 2016, *ApJ*, 828, 56

Challenges in Panchromatic Modelling with Next Generation Facilities
Proceedings IAU Symposium No. 341, 2018
M. Boquien, E. Lusso, C. Gruppioni & P. Tissera, eds.
doi:10.1017/S1743921319002667

The spectral energy distributions of active galactic nuclei

M. J. I. Brown[1], K. J. Duncan[2], H. Landt[3], M. Kirk[1], C. Ricci[4,5,6] and N. Kamraj[7]

[1]School of Physics & Astronomy, Monash University, Clayton, Victoria 3800, Australia
email: Michael.Brown@Monash.edu

[2]Leiden Observatory, Leiden University, NL-2300 RA Leiden, Netherlands

[3]Department of Physics, Centre for Extragalactic Astronomy, Durham University, South Road, Durham DH1 3LE, UK

[4]Núcleo de Astronomía de la Facultad de Ingeniería, Universidad Diego Portales, Av. Ejército Libertador 441, Santiago, Chile

[5]Kavli Institute for Astronomy and Astrophysics, Peking University, Beijing 100871, China

[6]Chinese Academy of Sciences South America Center for Astronomy, Camino El Observatorio 1515, Las Condes, Santiago, Chile

[7]Cahill Center for Astronomy and Astrophysics, California Institute of Technology, Pasadena, CA 91125, USA

Abstract. We present ongoing work on the spectral energy distributions (SEDs) of active galactic nuclei (AGNs), derived from X-ray, ultraviolet, optical, infrared and radio photometry and spectroscopy. Our work is motivated by new wide-field imaging surveys that will identify vast numbers of AGNs, and by the need to benchmark AGN SED fitting codes. We have constructed 41 SEDs of individual AGNs and 80 additional SEDs that mimic Seyfert spectra. All of our SEDs span 0.09 to 30 μm, while some extend into the X-ray and/or radio. We have tested the utility of the SEDs by using them to generate AGN photometric redshifts, and they outperform SEDs from the prior literature, including reduced redshift errors and flux density residuals.

Keywords. galaxies: active, (galaxies:) quasars: general, galaxies: Seyfert

---

1. Introduction

Photometric redshifts and AGN SED modelling will be critical for understanding the vast numbers of AGNs identified by wide-field surveys with new facilities such as eROSITA, ASKAP and LOFAR. It is also increasingly common to include AGN components in galaxy SED modelling, so empirical AGN SEDs could also prove critical for benchmarking the performance of SED modelling codes.

AGN SEDs have been developed by various groups over the past three decades, and some illustrative examples of quasar SEDs from the literature are provided in Figure 1. The SEDs in Figure 1 were developed with a variety of data, goals and methods, so some caution is required when comparing them. That said, there is an overall trend towards improved spectrophotometric accuracy, wavelength coverage and spectral resolution.

For this work we have produced AGN SEDs by combining X-ray, ultraviolet, optical, infrared and radio spectroscopy and photometry of individual objects (Brown *et al.* 2019). This approach can exploit the expanded wavelength range and spectrophotometric accuracy of ground-based telescopes and satellites from the past decade. However, it comes with the risk that variability and wavelength-dependent (extraction) aperture bias will

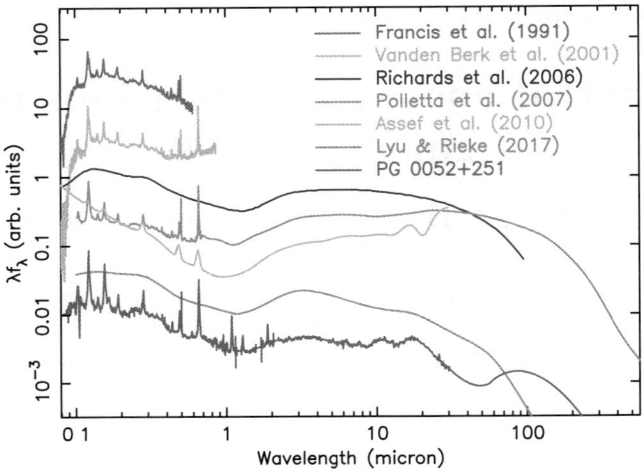

Figure 1. Quasar SEDs from the past three decades (Francis *et al.* 1991; Vanden Berk *et al.* 2001; Richards *et al.* 2006; Polletta *et al.* 2007; Assef *et al.* 2010; Lyu & Rieke 2017) along with our new SED for PG 0052+251. The trend towards greater spectral resolution and wavelength coverage is evident. Our PG 0052+251 SED has broad wavelength coverage while including near-infrared emission lines and 2mid-infrared silicate emission features.

produce unrealistic SEDs. To mitigate this risk we compare our SEDs with photometry and utilise the SEDs to determine AGN photometric redshifts.

2. Constructing AGN SEDs

To produce SEDs of individual AGNS we combine ground-based optical and near-infrared spectra with space-based X-ray, ultraviolet and infrared spectra. The availability (or lack) of near-infrared spectrophotometry strongly limits the building of SEDs, particularly as the near-infrared can include significant contributions from the AGN disk, torus and host galaxy. We mostly use published spectra or reduced spectra provided by archives, with only a few exceptions (NuSTAR, VLT XShooter, VLT SINFONI).

To rescale and verify spectrophotometry, and constrain SED models, we have measured aperture photometry across the ultraviolet to mid-infrared wavelength range using images from GALEX, Swift, SDSS, PanSTARRS, Skymapper, 2MASS and WISE. At longer wavelengths we have used published photometry from *Herschel*, WMAP, *Planck* and single-dish radio telescopes.

Most of the AGNs in our sample do not have far-infrared spectra, but the shape of far-infrared spectrum is sufficiently simple that it can often be adequately modelled with a grey-body curve. In the radio, we have modelled SEDs using power-laws and polynomials, which is adequate to model the observed flux densities with an accuracy of ~ 0.1 dex. Polynomials are used to interpolate over gaps in spectral coverage, including *Akari* to *Spitzer* and gaps in near-infrared spectra caused by atmospheric absorption.

To produce continuous SEDs, the individual spectra are initially scaled using the matched aperture photometry. The multiplicative scaling is then adjusted so neighbouring spectra have consistent flux densities. AGN variability and wavelength dependent (extraction) aperture bias does result in some SEDs being unphysical and inconsistent with the photometry, and these are rejected from our sample. For most of the remaining 41 AGNs, the multiplicative scalings range between 0.3 and 3 in the ultraviolet, and converge towards 1 with increasing wavelength. All of the SEDs span the 0.09 to 30 μm, but a subset extend into the X-ray, far-infrared and/or radio. Examples of several of our

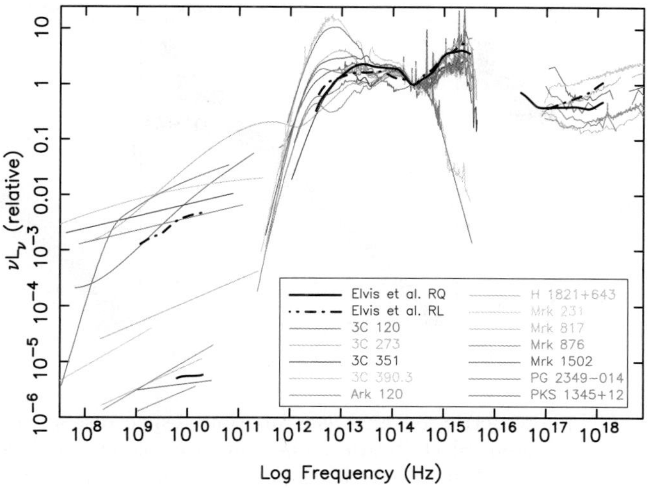

Figure 2. The AGN SEDs from our library with radio coverage, normalised at 1.25 μm, along with the Elvis *et al.* (1994) radio-loud and radio-quiet quasar SEDs for comparison. The diversity of AGN SEDs is apparent, including the broad range of radio luminosities and the diversity of SED shapes in the mid and far-infrared.

SEDs, along with the Elvis *et al.* (1994) radio-loud and radio-quiet quasar SEDs, are shown in Figure 2.

Our 41 SEDs for individual AGNs are biased towards luminous quasars and the central regions of nearby Seyferts, where the light is dominated by the AGN and aperture bias is thus reduced. To approximate the SEDs of Seyferts, including host galaxy light, we mix the SEDs of the central regions of Seyferts with galaxy SEDs taken from the Brown *et al.* (2014) sample, resulting in 80 additional SEDs spanning 0.09 to to 30 μm.

3. Photometric redshifts

We have tested the utility of our AGN SEDs by using them (and other SED libraries) to generate photometric redshifts (photo-zs) for X-ray selected AGNs in the Boötes. We compare photo-zs generated with our SEDs against those generated using the Brown *et al.* (2014) SEDs, which don't include powerful AGNs, and the AGN SEDs of Ananna *et al.* (2017), which build upon the Polletta *et al.* (2007) SEDs and were recently used to produce photo-zs for X-ray selected AGNs in Stripe-82. The photo-zs were determined using the EAZY code (Brammer *et al.* 2008) run on optical, near-infrared and mid-infrared photometry using the methods of Duncan *et al.* (2018).

Figure 3 illustrates the performance of the various SED libraries. Unsurprisingly, the Brown *et al.* (2014) galaxy SED library, which lacks powerful AGNs, performs poorly with $\sigma_{\rm NMAD} \sim 0.8 \times (1+z)$ for $z \sim 3$ quasars. Our new SEDs, in combination with the Brown *et al.* (2014) galaxy SEDs, produce significantly better quality photo-zs than the previous library alone and shows improvements over photo-zs using the Ananna *et al.* (2017) SEDs. Our photometric redshifts have a typical scatter of $\sigma_{\rm NMAD} = 0.096 \times (1+z)$ and flux density residuals are $\simeq 9\%$, with the exact value varying with wavelength. We attribute this improvement in part to the availability of new spectra, including from *Akari* and *Spitzer*. A complete and representative library of AGN SED templates will be vital for producing template-based photo-zs that are comparable in precision/reliability to those produced through the latest empirical methods (Duncan *et al.* 2019). Such redshifts will be necessary for as new AGN surveys push into parameter space that currently lacks the spectroscopic training samples required for empirical photo-z estimates.

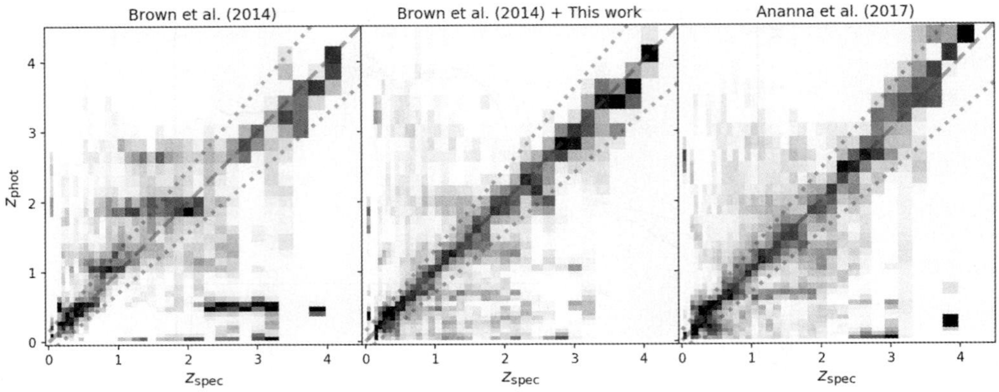

Figure 3. X-ray AGN photometric redshifts as a function of spectroscopic redshift, determined with EAZY while using different SED libraries. The plots are colour coded using the sum of the redshift probability density functions, so dark regions correspond to photometric redshifts with small uncertainties. Our new SEDs, in combination with the Brown *et al.* (2014) galaxy SEDs, produce the highest quality photometric redshifts, particularly for $z > 1.5$ quasars.

References

Assef, R. J., Kochanek, C. S., Brodwin, M., Cool, R., Forman, W., Gonzalez, A. H., Hickox, R. C., Jones, C., *et al.* 2010, *ApJ*, 713, 970

Ananna *et al.* 2017, https://ui.adsabs.harvard.edu/abs/2017ApJ...850...66A/abstract.

Brammer, G. B., van Dokkum, P. G., & Coppi, P. 2008, *ApJ*, 686, 1503

Brown, M. J. I., Moustakas, J., Smith, J.-D. T., da Cunha, E., Jarrett, T. H., Imanishi, M., Armus, L., Brandl, B. R., *et al.* 2014, *ApJS*, 212, 18

Brown *et al.* 2019, https://ui.adsabs.harvard.edu/abs/2019MNRAS.489.3351B/abstract.

Duncan, K. J., Brown, M. J. I., Williams, W. L., Best, P. N., Buat, V., Burgarella, D., Jarvis, M. J., Małek, K., *et al.* 2018, *MNRAS*, 473, 2655

Duncan, K. J, Sabater, J., Röttgering, H. J. A, Jarvis, M. J., Smith, D. J. B., Best, P. N., Callingham, J. R., Cochrane, R., *et al.* 2019, *A&A*, Forthcoming article

Elvis, M., Wilkes, B. J., McDowell, J. C., Green, R. F., Bechtold, J., Willner, S. P., Oey, M. S., Polomski, E., *et al.* 1994, *ApJS*, 95, 1

Francis, P. J., Hewett, P. C., Foltz, C. B., Chaffee, F. H., Weymann, R. J., Morris, S. L., 1991, *ApJ*, 373, 465

Lyu, J., Rieke, G. H., & Shi, Y. 2017, *ApJ*, 835, 257

Polletta, M., Tajer, M., Maraschi, L., Trinchieri, G., Lonsdale, C. J., Chiappetti, L., Andreon, S., Pierre, M., *et al.* 2007, *ApJ*, 663, 81

Richards, G. T., Lacy, M., Storrie-Lombardi, L. J., Hall, P. B., Gallagher, S. C., Hines, D. C., Fan, X., Papovich, C., *et al.* 2006, *ApJS*, 166, 470

Vanden Berk, D. E., Richards, G. T., Bauer, A., Strauss, M. A., Schneider, D. P., Heckman, T. M., York, D. G., Hall, P. B., *et al.* 2001, *AJ*, 122, 549

Discussion

David Rosario: Is there much variation in UV-optical SED shape?

Michael Brown: For the broadline quasars there's not that much variation in the continuum shape, although strong emission lines do contribute to the overall SED. There's significantly more variation in the mid-infrared, including the presence and absence of dust components.

David Rosario: What is the approach to the SED hybrids?

MICHAEL BROWN: We take the central SED of a Seyfert and then add contributions from a host galaxy SED. For example, 25% Seyfert and 75% host galaxy (defined at a wavelength of 0.6 μm).

THEMIYA NANAYAKKARA: What are the differences in performance between the empirical SEDs and models?

MICHAEL BROWN: When the new SEDs are fitted to AGN photometry they typically produce a reduced $\chi^2 \sim 1$, which is promising. This is, in part, due to the inclusion of strong emission lines, which are often difficult to include in SED models.

THEMIYA NANAYAKKARA: Can you explain the $z = 0$ photometric redshift outliers?

MICHAEL BROWN: All of the SED libraries produce some $z = 0$ photometric redshift outliers, and in some instances there are highly variable objects with photometry that wildly varies with wavelength, resulting in high χ^2.

Challenges in Panchromatic Modelling with Next Generation Facilities
Proceedings IAU Symposium No. 341, 2018
M. Boquien, E. Lusso, C. Gruppioni & P. Tissera, eds.

doi:10.1017/S1743921319003016

Panchromatic SED fitting codes and modelling techniques

Maarten Baes

Sterrenkundig Observatorium, Universiteit Gent, Krijgslaan 281 S9, 9000 Gent, Belgium
email: maarten.baes@ugent.be

Abstract. Modelling and interpreting the SEDs of galaxies has become one of the key tools at the disposal of extragalactic astronomers. Ideally, we could hope that, through a detailed study of its SED, we can infer the correct physical properties and the evolutionary history of a galaxy. In the past decade, panchromatic SED fitting, i.e. modelling the SED over the entire UV–submm wavelength regime, has seen an enormous advance. Several advanced new codes have been developed, nearly all based on Bayesian inference modelling. In this review, we briefly touch upon the different ingredients necessary for panchromatic SED modelling, and discuss the methodology and some important aspects of Bayesian SED modelling. The current uncertainties and limitations of panchromatic SED modelling are discussed, and we explore some avenues how the models and techniques can potentially be improved in the near future.

Keywords. galaxies: stellar content, galaxies: ISM, infrared: galaxies, ultraviolet: galaxies, methods: statistical, dust, extinction, radiative transfer

1. Introduction

The main goal of extragalactic astronomy is understanding the formation and evolution of galaxies throughout cosmic history. The global spectral energy distributions (SEDs) of galaxies are a key element in this grand challenge. Indeed, the different astrophysical processes that shape a galaxy, such as star formation, stellar evolution, chemical enrichment and processes in the interstellar medium, all leave an imprint on the SED. Ideally, we could hope that, through a detailed study of its SED, we can infer the physical properties and the evolutionary history of a galaxy. This detailed study is commonly referred to as SED modelling or SED fitting. For a general background on galaxy SED modelling we refer to the excellent reviews on this topic by Tinsley (1980); Walcher *et al.* (2011) and Conroy (2013).

Conceptually, SED modelling is relatively straightforward, and usually is done in two phases. The first phase consists of the construction of a theoretical or empirical parameterised SED model that is assumed to incorporate all the physical ingredients and processes that shape the SEDs of galaxies. The basis of all galaxy SED models are simple stellar populations (SSPs), which describe the SED of a coeval stellar population at a given age, metallicity and abundance pattern. Under the hood, any SSP model requires assumptions on the initial mass function, stellar evolution, and stellar spectral libraries, and different sets of SSP models are available in the literature (e.g., Bruzual & Charlot 1993; 2003; Leitherer *et al.* 1999; Maraston 2005; Conroy *et al.* 2009). From a given SSP library, a composite stellar population is built by combining different SSPs according to a chemical enrichment history (CEH) and star formation history (SFH). In practice, a parameterised galaxy SED model usually consists of a huge grid of individual SED templates, each of them corresponding to a specific set of values for the model parameters.

The second phase in the SED modelling consists of determining the parameter values of the model SED that reproduces the observed galaxy SED best. Again, this sounds relatively straightforward, but it is more complex in practice. Firstly, the theoretical model SEDs need to be converted to mock SED data that can immediately be compared to the observations, which involves redshifting, filtering through the spectral response curves of the instruments, and spatial redistribution over the point spread function of the instruments (Walcher *et al.* 2011). Secondly, determining which model SED reproduces the observed SED "best" is not a straightforward task with a unique answer. As for most optimisation problems, the simplest answer is usually some form of χ^2 optimisation, but this is not necessarily the "best" answer, especially when the data are noisy.

Until fairly recently, the observed SEDs of galaxies were often limited to the UVOIR (UV, optical and NIR) regime. At these wavelengths, the SED is dominated by stellar continuum emission, with usually minor contributions from ionised gas or an active galactic nucleus. The UVOIR SED could hence be modelled as a weighted sum of SSP SEDs, except for the disturbances caused by interstellar dust. Dust grains in the interstellar medium of galaxies are extremely efficient at absorbing and scattering UV and optical radiation. The net effect of dust absorption and scattering on the SED of galaxies is labelled attenuation. Correcting UVOIR galaxy SEDs for the effects of attenuation is nontrivial, as the shape of the attenuation curve depends on both the optical properties of the dust and the relative star-dust geometry (Baes & Dejonghe 2001; Tuffs *et al.* 2004; Trayford *et al.* 2017; Narayanan *et al.* 2018). Nearly all SED modelling codes assume a parameterised attenuation curve (e.g., Calzetti *et al.* 2000; Charlot & Fall 2000; Lo Faro *et al.* 2017; Buat *et al.* 2018).

The effect of dust attenuation on the SED of galaxies, and thus on the physical properties derived from SED fitting, are substantial. In nearby, star-forming galaxies, interstellar dust grains absorb about 30% of the radiative energy emitted by stars (Popescu & Tuffs 2002; Viaene *et al.* 2016; Bianchi *et al.* 2018). This absorbed energy is re-emitted at longer wavelengths, i.e., in the mid-infrared, far-infrared, and submm regimes. Except for some extreme cases where synchrotron emission from an AGN is important (e.g., Baes *et al.* 2010), thermal dust emission is the dominant contributor to the SED of galaxies in the infrared regime. In the past two decades, space missions such as WISE, Spitzer, and Herschel have opened up these regimes, and have observed large areas of the extragalactic sky. Major efforts have been undertaken to create databases with observed panchromatic SEDs† for thousands of nearby and distant galaxies (e.g., Scoville *et al.* 2007; Smith *et al.* 2012a; Liske *et al.* 2015; Vaccari 2016; Andrews *et al.* 2017; Clark *et al.* 2018). This availability of data across the entire UV–submm spectrum gives us much stronger constraints on both the stellar populations and the interstellar dust content in galaxies, and it is hence wise to use all of this information in the interpretation of galaxy SEDs. On the other hand, self-consistently modelling the entire UV–submm SED of galaxies, i.e., panchromatic SED modelling, also brings along some intellectual and technical challenges.

2. Panchromatic SED fitting codes

At UVOIR wavelengths, the SED of a galaxy can be regarded as the weighted sum of SEDs corresponding to different SSPs, with an additional attenuation correction factor. SED fitting then comes down to determining the optimal weights, which is a linear inversion problem that can be solved by classical inversion techniques. This approach is

† In this review we concentrate on the wavelength regime from UV to submm wavelengths, and we leave out the gamma-ray, X-ray, microwave and radio regimes. The UV–submm regime is reflects nearly all of the energy emitted by stars, either directly or reprocessed by dust absorption and re-emission. Outside this regime, non-stellar processes dominate the emission.

used by many spectral fitting codes including STARLIGHT (Cid Fernandes *et al.* 2005), STECKMAP (Ocvirk *et al.* 2006), VESPA (Tojeiro *et al.* 2007), ULySS (Koleva *et al.* 2009), and FIREFLY (Wilkinson *et al.* 2017). When we consider the full panchromatic SED of a galaxy, inversion techniques are not suitable anymore, because the SED is no longer a simple weighted sum of components when thermal dust emission is taken into account. Moreover, the number of free parameters to be considered is typically rather large, with many possible degeneracies between the parameters.

Nearly all panchromatic SED fitting codes operational today are based on Bayesian inference modelling. Bayes' theorem can be written as

$$P(\boldsymbol{\Theta}|\mathbf{D},\mathrm{M}) = \frac{P(\boldsymbol{\Theta}|\mathrm{M})\,P(\mathbf{D}|\boldsymbol{\Theta},\mathrm{M})}{P(\mathbf{D}|\mathrm{M})} \tag{2.1}$$

In this equation, $\boldsymbol{\Theta}$ corresponds to a set of parameters of the theoretical SED model M, and $\mathbf{D}$ is the observed SED data set. In words, Bayes' theorem states that the posterior distribution of the parameters $\boldsymbol{\Theta}$ can be found as the normalised product of the prior distribution of $\boldsymbol{\Theta}$ and the likelihood of the data, given this set of parameters. This formula enables us to draw rigorous constraints on the model parameters based on the observed SED, within the modelling framework M, without any specific assumptions about the shape of the posterior probability distribution. To determine the "best" parameters for a specific data set, the posterior distribution is marginalised over all parameters except one, which results in a series of one-dimensional posterior probability density functions,

$$P(\Theta_i|\mathbf{D},\mathrm{M}) = \int P(\Theta_i|\mathbf{D},\mathrm{M})\,\mathrm{d}\Theta_1 \cdots \mathrm{d}\Theta_{i-1}\,\mathrm{d}\Theta_{i+1} \cdots \mathrm{d}\Theta_m \tag{2.2}$$

For a given parameter, the median/mean and standard deviation of this marginalised distribution are taken as the best estimate for this parameter and the associated uncertainty. Specifically interesting about Bayesian inference modelling in the context of SED fitting is that instrumental systematics can be incorporated in the modelling in a consistent way. Also, full posterior distributions for the observables are available, and complex correlations between model parameters can be investigated. For a more detailed description of Bayesian inference in the context of SED modelling, we refer to Kauffmann *et al.* (2003); Han & Han (2014) and Chevallard & Charlot (2016).

Two aspects in Bayesian SED fitting deserve some special attention. The first one is the need for calculating the marginalised probability density functions for all parameters, which involves complex multi-dimensional integrations over the parameter space. The easiest approach to deal with this problem is to use a regular model grid, which makes the marginalisation rather trivial. Obvious disadvantages of a grid approach are that this method can become inefficient and very memory-intensive, in particular when many different parameters need to be considered. In this case, the grid often needs to be made rather coarse, which can result in poor accuracy and can lead to finite-resolution effects (see e.g., Nersesian *et al.* 2019; Decleir *et al.* 2019). Alternative sampling methods such as Markov Chain Monte Carlo or nested sampling can avoid these problems. These methods result in an increase in the efficiency and reliability of the computation of the marginalised distributions, as calculations in irrelevant parts of the parameter space are avoided, and the computational effort is concentrated on the important parts only (Conroy *et al.* 2009; Acquaviva *et al.* 2011). A challenge for such methods is that the theoretical SED needs to computable for *any* set of values of the parameters, which usually requires some kind of interpolation. Artificial neural networks are a promising technique to do this interpolation (Han & Han 2014).

A second important aspect is the choice of the model M and the prior distribution $P(\boldsymbol{\Theta}|\mathrm{M})$, which reflects our assumed knowledge about the model parameters without any knowledge of the particular data set **D**. In some cases, the likelihood distribution will be informative enough, such that the prior distribution has only a negligible impact on the resulting posterior distribution. In other cases, however, the data do not really constrain some model parameters, and the choice of the priors (and more generally, the choice of the model) can have a significant effect on the final results. Standard choices for the priors are often flat distributions in linear or logarithmic space, although it is clear that suitable priors should be preferred if possible. One advanced approach to obtain realistic priors is so-called hierarchical Bayesian SED modelling, which can be applied when modelling a sample of galaxies. In this case, the prior distributions are inferred from the data. This is achieved by parameterising the shape and position of the new prior distribution with hyperparameters, which control the distribution of the parameters (Kelly *et al.* 2012; Galliano 2018).

Over the past few years, many different Bayesian SED fitting codes have been conceived, and some of them were specifically designed to cover the entire UV–submm wavelength range. Two pioneering panchromatic SED fitting codes are MAGPHYS (da Cunha *et al.* 2008, 2015) and CIGALE (Noll *et al.* 2009; Boquien *et al.* 2019). These codes are grid-based Bayesian codes that fit the entire UV–submm SED of galaxies under the constraint that the energy absorbed by dust is balanced by the energy emitted at long wavelengths. Both codes have been applied extensively in the literature (e.g. Smith *et al.* 2012a; Driver *et al.* 2018, Małek *et al.* 2018). More recent codes that include advanced options on the prior distributions and/or the sampling methodology include BayeSED (Han & Han 2014), BEAGLE (Chevallard & Charlot 2016), Prospector (Leja *et al.* 2017) and BAGPIPES (Carnall *et al.* 2018). Comprehensive comparisons between different panchromatic galaxy SED codes have been presented by Pappalardo *et al.* (2016); Leja *et al.* (2017) and Hunt *et al.* (2019).

3. Uncertainties in panchromatic SED models

In spite of the progress that has been made over the past few years, there are still substantial uncertainties linked to SED modelling. In fact, all the different inputs in the SED modelling processes are prone to at least some degree of uncertainty.

In a series of papers, Conroy *et al.* (2009, 2010) and Conroy & Gunn (2010) discuss how uncertainties in stellar evolution and the IMF propagate to uncertainties in SSP models (see also Walcher *et al.* 2011). Among the most important uncertainties are particular late phases in stellar evolution (thermally pulsating AGB stars, blue stragglers, etc.), the effects of binary stellar evolution, and the high-mass slope of the IMF. They warn that the interpretation of the resulting uncertainties in the derived colours is highly non-trivial because many of the uncertainties are likely systematic, and possibly correlated with the physical properties of galaxies. As an indication, they find that stellar masses in the local Universe typically carry errors of about 0.3 dex with little dependence on luminosity or color, while this uncertainty increases to about 0.6 dex for bright red galaxies at $z \sim 2$.

Other components in panchromatic SED modelling also carry their uncertainties, and propagate them to the final model SEDs. The treatment of dust in particular is prone to many systematic uncertainties. The shape of the attenuation curve is one of these; most codes use either the empirical Calzetti *et al.* (2000) or Charlot & Fall (2000) prescriptions for the attenuation curve. Whether these prescriptions are realistic is subject to quite some debate, and it is definitely difficult to constrain the attenuation based on UVOIR data alone. In principle, strong constraints on the dusty ISM can be obtained from the infrared portion of the SED, and this information can be coupled to the attenuation obtained at UVOIR wavelengths. The relation between the amount of dust, the grain properties, and the attenuation is far from trivial, however, and strongly depends on the star-dust geometry, the structure of the ISM, and the importance of scattering.

Further complications arise from the fact that galaxy SEDs are anisotropic at short wavelengths (more attenuation for higher inclinations), whereas they are largely isotropic in the FIR/submm regime. This breaks the energy balance approach. Moreover, the actual physical dust models are subject to substantial uncertainties (Jones *et al.* 2017; Galliano *et al.* 2018, and references therein), and the implicit assumption that the dust properties are uniform across a galaxy is most probably a major simplification (Smith *et al.* 2012b; Mattsson & Andersen 2012; Relaño *et al.* 2018).

Finally, significant uncertainties in SED models are linked to how the different SSP and dust models are combined to a global SED model, and how this model is fitted to the data. As already indicated, most SED modelling codes are Bayesian in nature, and the choice of the model and the priors is crucial. A typical example is the choice of the SFH in the model. In many codes, the SFH is parameterised using simple analytical prescriptions (e.g., Lee *et al.* 2009; Pforr *et al.* 2012; Ciesla *et al.* 2016), which rule out a significant fraction of the full diversity of SFH shapes a priori. Using the Prospector code equipped with a non-parametric SFH module, Leja *et al.* (2018) demonstrate that the choice of prior rather than the photometric noise is the primary determinant of the size and shape of the posterior SFH. A similar, but probably less dramatic, effect is to be expected for the CEH. While Bayesian SED fitting codes can, in principle, include any parametric or non-parametric CEH, most codes assume a single metallicity for the entire composite population. The impact of this simplification has not been extensively explored (Conroy 2013).

Given all these uncertainties, simplifications and potential systematics, one could wonder whether we should believe the results of panchromatic SED modelling efforts. One way to answer this question, at least partially, is by applying panchromatic SED modelling to mock observables from hydrodynamically simulated galaxies. In a series of papers on this topic, Hayward & Smith (2015) and Smith & Hayward (2015, 2018) apply MAGPHYS to a suite of high-resolution hydrodynamical simulations. They conclude that most physical parameters are recovered fairly well in most cases, albeit with a number of caveats and systematics. In particular, recovering the SFH proved to be difficult, even though the mock photometry was fit reasonably well.

4. Towards more advanced SED templates

In order to bring panchromatic SED modelling to a new level, it is clear that more work is needed to decrease the uncertainties on all of these different aspects: stellar evolution, SSP models, dust models, attenuation curves, SFHs, CEHs, realistic priors, etc.

An obvious way to improve the handling of dust attenuation and emission is dust radiative transfer modelling. Only when dust radiative transfer is rigorously calculated in a realistic framework, one can be sure that attenuation and emission by dust are consistent, i.e., the dust energy balance is satisfied on global and local scales. The field of 3D dust radiative transfer has advanced enormously in recent years (for an overview, see Steinacker *et al.* 2013). Most progress has been made on Monte Carlo radiative transfer techniques, with novel acceleration techniques (Baes *et al.* 2011, 2016), advanced grids and associated traversal algorithms (Kurosawa & Hillier 2001; Camps *et al.* 2013; Saftly *et al.* 2013, 2014), and hybrid parallelisation strategies (Verstocken *et al.* 2017). As a result, grids of self-consistent radiative transfer models for AGNs (Stalevski *et al.* 2012, 2016; Siebenmorgen *et al.* 2015) and galaxies (Efstathiou *et al.* 2000; Law *et al.* 2018) have appeared in the literature, which can be used as the base components in SED modelling efforts.

These efforts could be taken a step further through the combination of high-resolution hydrodynamical modelling and 3D dust radiative transfer. Cosmological hydrodynamical simulations have only fairly recently achieved sufficient realism and statistics to provide a predictive model for the Universe (Vogelsberger *et al.* 2014; Schaye *et al.* 2015; Pillepich *et al.* 2018), and zoom simulations have managed to produce high-resolution galaxy models that reproduce the properties and scaling relations of galaxies in many aspects (Wang *et al.* 2015; Grand *et al.* 2017). Combined with the advances in 3D dust radiative transfer modelling, it is now possible to self-consistently calculate panchromatic mock observables for hydrodynamically simulated galaxies (Jonsson *et al.* 2010; Hayward *et al.* 2011; Domínguez-Tenreiro *et al.* 2014; Saftly *et al.* 2015; Goz *et al.* 2017; Rodriguez-Gomez *et al.* 2019).

Interestingly, 3D dust radiative transfer postprocessing can now be applied to entire populations of galaxies from cosmological simulations. Camps *et al.* (2016, 2018) applied the SKIRT radiative transfer code (Camps & Baes 2015) to almost half a million simulated galaxies from the suite of EAGLE simulations (Schaye *et al.* 2015). Such collections of panchromatic mock SEDs could form an ideal set of templates for panchromatic SED modelling. Indeed, for these templates, attenuation and thermal emission by dust is calculated in a self-consistent way, and in a realistic 3D geometry. Futhermore, the SFHs and CEHs of simulated galaxies are probably more realistic than the simple analytical prescriptions used in most panchromatic SED fitting codes. Finally, the current generation of cosmological hydrodynamical simulations manage to reproduce many observed population characteristics (stellar mass function, SFR function, etc.) and scaling laws (mass-metallicity relation, relation between stellar mass and angular momentum, etc.), such that the population of galaxies in a cosmological simulation could in principle be used to construct more informative priors on the model parameters.

5. Conclusions

Panchromatic UV–submm SED fitting has received an enormous boost in the past decade. In their 2011 review, Walcher *et al.* (2011) concluded that most panchromatic SED fitting codes are still in their testing phase, and argued that the large number of derived parameters and our still limited knowledge of their respective degeneracies and systematic uncertainties make it difficult to go a step further and fully use the full power of SED fitting. Since these conclusions were written, several new codes, all based on Bayesian inference modelling, have seen the light or have been profoundly upgraded, and our knowledge on the importance of realistic priors and noise models, and our methods to explore the parameter space and calculate posterior distributions have improved substantially. Without a doubt, there are still many uncertainties on many aspects involved in the SED modelling process, but these are gradually better understood, and their impact on the derived properties of galaxies are being investigated in detail. The current generation of cosmological hydrodynamical models, combined with 3D dust radiative transfer techniques, could be an interesting path to bring panchromatic modelling to a new level. In short: the future of panchromatic SED modelling is bright.

Acknowledgements

The author thanks the organisers of the PanModel2018 conference for a very interesting and entertaining conference, and particularly for the invitation to deliver an invited review on panchromatic SED fitting codes and modelling techniques. It was a honour to be asked. Wouter Dobbels and Peter Camps are acknowledged for comments on this manuscript.

References

Acquaviva, V., Gawiser, E., & Guaita, L. 2011, *ApJ*, 737, 47
Andrews, S. K., Driver, S. P., Davies, L. J. M., *et al.* 2017, *MNRAS*, 464, 1569
Aoyama, S., Hou, K.-C., Hirashita, H., Nagamine, K., & Shimizu, I. 2018, *MNRAS*, 478, 4905

Baes, M. & Dejonghe, H. 2001, *MNRAS*, 326, 733
Baes, M., Clemens, M., Xilouris, E. M., *et al.* 2010, *A&A*, 518, L53
Baes, M., Verstappen, J., De Looze, I., *et al.* 2011, *ApJS*, 196, 22
Baes, M., Gordon, K. D., Lunttila, T., *et al.* 2016, *A&A*, 590, A55
Baes, M., Trčka, A., Camps, P., *et al.* 2019, *MNRAS*, 484, 4069
Bekki, K. 2015, *MNRAS*, 449, 1625
Bianchi, S., De Vis, P., Viaene, S., *et al.* 2018, *A&A*, 620, A112
Boquien, M., Burgarella, D., Roehlly, Y., *et al.* 2019, *A&A*, 622, A103
Bruzual, G. & Charlot, S. 1993, *ApJ*, 405, 538
Bruzual, G. & Charlot, S. 2003, *MNRAS*, 344, 1000
Buat, V., Boquien, M., Małek, K., *et al.* 2018, *A&A*, 619, A135
Calzetti, D., Armus, L., Bohlin, R. C., *et al.* 2000, *ApJ*, 533, 682
Camps, P. & Baes, M. 2015, *Astron. Comput.*, 9, 20
Camps, P., Baes, M., & Saftly, W. 2013, *A&A*, 560, A35
Camps, P., Trayford, J. W., Baes, M., *et al.* 2016, *MNRAS*, 462, 1057
Camps, P., Trčka, A., Trayford, J., *et al.* 2018, *ApJS*, 234, 20
Carnall, A. C., McLure, R. J., Dunlop, J. S., & Davé, R. 2018, *MNRAS*, 480, 4379
Charlot, S. & Fall, S. M. 2000, *ApJ*, 539, 718
Chevallard, J. & Charlot, S. 2016, *MNRAS*, 462, 1415
Cid Fernandes, R., Mateus, A., Sodré, L., Stasińska, G., & Gomes, J. M. 2005, *MNRAS*, 358, 363
Ciesla, L., Boselli, A., Elbaz, D., *et al.* 2016, *A&A*, 585, A43
Clark, C. J. R., Verstocken, S., Bianchi, S., *et al.* 2018, *A&A*, 609, A37
Conroy, C. 2013, *ARA&A*, 51, 393
Conroy, C. & Gunn, J. E. 2010, *ApJ*, 712, 833
Conroy, C., Gunn, J. E., & White, M. 2009, *ApJ*, 699, 486
Conroy, C., White, M., & Gunn, J. E. 2010, *ApJ*, 708, 58
da Cunha, E., Charlot, S., & Elbaz, D. 2008, *MNRAS*, 388, 1595
da Cunha, E., Walter, F., Smail, I. R., *et al.* 2015, *ApJ*, 806, 110
Decleir, M., De Looze, I., Boquien, M., *et al.* 2019, *MNRAS*, 486, 743
Domínguez-Tenreiro, R., Obreja, A., Granato, G. L., *et al.* 2014, *MNRAS*, 439, 3868
Driver, S. P., Andrews, S. K., da Cunha, E., *et al.* 2018, *MNRAS*, 475, 2891
Efstathiou, A., Rowan-Robinson, M., & Siebenmorgen, R. 2000, *MNRAS*, 313, 734
Galliano, F. 2018, *MNRAS*, 476, 1445
Galliano, F., Galametz, M., & Jones, A. P. 2018, *ARA&A*, 56, 673
Goz, D., Monaco, P., Granato, G. L., *et al.* 2017, *MNRAS*, 469, 3775
Grand, R. J. J., Gómez, F. A., Marinacci, F., *et al.* 2017, *MNRAS*, 467, 179
Han, Y. & Han, Z. 2014, *ApJS*, 215, 2
Hayward, C. C. & Smith, D. J. B. 2015, *MNRAS*, 446, 1512
Hayward, C. C., Kereš, D., Jonsson, P., *et al.* 2011, *ApJ*, 743, 159
Hunt, L. K., De Looze, I., Boquien, M., *et al.* 2019, *A&A*, 621, A51
Jones, A. P., Köhler, M., Ysard, N., Bocchio, M., & Verstraete, L. 2017, *A&A*, 602, A46
Jonsson, P., Groves, B. A., & Cox, T. J. 2010, *MNRAS*, 403, 17
Kauffmann, G., Heckman, T. M., White, S. D. M., *et al.* 2003, *MNRAS*, 341, 33
Kelly, B. C., Shetty, R., Stutz, A. M., *et al.* 2012, *ApJ*, 752, 55
Koleva, M., Prugniel, P., Bouchard, A., & Wu, Y. 2009, *A&A*, 501, 1269
Kurosawa, R. & Hillier, D. J. 2001, *A&A*, 379, 336
Law, K.-H., Gordon, K. D., & Misselt, K. A. 2018, *ApJS*, 236, 32
Lee, S.-K., Idzi, R., Ferguson, H. C., *et al.* 2009, *ApJS*, 184, 100
Leitherer, C., Schaerer, D., Goldader, J. D., *et al.* 1999, *ApJS*, 123, 3
Leja, J., Johnson, B. D., Conroy, C., van Dokkum, P. G., & Byler, N. 2017, *ApJ*, 837, 170
Leja, J., Carnall, A. C., Johnson, B. D., Conroy, C., & Speagle, J. S. 2018, *ApJ*, 876, 3
Liske, J., Baldry, I. K., Driver, S. P., *et al.* 2015, *MNRAS*, 452, 2087
Lo Faro, B., Buat, V., Roehlly, Y., *et al.* 2017, *MNRAS*, 472, 1372
Małek, K., Buat, V., Roehlly, Y., *et al.* 2018, *A&A*, 620, A50

Maraston, C. 2005, *MNRAS*, 362, 799
Mattsson, L. & Andersen, A. C. 2012, *MNRAS*, 423, 38
McKinnon, R., Torrey, P., Vogelsberger, M., Hayward, C. C., & Marinacci, F. 2017, *MNRAS*, 468, 1505
McKinnon, R., Vogelsberger, M., Torrey, P., Marinacci, F., & Kannan, R. 2018, *MNRAS*, 478, 2851
Muzzin, A., Marchesini, D., Stefanon, M., *et al.* 2013, *ApJS*, 206, 8
Narayanan, D., Conroy, C., Davé, R., Johnson, B. D., & Popping, G. 2018, *ApJ*, 869, 70
Nersesian, A., Xilouris, E. M., Bianchi, S., *et al.* 2019, *A&A*, 624, A80
Noll, S., Burgarella, D., Giovannoli, E., *et al.* 2009, *A&A*, 507, 1793
Ocvirk, P., Pichon, C., Lançon, A., & Thiébaut, E. 2006, *MNRAS*, 365, 74
Pappalardo, C., Bizzocchi, L., Fritz, J., *et al.* 2016, *A&A*, 589, A11
Pforr, J., Maraston, C., & Tonini, C. 2012, *MNRAS*, 422, 3285
Pillepich, A., Springel, V., Nelson, D., *et al.* 2018, *MNRAS*, 473, 4077
Popescu, C. C. & Tuffs, R. J. 2002, *MNRAS*, 335, L41
Relaño, M., De Looze, I., Kennicutt, R. C., *et al.* 2018, *A&A*, 613, A43
Rodriguez-Gomez, V., Snyder, G. F., Lotz, J. M., *et al.* 2019, *MNRAS*, 483, 4140
Saftly, W., Camps, P., Baes, M., *et al.* 2013, *A&A*, 554, A10
Saftly, W., Baes, M., & Camps, P. 2014, *A&A*, 561, A77
Saftly, W., Baes, M., De Geyter, G., *et al.* 2015, *A&A*, 576, A31
Schaye, J., Crain, R. A., Bower, R. G., *et al.* 2015, *MNRAS*, 446, 521
Scoville, N., Aussel, H., Brusa, M., *et al.* 2007, *ApJS*, 172, 1
Serra, P., Amblard, A., Temi, P., *et al.* 2011, *ApJ*, 740, 22
Siebenmorgen, R., Heymann, F., & Efstathiou, A. 2015, *A&A*, 583, A120
Smith, D. J. B. & Hayward, C. C. 2015, *MNRAS*, 453, 1597
Smith, D. J. B. & Hayward, C. C. 2018, *MNRAS*, 476, 1705
Smith, D. J. B., Dunne, L., da Cunha, E., *et al.* 2012, *MNRAS*, 427, 703
Smith, M. W. L., Eales, S. A., Gomez, H. L., *et al.* 2012, *ApJ*, 756, 40
Stalevski, M., Fritz, J., Baes, M., Nakos, T., & Popović, L. Č. 2012, *MNRAS*, 420, 2756
Stalevski, M., Ricci, C., Ueda, Y., *et al.* 2016, *MNRAS*, 458, 2288
Steinacker, J., Baes, M., & Gordon, K. D. 2013, *ARA&A*, 51, 63
Tinsley, B. M. 1980, *Fundamentals of Cosmic Physics*, 5, 287
Tojeiro, R., Heavens, A. F., Jimenez, R., & Panter, B. 2007, *MNRAS*, 381, 1252
Trayford, J. W., Camps, P., Theuns, T., *et al.* 2017, *MNRAS*, 470, 771
Tuffs, R. J., Popescu, C. C., Völk, H. J., Kylafis, N. D., & Dopita, M. A. 2004, *A&A*, 419, 821
Vaccari, M. 2016, *The Universe of Digital Sky Surveys*, 42, 71
Verstocken, S., Van De Putte, D., Camps, P., & Baes, M. 2017, *Astron. Comput.*, 20, 16
Viaene, S., Baes, M., Bendo, G., *et al.* 2016, *A&A*, 586, A13
Vogelsberger, M., Genel, S., Springel, V., *et al.* 2014, *MNRAS*, 444, 1518
Walcher, J., Groves, B., Budavári, T., & Dale, D. 2011, *ApSS*, 331, 1
Wang, L., Dutton, A. A., Stinson, G. S., *et al.* 2015, *MNRAS*, 454, 83
Wilkinson, D. M., Maraston, C., Goddard, D., Thomas, D., & Parikh, T. 2017, *MNRAS*, 472, 4297
Zubko, V., Dwek, E., & Arendt, R. G. 2004, *ApJS*, 152, 211

Discussion

FABIO FONTANOT: How much can we hope to reduce the inferred error on recovered physical parameters obtained from panchromatic SED fitting?

MAARTEN BAES: This is very hard to estimate at this moment. We know from various studies that the choice of the prior is fundamental, and currently it seems unrealistic to claim uncertainties smaller than ~0.3 dex on stellar masses. Star formation histories are much more uncertain, and it would be interesting to test whether it is possible to reduce these uncertainties by building in priors as obtained from cosmological hydrodynamical simulations.

Alexa Villaume: Cosmological hydrodynamical simulations rely on subgrid prescriptions to include star formation. Is there something circular to using those to make templates for SED fitting to obtain information about star formation?

Maarten Baes: I believe that, indeed, there is some kind of circularity in this process, but this is unavoidable as long as we cannot model all processes in galaxy evolution ab initio. For most cosmological hydrodynamical simulations, the parameters of the subgrid models are calibrated based on observed population properties and scaling relations in the local Universe, which does not really put very strong constraints on the possible range of star formation histories. In general, the range of SFHs obtained from cosmological hydrodynamical simulations is broader than the set of parameterised SFHs often used in SED modelling.

Lapo Fanciullo: How is dust formation and evolution modelled in cosmological hydrodynamical simulations? What are the properties of the dust in the simulation?

Maarten Baes: In our SKIRT-EAGLE work (Camps *et al.* 2016, 2018, Trayford *et al.* 2017, Baes *et al.* 2019) we have adopted a fixed grain model in our post-processing radiative transfer modelling. More particularly, we have used the BARE_GR_S model from Zubko *et al.* (2004), and assumed a fixed dust-to-metal fraction of 30% for the diffuse cold gas. In principle, it is possible to include a dust evolution model into a cosmological hydrodynamical model (e.g., Bekki 2015, McKinnon *et al.* 2017, 2018, Aoyama *et al.* 2018). Unfortunately, the detailed physics involved in the different processes involved, like grain formation, coagulation, shattering, sputtering and drag forces, are still poorly constrained, and the coupling to a dust evolution model makes the hydrodynamical simulations extremely computationally demanding.

Hiroyuki Hirashita: Aoyama *et al.* (2018) also have a cosmological simulation with dust evolution, and with a grain size evolution. How good should the spatial resolution be to obtain a reliable SED?

Maarten Baes: In Camps *et al.* (2018) we have experimented with different models, and we found that a minimum of about 250 dust particles per galaxy in the simulation is needed for the dust radiative transfer to yield meaningful results. Obviously, the higher the spatial/mass resolution, the better.

Challenges in Panchromatic Modelling with Next Generation Facilities
Proceedings IAU Symposium No. 341, 2018
M. Boquien, E. Lusso, C. Gruppioni & P. Tissera, eds.

doi:10.1017/S174392131900262X

Evolutionary Population Synthesis model with binary stars – Yunnan-II model

F. Zhang[1,2], Z. Han[1,2] and L. Li[1,2]

[1]National Astronomical Observatories/Yunnan Observatory,
Chinese Academy of Sciences, Kunming, 650011, China

[2]Key Laboratory for the Structure and Evolution of Celestial Objects,
Chinese Academy of Sciences, Kunming, 650011, China
email: zhangfh@ynao.ac.cn

Abstract. By considering a modified version of the evolutionary population synthesis (EPS) model for stellar populations (SPs) comprising binary stars, the retrieved galaxy and HII-region parameters/properties differ from the case of neglecting binary stars. The retrieved age, stellar metallicity and mass of galaxies increase (e.g. ∼0.2 dex when using spectral fitting algorithm), whilst the star formation rate decreases (∼0.2 dex). The radiation fields from intermediate-age SPs with binary stars could be potentially important ionizing sources in HII regions. Under this possibility, the theoretical division between star forming galaxy and AGN on the diagnostic diagrams would move towards the up-right corner and the retrieved gaseous metallicity would decrease.

Our prediction for the birth rate of binary neutron stars in SPs ranges from 10^{-9} to $10^{-6}\,M_{\odot}^{-1}\,\mathrm{yr}^{-1}$ when the kick velocity is from 0 to 190 km s^{-1}.

Keywords. galaxies: stellar content, galaxies: fundamental parameters, galaxies: evolution, stars: neutron

1. Introduction

Observationally, binary stars have been found in many environments (Duchene & Kraus 2013): protostars, pre-main-sequence stars, young nearby associations, young massive star clusters (Sana *et al.* 2012), open clusters and Population II stars. Evolutionary population synthesis (EPS) models are a basic tool in studies of clusters and galaxies. So, we could not neglect the effect of binary stars in EPS model.

We have included the effect of binary stars in the Yunnan-II EPS model (Zhang *et al.* 2004, 2005, http://www1.ynao.ac.cn/∼zhangfh), in which different evolutions of binary stars are obtained via the rapid binary star evolution (BSE) algorithms of Hurley *et al.* (2002). Since then, the model is constantly updated by improving the algorithm, using up-to-date initial distributions for stellar populations (SPs) and the latest BSE code. BPASS is another EPS model considering the effect of binary stars (Eldridge & Tout 2008, http://bpass.auckland.ac.nz). In this model, the evolutions of binary stars are obtained by using the detailed Cambridge stellar evolution code STARS and the mass of the primary star in a binary system is within $5 \leqslant M_1/M_{\odot} \leqslant 120$. Recently, Hernandez-Perez & Charlot (2013) also have considered the effect of binary stars in their EPS model. They also used the BSE algorithms of Hurley *et al.* (2002), the main difference from our model is the adoption of different initial distributions for SPs.

Using the Yunnan-II EPS model, we studied the effects of binary stars on the retrieved parameters/properties of galaxies and HII regions, on the theoretical classification line between star forming galaxy (SFG) and AGN on the diagnostic diagrams. We propose that the radiation fields from intermediate-age (IMA) SPs with binary stars are in theory possible candidates for significant central ionizing sources (CISs) of classic HII regions. Ma *et al.* (2018) used the BPASS EPS model, in which the effect of binary stars are considered to explain the missing photon problem. Hernandez-Perez & Charlot (2014) used their EPS model to revisit the UV-upturn problem of elliptical galaxies.

Binary degenerated objects are important gravitational wave sources and relate to short $\gamma-$ray bursts. So binary stars have drawn more and more attention. We argue that the supernova (SN) kick velocity is a key factor in computing both the birth rate of binary neutron stars (BNSs) in SPs and the SPs' BNS birth rate.

2. Overview

First, we need to construct a SP comprising binary stars. Differing from single SP, which only needs initial mass function (IMF), the SP comprising binary stars needs the distributions of the initial primary- and secondary-star masses (M_1, M_2), the separation/period (P) and the eccentricity (e). For the IMF for primaries, $\phi(M_1)$, we used the prescriptions of Miller & Scalo (1979); Kroupa *et al.* (2001); Salpeter (1955) and Chabrier (2003). For the initial secondary-star mass distribution, we used two cases: (i) M_1 and M_2 are uncorrelated, $\phi(M_2)$ is the same as $\phi(M_1)$; (ii) M_1 and M_2 are correlated, the mass-ratio distribution ($q = M_2/M_1$) is a uniform or thermal form. For separation/period (P) distribution, we used a combination of power-law and constant separation distributions. Under this case, the binary fraction with $P < 100\,\mathrm{yr}$ is ~50%. For the eccentricity (e) distribution, we used a uniform or constant ($e = 0$) distribution.

Key ingredients in Yunnan-II EPS model are the stellar evolution tracks, the spectral library and the initial distributions of binary stars as outlined above. To obtain the stellar evolution tracks, we used the BSE algorithms of Hurley *et al.* (2002). For spectral library, we used BaSeL semi-empirical stellar spectral library (Lejeune *et al.* 1997, 1998).

3. Results

3.1. *Effects of binary stars*

Based on age, we divide the effects of binary stars on the SPs' model results and retrieved parameters/properties of galaxies and HII regions into three parts. The comparisons presented below are between the Yunnan-II models for SPs with and without binary stars.

At all ages ($t \sim 10^{6-10.1}$ yr), the inclusion of binary stars makes the SPs' colours bluer and mass-to-light ratio in K-band $(M/L)_{\mathrm{K}}$ larger by a factor of 0.2 dex at each metallicity Z. The reason is that binary interactions can produce some hotter stars, for example, naked helium stars, blue stragglers and subdwarf B stars (sdBs). The effects on the retrieved parameters of galaxies are displayed in two aspects: age/stellar-metallicity and stellar-mass. (i) The inclusion of binary stars would make the retrieved age and stellar metallicity of galaxies larger, consistent with the fact that the colours become red when the age or metallicity increases. (ii) The larger SPs' $(M/L)_{\mathrm{K}}$ naturally leads to the larger retrieved galaxies' stellar mass.

At late ages ($t \sim 10^9$ yr), the inclusion of binary stars raises the UV flux (by $\sim 10^2$ times at wavelength $\lambda \sim 1300$Å) at all metallicities (see panel A of Fig. 1). SdBs are the main reason for this flux increase. This result is consistent with the findings presented by Han *et al.* (2007) which have considered the presence of sdBs produced by binary evolution path to explain the UV-upturn phenomenon of elliptical galaxies.

At intermediate ages ($t \sim 10^{7-8}$ yr), the flux with $\lambda < 912$ Å is raised by the inclusion of binary stars (by $\sim 10^5$ times at $\lambda \sim 300$Å) at all metallicities (see panel B of Fig. 1). This effect is mainly caused by the naked helium stars, which temperature reaches up to 10^5K. We summarize our conclusions related to the parameters and properties of galaxies and HII regions below.

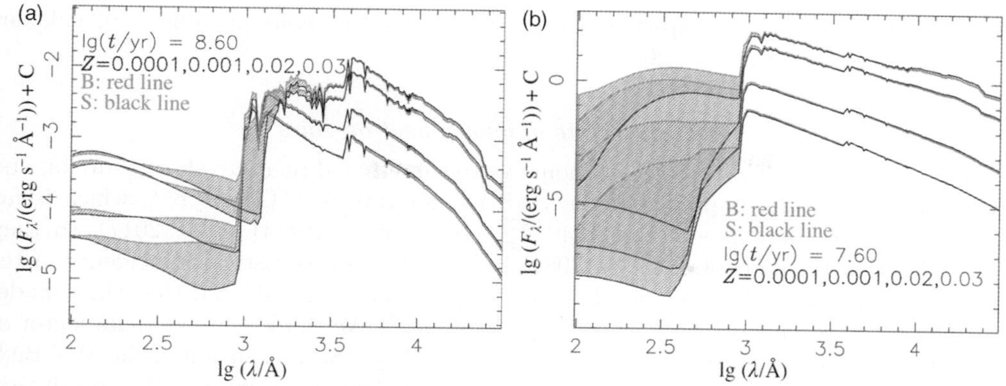

Figure 1. In each panel, 'B' and 'S' mean the cases considering binary stars and only single stars. The SPs' ISEDs at $\lg(t/\mathrm{yr}) = 8.6$ (panel A) and 7.6 (panel B).

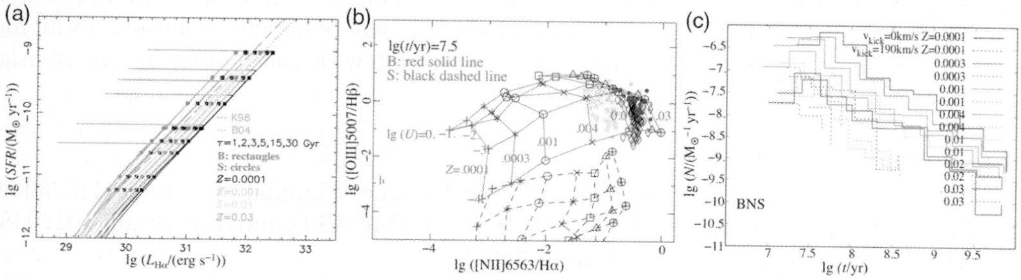

Figure 2. In each panel, 'B' and 'S' have the same meanings as in Fig. 1. (A) The relation between SFR and $L_{\mathrm{H}\alpha}$ for E, S0, Sa-d galaxies (corresponding to $\tau = 1, 2, 3, 5, 15$ and 30 Gyr, from top to bottom), also shown are the results of K98 and B04. (B) Diagnostic diagram lg([NII]6563/Hα) versus lg([OIII]5007/Hβ) when the radiation fields from SPs with $\lg(t/\mathrm{yr}) = 7.5$ are used as the CISs of HII regions and $\lg(U) = -4 \sim 0$, also shown are the observations of some spirals, dwarf-irregular galaxies and HII regions. (C) The BNS birth rate for SPs.

(i) *Star formation rate (SFR)*: the luminosity of the H_α recombination line (L_{H_α}) is usually used to derive SFR. The inclusion of binary stars makes the galaxies' model L_{H_α} larger at all metallicities (including E, S0, Sa-d galaxies, see panel A of Fig. 2), therefore making the retrieved SFR of galaxies smaller. For comparisons, we also plot the results of Kennicutt (1998, hereafter K98) and Brinchmann *et al.* (2004, hereafter B04). As the L_{H_α} increases with the number of ionizing photons $Q(\mathrm{H})$, the inclusion of binary stars makes the flux with $\lambda < 912$ Å larger and therefore $Q(\mathrm{H})$ larger (Zhang *et al.* 2012).

(ii) *The CIS' property of HII regions:* theoretical classification line between SFG and AGN on the diagnostic diagrams and gaseous metallicity of galaxies. In general, it is thought that the main CISs for classic H II regions are radiation fields emitted by O- and B-type stars or clusters of such stars. In the MAPPINGS photoionization code (Groves *et al.* 2008), we used the Yunnan-II EPS model results. When the radiation fields from IMA SPs with binary stars are used as the CISs of HII regions, we found the the model grid also can cover the regions occupied by spirals, dwarf-irregular galaxies and HII-regions (Zhang *et al.* 2015, and references therein) on the diagnostic diagrams. We concluded that the radiation fields from IMA SPs with binary stars can be in theory possible candidates for significant CISs of HII regions. The panel B of figure 2 presents our results for the predicted lg([NII]6563/Hα) versus lg([OIII]5007/Hβ) diagnostic diagram including binary stars. In the calculations, the hydrogen density n_H is fixed to 100 cm^{-3} while the ionization parameter ranges between $\lg(U) = -4$ and 0. With these assumptions, the theoretical classification line between SFG and AGN on the diagnostic diagrams goes toward the up-right corner, the derived gaseous metallicity of galaxies would decrease (Zhang *et al.* 2015).

3.2. *Birth rate of binary neutron stars*

Binary stars are important gravitational wave sources and relate to short γ-ray bursts. Belczynski *et al.* (2018) predicted the BNS merger rate is $48\,\mathrm{Gpc^{-3}yr^{-1}}$, which is far less than LISA/Virgo prediction ($1540^{+3200}_{-1200}\,\mathrm{Gpc^{-3}yr^{-1}}$, Abbott *et al.* 2017), so they concluded that the detection of GW170817 in NGC4993 is a statistical coincidence. After that, Chruslinska *et al.* (2018) used 19 sets of parameters and showed that their model prediction can reach to as high as $600^{+600}_{-300}\,\mathrm{Gpc^{-3}yr^{-1}}$. We think that the main factor of affecting the BNS birth rate is the SN kick velocity $V_{\rm kick}$. So we calculated the SPs' BNS birth rate as a function of time in the case of $V_{\rm kick} = 0$, 50 and $190\,\mathrm{km\ s^{-1}}$. The results are shown in panel C of Fig. 2 where we see that $V_{\rm kick}$ can largely affect the SPs' BNS birth rate by about two order of magnitudes between the cases of $V_{\rm kick} = 0$ and $190\,\mathrm{km\ s^{-1}}$ (see Zhang *et al.* 2018 for details). Jiang *et al.* (2019) used our SPs' BNS birth rate results in the case of $V_{\rm kick} = 50\,\mathrm{km\ s^{-1}}$ in their semi-analytic model of galaxy evolution, obtaining a the BNS merger rate of $\sim 600\,\mathrm{Gpc^{-3}yr^{-1}}$ consistent with the finding of Chruslinska *et al.* (2018).

Acknowledgements

This work was funded by the Chinese Natural Science Foundation (grants 11573062, 11973081, 11773065, 11521303 and 11390734), the YIPACAS Foundation (grant 2012048) and the Yunnan Foundation (grant 2011CI053).

References

Abbott, B., Abbott, R., Abbott, T., *et al.* 2017, *Phys. Rev. Lett.*, 119, 161101
Belczynski, K., Askar, A., Arca-Sedda, M., *et al.* 2018, *A&A*, 615, A91
Brinchmann, J., Charlot, S., White, S. D. M., *et al.* 2004, *MNRAS*, 351, 1151
Chabrier, G. 2003, *PASP*, 115, 763
Chruslinska, M., Belczynski, K., Klencki, J., Benacquista, M., *et al.* 2018, *MNRAS*, 474, 2937
Duchene, G. & Kraus, A. 2013, *ARA&A*, 51, 269
Eldridge, J. J., Izzard, R., & Tout, C. A. 2008, *MNRAS*, 384, 1109
Kennicutt, R. C., Jr 1998, *ARA&A*, 36, 189
Groves, B., Dopita, M., Sutherland, R., *et al.* 2008, *ApJS*, 176, 438
Han, Z., Podsiadlowski, Ph., Lynas-Gray, A., *et al.* 2007, *MNRAS*, 380, 1098
Hurley, J. R., Tout, C. A., & Pols, O. R. 2002, *MNRAS*, 329, 897
Hernandez-Perez, F. & Charlot, G. 2013, *MNRAS*, 431, 2612
Hernandez-Perez, F. & Charlot, G. 2014, *MNRAS*, 444, 2571
Jiang, Z., Wang, J., *et al.* 2019, *in prep.*
Kroupa, P., Aarseth, S., Hurley, J., *et al.* 2001, *MNRAS*, 321, 699
Lejeune, T., Cuisinier, F., & Buser, R. 1997, *A&AS*, 125, 229
Lejeune, T., Cuisinier, F., & Buser, R. 1998, *A&AS*, 130, 65
Ma, X., Hopkins, P., Garrison-Kimmel, S., *et al.* 2018, *MNRAS*, 478, 1694
Miller, G. E. & Scalo, J. M. 1979, *ApJS*, 41, 513
Salpeter, E. E. 1955, *ApJ*, 121, 161
Sana, H., de Mink, S. E., de Koteret, A., *et al.* 2012, *Sci.*, 444, 6
Zhang, F., Han, Z., Li, L., Hurley, J., *et al.* 2004, *A&A*, 415,117
Zhang, F., Han, Z., Li, L., Hurley, J., *et al.* 2005, *MNRAS*, 357,1088
Zhang, F., Li, L., Zhang, Y., *et al.* 2012, *MNRAS*, 421,743
Zhang, F., Li, L., Han, Z., Kang, X., *et al.* 2015, *MNRAS*, 447, L21
Zhang, F., Li, L., Han, Z., *et al.* 2018, *in prep.*

Challenges in Panchromatic Modelling with Next Generation Facilities
Proceedings IAU Symposium No. 341, 2018
M. Boquien, E. Lusso, C. Gruppioni & P. Tissera, eds.

doi:10.1017/S174392131900471X

HELP project - a dreamed-of multiwavelength dataset for SED fitting: The influence of used models for the main physical properties of galaxies

Katarzyna Małek[1,2], Veronique Buat[2], Denis Burgarella[2], Yannick Roehlly[2,3], Raphael Shirley[4,5] and the HELP team

[1]National Centre for Nuclear Research, ul.Pasteura 7, 02-093 Warszawa, Poland
email: katarzyna.malek@ncbj.gov.pl

[2]Aix Marseille Univ. CNRS, CNES, LAM Marseille, France

[3]Univ Lyon, Univ Lyon1, ENS de Lyon, CNRS, Centre de Recherche Astrophysique de Lyon UMR5574, F-69230, Saint-Genis-Laval, France

[4]Astronomy Centre, Department of Physics and Astronomy, University of Sussex, Falmer, Brighton BN1 9QH, UK

[5]Instituto de Astrofísica de Canarias, E-38205 La Laguna, Tenerife, Spain; Universidad de La Laguna, Dpto. Astrofísica, E-38206 La Laguna, Tenerife, Spain

Abstract. The Herschel Extragalactic Legacy Project (HELP) focuses to publish an astronomical multiwavelength catalogue of millions of objects over 1300 deg^2 of the Herschel Space Observatory survey fields. Millions of galaxies with ultraviolet–far infrared photometry make HELP a perfect sample for testing spectral energy distribution fitting models, and to prepare tools for next-generation data. In the frame of HELP collaboration we estimated the main physical properties of all galaxies from the HELP database and we checked a new procedure to select peculiar galaxies from large galaxy sample and we investigated the influence of used modules for stellar mass estimation.

Keywords. galaxies: fundamental parameters, infrared, methods: statistical, catalogs

1. Introduction

The primary objective of the Herschel Extragalactic Legacy Project (HELP project, Shirley *et al.* (2019)) founded by FP7 European Union is to provide homogeneously calibrated multiwavelength catalogues covering roughly 1300 deg^2 of the extragalactic Herschel Space Observatory surveys (HSO, Pilbratt *et al.* (2010)) at wide redshift range. Millions of galaxies with good coverage of ultraviolet–far infrared spectral range make HELP a perfect sample to prepare tools for next-generation data. The detailed description of a final master list creation of 170 million objects, selected at 0.36—4.5 μm from HSO, depth maps etc. can be found in Shirley *et al.* (2019). The catalogues supported by spectroscopic (if possible) or photometric redshift (Duncan *et al.* (2018)) will allow for colour-colour/colour-flux analysis, multi-wavelength spectral energy distribution (SED) fitting and many more statistical studies of the low-to-intermediate redshift galaxy population formation and evolution over cosmic time.

Tab. 1 shows the list of the HSO fields used for HELP project. It demonstrates that HELP not only created a huge multiwavelength, homogenized database, but also focuses

Table 1. Overview of 23 fields used for HELP project.

HELP field name	number of objects	area [deg^2]
AKARI-NEP	531 746	9.2
AKARI-SEP	844 172	8.7
Bootes	3 367 490	11
CDFS-SWIRE	2 171 051	13
COSMOS	2 599 374	5.1
EGS	1 412 613	3.6
ELAIS-N1	4 026 292	14
ELAIS-N2	1 783 240	9.2
ELAIS-S1	1 655 564	9.0
GAMA-09	12 937 982	62
GAMA-12	12 369 415	63
GAMA-15	14 232 880	62
HDF-N	130 679	0.67
Herschel-Stripe-82	50 196 455	363
Lockman-SWIRE	4 366 298	22
HATLAS-NGP	6 759 591	178
SA13	9 799	0.27
HATLAS-SGP	29 790 690	295
SPIRE-NEP	2 674	0.13
SSDF	12 661 903	111
xFLS	977 148	7.4
XMM-13hr	38 629	0.76
XMM-LSS	8 704 751	22
Total:	171 570 436	1270

both on wide and deep fields, with different area on the sky. This careful selection and the final data product can remove the barriers to multiwavelength data studies on the statistical level.

2. Data and short overview of the method

The European Large Area ISO Survey North 1 (ELAIS N1, 13.51 deg^2 area centred at $16^h10^m01^s$ $+54^o30^{'}36^{''}$, Oliver *et al.* 2000) was a pilot field for HELP. The HELP homogenized catalogue of ELAIS N1 includes 50 135 galaxies with good ultraviolet (UV)–far infrared (IR) measurements (quality criterion requires at least two optical – near IR measurements and at least two two of five Herschel measurements with signal to noise ratio $\geqq$2). We used the sample of 50 135 galaxies and we estimated the key physical parameters (i.e. stellar mass, star formation rate, dust luminosity) by fitting SED to all of them using Code Investigating GALaxy Emission (CIGALE, Burgarella *et al.* 2005, Noll *et al.* 2009, and Boquien *et al.* 2019).

CIGALE is designed to estimate the physical parameters by comparing modelled galaxy SEDs to observed ones. CIGALE conserves the energy balance between the dust-absorbed stellar emission and its re-emission in the IR. A more detailed description of the code can be found in Boquien *et al.* (2019).

All adopted parameters used for modules are presented in Table. 2. More detailed discussion of used parameters and description of addition quality tests for SED fitting procedure for ELAIS N1 field can be found in Małek *et al.* (2018). An exemplary fit of SED, showing typical photometric coverage of the spectra is shown in Fig. 1.

3. Impact of the dust attenuation law on the stellar mass

Based on the statistically significant sample of ~50 000 galaxies we check the influence of different dust attenuation recipes on the main physical parameters calculated for all HELP galaxies; stellar mass, star formation rate and dust luminosity. We perform the SED fitting of ELAIS N1 galaxies by assuming three different dust attenuation laws

Table 2. Main modules and input parameters used in CIGALE for the analysis of the high-z sample. The first column lists the CIGALE model, the second provides a brief description of the main parameters, and the third one shows the range of the selected values.

CIGALE module	**main parameter**	**description**
SFH delayed + additional burst	τ of the main stellar population model [Myr]	3 000
	τ of the late starburst population model [Myr]	10 000
	mass fraction of the late burst population	0.001–0.300
SSP: Bruzual & Charlot (2003)	initial mass function	Chabrier (2003)
dust attenuation: Charlot & Fall (2000)	A_V in the BCs	0.3–3.8
	power law slopes (BC and ISM)	–0.7
dust emission Draine & Li (2007)	minimum radiation field (U_{min})	5.0, 10.0, 25.0
	mass fraction of PAH	1.12, 2.5, 3.19
	power law slope dU/dM (U^{α})	2.0, 2.8
AGN emission: Fritz *et al.* (2006)	fractional contribution of AGN	0.0, 0.15, 0.25, 0.8

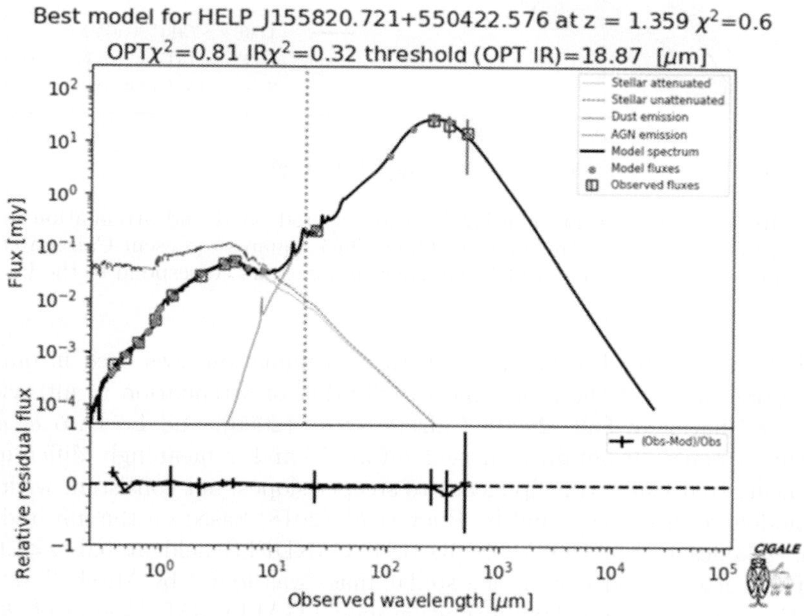

Figure 1. An example of SED fitting result. Open squares represent observed fluxes, while filled circles correspond to the model fluxes. The final model is plotted as a solid black line. The relative residual fluxes are plotted at the bottom of the spectra.

separately: Charlot & Fall (2000), widely used in the literature Calzetti *et al.* (2000), and Lo Faro *et al.* (2017) – dust attenuation recipe created in the framework of the HELP project for Ultra Luminous Infrared Galaxies at redshift > 2. This test allows us to analyze the impact of the assumed law on estimated physical parameters. We find that the attenuation law has an important impact on the stellar mass estimation (on average leading to disparities of a factor of 2), and we derived the relation between stellar mass estimates obtained by those three different attenuation laws. Found recipes (published in Małek *et al.* 2018) can help to homogenize estimated stellar masses from different attenuation laws, and allow to make more precise comparisons, sample selection or study of so called main sequence (stellar mass versus star formation rate relation) between different SED fitting procedures.

We check that the differences in obtained stellar masses are closely related to the shape of each attenuation law at near IR wavelengths. Fig. 2 shows relation between attenuation

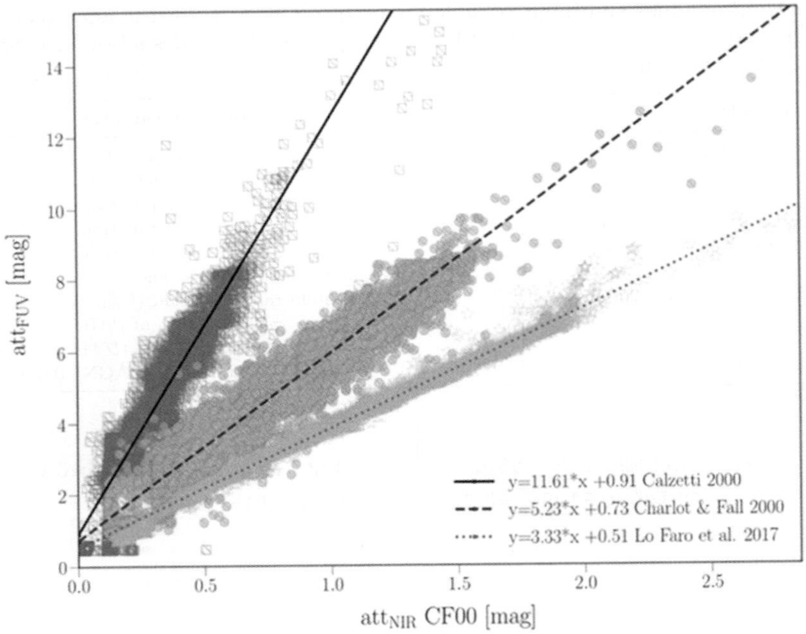

Figure 2. Relation between attenuation in near infrared band and attenuation in ultaviolet band for all three laws used in the analysis. Open black squares represent Calzetti *et al.* (2000) recipe, blue dots – Charlot & Fall (2000) law, and orange stars correspond to the Lo Faro *et al.* (2017) law.

in near IR band and far UV band for all three attenuation laws used in our analysis. This figure presents that the range and distribution of attenuation in ultraviolet band is similar for Charlot & Fall (2000), Calzetti *et al.* (2000), and Lo Faro *et al.* (2017), however the attenuation obtained in near infrared band is meaningly different. Similar result, showing that Calzetti recipe leads to steeper slopes, not consistent with radiation transfer models results, was found by Buat *et al.* (2018) based on the infrared complete sample of galaxies in the COSMOS 3D-HST CANDELS field at $0.6<z<1.6$. Similar impact of the attenuation law on the stellar mass was found by Mitchell *et al.* (2013) based on the semi-analytic galaxy formation model GALFORM (Cole *et al.* 2000).

Acknowledgement

Katarzyna Małek has been supported by the National Science Centre (grantUMO-2018/30/E/ST9/00082).

References

Boquien, M., Burgarella, D., Roehlly, Y., Buat, V., Ciesla, L., Corre, D., Inoue, A. K., & Salas, H. 2019, *A&A*, 622, A103

Buat, V., Boquien, M., Małek, K., Corre, D., Salas, H., Roehlly, Y., Shirley, R., & Efstathiou, A. 2018, *A&A*, 619, A135

Burgarella, D., Buat, V., & Iglesias-Páramo, J. 2005, *MNRAS*, 360, 1413

Bruzual, G. & Charlot, S. 2003, *MNRAS*, 344, 1000

Calzetti, D., Armus, L., Bohlin, R. C., Kinney, A. L., Koornneef, J., & Storchi-Bergmann, T. 2000,*ApJ*, 533, 682

Chabrier, G. 2003, *PASP*, 115, 763

Charlot, S. & Fall, S. M. 2000, *ApJ*, 539, 718

Cole S., Lacey C. G., Baugh C. M., & Frenk C. S. 2000, *MNRAS*, 319, 168

Draine, B. T. & Li, A. 2007, *ApJ*, 657, 810
Duncan, K. J., Brown, J. I., Williams, W. L., Best, P. N., Buat, V., Burgarella, D., Jarvis, M. J., Małek, K., *et al.* 2018, *MNRAS*, 473, 2655
Fritz, J., Franceschini, A., & Hatziminaoglou, E. 2006, *MNRAS*, 366, 767
Lo Faro, B., Buat, V., Roehlly, Y., Alvarez-Marquez, J., Burgarella, D., Silva, L., & Efstathiou, A. 2017, *MNRAS*, 472, 1372
Małek, K., Buat, V., Roehlly, Y., Burgarella, D., Hurley, P. D., Shirley, R., Duncan, K., Efstathiou, A., *et al.* 2018, *A&A*, 620, A50
Mitchell, P. D., Lacey, C. G., Baugh, C. M., & Cole, S. 2013, *MNRAS*, 435, 87
Noll, S., Burgarella, D., Giovannoli, E., Buat, V., Marcillac, D., & Muñoz-Mateos, J. C. 2009, *A&A*, 507, 1793
Oliver, S., Rowan-Robinson, M., Alexander, D. M., Almaini, O., Balcells, M., Baker, A. C., Barcons, X., Barden, M., *et al.* 2000, *MNRAS*, 316, 749
Pilbratt, G. L., Riedinger, J. R., Passvogel, T., Crone, G., Doyle, D., Gageur, U., Heras, A. M., Jewell, C., *et al.* 2010, *A&A*, 518, L1
Rowan-Robinson, M., Wang, L., Wardlow, J., Farrah, D., Oliver, S., Bock, J., Clarke, Ch., Clements, D., *et al.* 2014, *MNRAS*, 445, 3848R
Shirley, R., Roehlly, Y., Hurley, P.D., Buat, V., Campos Varillas, M., Duivenvoorden, S., Duncan, K.J., Efstathiou, A., *et al.* 2019, *A&A*, 490, 634S

Discussion

ADAM CARNALL: When you fit your galaxies with the Charlot & Fall (2000) dust law what do you assume for the slope?

KATARZYNA MAŁEK: The birth cloud slope is fixed to -0.7, the ISM slope is also fixed to –0.7. Those values were obtained by Charlot & Fall (2000).

MAARTEN BAES: You have demonstrated that the choice of the attenuation law is important for the $M_\star$ determination. But which one is the best one?

KATARZYNA MAŁEK: For most of the galaxies, the Charot & Fall attenuation law was giving the best fits (in terms of the lowest χ^2).

TOMOTSUGU GOTO: What are 30% of possible lensed objects which do not satisfy Rowan-Robinson *et al.* (2014) criteria?

KATARZYNA MAŁEK: Could be opt-IR mismatches.

Challenges in Panchromatic Modelling with
Next Generation Facilities
Proceedings IAU Symposium No. 341, 2018
M. Boquien, E. Lusso, C. Gruppioni & P. Tissera, eds.

doi:10.1017/S1743921319003065

Modelling the emission of passive galaxies at z∼3

C. D'Eugenio[1], E. Daddi[1], R. Gobat[2], V. Strazzullo[3], S. Jin[4,5]

[1]CEA, IRFU, DAp, AIM, Université Paris-Saclay, Université Paris Diderot, Sorbonne Paris Cité, CNRS, F-91191 Gif-sur-Yvette, France
email: chiara.deugenio@cea.fr

[2]Instituto de Física, Pontificia Universidad Católica de Valparaíso, Casilla 4059, Valparaíso, Chile

[3]Faculty of Physics, Ludwig-Maximilians-Universität, Scheinerstr. 1, 81679 Munich, Germany

[4]Instituto de Astrofiśica de Canarias (IAC), E-38205 La Laguna, Tenerife, Spain

[5]Universidad de La Laguna, Dpto. Astrofiśica, E-38206 La Laguna, Tenerife, Spain

Abstract. Early-type galaxies (ETGs) are of crucial importance to trace back the galaxy mass assembly across cosmic time, yet their formation and quenching remain remarkably elusive. The discoveries of massive, dead galaxies at ever-growing redshifts provided compelling evidence to push their formation up to redshift $> 4-5$ when the Universe was barely 1 Gyr old. In this talk I will present our results on the ages of a new sample of ETGs at $z \sim 3$, built by exploiting HST WFC3/G141 rest-frame optical/near-UV grism spectroscopy to study the nature of 10 passive galaxy candidates at $2.5 < z < 3.5$ in COSMOS.

This work is part of a PhD project aimed at quantifying the parent space density of distant *genuine* passive galaxies. I will also discuss the importance of multi-wavelength data in clarifying the degree of contamination by dusty star-forming galaxies affecting the color selection.

Keywords. Galaxies: evolution , Galaxies: elliptical and lenticular, cD, Galaxies: high-redshift

---

1. Introduction

The early formation and quenching of massive galaxies are still to be clarified by modern astrophysics. Passively Evolving Galaxies (PEGs) have been shown to host the bulk of the stellar mass in the local Universe (Baldry *et al.* (2004); Renzini & Peng 2015), thus representing good candidates to trace back the galaxy mass assembly. The larger their stellar mass, the earlier their formation. Specifically, archaeological studies of ETGs at $z \sim 0$ estimated the formation of the most massive galaxies to be at $z \gtrsim 4$ with Star Formation Histories (SFHs) characterized by strong and narrow peaks in Star Formation Rate (SFR) (e.g. Thomas *et al.* 2010). So, how fast is their mass growth and quenching? How many already passive galaxies can we expect to find as early as $z \sim 3$? Populations of compact quiescent galaxies have been found at progressively higher redshifts ($1 < z < 3$) (Franx *et al.* 2003; Cimatti *et al.* 2004; Kriek *et al.* 2006; Straatman *et al.* 2014). Nonetheless, they are described as rare objects at $z > 3$ by current number densities and mass functions at the high-mass end (Muzzin *et al.* 2013; Davidzon *et al.* 2017). These mass functions, however, heavily rely on galaxy colors to establish their passiveness, therefore a spectroscopic confirmation of photometric candidates is crucial to alleviate the uncertainty on the level of their star-formation.

While photometric identification of passive galaxy candidates is already probing $3 < z < 4$ range, spectroscopic confirmation of such objects is struggling at breaking the $z \sim 3$ barrier with current facilities. Individual high-z spectra where acquired for a few galaxies placed at z $\gtrsim$ 3 (Gobat *et al.* 2012; Glazebrook *et al.* 2017; Newman *et al.* 2018; Schreiber *et al.* 2018), the latter work showing MOSFIRE spectra of 4 objects, in the ZFOURGE catalog, with moderately evolved stellar populations (Balmer Breaks) at 3.2 < z < 3.7. The degree of contamination of the UVJ color selection by dusty star-forming galaxies was reported to be of 21% and the number density of quiescent galaxies of $(1.4 \pm 0.3) \times 10^{-5} Mpc^{-3}$, one dex higher than what hydrodynamical simulations and semi-analytic models predict (Qin *et al.* 2017a,b; Wellons *et al.* 2015; Davé *et al.* 2016) at $3 < z < 4$ and $log(M_\star/M_\odot) > 10.5$. Here we tackle this issue by studying the HST grism spectra of a pilot sample of 10 passive galaxy candidates at z ∼ 3.

2. Data and Methods

The sample consists of the most massive ($\log M_\star > 10.7$, $H_{AB} < 22$) 10 passive galaxy candidates with photometric redshifts within $2.5 < z_{phot} < 3.5$ in the COSMOS field. The selection as passive was performed on the basis of their rest-frame UVJ (Williams *et al.* 2009) colors, excluding galaxies with observed (B-z) and (z-K) colors (also known as the BzK selection) indicative of star-forming objects (Daddi *et al.* 2004). A fraction (50%) of the sample is sBzK-uncertain, i.e. formally classified as star-forming but with S/N in the B-band (and in some cases z-band) not sufficient to reject a passive nature. Additional criteria were the absence of 24μm detections in the McCracken *et al.* 2010 catalog and the lack of acceptable SED fits for dusty starburst solutions to broad-band photometry (from B to IRAC). Our HST program (P.I. E. Daddi) was designed to get WFC3/IR low resolution (R = 130) G141 slitless spectra, for a total of 17 orbits, including F160W ancillary imaging. The data were reduced and analized by means of the `grizli` software package†. The spectra were optimally extracted (Horne *et al.* 1986) and fitted with the composite stellar populations (CSPs) from Bruzual & Charlot 2003 templates, using four different SFHs to account for diverse modes of star-formation, namely allowing for a constant SFR, exponentially declining, delayed exponentially declining and a truncated SFH. We apply the Calzetti *et al.* 2000 extinction law and Salpeter IMF. Best fit templates are evaluated by their goodness of fit.

3. Results

We find 8 passive galaxies with moderately evolved stellar populations showing evidence of Balmer/4000 Åbreaks and low values of dust extinction, 3 of which are placed at z $\gtrsim$ 3 (Fig. 1). The determination of the significance for the secondary solution is currently work in progress. The remainder 2 candidates (ID266486 and ID581181) show best-fit solutions consistent with being star-forming interlopers. These two objects have 24 μm Spizter/MIPS $> 4\sigma$ detections in the Super-Deblended catalog by Jin *et al.* 2018, further supporting this picture. No further detections from 100 μm to 1.4 GHz (Spitzer/Herschel/SCUBA-2/AzTEC/MAMBO/VLA) are present for our sources. The fit results can be found in Table 1. The measured spectroscopic redshifts agree with the z_{phot} of the parent sample with the calibration by Strazzullo *et al.* 2015, given their uncertainty of $\frac{\delta z}{1+z} = 0.06$ (see Fig. 2).

4. Future Perspectives

A further improvement to be undertaken is the simultaneous fit of both spectra and photometry, fully exploiting the wavelength coverage of the COSMOS field. We will also check for any X-ray detections and ultimately derive the number density of the early passive population. Moreover, these galaxies offer an unique opportunity to extend the work of Gobat *et al.* 2018 on individual high-z quiescent galaxies. The goal will be to

† https://github.com/gbrammer/grizli

Table 1. Targets' characteristics and Simple Stellar Population equivalent best fit age. Stellar masses were derived using the Chabrier 2003 IMF.

ID	z	$\log(M_\star)$	t_{SSP} [Gyr]
20600	$2.841^{+0.021}_{-0.018}$	11.0	0.5
402009	$2.667^{+0.015}_{-0.002}$	11.1	0.6
266486	$2.666^{+0.046}_{-0.021}$	11.4	0.1
21292	$2.557^{+0.005}_{-0.005}$	11.2	0.7
93434	$3.124^{+0.003}_{-0.003}$	11.0	0.5
581181	$2.393^{+0.011}_{-0.000}$	11.0	0.6
316710	$2.998^{+0.002}_{-0.003}$	11.0	0.5
248503	$2.801^{+0.005}_{-0.002}$	10.9	0.3
227846	$2.665^{+0.007}_{-0.007}$	10.9	0.1
169603	$3.230^{+0.007}_{-0.006}$	10.7	0.3

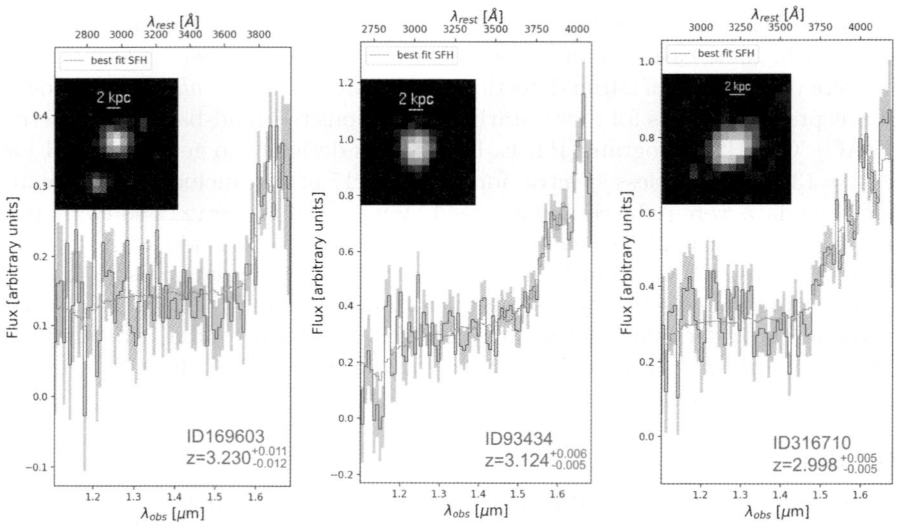

Figure 1. HST spectra are shown for our z∼3 objects. Red curves: best fit CSPs from BC03. Cutouts from HST/WFC3 F160W imaging are reported.

investigate whether the fast rise in the molecular gas fraction in ETGs from z ∼ 0 to z ∼ 1.8 steepens, remains flat or inverts when looking at higher lookback times (Fig. 3). The addition of photometric data in the 1.1mm regime (ALMA and NOEMA) will allow us to estimate the dust mass, M_{dust}, from the FIR flux taking advantage of the optically thin Rayleigh-Jeans tail of thermal dust emission as in Magdis *et al.* 2011 and Scoville 2013. This will allow us to estimate M_{dust} by fitting the 260 GHz fluxes with templates of dust grain emission (Draine & Li 2007) by varying the mean intensity of the radiation field $<U>$ and the dust temperature (T_{dust}) as in Béthermin *et al.* 2015; and to put a first constraint on the gas mass by assuming a reasonable gas-to-dust ratio (G/D) conversion (Galametz *et al.* 2011; Magdis *et al.* 2011; Magdis *et al.* 2012a; Gobat *et al.* 2012) and a gas-metallicity phase close to the solar value given the high stellar masses of our sample (see Table 1). The clarification of the high-z trend could have important implications in determining whether the quenching mechanisms that acted on these galaxies were efficient or not at removing the pre-existing fuel for star formation.

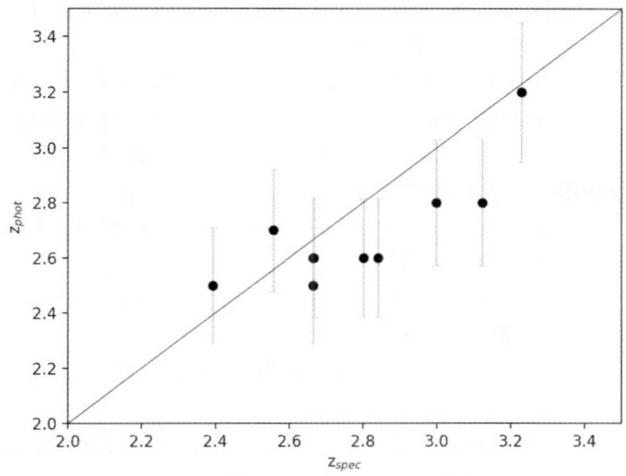

Figure 2. Comparison between spectroscopic and photometric redshifts of our sample, which confirm the appropriateness of our adopted photometric redshift calibration from Strazzullo *et al.* 2015. Degeneracies are not yet explored.

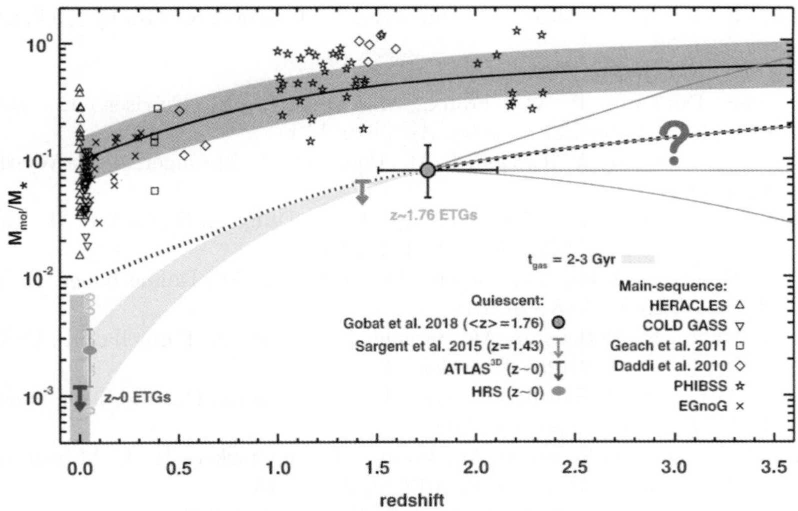

Figure 3. Evolution of molecular gas fraction $M_{mol}/M_\star$ as a function of redshift for main sequence and quiescent galaxies. The reference sample of 977 ETGs at $<z>\sim 1.76$ is marked as an orange dot. The solid line shows the the evolution of an average MS galaxy with stellar mass $5 \times 10^{10} M_\odot$. The red dashed curve shows the dust fraction evolution (traced by molecular gas) offset by a factor of 6 and for the median stellar mass of the Gobat *et al.* 2018 sample.

References

Baldry, I. K., Glazebrook, K., Brinkmann, J., Ivezić, Ž., Lupton R. H., Nichol, R. C., Szalay A. S., *et al.* 2004, *ApJ*, 600, 681–694

Bruzual, G. & Charlot, S. 2003, *MNRAS*, 344, 1000

Béthermin, M., Daddi, E., Magdis, G., Lagos, C., Sargent, M., Albrecht, M., Aussel, H., Bertoldi, F., *et al.* 2015, *A&A*, 573, A113

Calzetti, D., Armus, L., Bohlin, R. C., Kinney, A. L., Koornneef J., & Storchi-Bergmann, T. 2000, *ApJ*, 533, 682

Ceverino, D., Klessen, R. S., & Glover, S. C. O. 2018, *MNRAS*, 480, 4842

Chabrier, G. 2003, *PASP*, 115, 763

Cimatti, A., Daddi, E., Renzini, A., Cassata, P., Vanzella, E., Pozzetti, L., Cristiani, S., Fontana, A., *et al.* 2004, *Nature*, 430, 184
Franx, M., Labbé, I., Rudnick, G., van Dokkum, P. G., Daddi, E., Förster Schreiber, N. M., Moorwood, A., Rix, H.-W., Röttgering, H., *et al.* 2003, *ApJ*, 587, L79
Daddi, E., Cimatti, A., Renzini, A., Fontana, A., Mignoli, M., Pozzetti, L., Tozzi, P., Zamorani, G., *et al.* 2004, *ApJ*, 617, 746
Davidzon, I., Ilbert, O., Laigle, C., Coupon, J., McCracken, H.J., Delvecchio, I., Masters, D., Capak, P., *et al.* 2017, *A&A*, 605, A70
Davé, R., Thompson, R., & Hopkins, P. F. 2016, *MNRAS*, 462, 3265
Draine, B. T. & Li, A. 2007, *ApJ*, 657, 810
Galametz, M., Madden, S. C., Galliano, F., Hony, S., Bendo, G. J., & Sauvage, M. 2011, *A&A*, 532, A56
Glazebrook, K., Schreiber, C., Labbé, I., Nanayakkara, T., Kacprzak, G. G., Oesch, P. A., Papovich, C., Spitler, *et al.* 2017, *Nature*, 544, 71
Gobat, R., Strazzullo, V., Daddi, E., Onodera, M., Renzini, A., Béthermin, M., Dickinson, M., Carollo, M., Cimatti, A., *et al.* 2012, *ApJ*, 759, L44
Gobat, R., Daddi, E., Magdis, G., Bournaud, F., Sargent, M., Martig, M., Jin, S., Finoguenov, A., *et al.* 2018, *Nature Astronomy*, 2, 239
Horne, K. 1986, *PASP*, 98, 609
Jin, S., Daddi, E., Liu, D., Smolčić, V., Schinnerer, E., Calabrò, A., Gu, Q., Delhaize, J. *et al.* 2018, *ApJ*, 864, 56
Kennicutt, R. C. 1998, *ARAA*, 36, 189
Kriek, M., van Dokkum, P. G., Franx, M., Quadri, R., Gawiser, E., Herrera, D., Illingworth, G. D., Labbé, *et al.* 2006, *ApJ*, 649, L71
Qin, Y., Mutch, S. J., Duffy, A. R., Geil, P. M., Poole, G. B., Mesinger, A., & Wyithe, J. S. B. 2017a, *MNRAS*, 471, 4345
Qin, Y., Mutch, S. J., Poole, G. B., Liu, C., Angel, P. W., Duffy, A. R., Geil, P. M., Mesinger, A., Wyithe, J. S. B., *et al.* 2017b, *MNRAS*, 472, 2009
Magdis, G. E., Daddi, E., Elbaz, D., Sargent, M., Dickinson, M., Dannerbauer, H., Aussel, H., Walter, F., *et al.* 2011, *ApJ*, 740, L15
Magdis, G. E., Daddi, E., Béthermin, M., Sargent, M., Elbaz, D., Pannella, M., Dickinson, M., Dannerbauer, H., *et al.* 2012a, *Astrophys. J.*, 760, 6
McCracken, H. J., Capak, P., Salvato, M., Aussel, H., Thompson, D., Daddi, E., Sanders, D. B., Kneib, J.-P., *et al.* 2010, *ApJ*, 708, 202
Muzzin, A., Marchesini, D., Stefanon, M., Franx, M., McCracken, H. J., Milvang-Jensen, B., Dunlop, J. S., Fynbo, J. P. U., *et al.* 2013, *ApJ*, 777, 18
Newman, A. B., Belli, S., Ellis, R. S., & Patel, S. G.2018, *ApJ*, 862, 125
Renzini, A. & Peng, Y.-jie 2015, *ApJ*, 801, L29
Salpeter, E. E. 1955, *ApJ*, 121, 161
Schreiber, C., Glazebrook, K., Nanayakkara, T., Kacprzak, G. G., Labbé, I., Oesch, P., Yuan, T., Tran, K.-V., *et al.* 2018, *ApJ*, 618, A85
Scoville, N. 2013, *AAMA*,221, 03
Straatman, C. M. S., Labbé, I., Spitler, L. R., Allen, R., Altieri, B., Brammer, G. B., Dickinson, M., van Dokkum, P., *et al.* 2014, *ApJ*, 783, L14
Strazzullo, V., Daddi, E., Gobat, R., Garilli, B., Mignoli, M., Valentino, F., Onodera, M., Renzini, *et al.* 2015, *A&A*, 576, L6
Thomas, D., Maraston, C., Schawinski, K., Sarzi, M., & Silk, J. 2010, *MNRAS*, 404, 1775
Valentino, F., Daddi, E., Silverman, J.D., Puglisi, A., Kashino, D., Renzini, A., Cimatti, A., Pozzetti, L., Rodighiero, G., Pannella, M., Gobat, R., Zamorani, G., *et al.* 2017, *MNRAS*, 472, 4878
Williams, R. J., Quadri, R. F., Franx, M., van Dokkum, P., & Labbé, I. 2009, *ApJ*, 691, 1879–1895
Wellons, S., Torrey, P., Ma, C.-P., Rodriguez-Gomez, V., Vogelsberger, M., Kriek, M., van Dokkum, P., Nelson, E., Genel, S., *et al.* 2015, *MNRAS* 449, 361
Yan, R., Newman, J. A., Faber, S. M., Konidaris, N., Koo, D., & Davis, M. 2006, *ApJ*, 648, 281

Discussion

MANATARI: By what redshift should the passive galaxies form their stellar mass?

D'EUGENIO: Typically $z \sim 4$, but it depends on the underlying Star Formation History. Here we are sensitive to the last episode of star formation, so the bulk of the stellar mass could have formed even earlier.

CARNALL: Have you considered the implied OII fluxes you should see for your best-fitting solutions for your HST grim data? This could help you discriminate between dusty and passive.

D'EUGENIO: No, not yet but it is a nice suggestion. Thank you.

CEVERINO: A comment: The color selection may miss an important population of high-z, post-starburst galaxies with bluer colours, according to cosmological simulations (Ceverino, Klessen & Glover 2000).

D'EUGENIO: We are not looking for post-starburst galaxies but for galaxies with several hundred Myr of quiescence so, in this sense, the color selection might allow us to be conservative.

Challenges in Panchromatic Modelling with Next Generation Facilities
Proceedings IAU Symposium No. 341, 2018
M. Boquien, E. Lusso, C. Gruppioni & P. Tissera, eds.

doi:10.1017/S1743921319002734

Evolution histories of massive galaxies at $z \sim 2$ over the past 3 Gyr

T. Morishita[1], L. E. Abramson[2], T. Treu[3], G. B. Brammer[1,4], T. Jones[5], P. Kelly[6], M. Stiavelli[1], M. Trenti[7,8], B. Vulcani[9] and X. Wang[3]

[1]Space Telescope Science Institute, Baltimore, MD, USA
email: tmorishita@stsci.edu

[2]The Observatories of the Carnegie Institution for Science, Pasadena, CA, USA

[3]Department of Physics and Astronomy, UCLA, Los Angeles, CA USA

[4]Cosmic Dawn Centre, University of Copenhagen, Copenhagen, Denmark

[5]University of California Davis, Davis, CA , USA

[6]School of Physics and Astronomy, University of Minnesota, Minneapolis, MN, USA

[7]School of Physics, University of Melbourne, Parkville VIC, Australia

[8]ARC Centre of Excellence for All-Sky Astrophysics in 3 Dimensions, Parkville VIC, Australia

[9]INAF – Osservatorio Astronomico di Padova, Padova, Italy

Abstract. We study star formation and metallicity enrichment histories of 24 massive galaxies at $1.6 < z < 2.5$. Deep slitless spectroscopy + imaging data set collected from multiple *HST* surveys allows robust determination of their SEDs. Our new SED modeling with no functional assumptions on star formation histories revels that 1. most of the sample galaxies have already formed $> 50\%$ of their extant masses ~ 1.5 Gyr before the time of observed redshifts, with a trend where more massive galaxies form earlier, 2. most of our galaxies already have stellar metallicities compatible with those of local early-type galaxies, and 3. inferred metallicities are on average ~ 0.25 dex higher than observed gas-phase metallicities of star forming galaxies at the time of their formation. Continuation of low-level star formation, rather than abrupt termination of star forming activity, may explain the observed gap of metallicities.

Keywords. galaxies: evolution – galaxies: metallicity

1. Introduction

In the local universe, early-type galaxies dominate the massive end of the galaxy mass function, $\log M_*/M_\odot \gtrsim 11.5$ (Cole *et al.* 2001; Bell *et al.* 2003). Observations have, in fact, revealed that some galaxies are already massive and passively evolving at $z \gtrsim 2$ (e.g., Daddi *et al.* 2005; van Dokkum *et al.* 2008; Glazebrook *et al.* 2017). Given the short time since the Big Bang and their mass, their earlier star formation must be extremely intense, followed by a rapid cessation of their star formation activity, or quenching.

However, these episodes yet remain observationally indirect. What were their star formation histories like? How and why did they stop forming stars, especially at the peak time of the cosmic star formation? Are they already enriched in metallicity as local counterparts? Do any post-quenching processes play key roles over the following 10 Gyr? These are the central questions that the extra-galactic community has been trying to answer in the past decades.

A number of studies have investigated star formation histories of massive galaxies with different approaches, from archaeological studies of resolved stars in the local galaxies, detailed spectroscopic observations of low-z galaxies, to direct observation of high-z galaxies to reveal their progenitors. While each method has its strength, it is true that none of them is perfect to comprehensively understand the star formation history of galaxies. For example, spectroscopic information from the local galaxies is hampered up to several Gyrs with current observing facilities, which is too short for exploring star formation histories of massive galaxies, which were already dead at $z \sim 2$, or $\gtrsim 10$ Gyr ago.

To explore evolution histories in the earlier epoch, there has been a significant improvement with high-quality data of high-z galaxies. In this study, we target galaxies at high redshift, aiming at earlier evolution histories up to their formation redshift from fossil record obtained with low-spectral resolution yet high sensitivity *HST* spectrophotometric data set. The key of our SED modeling is to determine the best combination of those amplitudes that reproduces the observed data best (akin to Cid Fernandes *et al.* 2005; Panter *et al.* 2008), which gives us flexibility in reconstructed star formation and metallicity histories, leaving them free from any functional forms. Due to the large number of parameters ($N_{age} \times 2 + 1 + 1 = 22$), full exploration of parameter spaces is necessary for unbiased results. We accomplish this by adopting a Bayesian inference MCMC scheme (*emcee*; Foreman-Mackey *et al.* 2014), which is intensively tested with mock data set (for the detail see Morishita *et al.* 2018).

In this study, we collect 24 massive ($\log M_*/\mathrm{M}_\odot \gtrsim 11$) galaxies at $z \sim 2$ that have deep WFC3/G102 and G141 grism spectra coverage in their rest-frame 4000 Å, from multiple surveys (van Dokkum *et al.* 2013; Skelton *et al.* 2014; Schmidt *et al.* 2014; Treu *et al.* 2015; Kelly *et al.* 2016; Momcheva *et al.* 2016). The combination of grism spectra and wide broadband photometry (0.2 μm to 8.0 μm by *HST* and *Spitzer*) provides a unique opportunity to constrain not only the age but also the metallicity from the entire SED shape.

2. Results

While the result of individual fits and their detailed properties are shown in Morishita *et al.* (2018), we here focus on the global properties of our 24 galaxies. The left panel of Figure 1 shows their cumulative mass accumulation histories as a function of lookback time. The average trend is shown in the middle panel. We see that our galaxies on average form half of their observed mass by ~ 1.5 Gyr prior to the observation (which corresponds to $z \sim 2.5$-5). In the right panel, we see a positive correlation between the two parameters ($\log t_{\mathrm{half}}/\,\mathrm{Gyr} \propto 0.5 \log M_*/\mathrm{M}_\odot$). This means that the trend (downsizing) is observed in the early time of the universe, providing a further insight into their evolution at even earlier epoch, especially when those galaxies are in the star-forming phase. While it is possible our galaxies were once typical (but bright) Lyman break galaxies given the half mass (e.g., Verma *et al.* 2007), it should be reminded that there is also a population that are dusty and missed in most of current optical/NIR observations (e.g., those in Tao Wang's contribution).

In the left panel of Figure 2, we plot our galaxies in the stellar mass-stellar metallicity plane, for a comparison with massive galaxies at $z \sim 0$ (Gallazzi *et al.* 2005). Surprisingly, most of our galaxies are already above the local relation — median value of $\log Z_*/Z_\odot \sim 0.25$ with scatter of ~ 0.15 dex. While uncertainty of individual metallicity is still nonnegligible (~ 0.2 dex), the similar scatter to the local relation may indicate that metallicity evolution of the most massive galaxies completed by the time of $z \sim 2$. The consistent result was reported in Gallazzi *et al.* (2014) but for galaxies at $z \sim 0.7$.

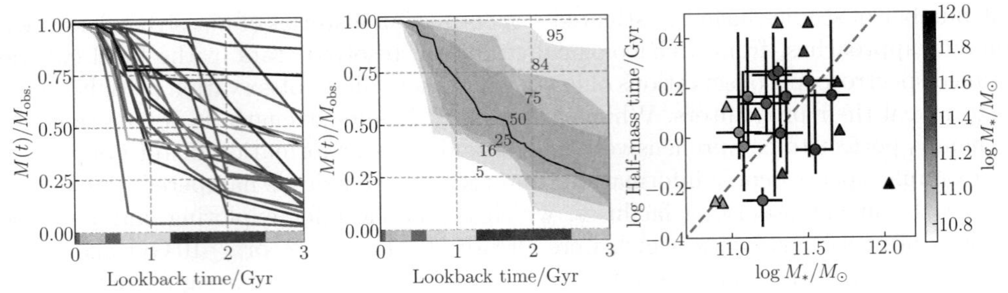

Figure 1. Left: Cumulative stellar mass evolution of individual galaxies. Color corresponds to the final stellar mass. **Middle:** Summary of individual cumulative stellar mass evolution, where contour boundaries and line correspond to 5/16/25/75/84/95th percentiles and median. **Right:** Half-mass time (lookback time from $t_{\rm obs}$ to the time when 50% of their current mass formed) as a function of observed stellar mass. Those with lower limits are shown with triangles.

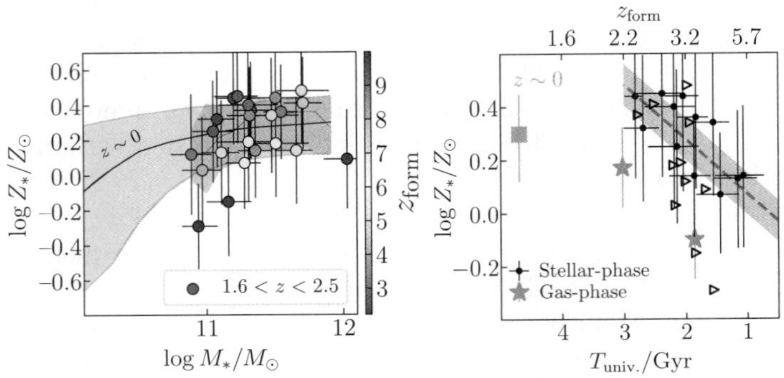

Figure 2. Left: Stellar mass-metallicity relation of 24 galaxies in this study (circles). The symbols are color-coded by formation redshift, $z_{\rm form}$. Running median (~ 0.25) is shown with 16-84th percentile range (red line with hatched region; $\sigma_{\log Z} \sim 0.15$). **Right:** Observed stellar metallicity as a function of the formation time, or $z_{\rm form}$ (black circles/triangles for those with an upper limit in age estimate). The linear fit with a slope (0.20 ± 0.08 dex/Gyr) and standard deviation ($\sigma \sim 0.16$ dex) is shown (red dashed line and hatched region).

Despite its tight relation in the plane, we see a trend for metallicity in Figure 2, where the color of metallicity as a function of formation redshift. In the right panel of Figure 2 we show the stellar metallicity but as a function of formation redshift, derived from the star formation history of individual galaxies. While some sample galaxies have a lower limit for their formation time (due to the low resolution power), the linear fit still shows an evolution of metallicity as a function of cosmic time, with a slope of $\log Z_*/dt \sim 0.20 \pm 0.08$ dex/Gyr. This reveals the rapid evolution of metallicity in the most massive galaxies, which is in fact consistent with other measurements but in e.g., star forming galaxies (e.g., Erb *et al.* 2006; Maiolino *et al.* 2008).

Interestingly, there is a nonnegligible gap (~ 0.25 dex) between the two different metallicities (stellar for ours and gas phase from Maiolino *et al.* (2008)) when compared at the time of formation. While the comparison is very challenging and may be hampered by systematics, this may hint at physical mechanisms for galaxy quenching. For example, Peng *et al.* (2015) demonstrated a chemical modeling by using the SDSS data for $z \sim 0$ galaxies and showed that strangulation is the main mechanism for quenching. While AGN feedback is one of popularly discussed quenching mechanisms, the gap should not be observed, as the feedback would blow the gas and leave no time for chemical enrichment from gas to stellar phase metallicities. Our result also supports the strangulation. In fact, there are several observations that reveal low-level star formation activity in quenched galaxies (Belli *et al.* 2017; Gobat *et al.* 2018). Direct comparison of both metallicities inside individual galaxies would provide more promising insight into this.

3. Conclusion

a. Our massive galaxies have already formed $> 50\%$ of their current mass by ~ 1.5 Gyr prior to the epoch of observation with a downsizing trend.

b. Stellar-phase metallicities of most of our galaxies are compatible with local values, indicating a rapid metallicity enrichment being associated with the early stellar mass formation.

c. By using the reconstructed SFHs and inferred metallicity, we revealed a rapid metallicity enrichment of this class of massive galaxies, at a rate of ~ 0.2 dex/Gyr in $\log Z_*/Z_\odot$ from $z \sim 5.5$ to $z \sim 2.2$.

d. The observed gap between the stellar phase metallicity and gas phase metallicity can be explained by continuation of low-level of star formation in quiescent systems. The scenario provides a picture where galaxy quenching happening more continuously, rather than abrupt/rapid termination, consistently but independently confirming our findings of individual star formation activity.

References

Bell, E. F., McIntosh, D. H., Katz, N., & Weinberg, M. D. 2003, *ApJS*, 149, 289
Belli, S., Newman, A. B., & Ellis, R. S. 2017, *ApJ*, 834, 18
Cid Fernandes, R., Mateus, A., Sodré, L., Stasińska, G., & Gomes, J. M. 2005, *MNRAS*, 358, 363
Cole, S., Norberg, P., Baugh, C. M., Frenk, C. S., Bland-Hawthorn, J., Bridges, T., Cannon, R., Colless, M., *et al.* 2001, *MNRAS*, 326, 255
Daddi, E., Renzini, A., Pirzkal, N., Cimatti, A., Malhotra, S., Stiavelli, M., Xu, C., Pasquali, A., *et al.* 2005, *ApJ*, 626, 680
Erb, D. K., Shapley, A. E., Pettini, M., Steidel, C. C., Reddy, N. A., & Adelberger, K. L. 2006, *ApJ*, 644, 813
Foreman-Mackey, D., Sick, J., & Johnson, B. 2014, doi:10.5281/zenodo.12157
Gallazzi, A., Bell, E. F., Zibetti, S., Brinchmann, J., & Kelson, D. D. 2014, *ApJ*, 788, 72
Gallazzi, A., Charlot, S., Brinchmann, J., White, S. D. M., & Tremonti, C. A. 2005, *MNRAS*, 362, 41
Glazebrook, K., Schreiber, C., Labbé, I., Nanayakkara, T., Kacprzak, G. G., Oesch, P. A., Papovich, C., Spitler, L. R., *et al.* 2017, *Natur*, 544, 71
Gobat, R., Daddi, E., Magdis, G., Bournaud, F., Sargent, M., Martig, M., Jin, S., Finoguenov, A., *et al.* 2018, Nature Astronomy, 2, 239
Kelly, P. L., Brammer, G., Selsing, J., Foley, R. J., Hjorth, J., Rodney, S. A., Christensen, L., Strolger, L.-G., *et al.* 2016, *ApJ*, 831, 205
Maiolino, R., Nagao, T., Grazian, A., Cocchia, F., Marconi, A., Mannucci, F., Cimatti, A., Pipino, A., *et al.* 2008, *A&A*, 488, 463
Momcheva, I. G., Brammer, G. B., van Dokkum, P. G., Skelton, R. E., Whitaker, K. E., Nelson, E. J., Fumagalli, M., Maseda, M. V., *et al.* 2016, *ApJS*, 225, 27
Morishita, T., Abramson, L. E., Treu, T., Brammer, G. B., Jones, T., Kelly, P., Stiavelli, M., Trenti, M., *et al.* 2018, arXiv e-prints, arXiv:1812.06980
Panter, B., Jimenez, R., Heavens, A. F., & Charlot, S. 2008, *MNRAS*, 391, 1117
Peng, Y., Maiolino, R., & Cochrane, R. 2015, *Natur*, 521, 192
Schmidt, K. B., Treu, T., Brammer, G. B., Bradač, M., Wang, X., Dijkstra, M., Dressler, A., Fontana, A., *et al.* 2014, *ApJL*, 782, L36
Skelton, R. E., Whitaker, K. E., Momcheva, I. G., Brammer, G. B., van Dokkum, P. G., Labbé, I., Franx, M., van der Wel, A., *et al.* 2014, *ApJS*, 214, 24
Treu, T., Schmidt, K. B., Brammer, G. B., Vulcani, B., Wang, X., Bradač, M., Dijkstra, M., Dressler, A., *et al.* 2015, *ApJ*, 812, 114

van Dokkum, P., Brammer, G., Momcheva, I., Skelton, R. E., Whitaker, K. E., & for the 3D-HST team, 2013, ArXiv e-prints, arXiv:1305.2140

van Dokkum, P. G., Franx, M., Kriek, M., Holden, B., Illingworth, G. D., Magee, D., Bouwens, R., Marchesini, D., *et al.* 2008, *ApJL*, 677, L5

Verma, A., Lehnert, M. D., Förster Schreiber, N. M., Bremer, M. N., & Douglas, L. 2007, *MNRAS*, 377, 1024

Discussion

David Rosario: Are the SFHs of the passive galaxies themselves consistent with the slow quenching scenario suggested by comparison to co-eval SFHs?

Takahiro Morishita: Many degeneracies and a lack of gas mass estimates make the check fairly unconstrained, but this is a valuable exercise.

Tomotsugu Goto: Sub-mm flux or 24-micron flux?

Takahiro Morishita: Will check.

Chiara D'Eugenio: How was the gas-phase metallicity of quiescent galaxies measured?

Takahiro Morishita: The gas-phase metallicity was derived from another sample. It would be interesting to have direct estimates for our galaxies.

Joel Leja: Have you tested using different priors for your non-parametric SFHs? The posteriors have a spiky shape which is likely related to the linear weighting system.

Takahiro Morishita: No.

Adam Carnall: How do you go about fitting 22 free parameters with MCMC? Have you looked at sensitivity to starting parameters?

Takahiro Morishita: We average our uncertainties over many randomized starting positions.

Challenges in Panchromatic Modelling with Next Generation Facilities
Proceedings IAU Symposium No. 341, 2018
M. Boquien, E. Lusso, C. Gruppioni & P. Tissera, eds.

doi:10.1017/S1743921319002680

Spatially resolved stellar mass buildup and quenching in massive disk galaxies over the last 10 Gyr revealed with spatially resolved SED fitting

Abdurro'uf[1,2] and Masayuki Akiyama[1]

[1]Astronomical Institute, Tohoku University, Aramaki, Aoba, Sendai 980-8578, Japan
email: abdurrouf@astr.tohoku.ac.jp

[2]Institute of Astronomy and Astrophysics, Academia Sinica, Taipei 10617, Taiwan

Abstract. Despite decreasing cosmic star formation rate density over the last 10 Gyr, the stellar mass (M_*) buildups in galaxies were still progressing during this epoch. About 50% of the current M_* density in the universe was built over the last ~ 8.7 Gyr. In this research, we investigated the stellar mass buildup and quenching of spatially resolved regions within massive disk galaxies over the last 10 Gyr. We apply the spectral energy distribution (SED) fitting method to SEDs of sub-galactic regions in galaxies to derive the spatially resolved distributions of SFR and M_* in the galaxies. This namely *pixel-to-pixel SED fitting* method is applied to massive disk galaxies at $0.01 < z < 0.02$ and $0.8 < z < 1.8$. We found that massive disk galaxies tend to build their M_* and quench their star formation progressively from the central region to the outskirts, i.e. *inside-out* stellar mass buildup and quenching.

Keywords. galaxies: evolution, galaxies: formation, galaxies: fundamental parameters, galaxies: spiral, galaxies: structure

1. Introduction

Galaxy redshift surveys over the last decades have revealed dozens of global galaxy scaling relations. One important scaling relation is a star formation main sequence (SFMS) which is a tight nearly linear relation between global star formation rates (SFRs) and stellar masses (M_*) of star-forming galaxies. This relation holds across a wide redshift range and evolves in terms of the slope and the normalization of its linear form.

The decreasing normalization of the SFMS with cosmic time (a factor of ~ 2 dex from $z \sim 6$ to 0 based on Speagle *et al.* (2014)) suggests a decreasing global specific SFR (sSFR$\equiv$SFR/M_*) which implies a decreasing rate of stellar mass buildup and star formation quenching in galaxies.

Besides the SFMS, the galaxy distribution is also bimodal on an optical color (e.g. $u - g$) versus M_* plane (i.e. color-mass diagram). In contrast to the early-type galaxies which predominantly reside in the red sequence on the color-mass diagram, late-type galaxies make a continuous population ranging from the blue cloud all the way to the red sequence, suggesting a slow quenching process in the late-type galaxies (Schawinski *et al.* (2014)). A slow quenching in late-type galaxies could be caused by a cut-off of cosmic gas supply (i.e. strangulation) into the galaxy or inefficient gas cooling due to some feedback mechanisms (Schawinski *et al.* (2014)). Once the cosmic gas supply of late-type galaxy is cut-off, star formation still proceeds and accumulates stellar mass using the remaining gas until the gas in the galaxy is used up. Once the gas is used up, star formation will stop and the galaxy eventually become a red late-type galaxy.

In this contributed paper, we discuss the spatially resolved distributions of SFR and M_* in massive disk galaxies and study their structural evolution over the last 10 Gyrs. We discuss the gradual processes of star formation quenching and stellar mass buildup in those galaxies.

2. Methodology

In this research, we use a spatially resolved SED fitting method, i.e. SED fitting technique applied to spatially resolved SED of a galaxy. The reason for using this method over an analysis using the integral field spectroscopy (IFS) data is that this method is applicable to sample galaxies in the local universe and at high redshift with nearly the same manner. The usage of a single method can reduce some systematic biases in comparison analysis between results obtained with the two samples. We chose massive ($\log(M_*) > 10.5$) face-on disk galaxies at $0.01 < z < 0.02$ and $0.8 < z < 1.8$ as the sample of galaxies in this research. The local and high redshift samples contain 93 and 152 galaxies, respectively. The local and high redshift samples are drawn from the MPA-JHU and 3D-HST (for the GOODS-South field), respectively.

The *pixel-to-pixel SED fitting* method used in this research can be divided into three main tasks: (1) construction of maps of the multiband fluxes of a galaxy from which spatially resolved SEDs of the galaxy are obtained, (2) construction of a set of model photometric SEDs with a large number of random parameters. The model SEDs are redshifted according to the redshift of the sample galaxy, (3) fitting of a spatially resolved SED with a set of model SEDs using the Bayesian statistics approach. Detailed description of this method is presented in Abdurro'uf & Akiyama (2017). In order to construct spatially resolved SED with rest-frame far-ultraviolet (FUV) to near infrared (NIR) coverage, we use 7 bands from GALEX and SDSS for the local sample and 8 bands from CANDELS and 3D-HST for the high redshift sample. In order to get spatially resolved SEDs with high S/N ratio, we bin neighboring pixels that have similar SED shapes. The binning method is described in Abdurro'uf & Akiyama (2017). We require all the spatially resolved SEDs to have a S/N larger than 10 in all filters.

The model SEDs are generated using Bruzual & Charlot (2003) stellar population synthesis (SPS) model assuming a Chabrier (2003) initial mass function (IMF), various stellar metallicities (Z) ranging from 0.004 to 0.05, exponentially declining star formation histories (SFHs), and Calzetti *et al.* (2000) dust extinction law. The Bayesian SED fitting that we adopted uses the Student's t distribution form for the likelihood function instead of Gaussian function. This likelihood function gives reasonable weight to the models with large χ^2, thereby allowing us to explore the models over whole parameter ranges. We use a flat prior over an assumed range for all model parameters.

3. Results and discussion

We investigated a linear scaling relation at a kpc scale between SFR surface density ($\Sigma_{\rm SFR}$) and M_* surface density (Σ_*), the so-called spatially resolved SFMS. This relation holds in both local and high redshift samples (Abdurro'uf & Akiyama 2017, 2018). Galaxies with lower global sSFR tend to have spatially resolved SFMS with lower normalization. Moreover, we found a tendency toward flattened spatially resolved SFMS relation in galaxies with lower global sSFR. Spatially resolved SFMS of massive star-forming (those residing on the global SFMS) disk galaxies at $0.8 < z < 1.8$ have a linear form with a slope of nearly unity over the entire Σ_* range without any 'flattening' tendency in the high Σ_* region (Abdurro'uf & Akiyama (2018)). This trend suggests a similar rate of stellar mass buildup over the entire region in those galaxies. The 'flattening' tendency at high Σ_* is observed in spatially resolved SFMS of galaxies with lower global sSFR than those of massive star-forming galaxies at $0.8 < z < 1.8$. The 'flattening' trend suggests that a quenching mechanism is taking place in sub-galactic regions at the high-mass end (predominantly located in the central regions of the galaxies). The spatially resolved SFMS relation shows that the amount of accumulated stellar mass and recent SFR trace each other even in a kpc scale within a galaxy. The origin of this relation has not been understood. Information on the spatially resolved gas surface density will be helpful to investigate the connection of this relation with the Kennicutt-Schmidt relation.

In order to investigate the internal structure of the galaxies in the sample, we derived the radial profiles of the surface densities of SFR ($\Sigma_{\rm SFR}(r)$), M_* ($\Sigma_*(r)$), and sSFR(r). We found that the average sSFR(r) of the massive star-forming (that reside on the global SFMS) disk galaxies at $0.8 < z < 1.8$ is roughly flat over the entire radial distances, suggesting a similar rate of stellar mass buildup over the entire region in these galaxies. The average sSFR(r) of the high redshift sample that reside below the global SFMS is lower over the entire radial distances compared to the average sSFR(r) of the star-forming galaxies with the central regions showing more significant suppression (Abdurro'uf & Akiyama (2018)). The sSFR(r) of all the local sample show suppression in the central region while roughly flat in the outskirts (Abdurro'uf & Akiyama (2017)). Overall, decreasing global sSFR is associated with suppression of the sub-galactic sSFR over the entire galaxy region with a sharper suppression in the central region compared to that in the disk. Similar trends were also observed in local galaxies by Belfiore *et al.* (2018) and Abramson *et al.* (2014). This trend suggests that a quenching process in a galaxy is propagating from the central region to outskirts, i.e. *inside-out quenching.*

In order to study the evolution of the internal structure of massive disk galaxies in a more quantitative manner, we tried to construct an evolutionary empirical model using the surface density radial profiles of the local and high redshift samples. First, we connect the high redshift and local samples by looking for a possible pair of progenitors and descendants. We did that using an evolutionary track on the sSFR versus M_* plane assuming an exponentially declining SFH with $\tau = [4.0 : 6.0]$ and an initial condition at $z = 2$ with $\log(\text{sSFR}(z = 2)) = [-8.6 : -8.4]$ and $\log(M_*(z = 2)) = [9.7 : 9.9]$ drawn from the scatter of the global SFMS at $z = 2$. Then progenitors and descendants are defined as galaxies in the high redshift and local samples that are passed by the model evolutionary track at the redshift of those galaxies, respectively. The number of progenitors and descendants are 20 and 14, respectively. The average $\Sigma_{\rm SFR}(r)$, $\Sigma_*(r)$, and sSFR(r) radial profiles of the progenitors and descendants are then calculated. Using the average $\Sigma_{\rm SFR}(r)$ of the progenitors and descendants as a boundary condition and assuming an exponentially declining SFH at each radial distance, we estimated the radial profile of the SFH, $\Sigma_{\rm SFR}(r, t)$. The $\Sigma_{\rm SFR}(r, t)$ and average $\Sigma_*(r)$ of the progenitors form a set of formula for the evolutionary empirical model of the surface density radial profiles. The predicted $\Sigma_*(r)$ and sSFR(r) at $z = 0$ by the empirical model are consistent with the average $\Sigma_*(r)$ and sSFR(r) of the descendants. Given that the empirical model is derived only using the average $\Sigma_{\rm SFR}(r)$ of progenitors and descendants, the consistency in $\Sigma_*(r)$ and sSFR(r) implies that the empirical modeling does make sense.

The empirical model for the evolution of $\Sigma_*(r)$ shows stellar mass buildup in the disk, while stellar mass accumulation in the central region is small. The $\Sigma_*(r, t)$ shows stellar mass buildup progressing from the central region toward the outskirts, i.e. *inside-out stellar mass buildup.* The $\Sigma_{\rm SFR}(r, t)$ and sSFR(r, t) show decreasing SFR and sSFR with cosmic time over the entire radial distances with larger suppression in the central region compared to that in the outskirts. This trend suggests a quenching process that propagates from the central region to the outskirts, i.e. *inside-out quenching.* However, the sSFR in the outskirts decreases mildly with cosmic time meaning that a quenching process also takes place in the outskirts. Using the empirical model and assuming log(sSFR) $= -10$ as a critical condition for quenching, we calculated the quenching timescale as a function of radius ($t_{\rm quench}(r)$). We found that the central region ($r \sim 1$ kpc) quenched by ~ 200 Myr from $z = 1.1$, while the outskirts quenched by ~ 5.2 Gyr from $z = 1.1$.

As suggested by $t_{\rm quench}(r)$, the quenching process progresses slowly in the disk. Although the disk slowly decreases its SFR, the stellar mass buildup is progressing in the disk. We see a simultaneous process of the *inside-out growth* and *inside-out quenching.* Both processes are connected in such a way that star formation was intense in the central region in the past (perhaps at the cosmic noon) but then quenched quickly. As star formation in the disk quenches slowly, the emerging picture is the stellar mass buildup that progresses *inside-out.*

These results agree with the slow quenching scenario of the late-type galaxies as suggested by Schawinski *et al.* (2014) and furthermore reveal the internal quenching process going on in the galaxies. It is possible that the disk galaxies observed in this research may experience a cut-off of cosmic gas supply (i.e. strangulation) since it can make SFR decreases slowly (perhaps exponentially) with time until the remaining gas in the galaxy is used up. The strangulation can happen once the halo mass reaches a critical mass ($\sim 10^{12} M_{\odot}$) above which a gas accretion to the halo enters a state of hot mode accretion (Birnboim & Dekel (2003)). The strangulation is also suggested as a primary mechanism for quenching in galaxies by Peng, Maiolino, & Cochrane (2015).

4. Conclusion

We investigated the spatially resolved distribution of Σ_* and $\Sigma_{\rm SFR}$ in massive face-on disk galaxies at $0.01 < z < 0.02$ and $0.8 < z < 1.8$. We derived the spatially resolved Σ_* and $\Sigma_{\rm SFR}$ in each galaxy using a method named *pixel-to-pixel SED fitting.* In this method, spatially resolved SEDs of a galaxy are fitted to a set of model SEDs using Bayesian statistics approach. We found a more fundamental SFMS relation at a kpc scale which is a nearly linear relation between $\Sigma_{\rm SFR}$ and Σ_*. We investigated the $\Sigma_{\rm SFR}(r)$, $\Sigma_*(r)$, and sSFR(r) radial profiles of the local and high redshift samples to study the structural evolution of massive disk galaxies and constructed an evolutionary empirical model. Overall, we observed simultaneous stellar mass buildup and quenching processes that progress *inside-out.*

References

Abdurro'uf & Akiyama M. 2017, *MNRAS*, 469, 2806

Abdurro'uf & Akiyama M. 2018, *MNRAS*, 479, 5083

Abramson, L. E., Kelson, D. D., Dressler, A., Poggianti, B., Gladders, M. D., Oemler, A., & Jr, Vulcani, B., *et al.* 2014, *ApJ*, 785, L36

Belfiore, F., Maiolino, R., Bundy, K., Masters, K., Bershady, M., Oyarzún, G. A., Lin, L., Cano-Diaz, M., *et al.* 2018, *MNRAS*, 477, 3014

Birnboim, Y. & Dekel, A. 2003, *MNRAS*, 345, 349

Bruzual, G. & Charlot, S. 2003, *MNRAS*, 344, 1000

Calzetti, D., Armus, L., Bohlin, R. C., Kinney, A. L., Koornneef, J., Storchi-Bergmann, T., *et al.* 2000, *ApJ*, 533, 682

Chabrier, G. 2003, *PASP*, 115, 763

Peng, Y., Maiolino, R., & Cochrane, R., 2015, *Nature*, 521, 192

Schawinski, K., Urry, C. M., Simmons, B. D., Fortson, L., Kaviraj, S., Keel, W. C., Lintott, C. J., Masters, K. L., *et al.* 2014, *MNRAS*, 440, 889

Speagle, J. S., Steinhardt, C. L., Capak, P. L., Silverman, J. D., *et al.* 2014, *ApJS*, 214, 15

Discussion

Tomo Goto: MS flattening is happening to all sub-samples. How does it get steeper after integration?

Abdurro'uf: The 'flattening' at high Σ_* end in the spatially resolved SFMS relation is caused by the suppression of sSFR in the central regions of the galaxies. The larger normalization difference in the global SFMS relation given only slight normalization difference in the spatially resolved SFMS is caused by the inclusion of the bulge component when calculating integrated SFR. The bulge component gives large contribution to the integrated M_* but only gives small contribution to the integrated SFR.

Challenges in Panchromatic Modelling with Next Generation Facilities
Proceedings IAU Symposium No. 341, 2018
M. Boquien, E. Lusso, C. Gruppioni & P. Tissera, eds.

doi:10.1017/S1743921319004770

Exploring the star formation histories of galaxies in different environments from MaNGA spectra

Maria Argudo-Fernández[1,2], Médéric Boquien[2], Shiyin Shen[3], Fangting Yuan[3], Jun Yin[3], Ruixiang Chang[3] and Lei Hao[3]

[1]Instituto de Física, Pontificia Universidad Católica de Valparaíso, Casilla 4059, Valparaíso, Chile

[2]Centro de Astronomía (CITEVA), Universidad de Antofagasta, Avenida Angamos 601 Antofagasta, Chile
email: maria.argudo@uantof.cl

[3]Shanghai Astronomical Observatory, Chinese Academy of Sciences, 80 Nandan Road, Shanghai 200030, China

Abstract. The star formation history (SFH) of galaxies allow us to investigate when galaxies formed their stars and assembled their mass. We can constrain the SFH with high level of precision from galaxies with resolved stellar populations, since we are able to discriminate between stars of different ages from the spectrum they emit. However, the relative importance of secular evolution (nature) over nurture is not yet clear, and separating the effects of interaction-driven evolution in the observed galaxy properties is not trivial. The aim of this study is to use MaNGA (Mapping Nearby Galaxies at APO) Integral Field Unit (IFU) data, in combination with multi-wavelength data, to constrain the SFH of nearby isolated galaxies. We present here the new techniques we are developing to constrain the SFH with high level of precision from Spectral Energy Distribution (SED) fitting. This study is part of a China-Chile collaboration program where we are applying these new techniques to investigate how galaxies formed and evolve in different environments.

Keywords. galaxies: general, galaxies: formation, galaxies: evolution, galaxies: interactions, galaxies: spiral, techniques: miscellaneous.

1. Introduction

The electromagnetic emission of galaxies is our main window into their formation and their evolution. Young stellar populations dominate the UV-optical energy budget of galaxies. As they evolve and die they produce dust and inject vast amounts of metals in the interstellar medium, which have a direct impact on the formation of new stars, shielding and cooling molecular clouds, but also reddening and dimming the UV-to-near-infrared spectrum. At the same time, galaxy can be red and dim simply because they are not forming stars anymore. In other words, the electromagnetic spectrum of galaxies encodes the past and current physical processes that have driven their evolution across cosmic times.

One of the major spectroscopic surveys in existence is the Sloan Digital Sky Survey (SDSS; York *et al.* 2000; Gunn *et al.* 2006; Aguado *et al.* 2019). It has provided us with spectra for a vast number of low redshift galaxies, yielding for each of them information about their star formation history, dust extinction, metallicity, stellar and ionized gas kinematics, etc. However, SDSS spectra are limited to a single spatial position at the center of galaxies (Strauss *et al.* 2002). The advent of the MaNGA (Mapping Nearby Galaxies at APO, Bundy *et al.* 2015) Integral Field Spectroscopy (IFS) survey is a giant leap forward. The survey is one of three core programs in the fourth-generation SDSS (SDSS-IV, Blanton *et al.* 2017). It provides us with a 3-dimensional view of galaxies (one spectral, two spatial) that encodes the processes driving their evolution, allowing us to go far beyond either the broadly used approaches of spatially-resolved photometry and spatially-unresolved spectroscopy.

What we present here is part of a China-Chile collaboration research program. Our aim is to gain a unique insight into how nearby galaxies form and evolve by comparing their observed properties with results from state-of-the-art chemical evolution models. MaNGA 3-D maps allow us to measure their physical properties (SFH, mass, metallicity, ionized gas and stellar continuum extinctions, etc.) with unprecedented precision and accuracy combining with multi-wavelength spectral energy distribution (SED) modeling from the ultraviolet to the infrared. This will therefore provides us new constraints on the assembly of galaxies across cosmic times. Because processes governing the evolution of galaxies can be strongly intertwined, we collaborate to address these issues putting our own tools and expertise in common, from fundamental spectral modeling to the interpretation of the results of these models to understand the effect of the environment on the evolution of galaxies. In particular, we show here the developed techniques and first results on the study of the evolution of spiral isolated galaxies.

2. Modelling of MaNGA galaxies

2.1. *Photometric and spectral modelling*

Spectral and photometric models are outstanding tools for measuring the physical properties of galaxies. Yet, they both have their own strengths and weaknesses. Spectra allow us to detect the presence of an AGN and access to the physical conditions of the ionized gas and of the stellar populations through emission and absorption lines and spectral indices. This yields, for instance, estimates of the extinction of the ionized gas, of the metallicity (both of the ionized gas and the stars), and constraints on the star-formation rate (SFR) on very short (Halpha, $\sim$10 Myr) and very long (D4000, $\sim$1 Gyr) timescales. However spectra are limited to a relatively narrow range in wavelengths and thus are only sensitive to a fraction of the baryonic components of galaxies. Conversely, photometric observations from the UV to the far-IR allow us to also probe young stellar populations that dominate the UV and the dust that reprocesses short wavelength radiation into the IR. Combined together, such a broad wavelength coverage provides us with the leverage we need to constrain the attenuation curve of the stellar continuum, the UV-based and IR-based SFR, and the stellar mass of galaxies. In other words, spectral and photometric data provide us with complementary and synergistic constraints on the physical properties of galaxies.

2.2. *CIGALE*

To perform the SED modelling combining all these data we use the state-of-the-art Code Investigating GALaxy Emission (CIGALE†, Boquien *et al.* 2018). CIGALE is a self-consistent SED modelling code from the far-ultraviolet to the radio domain, including the contributions from stellar populations of all ages, thermal and non-thermal gas emission, dust (both in absorption and emission: the energy absorbed by the dust in the UV-to-near-IR domain is re-emitted self-consistently in the mid- and far-IR), and optionally an active nucleus. It allows us to estimate the physical properties of the modelled

† https://cigale.lam.fr

targets (most importantly SFR over different timescales, burst strength, metallicities, dust attenuation and stellar masses). It is based on a Bayesian-like analysis method. Therefore, used in combination with spatially resolved multi-wavelength data to complement MaNGA observations, this will allow us to obtain better estimates of the SFR, the SFH, and other physical properties of the galaxies in our sample. In this study we use CIGALE to create a set of modelled spectra and use the same Bayesian-like analysis method to fit the observed MaNGA spectra and to extract the physical parameters of the SFH of the stellar populations.

2.3. *Application to a galaxy merger*

The UV and IR data are usually used to constrain the star formation rate and dust attenuation for mergers, which are tracers of the evolutionary stage of the systems. However, spatially resolved studies of galaxy mergers are restricted to very close galaxies, especially due to the lack spatial resolution in IR data. In Yuan *et al.* (2018), we show how the use of UV-to-IR broadband SED, in combination with MaNGA IFS, provide unique opportunity to study the star formation histories and dust attenuation at the tail and core parts of the merging galaxy Mrk 848 (see Figures 2 and 3 in Yuan *et al.* 2018). We find that dust derived from full spectra fitting is systematically lower than from UV to IR SED fitting (see Fig. 9 in Yuan *et al.* 2018). Additionally, using both photometry and spectroscopy SED modelling, we find a starburst younger than 100 Myr in core region, consistent with the scenario where the interaction-induced gas in-flow enhances the star formation in the center of galaxies.

3. Evolution of spiral isolated galaxies

We are interested into explore how galaxies build their disk when they grow in isolation, unaffected by external influences. From models, it is predicted that disks grow from inner to outer parts of galaxies, the so-called inside-out model of galaxy formation. As a consequence, we expect to observe gradients in stellar age and metallicity as a function of galaxy radii. The sample of isolated galaxies used in this study is selected from the SDSS-based catalog of Isolated Galaxies (SIG; Argudo-Fernández *et al.* 2015), composed of 3702 galaxies. Of the ∼200 SIG galaxies observed by MaNGA, we selected a sample of 18 almost face-on star-forming (SF) spiral galaxies. We focused on SF galaxies to assure that our results are not contaminating by quenching mechanisms (e.g., LINER or AGN emission). To identify the SF galaxies, we used Marvin (Cherinka *et al.* 2018) to create spatially-resolved BPT diagrams and selected the galaxies where more than the 90% of their spaxels are classified as star-forming (see the left panel of Fig. 1).

Considering that SIG galaxies in our sample are not subject to significant influence from close neighbours for significant time periods, we assume that their disk has homogeneously evolved form inside-out and we therefore divide each SIG galaxy in radial bins with separations of 3 arcsec (MaNGA spatial resolution is 2.5 arcsec). We finally use CIGALE, as explained in the previous section, to get the parameters of the SFH (see the right panel of Fig. 1). We find that, in general, the age t and τ gradients of spiral SIG galaxies are in agreement in the inside-out disk growth. However, the stellar metallicity (Z) and dust attenuation (A_v) gradients show more diverse results, which indicate that the evolution of the disk is more complex than expected (Argudo-Fernández *et al.* in prep.).

Acknowledgments

MAF acknowledges IAU Travel Grant for S341 and CAS-CONICYT N° 17002 for funding support to assist this conference. Funding for the Sloan Digital Sky Survey IV

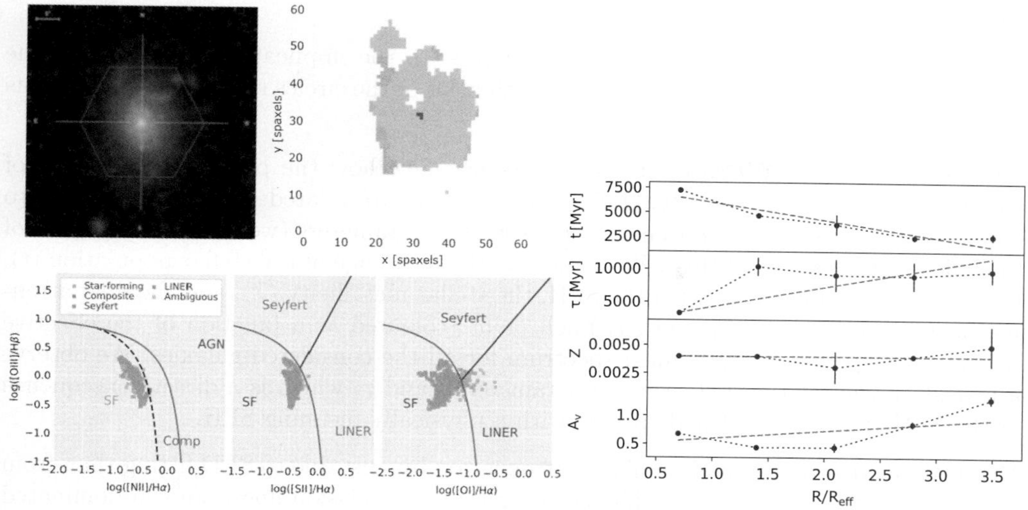

Figure 1. Results of the spectroscopic analysis for a SF SIG galaxy (Argudo-Fernández *et al.*, in prep.). *Left panel:* Spatially resolved BPT diagrams and map of the SIG galaxy of the upper left panel, colored as a function of the BPT classification according to the legend. *Right panel:* Bayesian-like estimation of SFH's parameters (age, τ), stellar metallicity (Z), and dust attenuation (A_v) in radial bins as a function of the effective radius R_{eff}.

has been provided by the Alfred P. Sloan Foundation, the U.S. Department of Energy Office of Science, and the Participating Institutions. SDSS-IV acknowledgessupport and resources from the Center for High-Performance Computing at the University of Utah. The SDSS web site is www.sdss.org.

References

Aguado, D. S., Ahumada, R., Almeida, A., Anderson, S. F., Andrews, B. H., Anguiano, B., Aquino Ortíz, E., Aragón-Salamanca, A., *et al.* 2019, *ApJS*, 240, 23

Argudo-Fernández, M., Verley, S., Bergond, G., Duarte Puertas, S., Ramos Carmona, E., Sabater, J., Fernández Lorenzo, M., Espada, D., *et al.* 2015, *A&A*, 578, A110

Blanton, M. R., Bershady, M. A., Abolfathi, B., Albareti, F. D., Allende Prieto, C., Almeida, A., Alonso-García, J., Anders, F., *et al.* 2017, *AJ*, 154, 28

Boquien, M., Burgarella, D., Roehlly, Y., Buat, V., Ciesla, L., Corre, D., Inoue, A. K., Salas, H., *et al.* 2018, arXiv:1811.03094

Bundy, K., Bershady, M. A., Law, D. R., Yan, R., Drory, N., MacDonald, N., Wake, D. A., Cherinka, B., *et al.* 2015, *ApJ*, 798, 7

Cherinka, B., Andrews, B. H., Sánchez-Gallego, J., Brownstein, J., Argudo-Fernández, M., Blanton, M., Bundy, K., Jones, A., *et al.* 2018, arXiv:1812.03833

Gunn, J. E., Siegmund, W. A., Mannery, E. J., Owen, R. E., Hull, C. L., Leger, R. F., Carey, L. N., Knapp, G. R., *et al.* 2006, *AJ*, 131, 2332

Strauss, M. A., Weinberg, D. H., Lupton, R. H., Narayanan, V. K., Annis, J., Bernardi, M., Blanton, M., Burles, S., *et al.* 2002, *AJ*, 124, 1810

York, D. G., Adelman, J., Anderson, J. E., Jr., *et al.* 2000, *AJ*, 120, 1579

Yuan, F.-T., Argudo-Fernández, M., Shen, S., Hao, L., Jiang, C., Yin, J., Boquien, M., and Lin, L., *et al.* 2018, *A&A*, 613, A13

Discussion

SADANORI OKAMURA: Could you please summarise the implication (meaning) of the plots where various parameters are shown with color of the circles representing the radius from the center of a galaxy.

ARGUDO FERNÁNDEZ MARIA: In the figure we show how the physical parameters of the SFH, derived from the CIGALE spectral fitting, are related each other. This is to check that there is not any degeneracy between the parameters (we create a large grid of modelled spectra to avoid this). We show how the mean age of the stellar population (t), the time-scale (τ) parameters of the SFH, the stellar metallicity (Z), and the dust attenuation (A_v) are related each other. Each point (coloured as a function of the effective radius) corresponds to one binned spectrum for all the considered galaxies. We observe a feature in the relation between the parameters t and τ, which is a direct consequence of the SFH, since we modelled spectra with a physically meaning SFH.

BUAT VERONIQUE: You find an higher attenuation by dust due to SED fitting than from spectral fits, it is at odds with a higher attenuation in emission lines, are you dominated by the spectral continuum?

ARGUDO FERNÁNDEZ MARIA: Yes, with the spectral fitting we focus on the continuum and the absorption lines. We mask the emission lines and create the grid of models without doing any assumption on the nebular emission.

TOMOTSUGU GOTO: Optical only SED fitting underestimate A_v. But MANGA is optical only?

ARGUDO FERNÁNDEZ MARIA: That is why our aim is to extend CIGALE to do photo + spectral fitting all together, to have more info on the UV and far-IR. But optical spectra is important because we can get information from many points in the SED, not only from a few points as usually occurs using photometry only, as we can get many information from the continuum.

Challenges in Panchromatic Modelling with Next Generation Facilities
Proceedings IAU Symposium No. 341, 2018
M. Boquien, E. Lusso, C. Gruppioni & P. Tissera, eds.

doi:10.1017/S1743921319002643

High-resolution radiation transfer modelling of barred galaxies

A. Nersesian[1,2,3], S. Verstocken[3], S. Viaene[3,4] and M. Baes[3]

[1]National Observatory of Athens, Institute for Astronomy, Astrophysics, Space Applications and Remote Sensing, Ioannou Metaxa and Vasileos Pavlou GR-15236, Athens, Greece
email: a.nersesian@noa.gr

[2]Department of Astrophysics, Astronomy & Mechanics, Faculty of Physics, University of Athens, Panepistimiopolis, GR15784 Zografos, Athens, Greece

[3]Sterrenkundig Observatorium Universiteit Gent, Krijgslaan 281 S9, B-9000 Gent, Belgium

[4]Centre for Astrophysics Research, University of Hertfordshire, College Lane, Hatfield, AL10 9AB, UK

Abstract. Dust radiative transfer simulations provide us with the unique opportunity to study the heating mechanisms of dust by the stellar radiation field. From 2D observational images we derive the 3D distributions of stars and dust. Our aim is to analyze the contribution of the different stellar populations to the radiative dust heating processes in the nearby face-on barred galaxies NGC 1365, M 83 and M 95. We wish to decompose the FIR-submm SED and quantify the fraction directly related to star formation. To model the complex geometries mentioned above, we used SKIRT, a state-of-the-art, 3D Monte Carlo radiative transfer code designed to simulate the absorption, scattering and thermal re-emission of dust in a variety of environments. We find that the contribution of the evolved stars (8 Gyr) to the dust heating is non-negligible ($\sim 35\%$) and can reach as high as 70%. We also find a tight correlation between the heating fraction by the unevolved stars ($\leqslant 100$ Myr) and the specific star formation rate.

Keywords. radiative transfer - ISM: dust, extinction - galaxies: individual: NGC 1365, M83, M95 - galaxies: ISM - infrared: ISM

1. Introduction

Dust grains are responsible for the attenuation and reddening effects at ultraviolet (UV) and optical wavelengths. Dust absorbs roughly 30% of the light emitted by stars and redistributes this light back in the infrared wavelengths (Popescu *et al.* 2002; Skibba *et al.* 2011; Viaene *et al.* 2016; Bianchi *et al.* 2018). The dust emission in this regime is often used to trace star formation activity (Calzetti *et al.* 2007, 2010; Kennicutt *et al.* 2009), however the contribution of the old stars to the dust heating can be non-negligible (e.g., Bendo *et al.* 2012, 2015) and needs to be considered while estimating the star formation rate (SFR) of a galaxy. We use radiation transfer (RT) modelling to create a realistic model of the radiation field and its interaction with dust, by quantifying the relative contribution of each stellar population to the dust heating. The advantage of using RT modelling in relation to other methods, such as pixel-by-pixel SED (spectral energy distribution) fitting, is that it takes into account the effect of non-local heating.

2. Modelling approach

We construct a self-consistent 3D model for a galaxy, with the Monte-Carlo RT code SKIRT (Baes *et al.* 2011; Camps & Baes 2015), taking into account all relevant physical

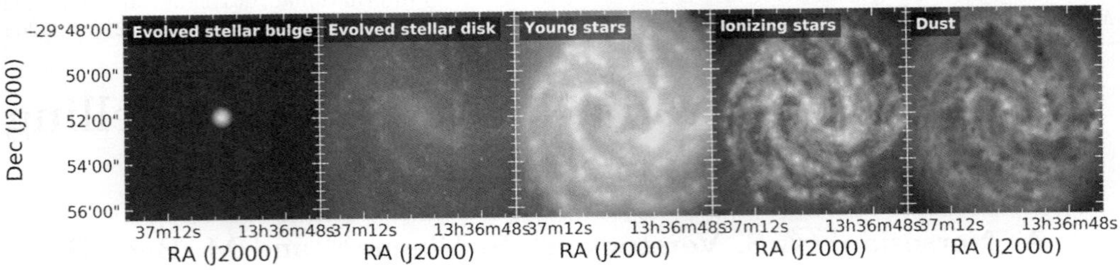

Figure 1. Input component maps of M 83. The colour coding is in log scale and reflects a normalised flux density.

processes such as scattering, absorption and thermal re-emission by dust. In this work we focus on three nearby face-on barred galaxies, NGC 1365, M 83 and M 95. The model for each galaxy consists of four stellar components and a dust component. We consider a central stellar bulge, and a disk populated by evolved and unevolved (young and ionizing) stars. We model the evolved and young stars using the Bruzual & Charlot (2003) single stellar populations (SSP) of ages 8 Gyr and 100 Myr respectively, while for the ionizing stars we adopt the SED templates from MAPPINGS III (Groves *et al.* 2008) assuming an age of 10 Myr.

Here we describe briefly the steps we followed to retrieve the stellar and dust geometries, for each galaxy. First we perform a bulge-disk decomposition using an IRAC 3.6 μm image, tracing the emission of evolved stars. We model the bulge with a flattened Sérsic profile and assume that the contamination from the unevolved stars and from dust emission in the central region are negligible. By subtracting the bulge from the disk we were able to obtain a map of the evolved stars in the disk. To get the distribution of the young and the ionizing stars in the disk we use the GALEX far-ultraviolet (FUV) image, and the MIPS 24 μm image combined with an Hα map respectively. We constrain the dust component in the disk through a FUV attenuation map, while for the dust composition we used the dust model THEMIS (Jones *et al.* 2017). The different components can be seen in Fig. 1.

To create the 3D distribution of the disk components we assign to each one of them an exponential profile of different scale heights based on previous estimates of the vertical extent of edge-on galaxies (De Geyter *et al.* 2014). We create a dust grid based on the dust component, through which the photons propagate in our simulations. We use a binary tree dust grid (Saftly *et al.* 2014) with approximately 3 million dust cells for each galaxy. Finally, three free parameters in our model are determined via a χ^2 optimisation procedure. These parameters are: the FUV luminosities of the young and ionizing stars and the dust mass. Initial guesses for these parameters are based on global SED fitting performed with CIGALE (Boquien *et al.* 2018) by (Nersesian *et al.* 2019) for 814 DustPedia galaxies (Davies *et al.* 2017).

3. Results

From our simulations it is possible to retrieve the absorbed energy in each dust cell originating from the different stellar populations in the model and thus to quantify the dust heating fraction from the unevolved stellar populations. An example of the dust heating map of M 83 is given in Fig. 2. The left panel shows the heating fraction map of a face-on view of M 83. The histogram in the right panel displays the heating fraction distribution within the dust cells, weighed by dust mass. For M 83 we find that 63% of the heating originates from the unevolved stars with the lowest values in the most central regions. Similarly, in NGC 1365, 66% of the heating originates from the unevolved stars, while for M 95 we find that the dust is mainly heated by the evolved stars with the

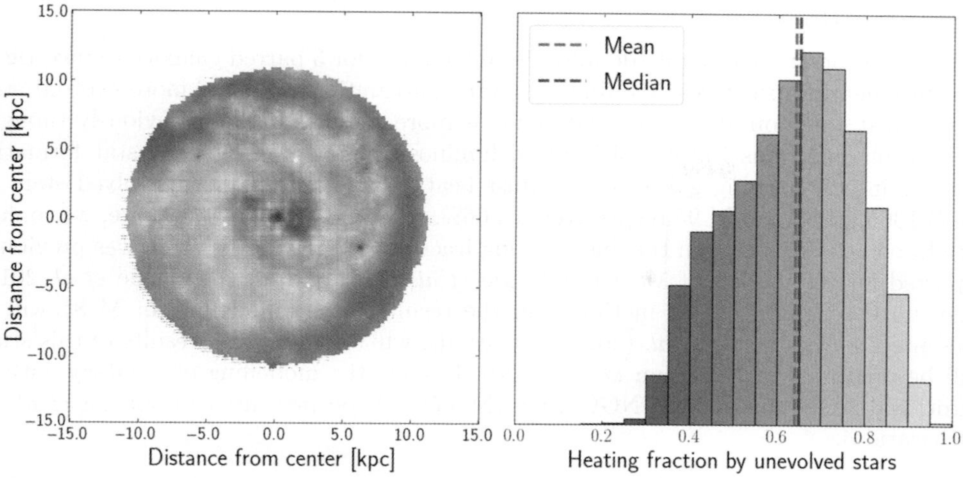

Figure 2. Left panel: Face-on view of the heating fraction by unevolved stars, as obtained from the 3D dust cell distribution, for M 83. Right panel: Distribution of the dust cell heating fractions, weighed by dust mass. The green dashed line is the mean value, while the purple dashed line is the median.

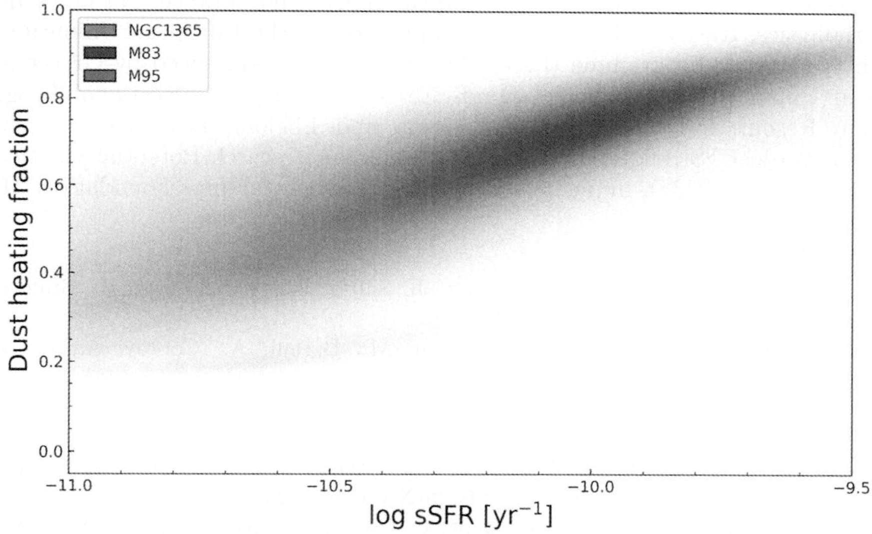

Figure 3. Relation between the sSFR and the relative contribution from the unevolved stars to the dust heating responsible for the total-infrared emission in the RT model of NGC 1365 (red data points), M 83 (blue data points) and M 95 (green data points).

contribution of the unevolved being only 32%. These are the fractions by which a typical far-infrared (FIR) SFR indicator should be corrected in these galaxies.

We also report a strong correlation between the dust heating fraction and the specific star formation rate (sSFR) calculated for each cell. We have used the IRAC 3.6 μm luminosities to estimate the stellar masses (Oliver *et al.* 2010) and the intrinsic FUV luminosities of the unevolved stars to estimate the star formation rates (Kennicutt & Evans 2012). In Fig. 3 we plot the heating fraction as was calculated for each cell and for all three galaxies, as a function of the sSFR. It is immediately evident that there is an increasing trend between the heating fraction and the sSFR. That, in principle, enables us to quantify the heating fraction based on sSFR measurements in other galaxies.

4. Conclusions

We have constructed highly detailed 3D RT models for 3 barred galaxies to investigate the dust heating processes. We find that the contribution from the more evolved stars to the dust emission at the infrared bands is more significant than previously thought. This result questions the use of infrared luminosity as a proxy for the star formation activity in star-forming galaxies. The dust heating fraction by the unevolved stars for NGC 1365, M 83 and M 95 respectively is: 66%, 63% and 32%. Furthermore, we confirm a tight correlation between the dust heating fraction and the sSFR which was previously reported for RT models of M 51 (De Looze *et al.* 2014) and M 31 (Viaene *et al.* 2017). The full description of our method and the results of the modelling of M 81 will be presented in Verstocken *et al.* (2020, submitted), while the detailed results of this study will be presented in Nersesian *et al.* (2020). Finally, the modelling of a galaxy with an additional AGN component, NGC 1068 (M 77) will be presented in Viaene *et al.* (in preparation).

Acknowledgements

DustPedia is a collaborative focused research project supported by the European Union under the Seventh Framework Programme (2007-2013) call (proposal no. 606847). The participating institutions are: Cardiff University, UK; National Observatory of Athens, Greece; Ghent University, Belgium; Université Paris Sud, France; National Institute for Astrophysics, Italy and CEA, France. supported by the European Union under the Seventh Framework Programme (FP7). This research is co-financed by Greece and the European Union (European Social Fund- ESF) through the Operational Programme ≪Human Resources Development, Education and Lifelong Learning≫ in the context of the project "Strengthening Human Resources Research Potential via Doctorate Research" (MIS-5000432), implemented by the State Scholarships Foundation (IKY).

References

Baes, M., Verstappen, J., De Looze, I., Fritz, J., Saftly, W., Vidal Pérez, E., Stalevski, M., Valcke, S., *et al.* 2011, *ApJS*, 196, 22

Bendo, G. J., Baes, M., Bianchi, S., Boquien, M., Boselli, A., Cooray, A., Cortese, L., De Looze, I., *et al.* 2015, *MNRAS*, 448, 135

Bendo, G. J., Boselli, A., Dariush, A., Pohlen, M., Roussel, H., Sauvage, M., Smith, M. W. L., Wilson, C. D., *et al.* 2012, *MNRAS*, 419, 1833

Bianchi, S., De Vis, P., Viaene, S., Nersesian, A., Mosenkov, A. V., Xilouris, E. M., Baes, M., Casasola, V., *et al.* 2018, *ArXiv e-prints*, [arXiv:1810.01208]

Boquien, M., Burgarella, D., Roehlly, Y., Buat, V., Ciesla, L., Corre, D., Inoue, A. K., Salas, H., *et al.* 2018, *ArXiv e-prints*, [arXiv:1811.03094]

Bruzual, G. & Charlot, S. 2003, *MNRAS*, 344, 1000

Calzetti, D., Wu, S.-Y., Hong, S., Kennicutt, R. C., Lee, J. C., Dale, D. A., Engelbracht, C. W., van Zee, L., *et al.* 2010, *ApJ*, 714, 1256

Calzetti, D., Kennicutt, R. C., Engelbracht, C. W., Leitherer, C., Draine, B. T., Kewley, L., Moustakas, J., Sosey, M., *et al.* 2007, *ApJ*, 666, 870

Camps, P. & Baes, M. 2015, *Astronomy and Computing*, 9, 20

Davies, J. I., Baes, M., Bianchi, S., Jones, A., Madden, S., Xilouris, M., Bocchio, M., Casasola, V., *et al.* 2017, *PASP*, 129, 044102

De Geyter, G., Baes, M., Camps, P., Fritz, J., De Looze, I., Hughes, T. M., Viaene, S., & Gentile, G. 2014, *MNRAS*, 441, 869

De Looze, I., Fritz, J., Baes, M., Bendo, G. J., Cortese, L., Boquien, M., Boselli, A., Camps, P., *et al.* 2014, *A&A*, 571, A69

Groves, B., Dopita, M. A., Sutherland, R. S., Kewley, L. J., Fischera, J., Leitherer, C., Brandl, B., van Breugel, W., *et al.* 2008, *ApJS*, 176, 438

Jones, A. P., Köhler, M., Ysard, N., Bocchio, M., & Verstraete, L. 2017, *A&A*, 602, A46
Kennicutt, R. C. & Evans, N. J. 2012, *ARA&A*, 50, 531
Kennicutt, R. C., Hao, C., Calzetti, D., Moustakas, J., Dale, D. A., Bendo, G., Engelbracht, C. W., Johnson, B. D., *et al.* 2009, *ApJ*, 703, 1672
Nersesian, A., Verstocken, S., Viaene, S., Baes, M., Xilouris, E. M., Bianchi, S., Casasola, V., Clark, C. J. R., *et al.* 2020, *A&A*, Forthcoming [doi:10.1051/0004-6361/201936176]
Nersesian A., Xilouris E., Bianchi S., Galliano, F., Jones, A. P., Baes, M., Casasola, V., Cassarà, L. P., *et al.* 2019, *A&A*, 624, A80
Oliver, S., Frost, M., Farrah, D., Gonzalez-Solares, E., Shupe, D.L., Henriques, B., Roseboom, I., Alfonso-Luis, A., *et al.* 2010, *MNRAS*, 405, 2279
Popescu, C. C., Tuffs, R. J., Völk, H. J., Pierini, D., & Madore, B. F. 2002, *ApJ*, 567, 221
Saftly, W., Baes, M., & Camps, P. 2014, *A&A*, 561, A77
Skibba, R. A., Engelbracht, C. W., Dale, D., Hinz, J., Zibetti, S., Crocker, A., Groves, B., Hunt, L., *et al.* 2011, *ApJ*, 738, 89
Verstocken, S., Baes, M., Nersesian, A., Viaene, S., Bianchi, S., Casasola, V., Clark, C. J. R., Davies, J. I., *et al.* 2020, *A&A*, submitted
Viaene, S., *et al.* 2020, in preparation
Viaene, S., Baes, M., Tamm, A., Tempel, E., Bendo, G., Blommaert, J .A. D. L., Boquien, M., Boselli, A., *et al.* 2017, *A&A*, 599, A64
Viaene, S., Baes, M., Bendo, G., Boquien, M., Boselli, A., Ciesla, L., Cortese, L., De Looze, I., *et al.* 2016, *A&A*, 586, A13

Discussion

V. Buat: You underlined an attenuation by dust from evolved stars for NIR data (M83 plot), what is the impact of this dust attenuation on the measure of stellar masses?

A. Nersesian: In the figure was depicted the relative contribution of the evolved and unevolved stars to the heating of dust across the wavelength range UV-optical, as determined from the RT model of M 83, and not the fraction of stellar luminosity that is absorbed or attenuated. Of course NIR data are affected by dust attenuation, but in the case of M 83 the effect to the stellar mass measurement is not significant.

F. Egusa: Residual maps look like to have some spiral-like structures? What would be the reason?

A. Nersesian: The spiral-like structures arise because the model has to some extent smoothed out the sharper features of the galaxy. But also has to do to the fact that multiple images, of different bands, were combined to create the input maps of the model.

F. Egusa: NGC 1365 and M 83 have ALMA data, which can be a tracer of molecular gas. Would this gas information help to improve the models?

A. Nersesian: Definitely, using gas maps to trace the SFR regions would help create a more accurate representation of those galaxies in our modelling.

T. Goto: Dust heating fraction has more values in NGC 1365. Why? Maybe AGN, inclination?

A. Nersesian: NGC 1365 has a very strong AGN emission in the center which eventually affects the heating fraction in the central regions of the galaxy.

S. Shen: Does your model include dust destruction? The evolved stars may have UV photons destructing dust grains.

A. Nersesian: No our model does not include dust destruction. However, we use observed images in the FIR to model the dust in the galaxies, which means that dust destruction is already accounted for.

Challenges in Panchromatic Modelling with Next Generation Facilities
Proceedings IAU Symposium No. 341, 2018
M. Boquien, E. Lusso, C. Gruppioni & P. Tissera, eds.

doi:10.1017/S1743921319002692

Which attenuation curves for star-forming galaxies?

Véronique Buat[1], David Corre[1], Médéric Boquien[2] and Katarzyna Małek[1,3]

[1]Aix Marseille Univ, CNRS, CNES, LAM Marseille, France
email: veronique.buat@lam.fr

[2]Centro de Astronomía (CITEVA), Universidad de Antofagasta, Avenida Angamos 601, Antofagasta, Chile

[3]National Centre for Nuclear Research, ul. Hoza 69, 00-681 Warszawa, Poland

Abstract. Dust attenuation shapes the spectral energy distributions of galaxies and any modelling and fitting procedure of their spectral energy distributions must account for this process. We present results of two recent works dedicated at measuring the dust attenuation curves in star forming galaxies at redshift from 0.5 to 3, by fitting continuum (photometric) and line (spectroscopic) measurements simultaneously with CIGALE using variable attenuation laws based on flexible recipes. Both studies conclude to a large variety of effective attenuation laws with an attenuation law flattening when the obscuration increases. An extra attenuation is found for nebular lines. The comparison with radiative transfer models implies a flattening of the attenuation law up to near infrared wavelengths, which is well reproduced with a power-laws recipe inspired by the Charlot and Fall recipe. Here we propose a global modification of the Calzetti attenuation law to better reproduce the results of radiative transfer models.

Keywords. galaxies: high-redshift–, dust : extinction –galaxies: ISM–infrared: galaxies

1. Introduction

Modelling the spectral energy distribution (SED) of galaxies is a method commonly used to derive physical parameters useful to quantify galaxy evolution like stellar masses or star formation histories. Dust plays a crucial role by strongly affecting and reshaping the spectral energy distribution (SED): it absorbs and scatters stellar photons, and thermally emits the absorbed energy in the infrared (IR) ($\lambda \sim 1 - 1000$ μm). The simplest way to model this dust effect to introduce effective attenuation curves which account for the complex blending of dust properties and relative geometrical distribution of stars and dust within a galaxy. The most commonly used recipes are the ones proposed by Calzetti *et al.* (1995) (hereafter C00) and Charlot & Fall (2000). These two recipes include a differential attenuation between young stars/nebular emission and older stellar populations. Since these original works, several authors proposed a modification of these fixed recipes (e.g. Buat *et al.* (2018); Kriek & Conroy (2013); Salmon *et al.* (2016); Lo Faro *et al.* (2017)). In this work we adopt flexible recipes based on both formalisms, the method is fully described in Buat *et al.* (2018). The general behaviour of the two formalisms, in their flexible form, is illustrated in Fig. 1. While both recipes give similar trends in the UV-V range, the shapes of the attenuation curves are systematically different in the V-NIR range with a flatter shape for the recipe derived from the Charlot & Fall (2000) model.

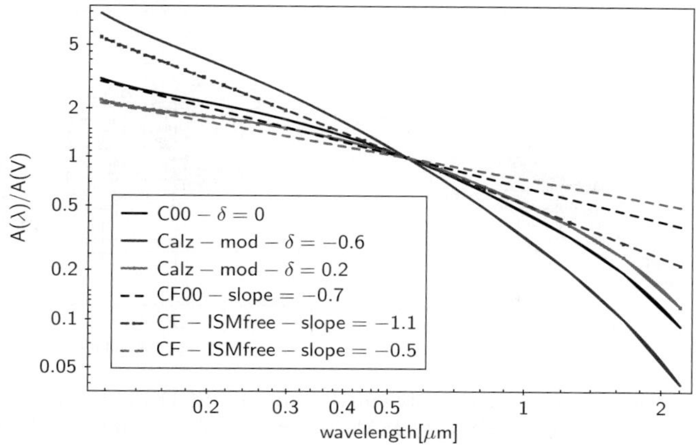

Figure 1. Flexible effective attenuation curves based on the Calzetti *et al.* original law (solid lines) and on the Charlot and Fall original power-laws (dotted lines, the curves refer only here to the attenuation in the interstellar medium (ISM) and not the birth clouds). Both recipes exhibit similar flexibility in the UV-V range. At V-NIR wavelengths the Calzetti-like recipes lead to much steeper slopes than the ones based on the power-laws introduced by Charlot and Fall

2. Determinations of dust attenuations laws in high redshift galaxies

We based our analysis on two different galaxy samples. The first one is presented in Buat *et al.* (2018). It consists of galaxies in the COSMOS field, detected with the PACS and SPIRE instruments of the *Herschel* satellite and for which low resolution spectra are available from the 3D-HST survey. The sample is made of 33 galaxies observed in 21 photometric bands from the NUV to the submm and with a Hα line measurement. The second galaxy sample was built by Corre (2018) and consists of 19 galaxies hosting a γ-ray burst (GRBHs), observed in at least 5 photometric bands starting in the UV rest frame and with Hα and Hβ line measurements.

The SEDs of all the galaxies were fitted with the CIGALE code which allows to fit simultaneously photometric and emission line fluxes (Boquien *et al.* (2018). Both measurements were very well fitted. The 3D-HST/COSMOS sample was fitted with both the modified C00 and Charlot and Fall recipes (Buat *et al.* (2018)). The GRBHs sample was only studied with the modified Calzetti recipe (Corre *et al.* (2018); Corre (2018)). For the purpose of the comparison between the two studies, we focus here on the results based on the modified Calzetti recipe.

The flexible recipe is defined as:

$$A(\lambda) = E(\mathrm{B-V})_{\mathrm{star}} k'(\lambda) \left(\frac{\lambda}{\lambda_{\mathrm{V}}}\right)^{\delta}, \tag{1}$$

$\delta = 0$ corresponds to the original C00 recipe. The nebular component is extinguished with a simple screen model, a Milky Way extinction curve and a color excess $E(\mathrm{B-V})_{\mathrm{line}}$. The ratio $E(\mathrm{B-V})_{\mathrm{star}}/E(\mathrm{B-V})_{\mathrm{line}}$ is defined as a free parameter in our recipe (it is fixed to 0.44 in the original C00 recipe).

The results are shown in Fig. 2. On the left plot, the color excesses for lines and stars are correlated and the two samples lead to consistent distributions. The average value of $E(\mathrm{B-V})_{\mathrm{star}}/E(\mathrm{B-V})_{\mathrm{line}}$ is found to be 0.56 ± 0.20, only slightly larger than the 0.44 value used by C00. The right panel of Figure 2 presents the correction δ to the slope of the attenuation law as a function of the attenuation, Av. The δ parameter spans a wide

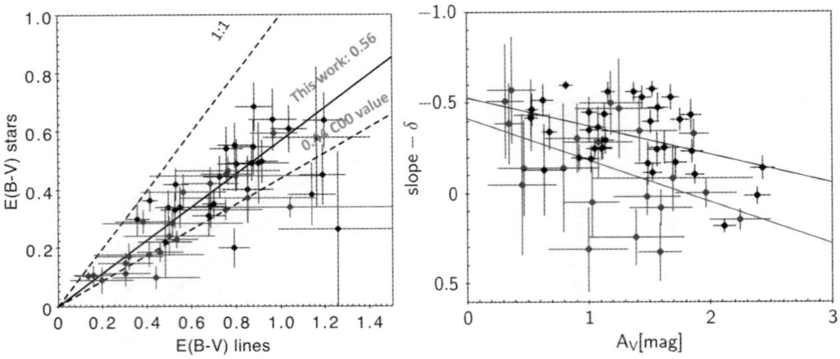

Figure 2. Attenuation characteristics obtained from the SED fitting analysis for the GRBH sample (blue symbols) and the 3D-HST/COSMOS sample (black symbols). Left: the color excess of the nebular emission is compared to the color excess of the stellar continuum, the average value of $E(\mathrm{B-V})_{\mathrm{star}}/E(\mathrm{B-V})_{\mathrm{line}}$ found for our samples and by C00 are also plotted as well as as the 1:1 relation. Right: the correction δ to apply to the original C00 attenuation curve is plotted against the global attenuation, represented by A_V

range of values, with most of the curves having steeper slopes than the original C00 law, corresponding to $\delta = 0$. The majority of the 3D-HST/COSMOS galaxies is well modelled by curves with $\delta < 0$, i.e. with higher levels of extinction at the blue wavelengths with respect to C00. Figure 2 also shows that, as Av increases the attenuation law flattens, consistently with previous observational and modelling studies (e.g. Salmon *et al.* (2016), Chevallard *et al.* (2013)). We stress that in the present study, the shape of the attenuation curve and the relative attenuation of young and old stars are taken free parameters which was not the case in previous published works.

3. Consistency of attenuation recipes with radiation transfer modelling

Radiation transfer models predict a flattening of the attenuation law when the global attenuation increase, and this flattening is seen up to the NIR. As shown in Fig.1, the Calzetti recipe, modified with the introduction of δ, does not allow for this flattening which is easier to get with power-laws modelling. We propose to modify the exponent δ in $\delta + \epsilon$ at wavelength longer than the V band as follows

$$\lambda < \lambda_\mathrm{V}, A(\lambda) = E(\mathrm{B-V})_{\mathrm{star}} k'(\lambda) \left(\frac{\lambda}{\lambda_\mathrm{V}}\right)^{\delta} \tag{2}$$

$$\lambda > \lambda_\mathrm{V}, A(\lambda) = E(\mathrm{B-V})_{\mathrm{star}} k'(\lambda) \left(\frac{\lambda}{\lambda_\mathrm{V}}\right)^{\delta+\epsilon} \tag{3}$$

The value of the correction ϵ can be inferred from the radiation transfer calculations of Seon & Draine (2016). Corre (2018) fitted the numerical results of their calculations with the Calzetti modified formalism, including the ϵ correction.

In Fig. 3 the values obtained of ϵ and δ are reported. The red curve is the result of a polynomial fit:

$$\epsilon = 0.35\ \delta^2 + 0.04\ \delta + 0.41. \tag{4}$$

The correction can be implemented either as this relation between δ and ϵ or as a single average value (Corre *et al.* 2019, in prep.).

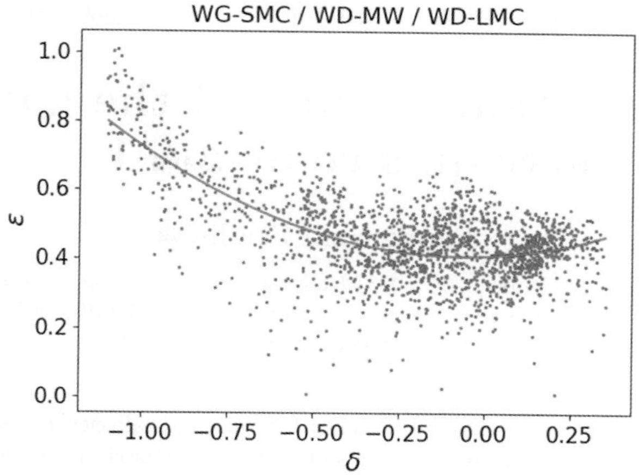

Figure 3. Correction of the slope of the attenuation law at wavelengths larger than the V band calculated with the models of Seon & Draine (2016). Several dust models are used, the blue points correspond to the results of the fits performed on the numerical results of Seon & Draine (2016). The red curve is the best fit (ϵ as a function of δ).

References

Boquien, M., Burgarella, D., Roehlly, Y., *et al.* 2018, arXiv:181103094B
Buat, V., Giovannoli, E., Heinis, S., *et al.* 2011, *A&A*, 533, A93
Buat, V., Boquien, M., Malek, K., *et al.* 2018, *A&A*, 619, A135
Calzetti, C., Armus, L., Bohlin, R. C., *et al.* 2000, *ApJ*, 533, 682
Charlot, S. & Fall, S. M. 2000, *ApJ*, 539, 718
Chevallard, J., Charlot, S., Wandelt, B. & Wild, V., *et al.* 2013, *MNRAS*, 462, 1415
Corre, D. 2018, *PhD thesis, Aix-Marseille Université*
Kriek, M. & Conroy, C. 2013, *ApJ*, 775, L16
Lo Faro, B., Buat, V., Roehlly. Y., *et al.* 2017, *MNRAS*, 472, 1372
Salmon, B., Papovich, C., Long, J., *et al.* 2016, *ApJ*, 827, 20
Seon, K.-I. & Draine, B. T. 2013, *ApJ*, 833, 201

Discussion

Gergo Popping: Do you see a correlation between extinction versus attenuation curve offset and SFR or main-sequence offset?

Veronique Buat: We found a flattening of the attenuation curve with SFR and no clear trend with sSFR. The offset between extinction and attenuation curves does not seem to be correlated to SFR but our sample remains very small and biased to extinction curves of LMC or MW types.

Tomotsugu Goto: What is the difference between modified Calzetti and CF2000? Which is better in the end?

Veronique Buat: The way to implement each of them in SED fitting is different. A single attenuation law (Calzetti-like) is easier to handle than an age dependent one. The most important issue is to keep some flexibility to represent the diversity of cases and to add a correction in NIR to the Calzetti-like recipe to fit dusty galaxies.

Yutaka Fujita: Did you study only Long GRBs

Veronique Buat: Yes, we did.

Challenges in Panchromatic Modelling with Next Generation Facilities
Proceedings IAU Symposium No. 341, 2018
M. Boquien, E. Lusso, C. Gruppioni & P. Tissera, eds.

doi:10.1017/S174392131900190X

Dust attenuation on and off the galaxy Main Sequence at $z \geqslant 1$

Annagrazia Puglisi

Laboratoire AIM-Paris-Saclay, CEA/DSM-CNRS-Université Paris Diderot, Irfu/Service d'Astrophysique CEA Saclay, Orme des Merisiers, 91191 Gif sur Yvette, France
email: annagrazia.puglisi@cea.frl

Abstract. Stellar masses are crucial ingredients for putting galaxies in the context of galaxy evolution and are commonly evaluated via Spectral Energy Distribution (SED)-fitting analyses which are hampered by dust attenuation. Observational constraints of attenuation in various galaxy classes provide key inputs for fitting a SED. I will present recent results about the attenuation properties of a sample of *Herschel*-selected galaxies at $0.7 \leqslant z \leqslant 1.6$ widely spanning the star-forming Main Sequence (MS). I will show that far-IR selected galaxies on the MS are well described with local attenuation recipes. Conversely, common recipes cannot recover the SFR of far-IR selected starburst galaxies well above the MS. The SFR of these outliers appears to be hidden by the ~90% in optically thick cores. These findings pose challenges for SED-fitting codes based on energy balance assumptions that might break in these peculiar sources.

Keywords. galaxies: evolution, galaxies: high-redshift, galaxies: star formation, dust, extinction, infrared, galaxies: interactions, galaxies: ISM

1. Introduction

The so-called Main Sequence is a tight relation that roughly 90% of star-forming galaxies follow in the stellar mass ($M_\star$) versus Star Formation Rate (SFR) plane (MS, Speagle *et al.* 2014). Its existence since very high redshifts along with its small scatter (~0.3 *dex*, Whitaker *et al.* 2014) has been often interpreted as an evidence that star formation events proceed smoothly with cosmic time and stochastic episodes (e.g. mergers) do not significantly contribute to galaxy growth. Nevertheless, outliers from this MS exist and might have a key role in the formation of massive elliptical galaxies despite contributing only ~10% to the cosmic SFR density (Rodighiero *et al.* 2011).

Locating a galaxy on the MS plane implies to measure accurate $M_\star$ and SFR. Traditionally, $M_\star$ is inferred through Spectral Energy Distribution (SED) fitting codes by comparing simple stellar population models to UV- to-NIR observations after applying a dust attenuation correction (Conroy 2013, e.g.). Measurements of SFR from UV or optical emission lines also require to correct for dust attenuation.

There are several dust attenuation indicators allowing us to correct the spectrum of a galaxy for this effect. For example, one can use the slope of the spectrum in the UV (the so-call UV-slope β, Calzetti *et al.* 1994) to evaluate the *reddening* of the stellar continuum $E_\star(B-V)$. To quantify the dust attenuation in HII regions, $E_{\rm neb}(B-V)$, one can directly measure nebular line ratios such as the Balmer Decrement (i.e., the $H\alpha/H\beta$ line ratio, BD). Finally, the so-called IRX=$L_{\rm FIR}/L_{\rm UV}$ ratio quantifies the total dust attenuation of a galaxy by comparing the bolometric emission of the dust ($L_{\rm IR}$) and the observed radiation at UV wavelengths ($L_{\rm UV,obs}$). These indicators appear to be correlated through empirical relationship which are mainly calibrated on local galaxies. Exploring the validity of dust attenuation correction in the high-z Universe is crucial given that star-forming galaxies show evolving properties as a function of redshift. Moreover, most studies are conducted on "typical" MS galaxies representative of the star-forming population. Systematic works exploring the dust attenuation properties of starbursting MS outliers are missing although it is likely to expect some variations in these objects given their large gas and dust content.

In this proceeding I will summarize our recent efforts to study the dust attenuation properties of $z \geqslant 1$ star-forming galaxies selected from *Herschel* far-IR observations in various extragalactic fields for having spectroscopy covering the nebular emission lines.

2. Dust attenuation on the $z \sim 1$ MS

In Puglisi *et al.* 2016 we analyzed the near-IR spectra of 79 *Herschel*-selected star forming galaxies at $0.7 \leqslant z \leqslant 1.5$, acquired from the 3D-HST survey (Brammer *et al.* 2012). The 3D-HST spectra allowed us to detect the Hα emission for the full sample, from which we measured spectroscopic redshifts and Hα fluxes. The sources were selected in the GOODS-Southern field to secure a high quality multi-wavelength data-coverage including 21 bands from then near-UV to the far-IR ($0.23\ \mu m \leqslant \lambda \leqslant 500\ \mu m$). This wide ancillary coverage allowed us to derive robust estimates of $M_\star$, $L_{\rm IR}$, $L_{\rm UV,obs}$ and β via SED fitting with the MAGPHYS software (da Cunha *et al.* 2008). These galaxies span a SFR and $M_\star$ range of $10 \leqslant SFR_{\rm IR+UV} \leqslant 230 M_\odot yr^{-1}$ and $3 \times 10^9 \leqslant M_\star \leqslant 3.5 \times 10^{11} M_\odot$ thus being representative of the MS population above $M_\odot \sim 3 \times 10^{10} M_\odot$.

The left panel in Figure 1 reports the ratio $SFR^{\rm uncorr}_{\rm H\alpha}/SFR^{\rm uncorr}_{\rm UV}$ as a function of $E_\star(B-V)$, as evaluated from the IRX$=L_{\rm IR}/L_{\rm UV}$ ratio. The solid line highlights the relation between the two axis with a nebular-to-continuum attenuation ratio $f = E_{\rm neb}(B-V)/E_{\rm star}(B-V) = 0.93$ that is the best-fit value from the data distribution. This value turns out to be larger than the local value (f=0.58 Calzetti *et al.* 2000). We tested this dust attenuation correction in the right panel of Fig. 1 where we compared the $SFR_{\rm IR+UV}$ with $SFR_{\rm H\alpha}$ corrected for dust attenuation using the prescription derived in this work. This dust attenuation correction yields a good agreement between the two SFR indicators, with a median ratio $SFR_{\rm H\alpha}/SFR_{\rm IR+UV} = 0.88$ and a median relative scatter $|SFR_{\rm IR+UV} - SFR_{\rm H\alpha}|\ /(1 + SFR_{\rm IR+UV}) = 0.38$ whereas using the local f-factor would lead to overestimate the $SFR_{\rm H\alpha}$ by a factor of ~3.3 above $SFR_{\rm IR+UV} \geq 50 M_\odot yr^{-1}$. In both panel, data-points are color-coded according to the $M_\star$. The two panels suggest a possible trend with the $M_\star$: using $f \sim 0.93$ would overestimate $SFR_{\rm H\alpha}$ for galaxies with $M_\star \sim 10^{10} M_\odot$.

3. Dust attenuation off the MS at $z \sim 1.6$

In Puglisi *et al.* 2017 we studied the dust attenuation of dusty starbursting galaxies above the MS at $z \sim 1.6$ (SBs hereafter). These SBs were drawn from a *Herschel*-selected sample and have near-IR spectroscopy from the FMOS-COSMOS survey allowing to detect the Hα and Hβ emission lines at $1.4 \leqslant z \leqslant 1.7$. SB sources analyzed in this work were selected as outliers from the MS, with $SFR \geqslant 4 \times SFR_{\rm MS}$, following Rodighiero *et al.* 2011.

Figure 2-left shows the dust attenuation on the Hα emission, $A_{\rm H\alpha}$, as a function of $M_\star$ for this sample. We use two methods for estimating $A_{\rm H\alpha}$ in this SB sample: the BD (cyan diamonds) and the IRX ratio (red circles). The BD yields an average value $A_{\rm H\alpha,BD} \sim 0.89 \pm 0.69$ mag, consistent with the value measured on the MS at the same redshift. The IRX gives $A_{\rm H\alpha,IRX} \sim 3.3 \pm 0.51$, much larger than the former. The trend seen in Fig. 2 suggests that the optical light (from which $A_{\rm H\alpha,BD}$ is measured) comes from relatively un-obscured lines of sight, whereas the bulk of the SF is almost completely obscured at optical wavelengths. The right panel of Fig. 2 illustrates this configuration,

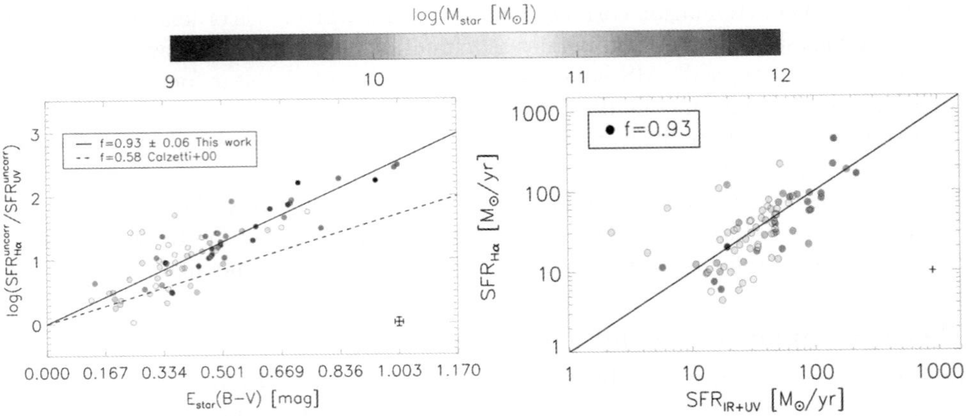

Figure 1. *Left:* Ratio of Hα to UV-based SFR (not corrected for dust extinction) as a function of $E_{\rm star}(B-V)$. The dashed line highlights the relation between these quantities adopting the local f-factor. The continuum line marks our estimate of the f -factor. *Right:* Comparison between $SFR_{\rm IR+UV}$ and $SFR_{\rm H\alpha}$ with the latter corrected for dust attenuation by using $f = 0.93$. The black solid line is the 1:1 correlation line. In both panels, the color-code corresponds to $M_{\star}$.

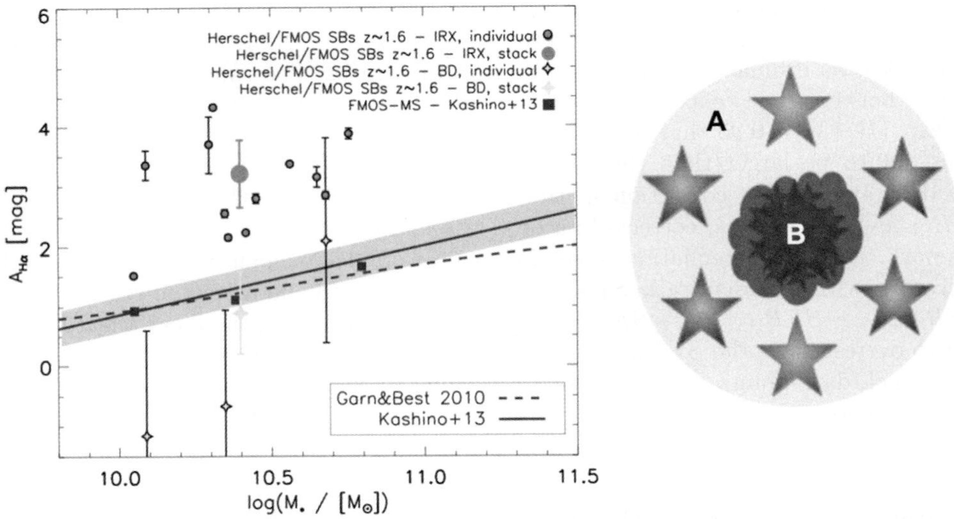

Figure 2. *Left:* Hα attenuation as a function of $M_{\star}$ for the twelve SBs at $z \sim 1.6$ presented in Puglisi *et al.* 2017. Red filled dots are measurements of dust attenuation from the IRX ratio whereas cyan diamonds are dust attenuation measurements as derived from the Balmer Decrement. The measurements are compared with the MS trend at $z \sim 1.6$ and in the local universe (solid and dashed lines, respectively). *Right:* Cartoon sketching our two-component dust model, with the optically-thin component A highlighted in yellow and the starbursting, optically thick region B in dark red.

in which a SB galaxy is modeled with an optically-thin part and an optically-thick region corresponding to the starburst core. Both components contribute to $SFR_{\rm tot}$ and to our integrated measurements, with their relative contribution depending on their attenuation. By assuming that all the dust is spatially concentrated in the starbursting core region and the optically-thin component is dust free, we estimated a lower limit for the attenuation of the thick core $A_{\rm H\alpha,thick} = 4.5$ mag. In this limiting situation the thick component has its maximum possible contribution to the optical emission lines of $\sim 33\%$. ALMA high-resolution imaging of the CO(5-4) emission for one of our sources confirms that L_{FIR} mainly arises from a region barely detected at UV/optical wavelengths (Silverman *et al.* 2018).

These results prompted us to use Magellan FIRE to obtain spectroscopy of starbursts in the near-IR rest-frame. Our aim is to study SBs at longer wavelengths thus suffering less by dust attenuation effects, at least in principle. The first results of this effort are presented in Calabrò *et al.* 2018. In this work by using Paschen-to-Balmer line ratios compared with far-IR dust attenuation indicators, we showed that SBs are consistent with a model in which young stars and dust are homogeneously mixed and are not consistent with attenuation laws assuming a foreground dust screen. With this mixed model the dust attenuation of the heavily obscured core reaches up to $A_{\mathrm{V}} \sim 30$ mag, with a median value $A_{\mathrm{V}} \sim 9$ mag. In these conditions observed optical/near-IR emission comes from surface regions, while the starburst core is still invisible at near-IR wavelengths.

4. Implications for SED-fitting studies

The results presented in Puglisi *et al.* 2017 and Calabrò *et al.* 2018 suggest that SBs are a separate class off galaxies with different properties with respect to the MS population. In particular, they show peculiar dust attenuation properties due to the large dust content and inhomogeneous dust distributions. These large dust content and the peculiar dust distribution may have dramatic effects on the multi-wavelength SED of these galaxies (see also Hainline *et al.* 2011). Geometrical effects in the dust distribution can produce blue spectra even for a large dust content or may reduce the far-IR emission also for red optical colors (Popping *et al.* 2017). This implies that spatially integrated observations at different wavelengths sample effectively physically disconnected regions (Puglisi *et al.* 2017). The question of whether energy balance assumptions can hold true in these extreme configurations has yet to be proven. Testing the impact of these peculiar dust attenuation properties on SED-estimated stellar population properties seems a crucial next step to firmly establish the role of starburst galaxies in galaxy evolution.

References

Brammer, G., van Dokkum, P. G., Franx, M., Fumagalli, M., Patel, S., Rix, H. W., Skelton, R. E., Kriek, M., *et al.* 2012, ApJs, 200, 13

Calabrò, A., Daddi, E., Cassata, P., Onodera, M., Gobat, R., Puglisi, A., Jin, S., Liu, D., *et al.* 2018, *ApJ*(Letters), 862, L22

Calzetti, D., Armus, L., Bohlin, R. C., Kinney, A. L., Koornneef, J., & Storchi-Bergmann, T. 2000, *ApJ*, 533, 682

Calzetti, D., Kinney, A. L., & Storchi-Bergmann, T. 1994, *ApJ*, 429, 582

https://ui.adsabs.harvard.edu/abs/2013ARA%26A..51..393C/abstract

da Cunha, E., Charlot, S., & Elbaz, D. 2008, *MNRAS*, 388, 1595

Hainline, L. J., Blain, A. W., Smail, I., Alexander, D., Armus, L., Chapman, & Ivison, R. https://ui.adsabs.harvard.edu/abs/2011ApJ...740...96H/abstract

Popping, G., Puglisi, A., & Norman, C. A. 2017, *MNRAS*, 472, 2315

Puglisi, A., Rodighiero, G., Franceschini, A., Talia, M., Cimatti, A., Baronchelli, I., Daddi, E., Renzini A., *et al.* 2016, *A&A*, 586, A83

Puglisi, A., Daddi, E., Renzini, A., Rodighiero, G., Silverman, J. D, Kashino, D., Rodríguez-Muñoz, L., Mancini C., *et al.* 2017, *ApJ*(Letters), 838, L18

Rodighiero, G., Daddi, E., Baronchelli, I., Cimatti, A., Renzini, A., Aussel, H., Popesso, P., Lutz D., *et al.* 2011, *ApJ*(Letters), 739, L40

Silverman, J., Daddi, E., Rujopakarn, W., Renzini, A., Mancini, C., Bournaud, F., Puglisi, A., Rodighiero G., *et al.* 2018, ArXiv e-prints, arXiv:1810.01595

Speagle, J. S., Steinhardt, C. L., Capak, P. L., & Silverman, J. D. 2014, *ApJs*, 214, 15

Whitaker, K. E., Franx, M., Leja, J., van Dokkum, P., Henry, Skelton, R. E., Fumagalli, M., Momcheva I., *et al.* 2014, *ApJ*, 795, 104

Challenges in Panchromatic Modelling with Next Generation Facilities
Proceedings IAU Symposium No. 341, 2018
M. Boquien, E. Lusso, C. Gruppioni & P. Tissera, eds.

doi:10.1017/S1743921319002710

Spatially resolved dust-to-gas mass ratios in nearby galaxies

Basilio Solís-Castillo[1,2] and Marcus Albrecht[1]

[1]Argelander Institute for Astronomy, University of Bonn,
Auf dem Hügel 71, D-53121 Bonn, Germany
email: bsolis@astro.uni-bonn.de

[2]Max-Planck Institute for Radio Astronomy, Auf dem Hügel 69, D-53121 Bonn, Germany

Abstract. We analyse the dust-to-gas mass ratio (DGR) in nearby galaxies on kiloparsec scales. We focus on their dependence on metallicity and the CO-to-H_2 conversion factor, α_{CO}. We use a sample of 25 nearby galaxies from SINGS and combine our data with CO (2-1) and H I observations from the HERACLES and THINGS surveys. We implement a Hierarchical Bayesian method to derive the dust mass via fitting the infrared data from 100 to 500 μm with a single modified blackbody. We find that the DGR-metallicity relation follows a power law and we study its strong dependency on the conversion factor α_{CO}. Our results indicate a strong connection between interstellar dust and gas. The resolved DGR-metallicity relation cannot be represented with a single power law. The scatter in this relation shows the strong impact of several processes that take place in every galaxy.

Keywords. evolution - dust, extinction - galaxies: ISM - infrared: ISM

1. Introduction

Interstellar dust plays a central role in the astrophysics of the interstellar medium (ISM), although it represents only ~1% of the total mass. Dust particles constitute an important agent in the fluid dynamics, chemistry, heating, cooling and ionisation balance of the ISM. Despite its importance, a comprehensive understanding of dust is still challenging, given the inherent complexity of its evolution in the ISM.

In the far-IR regime ($\lambda \gtrsim 60$ μm), the shape of the Spectral Energy Distribution (SED) coming from dust emission can be modelled as a modified blackbody with a frequency-dependent opacity. This model depends on three main components: dust mass, dust temperature, T, and emissivity index, β. Most observational analyses employ least-squares (χ^2) SED fits to estimate T and β. However, due to its degeneracy, these methods can lead to erroneous estimates and uncertainties in the flux measurements. Kelly *et al.* (2012) found that a Hierarchical Bayesian (hereafter, HB) fitting-technique can accurately estimate the dependence between SED parameters and therefore may be used to further understand grain evolution in the ISM. Additionally, Shetty *et al.* (2016) concluded that the HB methods explicitly treat the correlation between T and β among all pixels and accurately accounts for noise, while the normal χ^2-fit biased the distribution between these parameters towards an artificial anti correlation.

The evolution of the ISM is closely linked with metal enrichment. Elements present in the ISM are processed by gas and dust, providing the conditions to form more complex molecules. The relation between DGR and metallicity provides an important tool to study the evolutionary stage of a galaxy as it links the amount of metals bound in dust and in the gas phase. Therefore, it allows to put constrains on dust evolution models which predict the balance between dust formation, growth and destruction, and the total amount of metals. Rémy-Ruyer *et al.* (2014) analysed the relation between the DGR and metallicity for 126 galaxies and found that the DGR-metallicity relation cannot be represented by a single power law as in our galaxy, but rather a broken power law with a steeper trend for low metallicities. The DGR shows a large scatter for given metallicity throughout the metallicity range, indicating that metallicity is not the only driver for the DGR, and that environmental effects, variations in the star formation histories, dust properties and dust destruction efficiencies may play a role.

In the present work, we study the DGR-metallicity relations in nearby galaxies on kiloparsec scales.

2. Data and method

We use a sample of galaxies that are part of SINGS (Kennicutt *et al.* 2003) and combine it with H I data from THINGS (Walter *et al.* 2008) and VIVA (Chung *et al.* 2009), ^{12}CO $J = (2-1)$ from HERACLES; (Leroy *et al.* 2009), and far-infrared emission from KINGFISH (Kennicutt *et al.* 2011). The resulting sample consists of 25 galaxies. We perform a multi-wavelength analysis and convolve the data to the same resolution, using the lower SPIRE 500 μm resolution. All maps are trimmed to a common size and projected to a common grid with a pixel size of 36". Further details of the convolution and background subtraction can be found in Albrecht *et al.* (in preparation).

Data selection. We select all the pixels with a signal-to-noise (S/N) greater than 2 in the five bands that were used to fit the SED (100 to 500 μm) and in the CO maps. Additionally we exclude pixels with galactocentric radius, $r_{25} > 1.0$, with the exception of galaxy NGC 5055 where we use $r_{25} > 0.7$. The resulting pixels of our 19 galaxies correspond to physical regions that range from 0.28 to 2.6 kpc. We also use the gas-phase oxygen abundances and gradients given by Moustakas *et al.* (2010) for the SINGS sample and combine them with Schruba *et al.* (2018) work. Using the results of both works, we are able to derive a metallicity on every pixel of our galaxy sample.

Atomic and molecular gas. The atomic gas mass was derived using the H I observations from THINGS (Walter *et al.* 2008). We use the CO (2-1) integrated intensity to derive the molecular gas mass through the CO-to-H_2 conversion factor, α_{CO}. We convolve each moment 0 map with a PSF kernel following the procedure explained in Leroy *et al.* (2009). We study three possible scenarios for α_{CO}: a "standard value" of $\alpha_{CO} = 4.4\ M_{\odot}$ pc^{-2} (K km s^{-1})$^{-1}$ given by Bolatto *et al.* (2013), and a metallicity dependent α_{CO} according to Galametz *et al.* (2011) and Schruba *et al.* (2012).

Dust modelling. To provide a consistent determination of the dust masses throughout our sample, we use the model described in Albrecht *et al.* (in preparation). They use the specific intensity I_ν of the dust emission between 100 and 500 μm in galaxies to fit the SED with a single-component modified blackbody. For the SED fitting, we use a hierarchical bayesian approach, which define the joint posterior probability density function for every pixel of the galaxy and a hyperparameter vector for each galaxy through Monte Carlo sampling. From these hyperparameters we can derive mean values for the dust parameters: dust mass surface density, $\Sigma_{\rm dust}$, dust temperature, $T_{\rm dust}$, and emissivity index, $\beta_{\rm dust}$, for each pixel of the source.

3. Analysis and results

In Fig. 1 we show the DGR-metallicity relation for our 19 galaxy sample for three scenarios: on the left side, constant α_{CO}, and metallicity dependent according to Galametz *et al.* (2011) and Schruba *et al.* (2012), at the centre and right side respectively. The colours on the data points denote the dust temperature in K. We also show with a red line the power law that Sandstrom *et al.* (2013) found. Our best fit is shown as a blue dashed line. Spearman-rank order is shown as ρ_s. We can see that for a constant α_{CO}

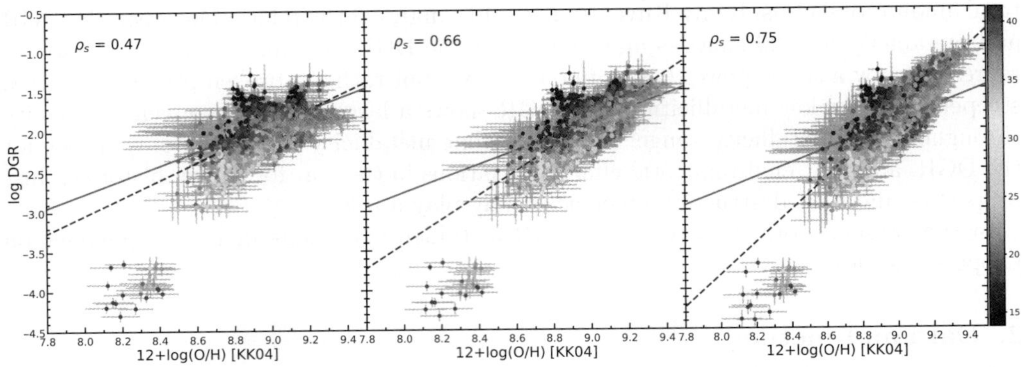

Figure 1. Dust-to-gas mass ratio (DGR) as a function of the metallicity according to the KK04 calibration for three scenarios: constant $\alpha_{\rm CO}$ (left panel), metallicity dependent $\alpha_{\rm CO}$ (centre and right panel). The color coding denotes the dust temperatures T_{dust} in K. The red line is the linear fit found by Sandstrom *et al.* (2013). The blue dashed line show our best linear fit.

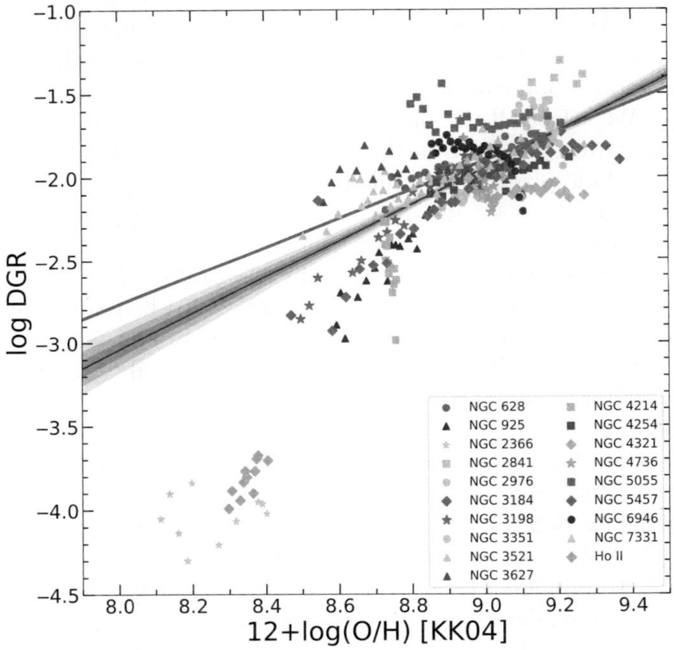

Figure 2. Dust-to-gas mass ratio (DGR) as a function of the metallicity according to the KK04 calibration for the galaxy sample. The red line is the fit found by Sandstrom *et al.* (2013). The blue line show our best linear fit and green shaded regions represent 1-, 2- and 3-σ deviation.

our results are similar with literature, although there are big differences for metallicity dependent $\alpha_{\rm CO}$. The low metallicity galaxies, represented by NGC 2366 and Holmberg II, do not follow the single power law, but a broken power law as Rémy-Ruyer *et al.* (2014) suggested, with a steeper slope for the low metallicity end. Additionally, we see that dust temperature is also correlated with metallicity, where higher dust temperatures are associated with low metallicity regions and low DGR. On the other hand, the DGR-metallicity relation using a stronger dependency of $\alpha_{\rm CO}$ (right panel) could be explained with a single power law covering the entire metallicity range. More low metallicity galaxies are needed for further analysis.

Figure 2 shows the DGR-metallicity relation for the galaxy sample, considering a constant $\alpha_{\rm CO}$. The binned values for every galaxy are shown with different symbols and colours. We do not find a correlation between the slope on the DGR-metallicity relation with metallicity and morphology of the galaxy. We see that the low metallicity regions located in the outskirts from galaxies NGC 925, NGC 3198 and NGC 5457 follow a different trend than more metal-rich galaxies, more similar to the low metallicity galaxies on the left end of the plot, which is consistent with the broken power law with a steeper slope. However, galaxies like NGC 3351, NGC 4321 and NGC 5055 tend to show a constant or a decreasing DGR with increasing metallicity, showing that DGR-metallicity relation is also dependent on the star formation rate, nucleus type and other internal processes that take place within the galaxies.

References

Aniano, G., Draine, B. T., Calzetti, D., Dale, D. A., Engelbracht, C. W., Gordon, K. D., Hunt, L. K., Kennicutt, R. C., *et al.* 2012, *ApJ*, 756, 138

Bolatto, A. D., Wolfire, M., & Leroy, A. K. 2013, *ARAA*, 51, 207

Chung, A., van Gorkom, J. H., Kenney, J. D. P., Crowl, H., & Vollmer, B. 2009, *AJ*, 138, 1741

Galametz, M., Madden, S. C., Galliano, F., Hony, S., Bendo, G. J., & Sauvage, M. 2011, *A&A*, 532, A56

Kelly, B. C., Shetty, R., Stutz, A. M., Kauffmann, J., Goodman, A. A., & Launhardt, R. 2012, *ApJ*, 752, 55

Kennicutt, Jr., R. C., Armus, L., Bendo, G., Calzetti, D., Dale, D. A., Draine, B. T., Engelbracht, C. W., Gordon, K. D., *et al.* 2003, *PASP*, 115, 928

Kennicutt, R. C., Calzetti, D., Aniano, G., Appleton, P., Armus, L., Beirão, P., Bolatto, A. D., Brandl, B., *et al.* 2011, *PASP*, 123, 1347

Leroy, A. K., Walter, F., Bigiel, F., Usero, A., Weiss, A., Brinks, E., de Blok, W. J. G., Kennicutt, R. C., *et al.* 2009, *AJ*, 137, 4670

Moustakas, J., Kennicutt, Jr., R. C., Tremonti, C. A., Dale, D. A., Smith, J. D. T., & Calzetti, D. 2010, *ApJS*, 190, 233

Rémy-Ruyer, A., Madden, S. C., Galliano, F., Galametz, M., Takeuchi, T. T., Asano, R. S., Zhukovska, S., Lebouteiller, V., *et al.* 2014, *A&A*, 563, A31

Sandstrom, K. M., Leroy, A. K., Walter, F., Bolatto, A. D., Croxall, K. V., Draine, B. T., Wilson, C. D., Wolfire, M., *et al.* 2013, *ApJ*, 777, 5

Shetty, R., Roman-Duval, J., Hony, S., Cormier, D., Klessen, R. S., Konstandin, L. K., Loredo, T., Pellegrini, *et al.* 2016, *MNRAS*, 460, 67

Schruba, A., Leroy, A. K., Walter, F., Bigiel, F., Brinks, E., de Blok, W. J. G., Kramer, C., Rosolowsky, E., *et al.* 2012, *AJ*, 143, 138

Schruba, A., Bialy, S., & Sternberg, A. 2018, *ApJ*, 862, 110

Walter, F., Brinks, E., de Blok, W. J. G., Bigiel, F., Kennicutt, Jr., R. C., Thornley, M. D., & Leroy, A. 2008, *AJ*, 136, 2563

Discussion

Tomotsugu Goto: Can you explain what are the differences amongst the curves plotted in the DGR-metallicity relation?

Basilio Solís-Castillo: The curves in the DGR-metallicity relation represent different models for global values and resolved studies of nearby galaxies, and our best fit. Major differences can be found from galaxy to galaxy due to different morphologies, grain composition and dust temperature.

Adam Carnall: Are you using a Bayesian hierarchical model to fit your galaxies or to combine the data from different galaxies in order to fit your linear relationship? It would be interesting to try SED fitting in a hierarchical Bayesian framework.

BASILIO SOLÍS-CASTILLO: I am using a hierarchical Bayesian model to fit the SED using FIR data. I also implemented a Bayesian method to make linear fitting in the relations I am studying.

TSUTOMI T. TAKEUCHI: One of the galaxy in you sample shows an increasing tendency of the DGR towards lower metallicity. Could you explain what happens to this galaxy?

BASILIO SOLÍS-CASTILLO: The galaxy that shows an increasing tendency in the DGR-metallicity relation is NGC 5055, and it is not the only one. This galaxy has a central AGN that is affecting the dust distribution and, therefore the DGR. Another important thing we can have to take into account is about the scatter in the metallicity gradient, which could also affect the relation.

LAPO FANCIULLO: Does your dust model (which you fit with a modified black body) assume the same dust optical constants (k_0, β) everywhere in the galaxy? If so (especially for the constant k_0), couldn't part of the DGR variation you see be actually an effect of spatial variation in dust optical constants inducing a bias on the fit?

BASILIO SOLÍS-CASTILLO: The dust modelling assumes a frequency dependent opacity (with a given k_0), giving as an output three different free parameters: dust mass surface density, emissivity index and dust temperature. So, on every kpc region there is a β and T, which is different to other regions of the same galaxy. This kind of resolved study allows us to fit pixel-by-pixel the dust model and help us to understand the internal processes of nearby galaxies.

Challenges in Panchromatic Modelling with Next Generation Facilities
Proceedings IAU Symposium No. 341, 2018
M. Boquien, E. Lusso, C. Gruppioni & P. Tissera, eds.

doi:10.1017/S174392131900139X

A MUSE inquiry into the physical processes taking place within the Abell 2667 Brightest Cluster Galaxy

E. Iani[1,2], G. Rodighiero[1,3], J. Fritz[4], G. Cresci[5], C. Mancini[1,3], P. Tozzi[5], L. Rodríguez-Muñoz[1], P. Rosati[6], G. B. Caminha[7], S. Berta[8], P. Cassata[1], A. Concas[9], A. Enia[1], D. Fadda[10], A. Franceschini[1], A. Liu[5], A. Mercurio[11], L. Morselli[9], P. G. Pérez-González[12,13], P. Popesso[9], G. Sabatini[14,15] and A. Zanella[2]

[1]Dipartimento di Fisica ed Astronomia, Università degli Studi di Padova, Vicolo dell'Osservatorio 3, I-35122 Padova, Italy
email: `edoardo.iani@phd.unipd.it`

[2]European Southern Observatory, Karl Schwarzschild Straße 2, D-85748 Garching, Germany

[3]INAF - Osservatorio Astronomico di Padova, Vicolo dell'Osservatorio 5, I-35122 Padova, Italy

[4]Instituto de Radioastronomía y Astrofísica, UNAM, Campus Morelia, A.P. 3-72, C.P. 58089, Mexico

[5]INAF - Osservatorio Astrofisico di Arcetri, Largo Enrico Fermi 5, I-50125 Firenze, Italy

[6]Dipartimento di Fisica e Scienze della Terra, Università degli Studi di Ferrara, Via Saragat 1, I-44122 Ferrara, Italy

[7]Kapteyn Astronomical Institute, University of Groningen, Postbus 800, 9700 AV Groningen, The Netherland

[8]Institut de Radioastronomie Millimétrique 300 rue de la Piscine, Domaine Universitaire 38406 Saint Martin d'Hères, France

[9]Excellence Cluster Universe, Boltzmannstraße 2, D-85748 Garching, Germany

[10]SOFIA Science Center, USRA, NASA Ames Research Center, M.S. N232-12 Moffett Field, CA 9403

[11]INAF - Osservatorio Astronomico di Capodimonte, via Moiariello 16, 80131 Napoli, Italy

[12]Departamento de Astronomía y Astrofísica, Universidad Complutense de Madrid, Av. Complutense s/n, C.P. 28040, Madrid, Spain

[13]Centro de Astrobiología (CAB, INTA-CSIC), Carretera de Ajalvir km 4, E-28850 Torrejón de Ardoz, Madrid, Spain

[14]Dipartimento di Fisica e Astronomia, Università degli Studi di Bologna, via Gobetti 93/2, I-40129 Bologna, Italy

[15]INAF - Istituto di Radioastronomia - Italian node of the ALMA Regional Centre (ARC), via Gobetti 101, I-40129 Bologna, Italy

Abstract. Brightest cluster galaxies (BCGs) residing in cool-core clusters are known to be the stage of intricate baryon cycle phenomena (e.g. gas inflows, AGN outflows, star formation feedback). The scenarios describing the observed properties of these galaxies are still controversial, suffering from limitations due to the spatial resolving power of the instruments, specifically for galaxies beyond the Local Universe. However, the dramatic improvements introduced by the integral-field unit instruments (e.g. MUSE) could shed light on the physical processes driving the evolution of these galaxies. We present an extensive analysis of the stellar and gas properties (i.e. kinematics, stellar mass, star formation rate) of the radio-loud BCG sitting at the centre of the X-ray luminous cool-core cluster Abell 2667 ($z = 0.23$), based on MUSE data. Our results indicate that the BCG is a massive elliptical, hosting an AGN that is possibly undergoing accretion of cold star-forming clouds of ICM or galactic cannibalism.

Keywords. galaxies: elliptical and lenticular, cD; galaxies: evolution; galaxies: kinematics and dynamics; galaxies: nuclei; (galaxies:) cooling flows

---

1. Introduction

Brightest cluster galaxies (BCGs) are the largest, most massive and luminous galaxies in the Universe. They host relatively old stellar populations and little on-going star formation activity, thus representing the brightest end of the luminosity function of early-type galaxies (ETGs). In dynamically relaxed galaxy clusters (i.e. no recent mergers with nearby clusters), BCGs sit almost at rest at the bottom of clusters potential wells and close to the peak of the thermal X-ray emission originated by the hydrostatical cooling of the hot (10^7–10^8 K) intracluster medium (ICM).

Despite their classification as ETGs, if specific environmental conditions are met, BCGs have been found to show exceptional properties over the whole electromagnetic spectrum, from the X-rays to the radio domain. In particular, BCGs residing in cool-core clusters, i.e. rich clusters with a relaxed morphology and showing a relevant drop (down to 10^4 K) in the X-ray temperature of the central ($r \lesssim 10 - 100$ kpc) ICM, are likely to host an active galactic nucleus (AGN) and have an intense on-going star formation activity. Optical nebulæ (e.g. Cowie *et al.* 1983; Cavagnolo *et al.* 2008) along with extended Lyα (e.g. Hu 1992) and far-ultraviolet continuum emissions (e.g. O'Dea *et al.* 2004; O'Dea *et al.* 2010) have been also detected around these BCGs. The co-existence of these phenomena and their tight relation with the properties of the galaxies environment (i.e. cool-core clusters) makes these BCGs a peculiar class of massive ellipticals, suggesting that they are the stage of very intricate baryon cycle processes. Different scenarios (e.g. AGN and/or star photoionisation, shocks, hot X-ray plasma penetrating cold gas) describing the origin and interplay between this phenomena, as well as the mechanisms regulating the BCGs AGN and star formation activity, have been proposed (e.g. Ferland *et al.* 2009; O'Dea *et al.* 2010). However, these scenarios are still a matter of debate, in particular for galaxies at intermediate-redshift where the spatial disentanglement between the different processes is limited by the resolving power of the instruments. In this regard, state-of-the-art integral-field unit instruments (e.g. MUSE) seem to give an answer to this problem, opening a new window on the description of galaxies inhabiting the far Universe.

In this context, we inquire the stellar and gas properties (i.e. kinematics, stellar mass, star formation rate) of the radio-loud (e.g. Kale *et al.* 2015) brightest cluster galaxy inhabiting the cool-core cluster (e.g. Rizza *et al.* 1998) Abell 2667 (see Fig. 1, left panel), a rich class-3 cluster, located in the southern celestial hemisphere at redshift $z = 0.2343$.

2. Data analysis and results

To unveil the complex dynamical structure of the BCG, we exploit the exquisite quality of the data from the ESO IFU instrument MUSE at VLT. MUSE observations of the Abell 2667 cluster were taken under the GTO program 094.A-0115(A) (P.I. J. Richard) for a total exposure time of 2h (4×1800s) and a seeing of $< 1''$ (FWHM). The raw-data are reduced by means of the `ESO-MUSE pipeline` (Weilbacher *et al.* 2014), and the `Zurich Atmosphere Purge` (`ZAP`, Soto *et al.* 2016).

We spatially resolve the BCG stellar kinematics with maps of stellar velocity and velocity dispersion by means of pPXF (Cappellari & Emsellem 2004) while the estimate of

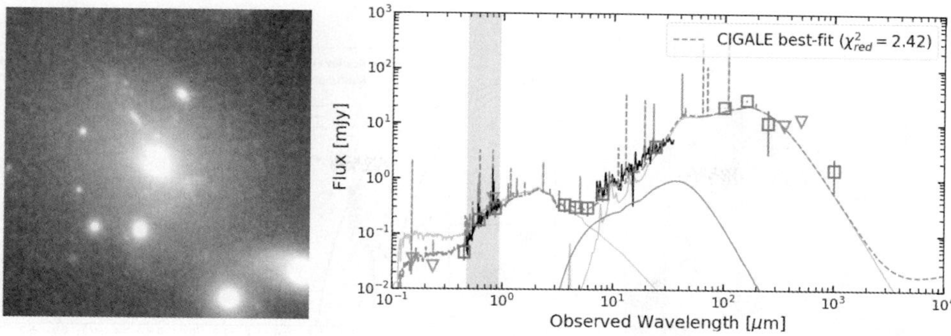

Figure 1. Left panel: *HST* RGB image (F450W + F606W + F814W) of the BCG in Abell 2667 (the cutout size is 12.8″ × 12.8″ , i.e. ∼50 kpc × 50 kpc, centered at RA(J2000.0) = $23^h51^m39.37^s$; Dec(J2000.0) = −26°05′2.7″). Right Panel: CIGALE best-fit SED (gray dashed-line) of the BCG photometric data (red open squares) and their 3σ error. The orange and cyan solid lines show the contribution of the unattenuated stellar populations and dust, respectively; while the red solid line highlights the AGN torus emission. In black, the MUSE (vertical shaded area 4750 − 9350 Å) and IRS spectra of the galaxy are presented too.

the galaxy total stellar mass, stellar mass assembly history and its current star formation are obtained from MUSE spectra thanks to SINOPSIS (Fritz *et al.* 2017).

In agreement with what typically observed in massive ellipticals, our results indicate that the BCG is a dispersionally supported system with no sign of a coherent stellar velocity pattern (e.g. rotation) and with a spatially flat velocity dispersion ($\sigma_0 \sim 216$ km s^{-1}). From the relation by Zubovas & King (2012) bounding the mass of supermassive black holes (M_{SMBH}) to the central velocity dispersion of elliptical galaxies in clusters, we estimate $M_{SMBH} \sim 3.8 \times 10^9$ M$_\odot$. According to the statistical analysis by Smolčić (2009) on the physical properties of AGNs, our estimate of M_{SMBH} suggests that the Abell 2667 BCG is a massive elliptical hosting a low-ionisation nuclear emission-line region (LINER).

We measure the galaxy total stellar mass† equal to $M_\star \sim 1.38 \times 10^{11}$ M$_\odot$. According to the BCG mass assembly history, ∼60% of $M_\star$ formed more than 5.6 Gyr ago. The value of $M_\star$ is confirmed by the analysis of the galaxy integrated photometric properties we perform thanks to the spectral energy distribution (SED) fitting code CIGALE (Noll *et al.* 2009; Boquien *et al.* 2018), using multi-wavelength data from rest-frame UV to IR (see Fig. 1, right panel). The SED fit indicates the BCG is hosting an obscured Type 2 AGN (as also pointed out by the analysis of *Chandra* data by Yang *et al.* 2018) whose emission does not contribute to the galaxy overall optical emission. The CIGALE analysis allows us to estimate the BCG star formation activity equal to‡ ∼14.5 M$_\odot$/yr, thus locating the galaxy firmly along the main-sequence of galaxies as defined by Speagle *et al.* 2014. The intense star formation activity of the BCG obtained from the SED fit is confirmed by the analysis of the emission lines in MUSE spectra. Specifically, MUSE detects the presence of a young stellar population (younger than 20 Myr and accounting for ∼0.6% of $M_\star$) all around the galaxy core (see Fig. 2, left panel). These young stars are distributed along a filamentary and clumpy structure that can be clearly observed in the *HST* data.

† The stellar mass we refer is composed both by stars that are still in the nuclear-burning phase, and remnants (e.g. white dwarfs, neutron stars and stellar black holes).

‡ The SFR we present is obtained from the conversion of the galaxy infrared luminosity into obscured star formation rate via the standard Kennicutt relation (Kennicutt 1998) for a Chabrier IMF, and adding the contribution of the unobscured component as in Santini *et al.* (2009).

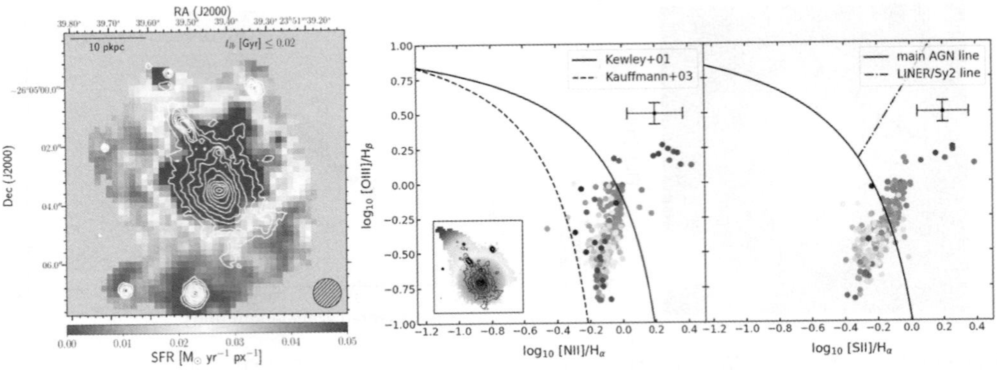

Figure 2. Left panel: SINOPSIS current SFR density map of the BCG with *HST* F450W contours superimposed. Right Panel: Spatially resolved BPT-[NII] (left) and BPT-[SII] (right) diagrams. Representative dots of the MUSE spaxels are colour-coded as a function of the projected spatial distance from the galaxy centre (see left bottom corner of the left diagram).

Since the Abell 2667 BCG has been proved to host an AGN, the results on the galaxy star formation rate obtained from emission lines have to be investigated in more detail as AGNs can contribute significantly to the gas ionisation and therefore to the total flux of the emission lines. To spatially disentangle the regions where the lines are emitted as a consequence of star formation than AGN, we resort to the empirical spectral diagnostic diagrams by Baldwin *et al.* (1981), the so-called BPTs, (see Fig. 2, right panel). According to BPTs, the very core of the BCG is consistent with a LINER while the gas in the galaxy outer regions seems to be ionised by processes that are compatible with both AGN activity and SF processes.

Ultimately, by studying the emission lines (specifically, H_β and [OIII]λ5007) spatial distribution in different velocity channels (e.g. McNamara *et al.* 2014), we detect two extended and kinematically decoupled gas components in the velocity range $\Delta v_{gas} = [100; 500]$ km s^{-1} and $\Delta v_{gas} = [-500; -100]$ km s^{-1}, respectively.

3. Discussion

From our analysis, Abell 2667 seems to be a clear example of how intricate the baryon cycle is in the centre of massive galaxy clusters. Similarly to the results obtained by Tremblay *et al.* (2018) for the nearby Abell 2597 cool-core cluster, we might have caught Abell 2667 in the process of experiencing a baryon cycle where a shrink of star-forming giant clouds (i.e. the clumps observed in the *HST* data) made of cold ICM (e.g. Li & Bryan 2014; Voit *et al.* 2015a) are accreting onto the BCG, refuelling it with gas thus turning on the central AGN and triggering its feedback (outflows). The combination of *HST* imaging and the detection of different velocity components of the ionised gas with respect to the rest-frame of the BCG, allow us to take into account also a scenario of galactic cannibalism. In this case, the clumps observed around the BCG could be stripped star-forming material left behind a small disc-galaxy (the major clump in the *HST* observations) that is merging along an inspiralling orbit and is undergoing disruption due to tidal interactions with the BCG and, possibly, ram-pressure stripping from the ICM. In this case, the galaxy AGN activity could be a pre-existent condition or could have been ignited as a consequence of accretion of material from the disc-galaxy.

References

Cowie, L. L., Hu, E. M., Jenkins, E. B., York, D. G., *et al.* 1983, *ApJ*, 272, 29
Cavagnolo, K. W., Donahue M., Voit, G. M., Sun M., *et al.* 2008, *ApJ*, 683, L107
Hu, E. M. 1992, *ApJ*, 391, 608

O'Dea, C. P., Baum, S. A., Mack J., Koekemoer, A. M., Laor A., *et al.* 2004, *ApJ*, 612, 131
O'Dea, K. P., Quillen, A. C., O'Dea, C. P., Tremblay, G. R., Snios, B. T., Baum, S. A., Christiansen K., Noel-Storr J., *et al.* 2010, *ApJ*, 719, 1619
Ferland, G. J., Fabian, A. C., Hatch, N. A., Johnstone, R. M., Porter, R. L., van Hoof, P. A. M., Williams, R. J. R., *et al.* 2009, *MNRAS*, 392, 1475
Kale, R., Venturi, T., Cassano, R., Giacintucci, S., Bardelli, S., Dallacasa, D., Zucca, E., *et al.* 2015, *A&A*, 581, A23
Rizza, E., Burns, J. O., Ledlow, M. J., Owen, F. N., Voges, W., Bliton, M., *et al.* 1998, *MNRAS*, 301, 328
Weilbacher, P. M., Streicher, O., Urrutia, T., Pécontal-Rousset, A., Jarno, A., Bacon, R., *et al.* 2014, *ASPC*, 485, 451
Soto, K. T., Lilly, S. J., Bacon, R., Richard, J., Conseil, S., *et al.* 2016, *MNRAS*, 458, 3210
Cappellari, M. & Emsellem, E. 2004, *PASP*, 116, 138
Fritz, J., Poggianti, B. M., Bettoni, D., Cava, A., Couch, W. J., D'Onofrio, M., Dressler, A., Fasano G., *et al.* 2017, *ApJ*, 848, 132
Zubovas, K. & King, A. R. 2012, *MNRAS*, 426, 2751
Smolčič, V. 2009, *ApJ*, 699, L43
Noll, S., Burgarella, D., Giovannoli, E., Buat, V., Marcillac, D., Muñoz-Mateos, J. C., *et al.* 2009, *A&A*, 507, 1793
Boquien, M., Burgarella, D., Roehlly, Y., Buat, V., Ciesla, L., Corre, D., Inoue, A. K., Salas, H., *et al.* 2018, *arXiv e-prints*
Yang, L., Tozzi, P., Yu, H., Lusso, E., Gaspari, M., Gilli, R., Nardini, E., Risaliti, G., *et al.* 2018, *ApJ*, 859, 65
Kennicutt, Jr. R. C. 1998, *ARA&A*, 36, 189
Santini, P., Fontana, A., Grazian, A., Salimbeni, S., Fiore, F., Fontanot, F., Boutsia, K., Castellano M., *et al.* 2009, *A&A*, 504, 751
Speagle, J. S., Steinhardt, C. L., Capak, P. L., Silverman, J. D., *et al.* 2014, *ApJS*, 214, 15
Baldwin, J. A., Phillips, M. M., Terlevich, R., *et al.* 1981, *PASP*, 93, 5
McNamara, B. R., Russell, H. R., Nulsen, P. E. J., Edge, A. C., Murray, N. W., Main, R. A., Vantyghem, A. N., Combes, F., *et al.* 2014, *ApJ*, 785, 44
Tremblay, G. R., Combes, F., Oonk, J. B. R., Russell, H. R., McDonald, M. A., Gaspari, M., Husemann, B., Nulsen, P. E. J., *et al.* 2018, *ApJ*, 865, 13
Li, Y. & Bryan, G. L. 2014, *ApJ*, 789, 153
Voit, G. M., Donahue, M., Bryan, G. L., McDonald, M., *et al.* 2015a, *Nature*, 519, 203

Discussion

YUTAKA FUJITA: Could star formation be triggered by AGN activity?

IANI: AGN activity triggering star formation is definitely a possible/compatible scenario. However, we do not have enough scientific evidences to rule out shocks or merging.

TOMOTSUGU GOTO: Can you explain the reason for the offset between the *HST* optical and MUSE data (i.e. H_β, [OIII]λ5007)?

IANI: The small offset ($< 0.2''$) is due to a residual mismatch between *HST* and MUSE data coordinates. The mismatch has been corrected in our subsequent analysis.

MAUD GALAMETZ: How about the other objects in the field-of-view? Are they kinematically related to the BCG?

IANI: According to their redshift, the closest sources (on the sky-plane) to the BCG are galaxies inhabiting the very same cluster but we have not investigated their kinematics yet.

Challenges in Panchromatic Modelling with Next Generation Facilities
Proceedings IAU Symposium No. 341, 2018
M. Boquien, E. Lusso, C. Gruppioni & P. Tissera, eds.

doi:10.1017/S1743921319003077

Pushing the technical frontier: From overwhelmingly large data sets to machine learning

Viviana Acquaviva

Physics Department, New York City College of Technology, 300 Jay Street, Brooklyn NY 11201
email: vacquaviva@citytech.cuny.edu

Abstract. This paper summarizes my thoughts, given in an invited review at the IAU symposium 341 "Challenges in Panchromatical Galaxy Modelling with Next Generation Facilities", about how machine learning methods can help us solve some of the big data problems associated with current and upcoming large galaxy surveys.

Keywords. Surveys, Big Data, Machine Learning

1. It's everywhere!

Machine learning and data mining methods have become ubiquitous in all fields of science, including Astronomy, in the last decade. Figure 1 shows the result of a search on the NASA-ADS archive service for papers that contain the words "Machine Learning" either in the title or in the abstract, in the last 12 years, in the Astronomy and Physics archives; the growth trend is obvious. But of course it's not just papers; during this time, we have seen the emergence of dedicated conferences and sessions in conferences (such as this one), scientific associations, perhaps best described in the Astrostatistics and Astroinformatics Portal (ASAIP) website, special initiatives (for example the COsmo INitiative), dedicated programming packages, such as AstroPy (Astropy Collaboration 2013, 2018) and AstroML (Vanderplas *et al.* 2012), books, ad-hoc review panels from the National Science Foundation, and, of course, jobs. The most recent ASAIP blog post (10/2018) estimates that 15–20% of faculty job ads on the AAS job register have a strong component in statistics, machine learning, or data science. But what is machine learning? And most importantly, what can we do with it?

2. What can we do with ML?

Machine learning is the art of teaching a machine to make informed decisions. Examples of its applications include recognizing and characterizing objects based on similarities and differences, picking up patterns, and distinguishing signal from noise. The following is a (certainly incomplete) list of important applications of machine learning methods to Astronomy.

Save time. This is perhaps the chief application of machine learning in a Universe as vast as ours. A typical example is the classification of galaxies according to their morphological type; de la Calleja & Fuentes (2004) is, to my knowledge, the first example, but there have been many more since then (e.g. Katebi *et al.* (2019) and references therein); figure 2 above shows an example from Dieleman *et al.* (2015). In this case, trained humans are the best classifiers; we have an uncanny ability to generalize just from a few examples,

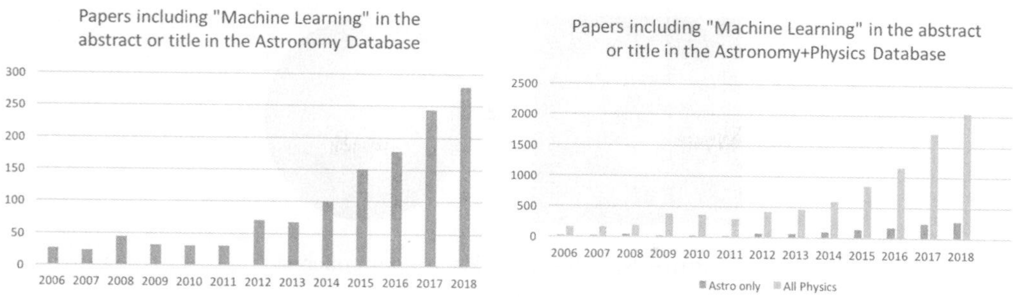

Figure 1. Machine learning related papers on the NASA/ADS archive from 2006 to today.

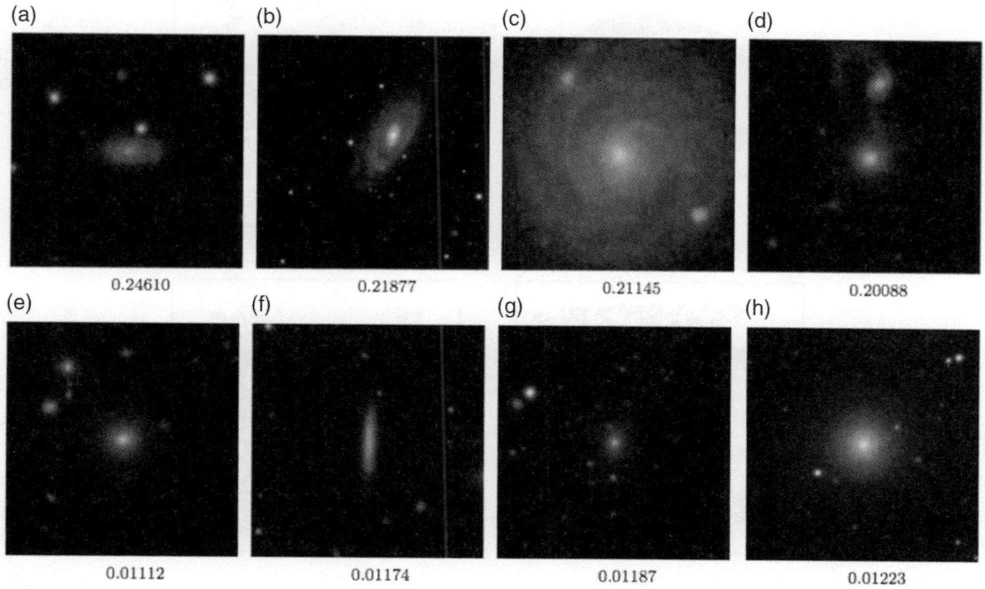

Figure 2. Dieleman *et al.* (2015) tackled the problem of galaxy morphology classification using deep learning.

and to weed out unimportant features. However, when the data sets become large (millions of objects), there are only two solutions: deploy the power of citizen science, with successful programs such as Galaxy Zoo (Lintott *et al.* 2008), or use machine learning. In a supervised approach, one would ask expert classifiers to provide "solutions" (estimated galaxy types) for a subset of objects, the learning set, and then teach the machine to learn from those examples; in an unsupervised approach, the machine determines its own set of galaxy types directly from the data, then assigns each object to a class (Hocking *et al.* 2018).

Provide a data-driven alternative to simplistic models. The alternative to machine learning is usually a model-based approach in which one creates intuition-based models of the objects, typically dependent on some input variables that are the parameters of the model. For example, when modeling galaxies, we use stellar population models paired with parameters such as stellar mass, stellar assembly history, metallicity, dust abundance and properties, and so on. The models are obviously an extreme simplification of complex objects such as real galaxies (in my own cheesy words: galaxies are not spheres, see Fig. 3). As a result, when we build likelihood functions and we infer

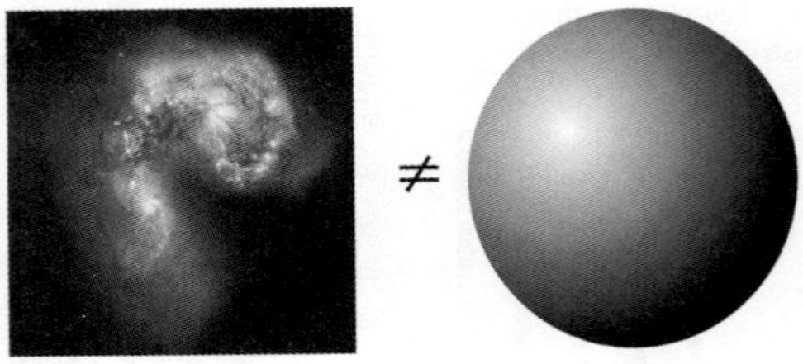

Figure 3. Alas, galaxies are not spheres.

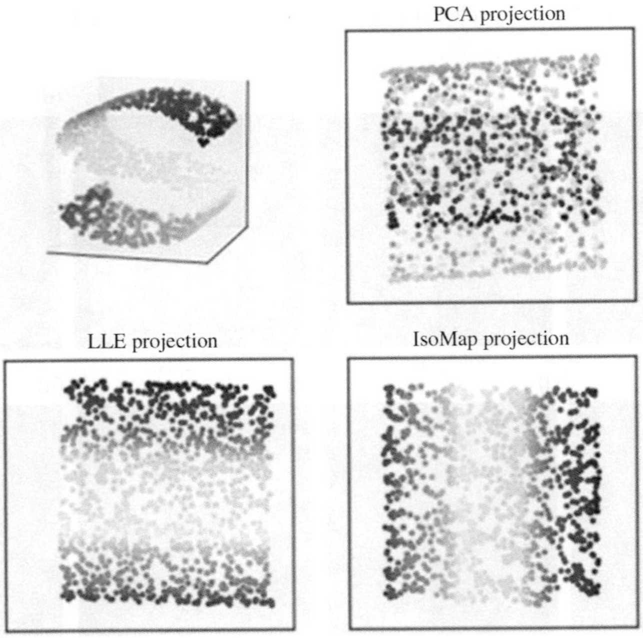

Figure 4. An example of 3D $\rightarrow$ 2D dimensionality reduction, with various degrees of information loss. Figure from the sklearn website.

probability distribution functions for our parameters, we bias our results by imposing our choice of parameters and priors. This model-related bias is something that machine learning can hope to circumvent, since the relation between input variables (observables) and output (measurements, or predictions) is purely data-driven.

Make problems more tractable. This is another field in which machine learning algorithms are quite amazing. One very simple but incredibly helpful application is dimensionality reduction, the process of identifying patterns in the data that allow one to reduce the size of the data space, making data easier to visualize, manipulate, and analyze, as shown in Fig. 4. Another great aspect is the increased resilience of ML methods to heterogeneous and sparse data. To stay within the regime of SED fitting, it is hard to create joint probability distributions to combine photometric and spectroscopic data, while for many ML methods, this problem just amounts to creating additional features.

Facilitate serendipitous discoveries. In 2017, a workshop called "Detecting the Unexpected" took place at the Space Telescope Science Institute. One of the most interesting talks (at least for me, following from afar) introduced me to the work of Norris (2017), who looked at the ten most important discoveries of the Hubble Space Telescope over the past decade. They pointed out that only a few key achievements, such as the use

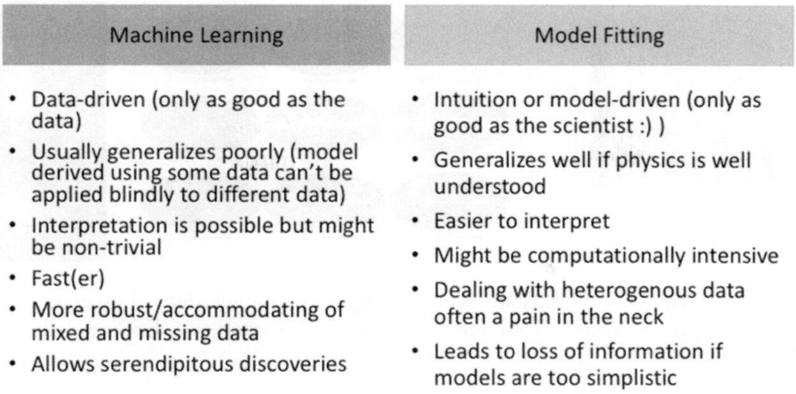

Figure 5. Some advantages and disadvantages of the two approaches. My claim is that **synergy** is often the best strategy.

of Cepheids to improve the accuracy of the measurement of the Hubble constant, were planned in advance, while other fundamental advances in Astronomy, from the discovery of Dark Energy to the Hubble Deep Fields, were the result of serendipity. This - like many other examples - shows that one of the most important ways to prepare ourselves for further discoveries is to plan for what we don't know yet. In this respect, machine learning and data mining methods are really powerful; ML algorithms for pattern recognition and outlier detection are in general superior to model fitting, because the latter assumes, by construction, that we know what we are looking for.

3. Machine Learning versus Model Fitting: Who wore it better?

While I do, in many ways, "believe the hype" about machine learning, and I think it is important for young and old scientists alike to become proficient with these methods, I think it is important to realize that these techniques are just another set of tools in an astronomer/data scientist's skill set. I've often joked that machine learning methods are the "Numerical Recipes" of this decade. They enable us to do some science that would be difficult to do otherwise; but as always, it's part of a scientist's job to understand how to "attack" a difficult problem, with or without machine learning. As a general and incomplete rule, I think that when we understand the physics but we don't have data, we use model fitting, and when we don't understand the physics but we have data, we use machine learning; the figure 5 below summarizes some more of my thoughts on either approach, but the conclusion is that it's the synergy between these two approaches that is most powerful, because it gives us a strong cross-check of our problem-solving architecture.

4. Why now?

I hope that the discussion above provides a fair assessment of why machine learning methods are useful in Astronomy. But another question arises about timing: after all, none of these problems or methods are new, so why have we seen such a strong increase in ML-related science in the last few years? The key to that, in my opinion, is that we have a *super* data problem (my way of saying it's not just *big* data). Sure, data are becoming bigger; if we use etendue as zeroth-order tracer of the data volume generated by a survey, we can see in the left panel of Fig. 6 that the Large Synoptic Survey Telescope (LSST) dwarfs other surveys by a factor of 100 or more. However, we also have *better* data, because of the extended wavelength coverage, depth, and resolution of galaxy surveys. We have a *wide* data problem (to borrow a term I first heard from Alyssa Goodman at

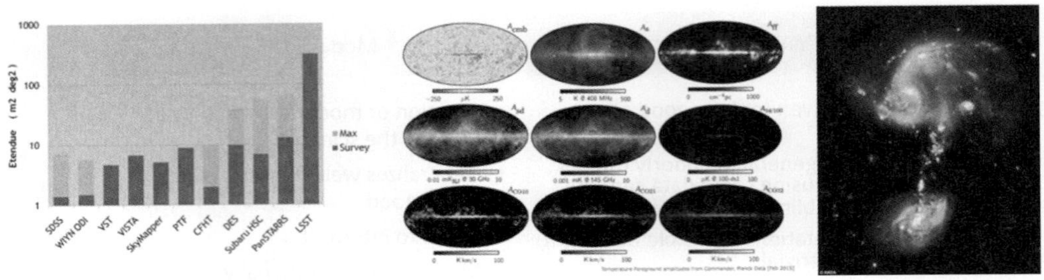

Figure 6. An exciting and terrifying scenario: we have bigger data, better data, wide data, and new data. Figure sources, from left to right: LSST website, ESA website and Planck Collaboration (2016), and the Hubble website.

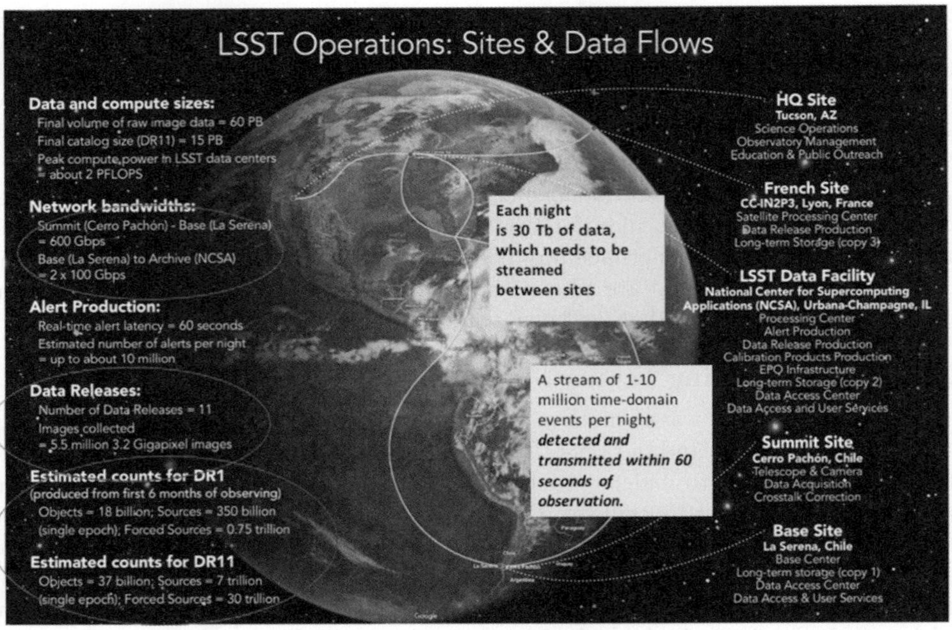

Figure 7. The "Big Data" problem of LSST. From the LSST website; annotations are mine.

a 2015 talk), where we have multiple representations of the same object, contributing information from different parts of the spectrum. And we have a *new* data problem, where we are bound to discover things that we didn't think could exist, like hot Jupyters or super massive stars, or to unveil relationships that we don't yet understand. It is in dealing with wide and new data that, in my opinion, machine learning methods really shine, and the timing is now because of upcoming observatories like LSST, the James Webb Space Telescope, and Euclid.

5. The example of LSST

The Large Synoptic Survey Telescope provides a great example of the type of big data problem that astronomers are facing, and it's right around the corner, with a projected start date of 2023 (and an impressive record of being on schedule). Even the first of eleven data releases (DRs), anticipated to happen just six months after the beginning of science operations, will contain data for *eighteen billion* galaxies; DR11 will contain data for 37 billion galaxies, and will amount to the equivalent of 5.5 billion

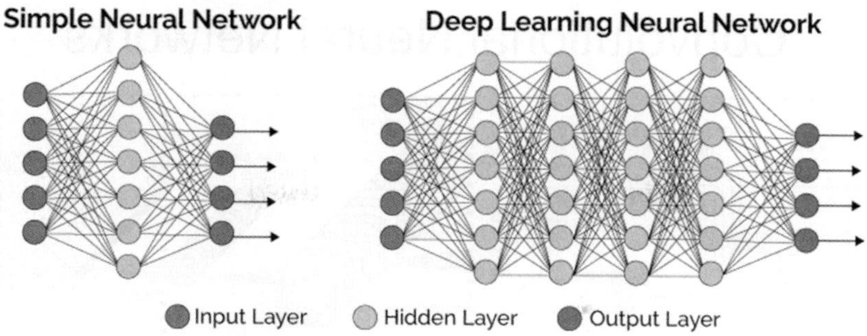

Figure 8. Deep Learning algorithms (right) use multiple layers to characterize the relation between input and output. Figure from this post.

Gigapixel-size images. That's a lot of huge pictures! The problem of transferring and storing these data is only exacerbated by the fact that each night's observations will amount to 30 Tb of data; within those, 1–10 million time-domain events will need to be streamed within 60 seconds, because real-time detection of unusual and serendipitous phenomena is essential; see Fig. 7. In one word: Ouch!

6. What's next?

This section summarizes my thoughts about what are going to be the most pressing issues that we will face in the next five years or so, and what algorithms and techniques will, in my opinion, become most important in the "astronomy x data science" world. Of course, it's a limited, biased view - if I were a ML algorithm, I would have low precision and low recall!

5.1. Too much information, too little space. One of the big issues we will face is going to be data storage. On the one hand, this applies to the raw data - for example, while searching and downloading the SDSS archive is a relatively manageable task for anyone with a laptop and a decent internet connection, the corpus of LSST data will certainly make this impossible. As a result, data compression, representation, and visualization techniques will become increasingly important. Examples of them range from the simple Principal Component Analysis to nonlinear techniques such as Self Organized Maps, t-distributed Stochastic Neighbor Embedding, Auto Encoders, and many others, some of which are described here.

However, the tricky part is that different science applications will need different types of data, so the information loss associated to any technique may be perfectly acceptable for some projects, but not for others. The problem becomes even more pressing if one also considers *derived* products: catalogs, estimated parameters with their probability distributions, and so on. I found it telling that in my field of galaxy SED fitting, we spent a good fraction of the last decade arguing in favor of going beyond the best-fit model and using more sophisticated methods, such as various flavors of Markov Chain Monte Carlo methods, to explore the parameter spaces and generate full probability distributions. Instead, by the last CANDELS meeting (April 2018 at UC Riverside) the discussion had largely shifted towards what is the best way to make the products available, since full probability distributions are expensive to compute and store. I don't really have a proposed solution, but I argue that we will all collectively need to discuss what products should be made available, and to develop/learn smart tools for data representation.

5.2. Deep Learning FTW. My prediction is that we will see a proliferation of deep learning methods (Fig. 8), a subset of machine learning methods in which the input/output relation is modeled through a series of intermediate steps (layers) that

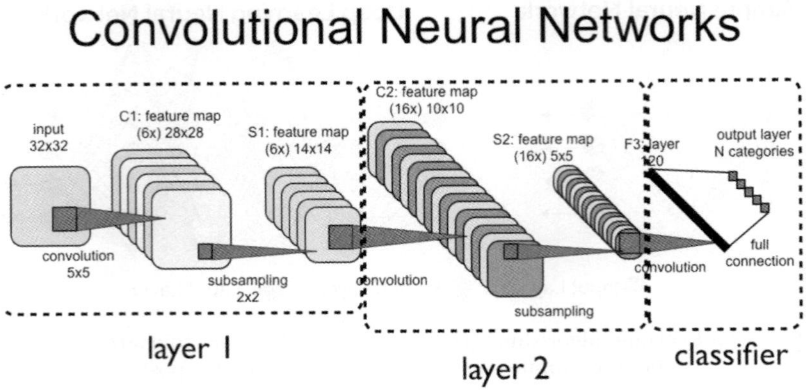

Figure 9. A scheme of a convolutional neural network. The network includes convolution layers, in which complicated patterns are broken down into simple ones, creating data-driven features, pooling layers that are used for subsampling/dimensionality reduction, and a final fully connected layer that predicts the output. Figure courtesy of Eugenio Colurciello.

make them really flexible and powerful in capturing features from raw data, moderating the need for feature selection and engineering, and modeling complex relationships.

The following are three types of neural networks that have already been shown to be very useful in Astronomy, and I expect that we'll hear a lot more about them in the next five years.

Convolutional Neural Networks. Convolutional Neural Networks (CNNs) are growing in popularity in many areas of Astronomy, typically as a means of analyzing 2D image data (*e.g.* Dieleman *et al.* (2015); Tuccillo *et al.* (2017); Petrillo *et al.* (2017); Fussell & Moews (2018); Huertas-Company *et al.* (2018)). One powerful feature of CNNs is that they alternate different type of layers: convolution layers, used for feature extraction, and pooling layers, used for subsampling/dimensionality reduction. The final layer is a fully connected neural network, which generates the output. This structure, shown in Fig. 9 above, enables them to autonomously extract significant features from images, reducing the need for feature selection and feature engineering, and making them suitable to process astronomical images, in which information is often present in the form of gradients, breaks, and spatial correlations. A very nice, detailed discussion of how CNNs can master morphology classifications can be found in this blog post.

Generative Adversarial Networks. Generative Adversarial Networks (Goodfellow *et al.* 2014) are built as a double system where one network attempts to build realistic examples of the objects, and the other one tries to detect whether they are fake or real. While the first applications were fraud detection, GANs have now been applied to a variety of problems, for example generating art in the style of a specific painter (Zhu *et al.* 2017); see Fig. 10 above. This blog post has a wonderful collection of examples. In Astronomy, GANs' recent claims to fame have been "upsampling" image resolution (Ledig *et al.* 2016; Schawinski *et al.* 2017), and generating synthetic galaxy images that look realistic at the object and at the population level (Fussell & Moews 2018), as shown in Fig. 11. Both of these approaches have profound consequences for Astronomy, because they potentially lead to reduced observing times and to scaling down the processing time for large galaxy evolution simulations.

Recurrent Neural Networks. Recurrent neural networks, whose structure is shown in Fig. 12, have the characteristic of using not just the inputs, but the output of previous queries as part of their output generation. This characteristic makes them very suitable for all sorts of time series (time domain) problems, in which we care about not just the

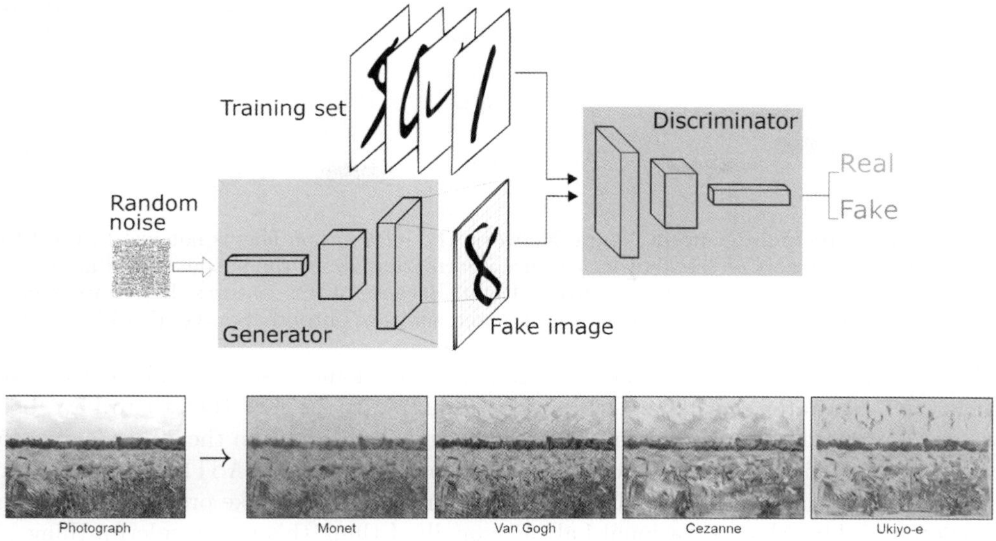

Figure 10. Top: Generative Adversarial Networks (GANs) learn how to generate realistic images thanks to their dual generator/discriminator structure. From Skymind. Bottom: an example of style transfer, where the GAN processes the input photograph to generate art in the style of four different painters. From Zhu *et al.* (2017).

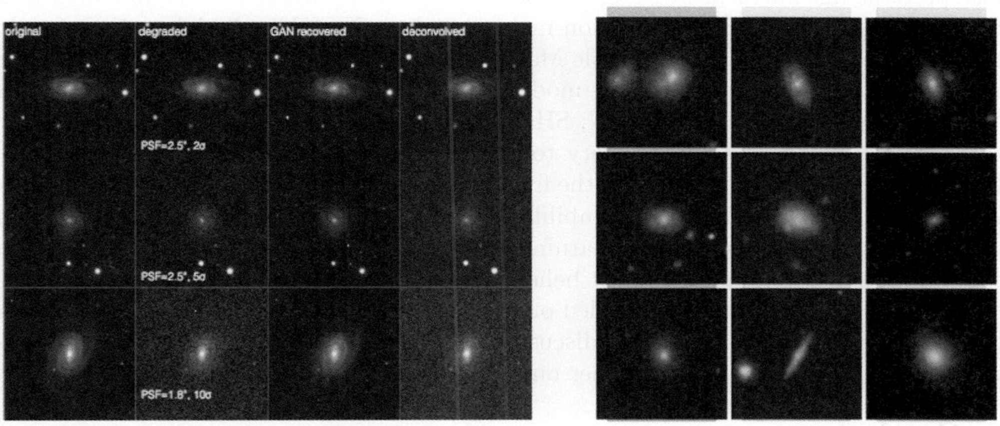

Figure 11. Left: An example of how GANs can be used to recover high-resolution galaxy images, from Schawinski *et al.* (2017) From left to right: the original SDSS image, the degraded image with a worse PSF and higher noise level (indicated), the image as recovered by the GAN, and for comparison, the result of a deconvolution. Right: Generating simulated galaxy images with GANs; the first two columns show synthetic objects, the last one real SDSS images. From v1 of Fussell & Moews (2018); further improvements in removing artifact features were solved by these authors through the development of a two-stage GAN approach.

characteristic of a single input instance but about a *sequence* of them. This is very helpful, among other cases, in outlier detection and hardware failure detection (Naul *et al.* 2018; Zhang & Zou 2018).

5.3. (Sorry,) We need to talk to non-astronomers. An important consequence of the widespread impact of machine learning techniques in Astronomy is the fact that we need to develop interdisciplinary collaborations - for real! While this is in general always a good idea, I finally see it happening and it's in part due to the fact that data

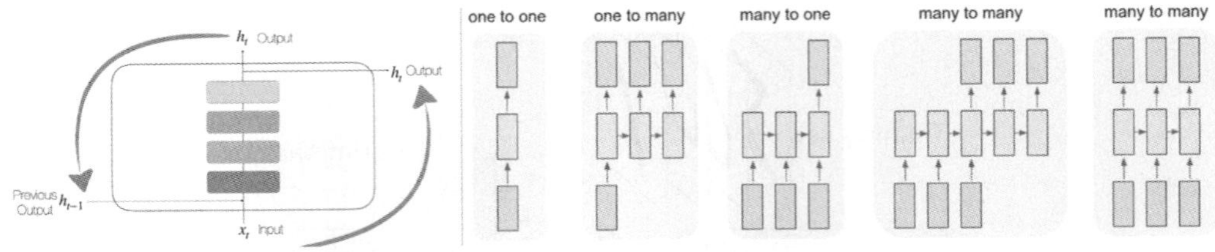

Figure 12. Left: In a Recurrent Neural Network, the information flow is not just forward-fed, but the unit processors of the network can use information about the output of the network in previous steps. Figure by Bruno Goncalves). Right: Because of this features, RNNs are good at modeling relationships between series of inputs and series of outputs. Source: this blog post.

science seems to provide a common language for astronomers, statisticians, mathematicians, and computer scientists. One positive example is the fact that there have been several Astronomy/Physics challenges on the Kaggle website, from the 2014 Galaxy Zoo challenge and Higgs boson challenge to recent ones like the PLASTICC challenge for light curves classification in LSST data and the ongoing earthquake prediction challenge promoted by Los Alamos National Lab. Personally, I think this is a wonderful thing.

5.4. What the heck are we doing? Model interpretability/Explainability. One drawback of powerful deep learning methods (but also of simpler models with heavily engineered features) is that they are difficult to interpret, hindering the process of producing generalizable insights. As a result, the last few years have seen the advent of interpretable machine learning (e.g. Molnar 2018), a discipline that seeks to increase the transparency of the machine's decision-making process. Useful methodology tools for this task are LIME (Locally Interpretable Model-Agnostic Explanations, Tulio Ribeiro *et al.* 2016), which uses surrogate simpler models to map the input-output relationships of an "opaque" algorithm such as a CNN; SHAP (SHapley Additive exPlanations, Lundberg *et al.* 2018), which uses game theory to understand the interactions between different features (players) and their roles in the final decision, and Recourse Analysis Ustun *et al.* 2018, which aims to build interpretability by understanding what feature modifications would change the outcome of a learning algorithm. The topic is controversial in the computer science community, and I believe that the trade-off between predictive power and interpretability should be decided on a discipline-by-discipline, or even case-by-case basis; see Lipton (2016) for a great discussion. Nonetheless, I imagine that many of these will be helpful in our quest to further our understanding of the physics of galaxies.

7. Conclusions

Machine learning and data mining methods provide an exciting opportunity to deal with our "super data" problems. Here are some ideas on how to leverage them:

• **Teach/learn this stuff early and often**. There is no doubt that any Astronomy graduate student can benefit from learning about data science methods, and there are infinite possibilities to do so. Many resources, such as classes and books, are available for free online; there are also dedicated initiatives such as the LSST data science fellowship.

• **(Encourage mentees to) Code in Python.** This is essential, I think, for two reasons: one is convenience, because so many of the dedicated packages and support networks are available for Python, and second, because learning Python (and data science tools) will keep options open in case a career change is desired or necessary. I should add that some database experience and distributed computing experience is also quite desirable for this purpose.

• **(Encourage mentees to) Talk to people and go to conferences outside your field**. This year, the most inspiring talks I have heard were given in non-academic settings. I want to acknowledge these people here: Mikio Braun (Zalando) and Adil Aijaz (Split Software) at Strata Data NYC, Manojit Nandi (Rocketrip) at PyGotham, Stephanie Yang (Foursquare) at a NYC Women in Machine Learning and Data Science meetup, and Francesca Lazzeri (Microsoft) at DataX NYC. This is of course driven by my own interests, but I think it's generally true that hearing about different topics and gathering fresh perspectives is exhilarating, and I have found these environments generally welcoming (not to mention the networking value).

- **(Encourage mentees to) Publish your code and make your results reproducible.** As the use of computational techniques becomes increasingly essential and algorithms become possibly more opaque, sharing code, even if it's not pretty, is the main way of keeping each other accountable and learning from each other. Plus, having an online presence on github is a good idea in view of career advancement, both in academia and in industry.
- **Rest assured that science is science.** Machine learning is not a magic wand; it's just a way of enhancing the tool set that we have. In fact, I think that it is even more essential to learn how to formalize, attack, solve, and test our solutions to problems now that we have more ways to solve them. In other words, the scientific method stands high and strong, and intuition and rigor continue, even more than before, to be the pillars of advancement in science.

References

Astropy Collaboration. 2013, *Astron. Astrophys.*, 558, A33
—. 2018, *Astron. J.*, 156, 123
de la Calleja, J. & Fuentes, O. 2004, *Mon. Not. R. Astron. Soc.*, 349, 87
Dieleman, S., Willett, K. W., & Dambre, J. 2015, *Mon. Not. R. Astron. Soc.*, 450, 1441
Fussell, L. & Moews, B. 2018, ArXiv e-prints, 1811.03081
Goodfellow, I. J., Pouget-Abadie, J., Mirza, M., Xu, B., Warde-Farley, D., Ozair, S., Courville, A., & Bengio, Y. 2014, ArXiv e-prints, 1406.2661
Hocking, A., Geach, J. E., Sun, Y., & Davey, N. 2018, *Mon. Not. R. Astron. Soc.*, 473, 1108
Huertas-Company, M., Primack, J. R., Dekel, A., Koo, D. C., Lapiner, S., Ceverino, D., Simons, R. C., Snyder, G. F., Bernardi, M., Chen, Z., Domínguez-Sánchez, H., Lee, C. T., Margalef-Bentabol, B., & Tuccillo, D. 2018, *Astrophys. J.*, 858, 114
Katebi, R., Zhou, Y., Chornock, R., & Bunescu, R. 2019, *Mon. Not. R. Astron. Soc.*, 486, 1539
Ledig, C., Theis, L., Huszar, F., Caballero, J., Cunningham, A., Acosta, A., Aitken, A., Tejani, A., Totz, J., Wang, Z., & Shi, W. 2016, arXiv e-prints, arXiv:1609.04802
Lintott, C. J., Schawinski, K., Slosar, A., Land, K., Bamford, S., Thomas, D., Raddick, M. J., Nichol, R. C., Szalay, A., Andreescu, D., Murray, P., & Vandenberg, J. 2008, *Mon. Not. R. Astron. Soc.*, 389, 1179
Lipton, Z. C. 2016, arXiv e-prints, arXiv:1606.03490
Lundberg, S. M., Erion, G. G., & Lee, S.-I. 2018, ArXiv e-prints
Molnar, C. 2018, Interpretable Machine Learning (https://christophm.github.io/interpretable-ml-book/), https://christophm.github.io/interpretable-ml-book/
Naul, B., Bloom, J. S., Pérez, F., & van der Walt, S. 2018, Nature Astronomy, 2, 151
Norris, R. P. 2017, *Publications of the Astron. Soc. of Australia*, 34, e007
Petrillo, C. E., Tortora, C., Chatterjee, S., Vernardos, G., Koopmans, L. V. E., Verdoes Kleijn, G., Napolitano, N. R., Covone, G., Schneider, P., Grado, A., & McFarland, J. 2017, ArXiv e-prints, 1702, arXiv:1702.07675
Planck Collaboration. 2016, *A&A*, 594, A10
Schawinski, K., Zhang, C., Zhang, H., Fowler, L., & Santhanam, G. K. 2017, *Mon. Not. R. Astron. Soc.*, 467, L110
Tuccillo, D., Huertas-Company, M., Decencière, E., & Velasco-Forero, S. 2017, in IAU Symposium, Vol. 325, Astroinformatics, ed. M. Brescia, S. G. Djorgovski, E. D. Feigelson, G. Longo, & S. Cavuoti, 191–196

Tulio Ribeiro, M., Singh, S., & Guestrin, C. 2016, ArXiv e-prints, 1602.04938
Ustun, B., Spangher, A., & Liu, Y. 2018, ArXiv e-prints, 1809.06514
Vanderplas, J., Connolly, A., Ivezić, Ž., & Gray, A. 2012, in Conference on Intelligent Data Understanding (CIDU), 47 –54
Zhang, R. & Zou, Q. 2018, Journal of Physics: Conference Series, 1061, 012012
Zhu, J.-Y., Park, T., Isola, P., & Efros, A. A. 2017, arXiv e-prints, arXiv:1703.10593

Discussion

David Rosario: With deep learning, there are examples of "emergent behaviour" where patterns and relationships are generated that were not represented in the training data. How could one identify such classes and either guard against or learn from them.

Viviana Acquaviva: This can happen. Cross-validation and expert understanding of the data and the scientific problem are essential to understand emergent effects.

Challenges in Panchromatic Modelling with Next Generation Facilities
Proceedings IAU Symposium No. 341, 2018
M. Boquien, E. Lusso, C. Gruppioni & P. Tissera, eds.

doi:10.1017/S1743921319001418

Modeling with the crowd: Optimizing the human-machine partnership with Zooniverse

Hugh Dickinson[1], Lucy Fortson[1], Claudia Scarlata[1], Melanie Beck[1] and Mike Walmsley[2]

[1]School of Physics and Astronomy, University of Minnesota, 116 Church Street SE, Minneapolis, MN 55455, USA

[2]Oxford Astrophysics, Denys Wilkinson Building, Keble Road, Oxford OX1 3RH, UK

Abstract. LSST and Euclid must address the daunting challenge of analyzing the unprecedented volumes of imaging and spectroscopic data that these next-generation instruments will generate. A promising approach to overcoming this challenge involves rapid, automatic image processing using appropriately trained Deep Learning (DL) algorithms. However, reliable application of DL requires large, accurately labeled samples of training data. Galaxy Zoo Express (GZX) is a recent experiment that simulated using Bayesian inference to dynamically aggregate binary responses provided by citizen scientists via the Zooniverse crowd-sourcing platform in real time. The GZX approach enables collaboration between human and machine classifiers and provides rapidly generated, reliably labeled datasets, thereby enabling online training of accurate machine classifiers. We present selected results from GZX and show how the Bayesian aggregation engine it uses can be extended to efficiently provide object-localization and bounding-box annotations of two-dimensional data with quantified reliability. DL algorithms that are trained using these annotations will facilitate numerous panchromatic data modeling tasks including morphological classification and substructure detection in direct imaging, as well as decontamination and emission line identification for slitless spectroscopy. Effectively combining the speed of modern computational analyses with the human capacity to extrapolate from few examples will be critical if the potential of forthcoming large-scale surveys is to be realized.

Keywords. Surveys, Morphology, Citizen Science, Machine Learning

1. Astronomy in the age of Big Data

The coming decade promises an unprecedented expansion in the volume and diversity of high-quality data that will be available to the astronomical community. Forthcoming instruments including the Large Size Synoptic Telescope (LSST) (Ivezić *et al.* 2018) or *Euclid* (Laureijs *et al.* 2011) will provide galaxy catalogues containing billions of sources. The potential scientific yield of these datasets is correspondingly unprecedented as long as they can be rapidly and effectively analysed.

To be robust, traditional automated analyses of extensive astrophysical datasets have remained somewhat rudimentary. For example, in the context of galaxy morphology, commonly applied techniques for automatic classification use the radial light profile of galaxies to estimate the relative contributions from distinct bulge and disk components (e.g. Simard *et al.* 2011). While such methods have largely solved the problem of *coarse* morphological classification for large populations of galaxies, they cannot identify more subtle features like stellar shells, spiral arms, bars and clumps. Indeed, automatic detection of galaxy substructure is notoriously difficult, which often limits related studies to small, visually inspected samples (e.g. Nair & Abraham 2010).

To visually inspect and morphologically categorize large numbers ($\sim 10^5$) of galaxy images, the *Galaxy Zoo* project (hereafter GZ; Lintott *et al.* 2008) used a web-based classification interface to recruit over 350,000 volunteers from the public. Unfortunately, even the impressive resources mobilized by GZ will not be sufficient to process the enormous volumes of data that forthcoming instruments will deliver. To meet this challenge, robust new automatic tools that provide at least human-equivalent performance when applied to complex astrophysical analyses are urgently required.

Modern machine learning algorithms are designed specifically to process and analyse large data volumes associated with Big Data. Given sufficient training data, DL techniques can achieve human-like performance across a wide variety of image analysis tasks including morphological classification of galaxies (e.g. Dieleman *et al.* 2015; Domínguez Sánchez *et al.* 2018). The main obstacle for applying DL in any context is the requirement for abundant labeled training data. In this proceeding, we outline methods that leverage citizen science to generate accurate data labels in sufficient volume to train robust DL algorithms. We use examples based on GZ and use results of Galaxy Zoo Express (hereafter GZX; Beck *et al.* 2018) and planned extensions thereof to illustrate the techniques we discuss. We address three crucial questions:

- How can the compromise between label accuracy and volunteer effort be optimized?
- How can we avoid volunteers labelling mundane data that are well characterized by DL models?
- How can the unique capabilities of humans and DL best be combined?

2. Optimizing Accuracy in Citizen Science

Aggregation of annotations that are provided by different citizen scientists *for the same subject* is essential to mitigate individual subjectivity and compensate for outliers and catastrophic misclassifications (see e.g. Lintott *et al.* 2008; Willett *et al.* 2013; Hart *et al.* 2016, for some GZ-related examples). While aggregation can be performed *a posteriori* using a fixed number of independent classifications per subject, this approach is generally inefficient because stable consensus between classifiers is often achieved before the prescribed number of annotations have been gathered. The *SpaceWarps* project (Marshall *et al.* 2016) developed the *SpaceWarps Analysis Program* (SWAP) for binary-valued classification tasks. SWAP dynamically computes Bayesian estimates for the probability $P_{\rm correct}$ that the current consensus label is correct, based on calibrated reliabilites of the volunteers that provided annotations. SWAP provides a quantitative means to optimize the number of annotations collected per subject by accepting the consensus once $P_{\rm correct}$ exceeds an appropriately chosen threshold. When the consensus is accepted the subject is said to be "retired". GZX applied SWAP to the morphological classification of galaxies by post-processing binary-valued volunteer responses from Galaxy Zoo 2 (hereafter GZ2; Willett *et al.* 2013) to the question "Is the galaxy simply smooth and rounded, with no sign of a disk?" Comparing the *dashed blue* and *solid blue* curves in the left panel of Figure 1 reveals that application of the SWAP algorithm increased the subject retirement rate by a factor of ~ 5. The increased rate reflects the fact that most galaxies achieve stable consensus using far fewer than the default 40 annotations required by GZ. Indeed, application of SWAP reduced the overall volunteer effort by a factor of 7 while replicating $> 95\%$ of the original GZ2 labels for the smooth versus disk classification.

The SWAP algorithm is designed to accelerate consensus for binary classifications, which restricts its domain of applicability to a subset of astrophysical use cases. In the context of galaxy morphology, many analyses require identification and labeling of specific galactic substructures, corresponding to specific locations or regions within a two-dimensional image. Optimal retirement algorithms for two-dimensional annotations

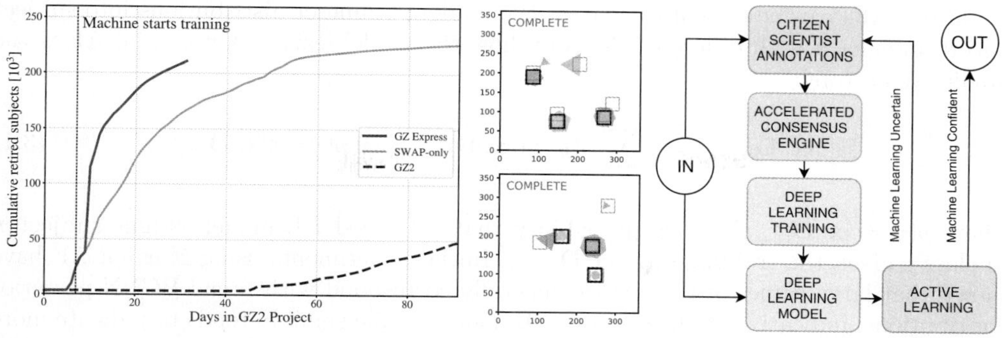

Figure 1. Optimizing the utility of citizen scientists' annotations using. *Left:* Adapted from Beck *et al.* (2018). The *dashed* curve illustrates the cumulative retirement rate in the original GZ2 project. GZ2 required 40 annotations in order to retire each subject. The *solid blue* curve illustrates the same quantity using the SWAP algorithm with a retirement threshold $P_{\rm correct}$ that replicates $> 95\%$ of the original GZ2 labels. The *solid red* curve illustrates the cumulative retirement rate using a simple random forest algorithm trained using a subset of the consensus classifications provided by SWAP within the first 12 days of volunteer activity. *Middle:* Results demonstrating the effectiveness of an optimal aggregation algorithm for two-dimensional annotations. Volunteers are instructed to mark shapes with more than four sides. *Thick black* boxes illustrate the consensus after annotation by four volunteers. The consensus annotations for both subjects are accurate. The coloured boxes indicate the annotations provided by individual volunteers. Evidently, the consensus algorithm is robust against the noisy or spurious *green*-coloured annotations. *Right:* Data flow schematic illustrating an infrastructure designed to enable efficient human-machine collaboration for the analysis of extensive astronomical datasets. The schematic unites a GZX-like infrastructure involving citizen science (*purple*), optimal subject retirement (*green*) and machine learning (*blue*) components with an active learning engine (*red*) designed to optimise DL model training.

have been developed in the context of commercial crowdsourcing (e.g. Branson *et al.* 2017) and we are currently adapting them to the Zooniverse (www.zooniverse.org) infrastructure that hosts GZ. The right panel in Figure 1 illustrates the effectiveness of one such algorithm using a simple test project in which volunteers were asked to mark all shapes with more than four sides. Reliable consensus was achieved after only four annotations for $> 96\%$ of the test subjects. The results also demonstrate the algorithm's robustness against outliers and catastrophic mistakes.

3. Minimizing wasted volunteer effort

Volunteer effort in citizen science is a valuable resource. It is not rational or ethical to waste volunteer time annotating subjects that DL algorithms can accurately characterize. A more sensible approach would focus volunteer attention on data that machine learning cannot handle, and use the resultant annotations as training data to refine the DL model. *Active Learning* (AL) is a general machine learning technique that incrementally updates a small initial training set using an *acquisition function* (AF) to request annotations for subjects that, if annotated, would be most informative for the DL model in its current state. Subjects about which the model is currently uncertain are a natural choice. However, many conventional DL algorithms are incapable of quantifying the uncertainty associated with their model. Bayesian Convolutional Neural Networks (BCNNs; e.g. Gal & Ghahramani 2015) are DL algorithms that are tailored for image analysis and which do provide an estimate of the model uncertainty required by AL in the form of approximate posterior distributions over the BCNN parameters.

Various AFs can be chosen to convert the model parameter distributions into subject uncertainty, each with different effects on how the model learns. A potential AF is the predictive entropy:

$$\mathbb{H}[y|\mathbf{x}, \mathbf{D}_{train}] = -\sum_{c} p(y=c|\mathbf{x}, \mathbf{D}_{train}) \log p(y=c|\mathbf{x}, \mathbf{D}_{train}) \quad (3.1)$$

where $p(y=c|\mathbf{x}, \mathbf{D}_{train})$ is the probability that a trained DL model assigns subject $\mathbf{x}$ to class c given the training dataset $\mathbf{D}_{train}$. Initial experiments using $\mathbb{H}$ as an AF have shown that large values of this metric generally correspond with broad BCNN posterior distributions, indicative of an uncertain DL model, while small $\mathbb{H}$ tends to indicate more concentrated BCNN posteriors. Moreover, $\mathbb{H}$ is also correlated with the accuracy of the BCNN model, with smaller values indicating a correct DL prediction is more likely.

4. Combined Analysis by Humans and Machines

GZX explored the potential of human-generated classifications as training data for machine learning algorithms by using SWAP-accelerated consensus labels to train a simple random forest (RF) classifier. The *red* line in Figure 1 reveals that the adequate initial training of the RF was possible within 8 days of volunteer activity. After 12 days, the SWAP provided over 40,000 labelled training images and the RF model was deemed sufficiently accurate to retire additional subjects based on its classifications. Using the RF classifier, over 200,000 subjects were retired within 27 days, increasing the overall classification rate by a factor of 10 while requiring $\sim 13 \times 10^6$ fewer volunteer anootations with respect to GZ2. The RF classifications replicated $> 93\%$ of the original GZ2 labels.

The right panel of Figure 1 illustrates a data flow schematic that unites and generalizes the infrastructure for accelerated consensus and machine learning training implemented in GZX. The schematic also includes an AL engine designed to optimize the DL training efficiency. We envision a system in which a small fraction of data are labeled by professional astronomers and citizen scientists. These labeled data are then used to train DL algorithms to perform specific analysis tasks. Partially trained DL algorithms classify unlabeled data and an active learning engine is used to estimate model uncertainty and request additional human annotations for poorly classified subjects. This design benefits from humans' unique ability to extrapolate reliably from very few examples and enables continual refinement of machine learning models as new and potentially unanticipated data classes are encountered. We reiterate that sophisticated automatic analyses will become essential tools for astronomy within the next few years and assert that infrastructures similar to that we have described will be invaluable to train, calibrate and validate such analyses.

References

Ž. Ivezić, J. A. Tyson, E. Acosta, R. Allsman, S. F. Anderson, J. Andrew, J. R. P. Angel, T. S. Axelrod *et al.*, 2008. ArXiv e-prints 0805.2366

R. Laureijs, J. Amiaux, S. Arduini, J. L. Augùeres, J. Brinchmann, R. Cole, M. Cropper, C. Dabin *et al.*, 2011. ArXiv e-prints 1110.3193

L. Simard, J. T. Mendel, D. R. Patton, S. L. Ellison & A. W. McConnachie, 2011. *ApJS* 196, 11

P. B. Nair & R. G. Abraham, 2010. *ApJS* 186, 427

C. J. Lintott, K. Schawinski, A. Slosar, K. Land, S. Bamford, D. Thomas, M. J. Raddick, R. C. Nichol *et al.*, 2008. *MNRAS* 389, 1179

S. Dieleman, K. W. Willett & J. Dambre, 2015. *MNRAS* 450, 1441

H. Domínguez Sánchez, M. Huertas-Company, M. Bernardi, D. Tuccillo & J. L. Fischer, 2018. *MNRAS* 476, 3661

M. R. Beck, C. Scarlata, L. F. Fortson, C. J. Lintott, B. D. Simmons, M. A. Galloway, K. W. Willett, H. Dickinson *et al.*, 2018. *MNRAS* 476, 5516

K. W. Willett, C. J. Lintott, S. P. Bamford, K. L. Masters, B. D. Simmons, K. R. V. Casteels, E. M. Edmondson, L. F. Fortson *et al.*, 2013. *MNRAS* 435, 2835

R. E. Hart, S. P. Bamford, K. W. Willett, K. L. Masters, C. Cardamone, C. J. Lintott, R. J. Mackay, R. C. Nichol *et al.*, 2016. *MNRAS* 461, 3663

P. J. Marshall, A. Verma, A. More, C. P. Davis, S. More, A. Kapadia, M. Parrish, C. Snyder *et al.*, 2016. *MNRAS* 455, 1171

S. Branson, G. V. Horn & P. Perona, 2017. In 2017 IEEE Conference on Computer Vision and Pattern Recognition (CVPR). 6109–6118

Y. Gal & Z. Ghahramani, 2015. ArXiv e-prints 1506.02158

Discussion

KATARZYNA MAŁEK: What should we do when SWAP might not be able to classify more than a few classes of galaxies/objects? Should we then use unsupervised machine learning?

HUGH DICKINSON: SWAP is designed to provide a quantitative estimate of the reliability of the consensus label after several citizen scientists have provided an annotation. If the probability of the consensus label indicates uncertainty (probability ~ 0.5) after many volunteer classifications, then it may be that the particular image is particularly ambiguous. If this is true for a large fraction of your dataset, you may want to check that professional astronomers can provide a consistent classification for a representative sample of your data. If so, then unsupervised learning is an option, as is changing the question asked of citizen scientists - they may be struggling because they don't understand what is being asked of them.

DAN TARANU: In large datasets with millions of "strange"/unclassifiable objects, how do you choose which ones should be sent back to astronomers for a second look? This was basically answered in the previous question as "by finding clusters of outlier", so follow-up: what if there are many small clusters of tens of rare objects?

HUGH DICKINSON: There are several approaches being explored. Perhaps the most promising approach involved iterative refinement of initial clusters by human volunteers. By learning an embedding of the data space containing the poorly classified data clusters, nearby clusters can be presented to volunteers, whose classifications can be used to split, unify or purify each cluster. After several iterations we hope that very small clusters will merge enabling professional astronomers to investigate the larger cluster's unique characteristics.

Challenges in Panchromatic Modelling with Next Generation Facilities
Proceedings IAU Symposium No. 341, 2018
M. Boquien, E. Lusso, C. Gruppioni & P. Tissera, eds.

doi:10.1017/S1743921319002187

Deep learning for galaxy mergers in the galaxy main sequence

William J. Pearson[1,2], Lingyu Wang[1,2], James Trayford[3], Carlo E. Petrillo[2] and Floris F. S. van der Tak[1,2]

[1]SRON Netherlands Institute for Space Research
Landleven 12, 9747 AD, Groningen, The Netherlands
email: w.j.pearson@sron.nl

[2]Kapteyn Astronomical Institute, University of Groningen
Postbus 800, 9700 AV Groningen, The Netherlands

[3]Leiden Observatory, Leiden University, P.O. Box 9513, 2300 RA Leiden, The Netherlands

Abstract. Starburst galaxies are often found to be the result of galaxy mergers. As a result, galaxy mergers are often believed to lie above the galaxy main sequence: the tight correlation between stellar mass and star formation rate. Here, we aim to test this claim.

Deep learning techniques are applied to images from the Sloan Digital Sky Survey to provide visual-like classifications for over 340 000 objects between redshifts of 0.005 and 0.1. The aim of this classification is to split the galaxy population into merger and non-merger systems and we are currently achieving an accuracy of 92.5%. Stellar masses and star formation rates are also estimated using panchromatic data for the entire galaxy population. With these preliminary data, the mergers are placed onto the full galaxy main sequence, where we find that merging systems lie across the entire star formation rate - stellar mass plane.

Keywords. methods: data analysis, galaxies: evolution, galaxies: interactions, galaxies: starburst, infrared: galaxies

1. Introduction

Galaxy mergers are highly important for fully understanding how galaxies form and evolve, underpinning hierarchical growth models. Also, galaxy mergers are believed to be the driving force behind some of the brightest infrared objects known (Sanders *et al.* 1996). With bright infrared emission often comes high star formation rates (SFRs), hence the argument that most mergers go through a starburst phase (e.g. Schewizer 2005).

More recent studies have begun to show that merging systems may not always cause starbursts. Knapen *et al.* (2015) have found that the increase in SFR in merging galaxies is at most a factor of two in the local universe, with the majority of galaxies showing no evidence of an enhance SFR or showing evidence of merger driven quenching. However, mergers can still cause starbursts with the majority of starbursts being caused by galaxy mergers (Luo *et al.* 2014; Knapen *et al.* 2015; Cortijo-Ferrero *et al.* 2017).

The main sequence of star forming galaxies (MS) is a tight correlation between the stellar mass ($M_\star$) and SFR of star forming galaxies (e.g. Brinchmann *et al.* 2004, Noeske *et al.* 2007; Speagle *et al.* 2014). The MS has been shown to be present out to at least $z = 6$, although the presence, or not, of a high mass turnover is still debated (e.g. Tomczak *et al.* 2016; Pearson *et al.* 2018). Whatever the form it takes, the majority of galaxies are found to lie on the MS, with high SFR objects and starbursts above and quenched or quiescent objects below. Thus, if all merging galaxies are starbursts, all merger galaxies should lie above the MS, otherwise they will be scattered around the SFR-$M_\star$ plane.

Deep learning neural networks are a rapidly growing technique that can be used to classify galaxies. These techniques use layers of non-linear mathematical functions that are inspired by the functioning of biological networks of neurons. Once properly trained, a neural network can classify thousands of galaxies in a fraction of the time it would take a human, or team of humans, to perform the same task. Deep neural networks are increasingly used in astronomy to perform tasks such as determining galaxy morphology (e.g. Huertas-Company *et al.* 2015), detecting lensed systems (e.g. Petrillo *et al.* 2017) and detecting galaxy mergers (e.g. Ackermann *et al.* 2018).

2. Stellar Masses and Star Formation Rates

To populate the SFR-$M_\star$ plane, we need SFRs and $M_\star$s for the objects. For the SDSS data release 7 galaxies with spectroscopic redshifts between 0.005 and 0.1, the five SDSS bands (ugriz) were run through CIGALE (Noll *et al.* 2009; Boquien *et al.* 2018) using the same configuration as Pearson *et al.* (2018) but removing the active galactic nuclei component to generate $M_\star$, and U-V and V-J colour estimates. The SFR estimates were derived from the SDSS Hα spectral lines but still need to be corrected for extinction.

We intend to improve the $M_\star$ and SFR estimates, from CIGALE, over the region of SDSS that coincides with the Galaxy And Mass Assembly survey (GAMA; Driver *et al.* 2009). The GAMA region contains data from the ultra-violet (UV) to the far-infrared (FIR), allowing better constraints on the stellar mass and the thermal dust emission than just using the five SDSS bands. We will improve the FIR data in the GAMA fields by following Pearson *et al.* (2017) to de-blend the *Herschel* SPIRE (Griffin *et al.* 2010) data. This involves running CIGALE with the UV to near-infrared data to predict the flux density in the three SPIRE bands and using these predictions as a prior in XID+ (Hurley *et al.* 2017) to de-blend the SPIRE images. With more reliable *Herschel* SPIRE data from this technique, we can generate more reliable SFR estimates.

3. Identifying Mergers

To identify merging systems, a convolutional neural network (CNN) was employed. This type of neural network uses a series of kernels to identify features in the input image and create classifications, in this case if the galaxy is merging or not. The network used is based on the Dieleman *et al.* (2015) architecture with some changes to the parameters of the four convolutional layers and two fully connected layers, as well as changing the input to be a 64×64 pixel, 3 colour channel image. To train the network, the 3003 Sloan Digital Sky Survey (SDSS; York *et al.* 2000) objects classified as merging by Darg *et al.* (2010a,b) were used along with 3003 randomly selected galaxies that had Galaxy Zoo merger classification rate of less than 0.2 and fell within the same redshift range (Lintott *et al.* 2011; Ackermann *et al.* 2018). After training, this resulted in an accuracy (percentage of objects that are correctly classified) of 92.5%. The network was then applied to over 340 000 SDSS galaxies with redshifts between $z = 0.005$ and $z = 0.1$, categorising them in approximately two hours. Approximately 19% of these objects are preliminarily classified as being merging systems, twice the merger fraction seen in other observational works (Man *et al.* 2016).

4. The Main Sequence of Starforming Galaxies

To find the trend of the MS for this data set, we follow Pearson *et al.* (2018). The quiescent galaxies are split off using a UVJ colour cut and the 90% mass completeness limit determined empirically by following Pozzetti *et al.* (2010) and using predicted Ks-band magnitudes from CIGALE. A simple power law is fitted to the remaining objects using a Markov Chain Monte Carlo routine to simultaneously fit the slope, normalisation and

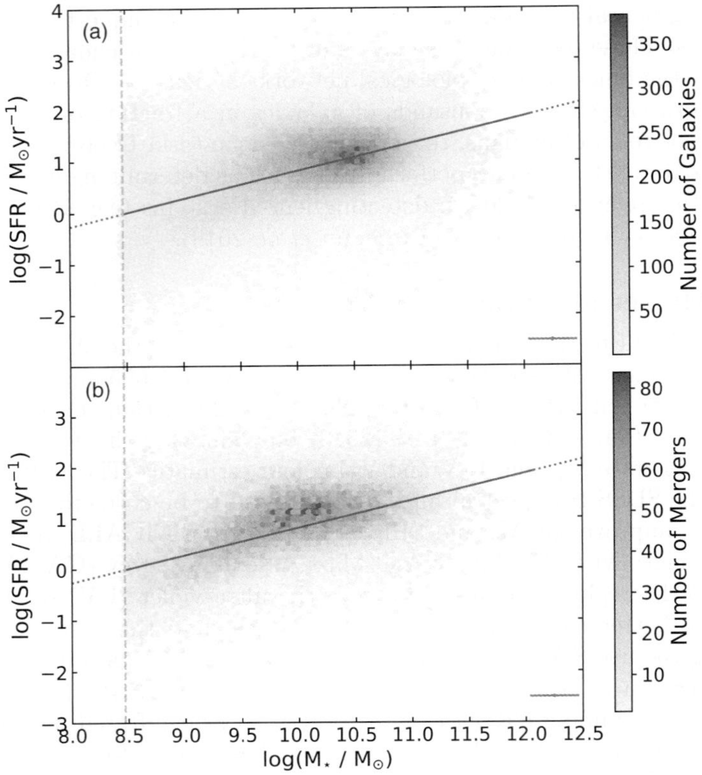

Figure 1. $M_\star$-SFR plane as a number density plot from low (light) to high (dark) with (a) all SDSS objects with detections (at any confidence level) of Hα and Hβ lines above the mass limit (vertical dashed line) and (b) only the objects from (a) that are identified as mergers. The solid line is the main sequence trend, with extrapolations to higher and lower mass as a dotted line. As can be seen, the merging systems closely follow the entire galaxy population, perhaps with a slight rise, and are not preferentially starbursting or quenched.

scatter of the MS, taking into account the observational errors of both the SFR and $M_\star$. The resulting trend is plotted in Fig. 1 as a solid line.

To determine if all mergers are starbursts, we present two plots. Fig. 1a shows the SFR-$M_\star$ plane populated with all SDSS galaxies that have detections (at any confidence level) of Hα and Hβ lines above our mass completeness limit. As can be seen the MS trend runs through the high number density regions. In Fig. 1b we also plot the SFR-$M_\star$ plane populated only objects in Fig. 1a that our CNN identifies as mergers. As can be seen, there is no excess of galaxies above the MS, as would be expected if every merger caused a starburst at the time of observation. Equally, there is no clumping of objects below the MS, showing no evidence that all mergers trigger quenching. Indeed, the merging systems follow the distribution of the entire galaxy population, perhaps with a slight rise in SFR for the merger population. With this preliminary result, there is no evidence that all merging systems are starbursts.

5. Ongoing Work

We are currently working to train a neural network with simulated SDSS images from the EAGLE simulation (Schaye *et al.* 2015). This network will be applied to the observed SDSS images, allowing us to test for possible biases and incompleteness in human classification: galaxies that are classed by humans as merger or non-merger but are given the opposite class by the EAGLE trained network. This also has applications for the upcoming large surveys, such as *Euclid* (Laureijs *et al.* 2011) and the Large Synoptic Survey Telescope (LSST Science Collaboration *et al.* 2009). If this technique proves to function as well as, or better than, an observationally trained network, it will reduce the need to first create a sample of galaxies for humans to classify for training a network. The simulation trained network can immediately be applied to the data from the large surveys and create data sets for science quickly.

Similarly, the simulated images can be passed through the observation trained network. This will allow us to check that the simulation data are representative of the observed universe. Objects from simulations that are classified as mergers but the observation trained network classifies as non-mergers can be identified. Understanding why these objects are rejected as mergers by observations could provide insight into shortcomings of the simulations and inform future work simulating our Universe.

References

Ackermann, S., Schawinski, K., Zhang, C., Weigel, A., & Turp, M. 2018, *MNRAS*, 479, 415
Boquien, M., Burgarella, D., Roehlly, Y., *et al.* 2018, *ArXiv* arXiv:1811.03094
Brinchmann, J., Charlot, S., White, S. D. M., *et al.* 2004, *MNRAS*, 351, 1151
Cortijo-Ferrero, C., González Delgado, R. M., Pérez, E., *et al.* 2017, *A&A*, 607, A70
Darg, D. W., Kaviraj, S., Lintott, C. J., *et al.* 2010a, *MNRAS*, 401, 1552
Darg, D. W., Kaviraj, S., Lintott, C. J., *et al.* 2010b, *MNRAS*, 401, 1043
Dieleman, S., Willett, K. W., & Dambre, J. 2015, *MNRAS*, 450, 1441
Driver, S. P., Norberg, P., Baldry, I. K., *et al.* 2009, *Astronomy and Geophysics*, 50, 5.12
Griffin, M. J., Abergel, A., Abreu, A., *et al.* 2010, *A&A*, 518, L3
Huertas-Company, M., Gravet, R., Cabrera-Vives, G., *et al.* 2015, *ApJS*, 221, 8
Hurley, P. D., Oliver, S., Betancourt, M., *et al.* 2017, *MNRAS*, 464, 885
Knapen, J. H. & Cisternas, M. 2015, *ApJ*, 807, L16
Knapen, J. H., Cisternas, M., & Querejeta, M. 2015, *MNRAS*, 454, 1742
Laureijs, R., Amiaux, J., Arduini, S., *et al.* 2011, *ArXiv* arXiv:1110.3193
Lintott, C., Schawinski, K., Bamford, S., *et al.* 2011, *MNRAS*, 410, 166
LSST Science Collaboration, Abell, P. A., Allison, J., *et al.* 2009, *ArXiv* arXiv:0912.0201
Luo, W., Yang, X., & Zhang, Y. 2014, *ApJ*, 789, L16
Man, A. W. S., Zirm, A. W., & Toft, S. 2016, *ApJ*, 830, 89
Noeske, K. G., Weiner, B. J., Faber, S. M., *et al.* 2007, *ApJ*, 660, L43
Noll, S., Burgarella, D., Giovannoli, E., *et al.* 2009, *A&A*, 507, 1793
Pearson, W. J., Wang, L., Hurley, P. D., *et al.* 2018, *A&A*, 615, A146
Pearson, W. J., Wang, L., van der Tak, F. F. S., *et al.* 2017, *A&A*, 603, A102
Petrillo, C. E., Tortora, C., Chatterjee, S., *et al.* 2017, *MNRAS*, 472, 1129
Pozzetti, L., Bolzonella, M., Zucca, E., *et al.* 2010, *A&A*, 523, A13
Sanders, D. B. & Mirabel, I. F. 1996, *ARA&A*, 34, 749
Schaye, J., Crain, R. A., Bower, R. G., *et al.* 2015, *MNRAS*, 446, 521
Schweizer, F. 2005, in Astrophysics and Space Science Library, Vol. 329, Starbursts: From 30 Doradus to Lyman Break Galaxies, ed. R. de Grijs & R. M. González Delgado, 143
Speagle, J. S., Steinhardt, C. L., Capak, P. L., & Silverman, J. D. 2014, *ApJS*, 214, 15
Tomczak, A. R., Quadri, R. F., Tran, K.-V. H., *et al.* 2016, *ApJ*, 817, 118
York, D. G., Adelman, J., Anderson, Jr., J. E., *et al.* 2000, *AJ*, 120, 1579

Discussion

FABIO FONTANOT: What is the fraction of merging galaxies as a function of stellar mass?

William J. Pearson: This is not something we have looked at yet but it would be interesting to see. (A quick look after the talk shows that the merger fraction decreases as stellar mass increases).

Minju Lee: It may be worth to separate the sample into two groups of gas-rich and gas-poor to test if gas-rich mergers cause SB while gas-poor don't.

William J. Pearson: Yes, it would indeed be interesting to split by gas-rich/poor to see where they lie. I have not done this yet but would be interesting.

Challenges in Panchromatic Modelling with Next Generation Facilities
Proceedings IAU Symposium No. 341, 2018
M. Boquien, E. Lusso, C. Gruppioni & P. Tissera, eds.

doi:10.1017/S1743921319002576

Automatic classification of sources in large astronomical catalogs

Agnieszka Pollo[1,2], Aleksandra Solarz[2], Małgorzata Siudek[3,4,2], Katarzyna Małek[2,5], Maciej Bilicki[6,4], Tomasz Krakowski[2], Tsutomu Takeuchi[7] and the VIPERS team[8]

[1]Astronomical Observatory of the Jagiellonian University, ul. Orla 171, 30-001 Cracow, Poland
email: agnieszka.pollo@gmail.com

[2]National Center for Nuclear Research,ul. A. Sołtana 7, 05-400 Otwock, Poland

[3]IFAE, The Barcelona Institute of Science and Technology, 08193 Bellaterra (Barcelona), Spain

[4]Center for Theoretical Physics, PAS, al. Lotników 32/46, 02-668, Warsaw, Poland

[5]Aix Marseille Univ. CNRS, CNES, LAM Marseille 13388, France

[6]Leiden Observatory, Leiden University, P.O. Box 9513, NL-2300 RA Leiden, The Netherlands

[7]Nagoya University, Furo-Cho, Chikusa-ku, Nagoya 464-8602, Japan

[8]listed at the end of this proceedings

Abstract. In this paper we address two questions related to data analysis in large astronomical datasets, and we demonstrate how they can be answered making use of machine learning techniques. The first question is: how to efficiently find previously unknown or rare objects which can be expected to exist in big data samples? Using the largest existing extragalactic all-sky survey, provided by the WISE satellite, we demonstrate that, surprisingly, *supervised* classification methods can come to aid. The second question is: having a sufficiently large data sample, how can we look for new optimal classification schemes, possibly finding new and previously unknown classes and subclasses of sources? Based on the VIPERS cutting-edge galaxy catalog at redshift $z > 0.5$, we demonstrate that *unsupervised* classification methods can give unexpected but physically well-motivated results.

Keywords. surveys, galaxies: statistics, quasars: general

1. Introduction

Currently, astronomy is dealing with increasingly larger data samples, and even bigger data, like those from the Large Synoptic Survey Telescope (LSST), are coming soon. Such data open new possibilities: both for discoveries of previously unknown rare classes of objects and for understanding the global properties of the known sources, as they provide samples allowing for much more refined statistical treatment. This new situation on the market of astronomical data, coupled with increasing capabilities of modern computers, resulted in a boost of different machine learning and data mining methods applied in the field of astronomy over the last few years.

Very broadly speaking, machine learning based classification methods can be divided into two main families: *supervised* methods, where we know a priori what sources we expect to find and we use some known datasets to train algorithms to look for them, and *unsupervised* methods which look for separate clusters or groups in the data based on similarity of their properties in a given feature space. In this paper, we present two examples of applications of *semi-supervised* (a hybrid between supervised and unsupervised) and

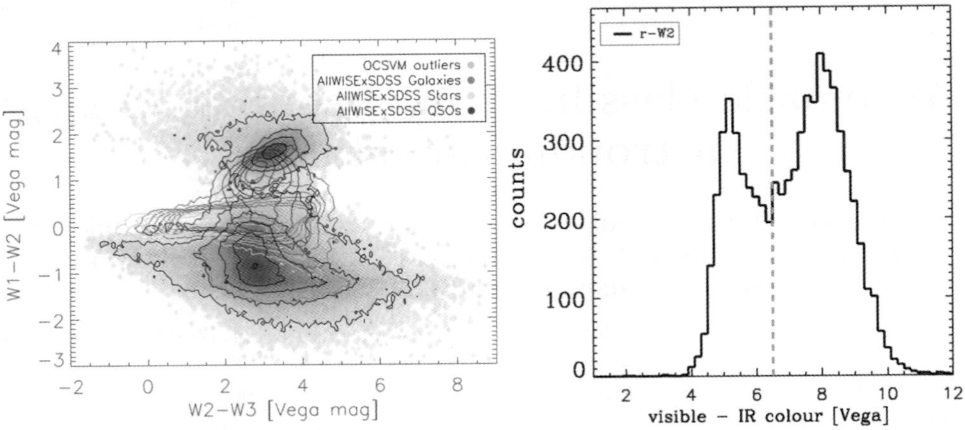

Figure 1. Classification of the AllWISE sources by the OCSVM method. *Left panel*: infrared color-color $W1 - W2$ vs $W2 - W3$ distribution of AllWISE sources: in addition to objects following the pattern of known galaxies, stars and AGN, a large number of anomalous sources was found. *Right panel*: histogram of the visible-infrared $r - W2$ color of ~7,000 anomalous objects identified as genuine astrophysical sources with counterparts in the photometric part of the SDSS catalog; the preliminary analysis indicates that they form two groups of dusty (redder) and non-dusty (bluer) AGN.

unsupervised machine learning techniques to search for new unknown classes of sources and new classification schemes for known (and unknown) sources.

2. Supervised methods in search for novel sources

The largest presently existing all-sky extragalactic catalog was created based on the near- to mid-infrared data gathered by the Wide-field Infrared Survey Explorer (WISE, Wright *et al.* 2010) in four passbands ($W1$, $W2$, $W3$, $W4$) centered at 3.4, 4.6, 12 and 23 μm, respectively. It contains over 747 million sources in its AllWISE data release. We can reasonably expect that among such a wealth of data new, rare and previously unclassified categories of objects should be contained. The question is – how to find them?

Solarz *et al.* (2017) applied a *semi-supervised* machine learning method, called one-class support vector machines (OCSVM, Schölkopf *et al.* 2000), in order to separate all *known* classes of objects from those which do not fit to any known pattern in the considered parameter space. A model of expected data was trained based on 2.6 million sources present in the spectroscopic SDSS DR13 database, and found also in AllWISE. As shown in the left panel of Fig. 1, among the AllWISE data with no spectroscopic SDSS counterparts, the OCSVM algorithm found a large number of sources fitting well the pattern corresponding to known stars, galaxies and quasars. However, it also identified two big groups of *anomalous* sources. The largest among these two groups was found to contain mainly artifacts, such as objects with spurious photometry due to blending. It should be stressed that these artifacts were not detected previously by any other cleaning algorithms. Thus, the OCVSM method was proven to be an efficient tool for cleaning large catalogs. However, even more importantly, a second, smaller group of anomalies appeared to contain ~40,000 real sources of genuine astrophysical interest, among them a sample of heavily reddened Active Galactic Nuclei (AGN) / quasar candidates distributed uniformly over the sky and in a large part absent from other WISE-based AGN catalogs. The $r - W2$ color distribution of 7,000 among these sources having counterparts in the photometric part of the SDSS survey is presented in the right panel of Fig. 1; two groups

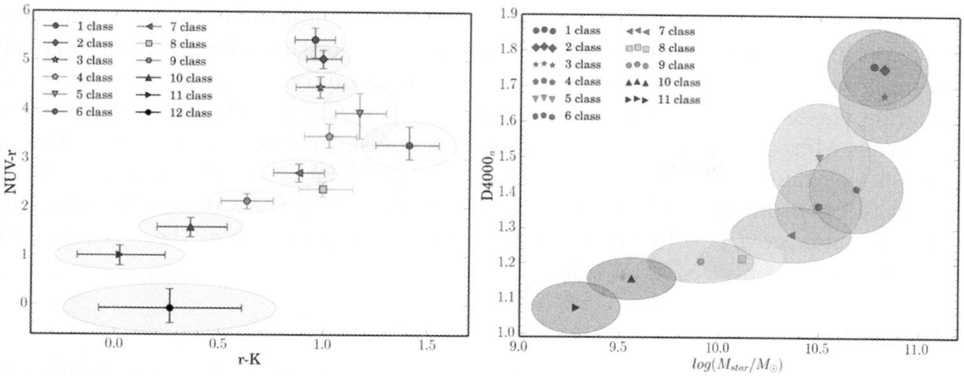

Figure 2. Classification of the VIPERS galaxies by the FEM algorithm. *Left panel*: projection of the FEM classes on the $NUV - r$ vs $r - K$ color-color diagram. The error bars correspond to the first and third quartile of the distribution, while the semi-axes of the ellipses to the median absolute deviation. *Right panel*: median value of the $D4000$ spectral feature as a function of a median stellar mass of the FEM classes. Again, the two semi-axes of the ellipses correspond to the median absolute deviation of the distributions.

of "red" and "blue" sources presumably correspond to obscured and unobscured AGN, respectively. The follow-up observations of these sources are now on-going.

3. Unsupervised methods in search for new classification schemes

The VIMOS Public Extragalactic Redshift Survey (VIPERS; Scodeggio *et al.* 2018) contains ~90,000 spectroscopically measured galaxies at $z \geqslant 0.5$, covering a large volume of ~5x$10^7 h^{-3}$Mpc^3 with an effective spectroscopic sampling (~46%), which makes it a state-of-the-art equivalent of local surveys but at $z \sim 1$.

Siudek *et al.* (2018a) applied a Fisher Expectation-Maximization (FEM) *unsupervised* algorithm (Bouveyron & Brunet 2018) to classify VIPERS galaxies in a parameter space of 12 rest-frame magnitudes and spectroscopic redshift. The FEM algorithm has automatically distinguished 12 classes: 11 classes of galaxies at $0.5 \leqslant z \leqslant 1.2$ and an additional class of outliers consisting mostly of broad-line AGNs. The first broad division was into red (passive), green (intermediate), and blue (star-forming) galaxy populations. A further sub-division yielded three red, three green, and five blue galaxy classes. A joint analysis based on standard statistical criteria (BIC, AIC, ICL), a flow chart of the subsequent divisions performed by the algorithm in the consecutive steps, and a posteriori checks of physical properties of galaxies in resultant FEM classes, allowed for the conclusion that the 11 galaxy classes obtained reflect indeed different galaxy subpopulations, and that the transition of galaxy properties between subsequent classes is not continuous. In other words, the FEM classes can be treated as physically different subcategories of galaxies. The differences between classes are seen not only in galaxy rest-frame magnitudes or colors, but they are also reflected by the properties which were not used for classification neither directly nor indirectly, like spectral lines, shapes or sizes. For example, one among the classes of passive galaxies contains galaxies significantly more compact than two other red classes. As shown in Fig. 2, FEM classes follow the sequence from the earliest to the latest types, which is reflected both by their colors (left panel), and physical as well as spectroscopic properties (right panel).

Can such a fine classification be reproduced based on photometric data only? The question is timely, having in mind huge photometric sky surveys which are expected in the near future, like the LSST. Siudek *et al.* (2018b) extended the previous FEM classification of VIPERS galaxies this time making use of *photometric* redshifts only, and corresponding rest-frame magnitudes based on these photometric redshifts and apparent magnitudes. Under these conditions, the FEM algorithm effectively distinguished three main red, intermediate, and blue galaxy classes, and the agreement with the corresponding classes based on the spectroscopic data was very high: 92%, 84% and 96%, respectively. Moreover, most of the subclasses were also successfully recovered. The only exception was one of the star-forming classes, containing dusty star-forming galaxies, which in the photometric classification merged with other groups of star-forming galaxies. It was thus proven to be the most sensitive to the accuracy of the recovery of the rest-frame color properties of galaxies.

4. Summary

As we have shown, both supervised and unsupervised machine learning methods give great promise for providing proper understanding of source properties in the future sky surveys. Combining these methods will become a powerful tool for future data analysis. It can allow first for cleaning the data and then for finding novel sources which do not correspond to any known pattern – a semi-supervised algorithm such as OCSVM can add flexibility and reliability to automated source separation procedures. In the next step, unsupervised methods like FEM can be used for efficient and robust classification of already known and, even more importantly, novel samples of sources.

Acknowledgements

This research has been partially supported by the Polish NCN grants 2017/26/A/ST9/00756; 2016/23/N/ST9/02963; 2018/30/M/ST9/00757 and MNiSW grant DIR/WK/2018/12.

References

Bouveyron, C. & Brunet, C. 2012, *Statistics and Computing*, 22, 301
Schölkopf, B., Williamson, R., Smola, A., *et al.* 2000, *Adv. Neural Inf. Process. Syst.*, 582–588
Scodeggio, M., Guzzo, L., Garilli, B., *et al.* 2018, *Astronomy & Astrophysics*, 609, A84
Siudek, M., Małek, K., Pollo, A., *et al.* 2018a, *Astronomy & Astrophysics*, 617, A70
Siudek, M., Małek, K., Pollo, A., *et al.* 2018b, arXiv:1805.09905
Solarz, A., Bilicki, M., Gromadzki, M., *et al.* 2017, *Astronomy & Astrophysics*, 606, A39
Wright, E. L., Eisenhardt, P. R. M., Mainzer, A. K., *et al.* 2010, *Astronomical Journal*, 140, 6, 1868

Discussion

Fumi Egusa: How are different classes different? Are they really discrete groups or part of a continuously/smoothly changing properties?

Agnieszka Pollo: Different classes selected by the FEM algorithm seem to be indeed well separated in the multidimensional feature space. In particular, most galaxies have very clear class assignment with high probability of belonging to a given class and low second best class membership probability; different classes gather galaxies of different physical properties: for instance, one of the red classes gathers galaxies with UV upturn; one of the intermediate classes contains very dusty objects. Thus, the change in properties between different classes is not smooth.

Fumi Egusa: Comparison with $z = 0$ would be interesting.

Agnieszka Pollo: Yes, indeed, and we are working on such a comparison right now.

The VIPERS Team: U. Abbas[5], C. Adami[6], S. Arnouts[6], J. Bel[6,8], M. Bolzonella[9], D. Bottini[18], E. Branchini[10,11,12], A. Cappi[9,13], J. Coupon[14], O. Cucciati[15,9], I. Davidzon[6,9], G. De Lucia[16], S. de la Torre[6], P. Franzetti[18], B. Garilli[18], B. R. Granett[8], L. Guzzo[8,17], O. Ilbert[6], A. Iovino[8], J. Krywult[30] V. Le Brun[6], O. Le Fèvre[6], D. Maccagni[18], F. Marulli[15,19,9], H. J. McCracken[20], M. Polletta[18], L. A. .M. Tasca[6], R. Tojeiro[23], D. Vergani[24,9], A. Zanichelli[25], A. Burden[23], C. Di Porto[9], A. Marchetti[26,8], C. Marinoni[27,12,28], L. Moscardini[15,19,9], J. A. Peacock[29], M. Scodeggio[18], G. Zamorani[9]

[5] INAF - Osservatorio Astronomico di Torino, 10025 Pino Torinese, Italy,

[6] Aix Marseille Universit'e, CNRS, LAM (Laboratoire d'Astrophysique de Marseille) UMR 7326, 13388, Marseille, France;

[7] Canada-France-Hawaii Telescope, 65–1238 Mamalahoa Highway, Kamuela, HI 96743, USA,

[8] INAF - Osservatorio Astronomico di Brera, Via Brera 28, 20122 Milano, via E. Bianchi 46, 23807 Merate, Italy,

[9] INAF - Osservatorio Astronomico di Bologna, via Ranzani 1, I-40127, Bologna, Italy,

[10] Dipartimento di Matematica e Fisica, Università degli Studi Roma Tre, via della Vasca Navale 84, 00146 Roma, Italy,

[11] INFN, Sezione di Roma Tre, via della Vasca Navale 84, I-00146 Roma, Italy,

[12] INAF - Osservatorio Astronomico di Roma, via Frascati 33, I-00040 Monte Porzio Catone (RM), Italy,

[13] Laboratoire Lagrange, UMR7293, Université de Nice Sophia Antipolis, CNRS, Observatoire de la Côte dAzur, 06300 Nice, France,

[14] Astronomical Observatory of the University of Geneva, ch. d'Ecogia 16, 1290 Versoix, Switzerland,

[15] Dipartimento di Fisica e Astronomia - Alma Mater Studiorum Università di Bologna, viale Berti Pichat 6/2, I-40127 Bologna, Italy,

[16] INAF - Osservatorio Astronomico di Trieste, via G. B. Tiepolo 11, 34143 Trieste, Italy,

[17] Dipartimento di Fisica, Università di Milano-Bicocca, P.zza della Scienza 3, I-20126 Milano, Italy,

[30] INAF - Istituto di Astrofisica Spaziale e Fisica Cosmica Milano, via Bassini 15, 20133 Milano, Italy,

[19] INFN, Sezione di Bologna, viale Berti Pichat 6/2, I-40127 Bologna, Italy,

[20] Institute d'Astrophysique de Paris, UMR7095 CNRS, Université Pierre et Marie Curie, 98 bis Boulevard Arago, 75014 Paris, France,

[21] Universitätssternwarte München, Ludwig-Maximillians Universität, Scheinerstr. 1, D-81679 München, Germany,

[22] Max-Planck-Institut für Extraterrestrische Physik, D-84571 Garching b. München, Germany,

[23] Institute of Cosmology and Gravitation, Dennis Sciama Building, University of Portsmouth, Burnaby Road, Portsmouth, PO1 3FX,

[24] INAF - Istituto di Astrofisica Spaziale e Fisica Cosmica Bologna, via Gobetti 101, I-40129 Bologna, Italy,

[25] INAF - Istituto di Radioastronomia, via Gobetti 101, I-40129, Bologna, Italy,

[26] Università degli Studi di Milano, via G. Celoria 16, 20130 Milano, Ital,

[27] Aix Marseille Université, CNRS, CPT, UMR 7332, 13288 Marseille, France,

[28] Université de Toulon, CNRS, CPT, UMR 7332, 83957 La Garde, France,

[29] SUPA, Institute for Astronomy, University of Edinburgh, Royal Observatory, Blackford Hill, Edinburgh EH9 3HJ, UK

[30] Institute of Physics, Jan Kochanowski University, ul. Swietokrzyska 15, 25-406 Kielce, Poland

Challenges in Panchromatic Modelling with Next Generation Facilities
Proceedings IAU Symposium No. 341, 2018
M. Boquien, E. Lusso, C. Gruppioni & P. Tissera, eds.

doi:10.1017/S1743921319002515

Predicting the global far-infrared emission of galaxies

Wouter Dobbels and Maarten Baes

Sterrenkundig Observatorium, Universiteit Gent, Krijgslaan 281, B-9000 Gent, Belgium
email: wouter.dobbels@ugent.be

Abstract. Dust absorbs stellar emission and reradiates this energy in the far-infrared (FIR). FIR observations hence give us a direct view of the dust, and allow us to study its properties. Unfortunately, FIR observations are only available for a small subset of galaxies. In this work, we estimate the global FIR emission from global UV-NIR observations. We show that a machine learning method clearly outperforms a SED modelling approach. For each galaxy, we not only predict the FIR flux across the 6 Herschel bands, but also estimate individual uncertainties. We inspect the worst predictions, and investigate how the machine learning predictor generalizes on new data. Our predictor can be used as a virtual observatory, which is especially useful now that there is still no confirmed next-generation FIR telescope.

Keywords. infrared: galaxies, infrared: ISM, galaxies: ISM

1. Introduction

When studying galaxies through broadband photometry, two main components arise. At ultraviolet (UV) to near-infrared (NIR) wavelengths, the stellar emission dominates. This emission is partially absorbed by dust, heating the dust. The dust then emits thermally at longer wavelengths. Very small grains are stochastically heated, and dominate at mid-infrared (MIR) wavelengths. Larger grains are in thermal equilibrium, and they emit at FIR wavelengths. We can learn some things about the dust through UV-NIR attenuation (e.g., Charlot & Fall 2000; De Geyter *et al.* 2014; Leja *et al.* 2017), however there is some degeneracy between attenuation, stellar age and metallicity (Bell & de Jong 2001). The FIR can be used to break this degeneracy, and gives a direct, more accurate estimate of properties such as dust mass and effective temperature (see Conroy 2013). By assuming energy balance, the total IR luminosity L_{IR} equals the total attenuated at shorter wavelengths. Hence, FIR observations are not only useful to study dust, but help constrain other properties as well (Leja *et al.* 2017).

Unfortunately, the FIR is hard to observe. Ground observations are not possible, cooling is required to limit the thermal noise of the instrument itself, and the angular resolution is a lot worse than for optical telescopes. The Herschel Space Observatory stopped operations in 2013 after cooling ran out, and a follow-up mission is still not confirmed. Meanwhile, excellent optical and NIR telescopes are upcoming, including JWST, LSST, Euclid and WFIRST. For many galaxies, both nearby and at larger redshift, we have UV-NIR observations but no FIR observations.

The goal of this work is to predict the FIR from UV-NIR observations. We know the two are related through the dust energy balance. However, even if the total attenuation can be recovered perfectly, it only constrains L_{IR} and not the shape of the infrared SED. Dust emission and attenuation are usually modelled independently, but this of course doesn't mean that the two are unrelated.

To achieve this goal, our main result uses machine learning techniques. This ensures that we learn an optimal mapping from input (UV-NIR) to output (FIR), given a large enough training dataset. However, we start with a more traditional SED modelling approach.

2. Dataset and SED modelling

In order to test our methods, we require a set of galaxies with UV to FIR observations. For this, we use two samples. The first is H-ATLAS DR1 (Eales *et al.* 2010; Valiante *et al.* 2016), which covers 3 equatorial fields also observed by the GAMA survey (Driver *et al.* 2011). After limiting these to $z < 0.1$, this sample contains 3 618 galaxies. The second sample is DustPedia (Davies *et al.* 2017; Clark *et al.* 2018), a sample of local ($z < 0.01$) galaxies observed by Herschel. After removing galaxies with contamination and galaxies with insufficient datapoints (minimum of 4 UV-NIR and 3 FIR observations), we have 715 galaxies remaining. For now, the two datasets are combined into a total sample of 4 333 galaxies.

The FIR can be estimated from UV-NIR through Bayesian SED modelling, using stellar population synthesis (SPS). For this, we used the SED fitting code CIGALE (Boquien *et al.* 2018), with a grid of nearly 80 million models (Nersesian et al., in prep). The models are only fit to UV-NIR observations. From this, we extract a Bayesian estimate of each of the 6 Herschel bands (70 μm – 500 μm).

We could compare these predictions against the raw Herschel observations. However, the raw data do not always lead to a realistic SED, and include missing fluxes, negative fluxes (due to sky subtraction) and some outliers (resulting in unrealistic peaks or troughs). We can do better by making use of all information: a full UV-FIR SED fit. For this we use a similar setup as the short wavelength CIGALE fit, but this time using UV-FIR data. We extract the Bayesian estimates for the 6 Herschel bands, which are now much better constrained due to the corresponding observed data points. These serve as our ground truth FIR fluxes: a Bayesian estimate for a realistic SED which makes use of all available datapoints.

We can now compare the predicted values (from the UV-NIR CIGALE fit) to the ground truth values (from the UV-FIR CIGALE fit). Instead of directly comparing the fluxes, we first divide by the WISE 3.4 μm flux, in order to remove the dependency on distance and total luminosity. Then, we take the base 10 logarithm due to the large dynamic range. The comparison of these log normalized fluxes ($\log(F_{\mathrm{FIR}}/F_{3.4})$) is shown in Fig. 1 for each of the 6 Herschel bands.

We find a total root mean squared error (RMSE) of 0.39 dex across the 6 bands. There is a clear correlation between prediction and target, but there are definitely some biases, especially at longer wavelengths. There are two main problems. First: it is hard to estimate the total absorbed (and hence emitted) energy budget from UV-NIR wavelengths, due to the degeneracy discussed earlier. Second: even when this energy budget is known, we don't know the shape of the FIR, since the emission parameters are varied independently of the absorption parameters. The result of these predictions depend on the grid of models chosen. Leja *et al.* (2017) show that better predictions are possible by using more constrained priors. However, this choice of prior depends on the sample, and limits the flexibility.

3. Machine learning

Due to the model limitations described above, it is possible to do better. Machine learning ensures that we learn an optimal mapping between the input (UV-NIR fluxes) and output (FIR fluxes), at least for the training set. The method can then be evaluated

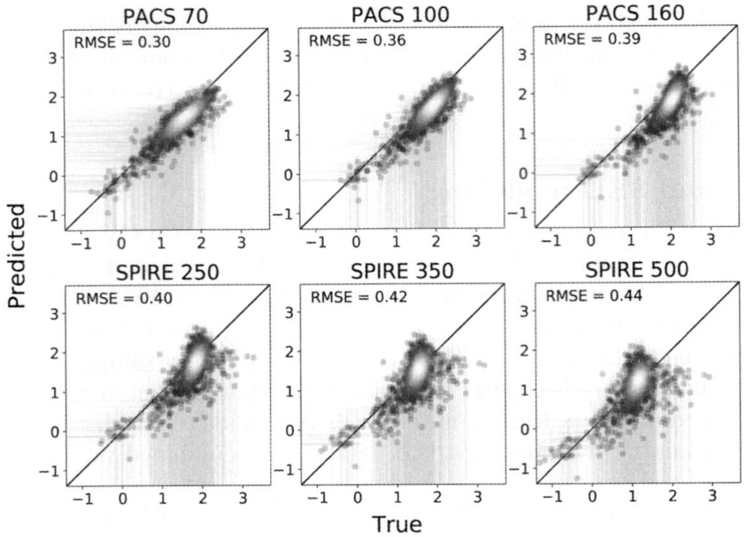

Figure 1. Comparing the SED modelling predictions against the ground truth. Both predicted and true values are Bayesian estimates of the FIR fluxes from CIGALE, and show the log normalized fluxes ($\log(F_{\mathrm{FIR}}/F_{3.4})$). However, the predicted values only make use of UV-NIR data, while the true values make use of UV-FIR data.

on unseen data: the test set. We (randomly) set aside 25% of the data as a test set. For the input, we again have problems when using the raw fluxes. In particular, machine learning has trouble using missing values and large uncertainties in the input. Hence, we use the UV-NIR CIGALE fit and extract the Bayesian estimates of the 14 UV-NIR fluxes that were observed. Of course, no Herschel observations are used in order to make a prediction. We use a 3-layer neural network with an Adam optimizer (Kingma & Ba 2014), using warm restarts. We pick the model that performed best on a cross-validation set (split from the training set).

The predictions on the test set are shown in Fig. 2. The total RMSE is now 0.20 dex, compared to 0.39 dex for the SED modelling. The result does not show significant biases. Note that we also estimated uncertainties on the predictions. For this we used the methodology of Gurevich & Stuke (2017): we train a second neural network as an uncertainty estimator, by optimizing a joint loss function. This allows us to identify which predictions can be trusted and which can not. When inspecting the worst predictions, we see that these can often be explained. Sometimes, the WISE 12 μm and WISE 22 μm bands are missing, and the Bayesian estimate of these values can mislead the neural network. Some galaxies also have unusual SED shapes, with possible AGN contribution, and did not resemble the training set.

Since machine learning tends to only work on samples that resemble the training set, we investigated how the algorithm performs on different data. When testing on a H-ATLAS sample of $0.1 < z < 0.5$ (training was for $z < 0.1$), we find a total RMSE of 0.27. This is a satisfying result, meaning that there is some room for extrapolation. When training only on H-ATLAS and testing on DustPedia, the total RMSE is 0.40. There are several explanations for this rather large RMSE value. First of all, the two data sets were each reduced in their own way. Moreover, DustPedia is a more local sample and contains less strict criteria on FIR S/N. Especially at longer wavelengths, we find that DustPedia includes galaxies with lower FIR fluxes than available in the H-ATLAS set. Keeping these differences in mind, the neural network also produces satisfying results. The individual

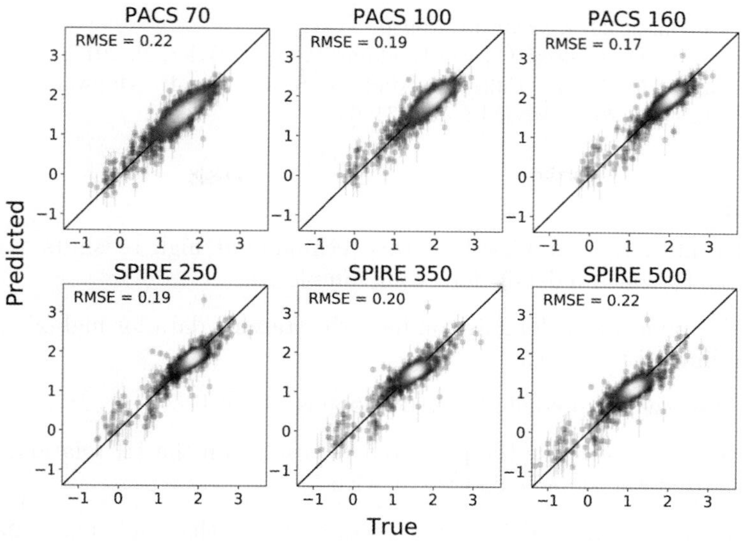

Figure 2. Similar to Fig. 1, but this time showing the neural network predictions against the Bayesian UV-FIR ground truth.

outliers can be inspected, and again it is possible to see in which scenarios the prediction is not trustworthy.

Overall, this neural network gives us an interesting tool to study the FIR. It is possible to explore which UV-NIR properties lead to a certain FIR shape, without being limited to a sample of real galaxies: we can predict for arbitrary UV-NIR input. Our model clearly outperforms SED modelling approaches. Through a predicted uncertainty, we know when the predictions can not be trusted (e.g. when there is large uncertainty on the input). Since we are able to predict a full FIR spectrum from UV-NIR, we are also able to predict related dust properties, such as dust mass and temperature. Since there are currently no confirmed next generation FIR telescopes, we hope these predictions can exploit the excellent upcoming optical/NIR telescopes in order to bridge this gap.

References

Bell, E. F. & de Jong, R. S. 2001, *ApJ*, 550, 212

Boquien, M., Burgarella, D., Roehlly, Y., Buat, V., Ciesla, L., Corre, D. Inoue, A. K., Salas, H., *et al.* 2018, arXiv:1811.03094

Charlot, S. & Fall, S. M. 2000, *ApJ*, 539, 718

Ciesla, L., Boselli, A., Elbaz, D., Boissier, S., Buat, V., Charmandaris, V. Schreiber, C., Béthermin, M., *et al.* 2016, *A&A*, 585, A43

Clark, C. J. R., Verstocken, S., Bianchi, S., Fritz, J., Viaene, S., Smith, M. W. L., Baes, M., Casasola, V., *et al.* 2018, *A&A*, 609, A37

Conroy, C. 2013, *ARAA*, 51, 393

Davies, J. I., Baes, M., Bianchi, S., Jones, A., Madden, S., Xilouris, M., Bocchio, M., Casasola, V., *et al.* 2017, *PASP*, 129, 044102

De Geyter, G., Baes, M., Camps, P., Fritz, J., De Looze, I., Hughes, T. M., Viaene, S., Gentile, G., *et al.* 2014, *MNRAS*, 441, 869

Driver, S. P., Hill, D. T., Kelvin, L. S., Robotham, A. S. G., Liske, J., Norberg, P., Baldry, I. K., Bamford, S. P., *et al.* 2011, *MNRAS*, 413, 971

Eales, S., Dunne, L., Clements, D., Cooray, A., De Zotti, G., Dye, S., Ivison, R., Jarvis, M., *et al.* 2010, *PASP*, 122, 499

Gurevich, P. & Stuke, H. 2017, arXiv:1707.07287

Jones, A. P., Köhler, M., Ysard, N., Bocchio, M., & Verstraete, L. 2017, *A&A*, 602, A46

Kingma, D. P. & Ba, J. 2014, arXiv:1412.6980
Leja, J., Johnson, B. D., Conroy, C., van Dokkum, P. G., & Byler, N. 2017, *ApJ*, 837, 170
Valiante, E., Smith, M. W. L., Eales, S., Maddox, S. J., Ibar, E., Hopwood, R., Dunne, L., Cigan, P. J., *et al.* 2016, *MNRAS*, 462, 3146

Discussion

CALISTRO RIVERA: How applicable is this technique at high redshifts, e.g. to predict ALMA fluxes? To which redshift does your sample go?

DOBBELS: It's applicable as long as you have the training data for high-z and ALMA. In my work I'm limited to $z < 0.5$.

GOTO: It seems like the predicted vs true slope is below 1?

DOBBELS: By eye it looks like the predicted vs true lie on the 1:1 relation, but I can do a fit and hope that the slope is consistent with 1.

BUAT: From the resulting SEDs you showed, it seems that the energy balance is not always preserved. Does this mean that your estimates are not completely physically consistent?

DOBBELS: It is hard to estimate the total absorbed energy (more attenuation and older stars can be confused) from UV-NIR. The same happens when you fit CIGALE to UV-NIR: there will be significant scatter in the predicted $L_{\rm FIR}$ around the true values. Moreover, energy balance does not hold for a single viewing angle: edge-on and face-on have the same $L_{\rm FIR}$ (isotropic), but the edge-on is more attenuated.

LEE: What is the best strategy to choose the number of samples to train?

DOBBELS: This is mainly driven by the test set, where we want decent statistics. I have 4000 total samples, and we used 1000 for testing (25%). I then look at the true vs predicted plot, and this seems to be well sampled. This can be quantified by taking a large number of bootstrap samples of the test set. If the bootstrap distribution of MSE is narrow, you have enough test samples. If you have 10 million total samples, 1% test size is probably enough.

Challenges in Panchromatic Modelling with Next Generation Facilities
Proceedings IAU Symposium No. 341, 2018
M. Boquien, E. Lusso, C. Gruppioni & P. Tissera, eds.

doi:10.1017/S1743921319002813

Mentari: A pipeline to model the galaxy SED using semi analytic models

Dian Triani[1,2], Darren Croton[1,2] and Manodeep Sinha[1,2]

[1]Centre for Astrophysics and Supercomputing, Swinburne University of Technology
PO Box 218, Hawthorn VIC 3122, Australia
email: dtriani@swin.edu.au

[2]ARC Centre of Excellence for All Sky Astrophysics in 3 Dimensions (ASTRO 3D)

Abstract. We build a theoretical picture of how the light from galaxies evolves across cosmic time. In particular, we predict the evolution of the galaxy spectral energy distribution (SED) by carefully integrating the star formation and metal enrichment histories of semi-analytic model (SAM) galaxies and combining these with stellar population synthesis models which we call mentari. Our SAM combines prescriptions to model the interplay between gas accretion, star formation, feedback process, and chemical enrichment in galaxy evolution. From this, the SED of any simulated galaxy at any point in its history can be constructed and compared with telescope data to reverse engineer the various physical processes that may have led to a particular set of observations. The synthetic SEDs of millions of simulated galaxies from mentari can cover wavelengths from the far UV to infrared, and thus can tell a near complete story of the history of galaxy evolution.

Keywords. galaxies: evolution - galaxies: stellar content - galaxies.

1. Introduction

The study of galaxy evolution has a long history, both theoretically and empirically. The general view of how the large-scale structure in the Universe grows includes an understanding of the virialization of dark matter and baryons, gas cooling, star formation, feedback, and mergers. Galaxy properties are driven by various possible scenarios of how such nonlinear processes occur and compete. These properties include the stellar mass, star formation history (SFH), metallicity, initial mass function (IMF), and physical state and quantity of dust and gas (Conroy 2013).

From the theoretical side, semi-analytic models (SAMs) have been widely used to predict the evolution of galaxies. Such models are run as a post-processing step on dark matter only N-body simulations. SAMs consist of a set of coupled differential equations, both empirically and/or physically driven, that describe each physical process believed important. Compared to a hydrodynamical simulation that solves the equations of gravity and fluid dynamics of each particle, SAMs are significantly more computationally efficient and faster at exploring the parameter space.

On the observation side, the spectral energy distribution (SED) of stars and interstellar medium in galaxies encode many of the fundamental aspects of galaxies that are key to studying their evolution. A large number of surveys using different methods have been conducted to measure the SEDs of galaxies, from wide field surveys of the local Universe such as the Sloan Digital Survey (SDSS, York *et al.* 2000), and the Two-degree Field Galaxy Redshift Survey (2dFGRS, Coless *et al.* 2001), to the deep-field survey of

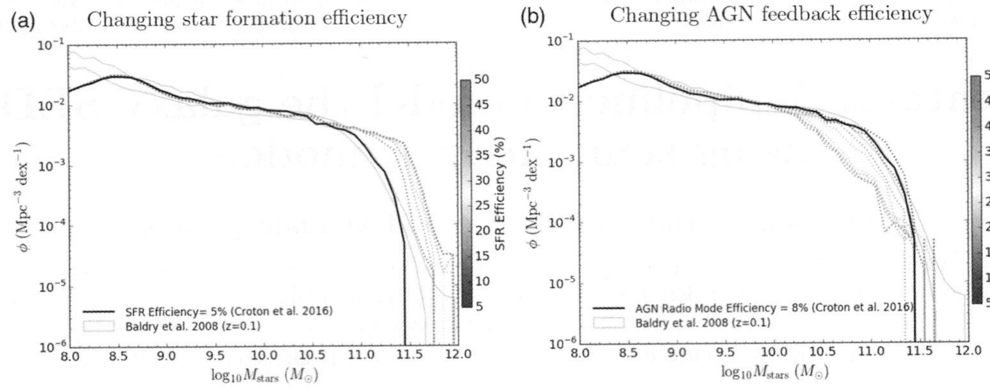

Figure 1. Effect of changing SAGE parameter in the stellar mass function.

the high-redshift universe such as the Deep Extragalactic Evolutionary Probe (DEEP2, Davis *et al.* 2003).

To infer the properties of galaxies from the SED, an early approach involved trial-and-error techniques by combining stellar spectra until they matched the observations (Spinrad & Taylor 1971). Following this, an automated fitting technique that incorporated physical constraints was developed (Faber 1972). These methods actually matured into what has become the standard technique today, known as stellar population synthesis (SPS) modeling; it uses stellar evolution theory to constrain the range of possible stellar populations at a given age and metallicity (Tinsley 1968). Currently, a number sophisticated models of galaxy spectral evolution have been developed for public use (e.g., Bruzual & Charlot 2003; Conroy & Gunn 2010).

This work combines the physically motivated star formation and metallicity histories from a SAM with the widely-applied method of stellar population synthesis. The predicted SED covers wavelengths from the far UV to infrared, will eventually include dust, gas and AGN physics, and thus will be able to tell a complete history of each galaxy and identify their fundamental properties.

2. Semi-analytic model

The galaxy model we use is called Semi-Analytic Galaxy Evolution, or SAGE for short (Croton *et al.* 2006, 2016). SAGE is a modular code written in C and available at https://github.com/darrencroton/sage. We run SAGE on the Millennium simulation of Springel *et al.* (2005). This cosmological simulation follows the growth and interaction of dark matter particles in a box of length $500h^{-1}$ Mpc. There are 2160^3 particles, each with mass $8.6 \times 10^8 h^{-1} M_\odot$. A dark matter halo is defined to have a minimum of 20 particles. The simulation assumes the standard WMAP1 cosmological parameters (Spergel *et al.* 2003).

SAGE consists of coupled equations that track the evolution of baryons inside dark matter halos, including the process of gas cooling, star formation, feedback, mergers, starbursts, and the effect of early Universe reionization. The details of the model can be found in Croton *et al.* (2016). When parameterizing the model, our main objective is to reproduce the stellar mass function at redshift zero. For producing stellar mass, we use the initial mass function of Chabrier (2003). We give an example of how changing the star formation and AGN feedback efficiency can significantly change the stellar mass function in Fig. 1. For this example, we run SAGE on the mini version of Millennium simulation, made for testing purposes. The mini version is identical to the full version but in a smaller box of side length $62.5h^{-1}$ Mpc. We use the stellar mass function presented in

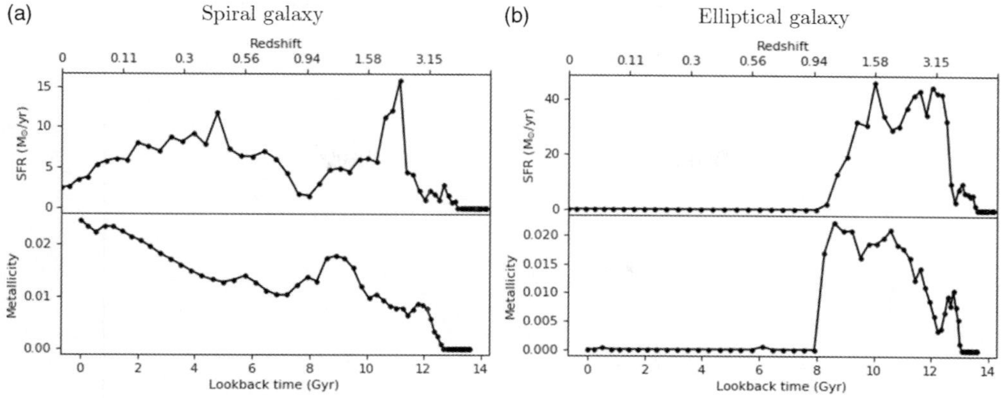

Figure 2. Example star formation and metallicity histories of simulated SAGE galaxies (details are given in the text).

Baldry, Glazebrook & Driver (2008) as the observational benchmark in setting the model parameters.

We compute the star formation and metallicity histories of 36 thousand galaxies from $z = 127$ to $z = 0$ using `SAGE`. Fig. 2 and shows an example of the star formation and metallicity history of a Milky Way like galaxy (morphologically spiral, stellar mass $5.86 \times 10^{10} M_{\odot}$, star formation rate $1.6 M_{\odot} yr^{-1}$) and a typical elliptical galaxy with no current star formation rate and present stellar mass of $9.3 \times 10^{10} M_{\odot}$ in the simulation. We can see the history of the spiral galaxy consists of several starbursts triggered by merger events of varying significance, and it is still undergone the star formation activity at z=0. Conversely, the elliptical galaxy had been dead since a lookback time of 8 Gyr. Using SAGE to build the star formation and metallicity histories of a large number of galaxies gives us the flexibility to customize our subject of interest. For example, it allows us to select galaxies with certain properties or history, explore the star formation activity occurring only in disk or bulge, or even look only at the intra-cluster stars around massive galaxies. The idea is to enable us to build a library of theoretical galaxies to be referenced against their observational counterparts.

3. SED construction pipeline

`Mentari` is an SED generation pipeline that takes semi-analytic galaxy model data as its input. The code is python-based and although still under development, has been made publicly available at https://github.com/dptriani/mentari. It has four simple functions.

• `read_properties(redshift, firstfile, lastfile, directory, filename)`. This function reads the properties of simulated galaxies from the `SAGE` model output.

• `build_history(redshift, firstfile, lastfile, directory, filename)`. This function uses the merger tree to the correct star formation and metallicity history for each galaxy from the desired redshift. Each merger in a tree needs careful treatment because when one occurs, the old stars from the satellite galaxy will merge to the central, while the central may additionally undergo a starbursttriggered by the merger. This function will separate old stars and newly formed stars and make sure they are assigned the correct age.

• `generate_SED(lookbacktime, mass, metallicity)`. This function generates the SED from the star formation and metallicity history calculated in `build_history`.

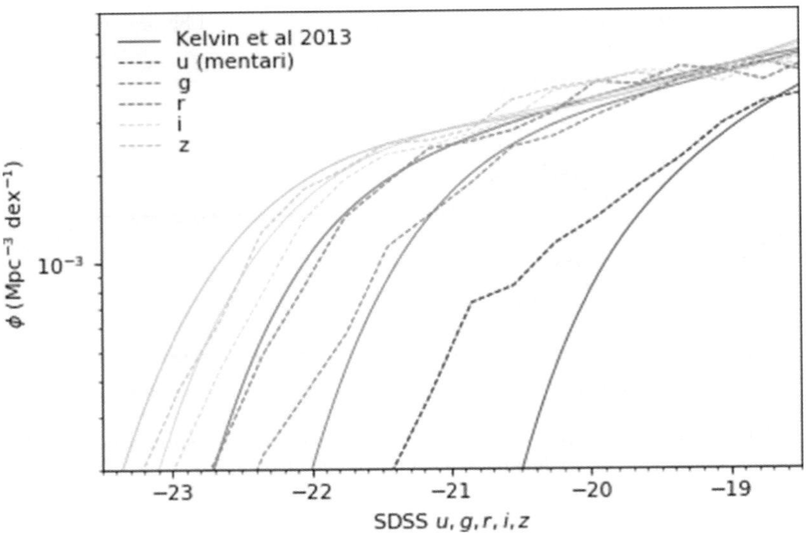

Figure 3. Luminosity functions of simulated galaxies (thick line) and observed galaxies from Kelvin *et al.* 2014. Shown are the SDSS ugriz filters, as indicated in the legend.

• `mab(wavelength, spectra, filter_list, z)`. This function computes the AB magnitudes for the listed filters at the specified redshift. It will produce apparent magnitudes when the redshift is greater than zero, otherwise it assumes the user wants the absolute magnitude.

To construct the SED, `mentari` reads the star formation history and metallicity evolution directly from `SAGE` and uses the Bruzual & Charlot (2003) stellar population synthesis library, although the pipeline will be flexible and allow other libraries. This library contains spectra per unit mass of single stellar populations at ages between 1×10^5 and 2×10^{10} years with metallicities that range from $0.005Z_\odot$ to $5Z_\odot$. The spectra curently cover panchromatic wavelengths from $91\mathring{A}$ to 160μm. For the stellar evolution tracks and initial mass function we use Padova 1994 (Alongi *et al.* 1993; Bressan *et al.* 1993; Fagotto *et al.* 1994a,b; Girardi *et al.* 1996) and Chabrier (2003), respectively.

We provide a set of filters in the pipeline to convolve the SED to AB magnitude. We can create photometric datasets of galaxies at any redshift that can then be compared with observations. Using `mentari`, we construct the SED and the SDSS *ugriz* photometry for 36 thousand galaxies from our model at z=0. We compare with the luminosity function from the GAMA survey (Kelvin *et al.* 2014). This is shown in Fig. 3.

`Mentari` is currently still under development. The future enhancements will include additional stellar population models; more sophisticated dust modelling, especially for the IR part of the spectrum; quasar and radio mode AGN components to the composite SED; emission and absorption lines in the SED; and H_I and H_2 gas components. The modular nature of `mentari` will mean that such enhancements can be included incrementally as the technical resources and scientific needs dictate.

References

Alongi, M. *et al.* 1993, *A&AS*, 97, 851
Bressan, A., Fagotto, F., Bertelli, G., Chiosi, C., *et al.* 1993, *A&AS*, 100, 647
Bruzual, G. & Charlot, S. 2003, *MNRAS*, 344, 1000
Baldry, I. K., Glazebrook, K., & Driver, S. P. 2008, *MNRAS*, 388, 945
Chabrier, G. 2003, *PASP*, 115, 763
Coless, M. *et al.* 2001, *MNRAS*, 328, 1039

Conroy, C. 2013, *ARA&A*, 51, 393
Conroy, C. & Gunn, J. E. 2010, *ApJ*, 712, 833
Croton, D. J. etal 2006, *MNRAS*, 365, 11
Croton, D. J. *et al.* 2016, *ApJS*, 222, 22
Davis, M. *et al.* 2003, *Proc. SPIE*, 4834, 161
Faber, S. M. 1972, *A&A*, 20, 361
Fagotto, F., Bressan, A., Bertelli, G., Chiosi, C., *et al.* 1994a, *A&AS*, 104, 365
Fagotto, F., Bressan, A., Bertelli, G., Chiosi, C., *et al.* 1994b, *A&AS*, 105, 29
Girardi, L., Bressan, A., Chiosi, C., Bertelli, G., Nasi, E., *et al.* 1996, *A&AS*, 117, 113
Kelvin, L. S. *et al.* 2014, *MNRAS*, 439, 1245
Spergel, D. N. *et al.* 2003, *ApJS*, 148, 175
Spinrad, H. & Taylor, B. J. 1971, *ApJS*, 22, 445
Springel, V. *et al.* 2005, *Nature*, 435, 629
Tinsley, B. M. 1968, *ApJ*, 754, L29
York, D. G. *et al.* 2000, *AJ*, 120, 1579

Discussion

FABIO FONTANOT (INAF-OATS): Which kind of IR templates are you implementing to model the IR re-emission?

ANSWER: No one, only pure stellar (Bruzual & Charlot '03)

Challenges in Panchromatic Modelling with Next Generation Facilities
Proceedings IAU Symposium No. 341, 2018
M. Boquien, E. Lusso, C. Gruppioni & P. Tissera, eds.

doi:10.1017/S1743921319002370

Variations of the stellar Initial Mass Function in semi-analytic models: Implications for the mass assembly of galaxies in the GAEA model

Fabio Fontanot

INAF - Astronomical Observatory of Trieste, via G.B. Tiepolo 11, I-34143 Trieste, Italy
email: fabio.fontanot@inaf.it

Abstract. A wealth of observations recently challenged the notion of a universal stellar initial mass function (IMF) by showing evidences in favour of a variability of this statistical indicator as a function of galaxy properties. I present predictions from the semi-analytic model GAEA (GAlaxy Evolution and Assembly), which features (a) a detailed treatment of chemical enrichment, (b) an improved stellar feedback scheme, and (c) implements theoretical prescriptions for IMF variations. Our variable IMF realizations predict intrinsic stellar masses and mass-to-light ratios larger than those estimated from synthetic photometry assuming a universal IMF. This provides a self-consistent interpretation for the observed mismatch between photometrically inferred stellar masses of local early-type galaxies and those derived by dynamical and spectroscopic studies. At higher redshifts, the assumption of a variable IMF has a deep impact on our ability to reconstruct the evolution of the galaxy stellar mass function and the star formation history of galaxies.

Keywords. galaxies: evolution - galaxies: fundamental parameters - galaxies: stellar content

1. Introduction

The IMF quantifies the number of stars formed per stellar mass bin in a given episode of star formation. As such, it represents a key parameter to characterise galaxy evolution from photometric surveys. The IMF has long been assumed to have a universal shape, mainly because of near-invariant results obtained for the closest star forming regions in our Galaxy (see e.g. Kroupa 2001). However, recent observational developments put this notion in question. These evidences came mainly trough two separated channels. A systematic excess of the mass-to-light ratios derived using integral field stellar kinematics with respect to photometric values based on a universal IMF has been reported by many authors (see e.g. Cappellari *et al.*, 2012). Moreover, high-resolution spectroscopy of spectral features sensitive to IMF variations (due of their dependence on stellar surface gravity and/or effective temperature) suggests an excess of low- relative to intermediate-mass stars, compared to the universal IMF, in massive early-type galaxies (ETGs, see e.g. Conroy *et al.*, 2013 and La Barbera *et al.* 2013). In both cases, the reported excess increases with increasing galaxy stellar mass ($M_\star$) and/or velocity dispersion. It is worth stressing that the spectroscopic probe clearly favours a *bottom-heavy* scenario (i.e. an excess of low-mass stars with respect to the universal IMF), but it samples only to the low-mass end of the IMF ($<$1 $M_\odot$). On the other hand, the dynamical probe is sensible to the whole shape of the IMF, but cannot distinguish between a *top-heavy* (i.e. an excess of high-mass stars with respect to the universal IMF) or a *bottom-heavy* scenario. The observational picture is further complicated by the radial IMF gradients reported for local

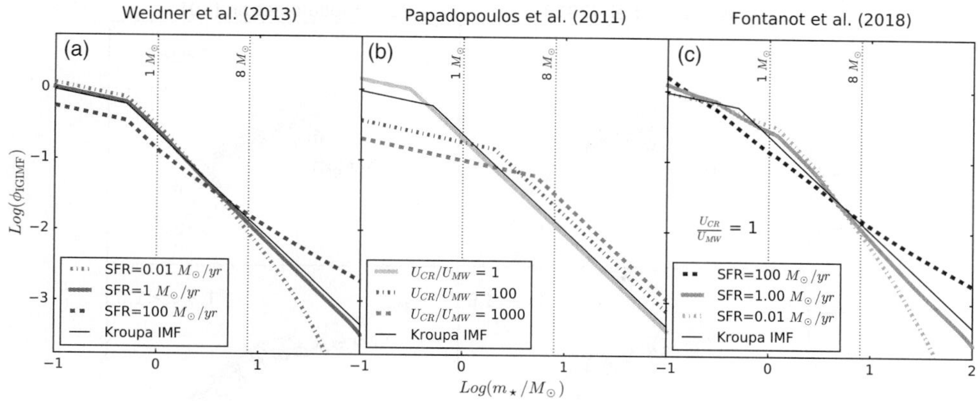

Figure 1. Variable IMF implementations discussed in this talk (in all panels the thin solid line represents the universal IMF in the solar neighbourhood as in Kroupa 2001). *Panel (a):* evolution of the IMF shape as a function of SFR, according to the IGIMF theory Weidner *et al.* 2013. *Panel (b):* evolution of the IMF shape as a function of cosmic ray energy densities, using the numerical results from Papadopoulos *et al.* (2011). *Panel (c):* composite IMF evolution as a function of SFR and cosmic ray energy densities (only the $U_{\rm CR} = U_{\rm MW}$ case is shown) as proposed in Fontanot *et al.* (2018b). In all panels, each IMF is normalized to 1 $M_\odot$ in the stellar mass interval 0.1-100 $M_\odot$.

ETGs using integral field unit spectroscopy (see e.g. Sarzi *et al.* 2018). These findings suggest that the low-mass stars excess increases in the innermost regions of these galaxies. However, these trends stand in apparent contradiction with the well-known metallicity gradients in ETGs: the increasing α-enhancement towards the centre seems to require a larger contribution from Type-II SNe, thus a larger fraction of high-mass stars.

2. Variable IMF implementations in GAEA

In my contribution, I present recent results obtained in the framework of the semi-analytic model GAEA (GAlaxy Evolution and Assembly; Hirschmann, De Lucia & Fontanot, 2016). This state-of-the-art model features key improvements in the treatment of (a) chemical enrichment (i.e. it follows the differential evolution of AGB stars, Type-Ia and Type II SNe) and (b) stellar feedback (i.e. it implements an ejective feedback prescription that combines results of high-resolution hydrodynamical simulations with energy conservation arguments). This model is able to provide a self-consistent picture of galaxy evolution, reproducing the evolution of the galaxy stellar mass function (GSMF) and cosmic star formation rate (SFR) up to $z \sim 8 - 10$ (Fontanot, Hirschmann & De Lucia, 2017b).

In a series of recent papers, we include in GAEA prescriptions that link the variability of the IMF to physical properties of model galaxies. In particular, in Fontanot *et al.* (2017a) we test the integral galaxy-wide IMF theory (IGIMF - Weidner *et al.* 2013 - Fig. 1, *panel (a)*): this analytic model predicts an evolution of the high-mass-end slope of the IMF as a function of the global SFR of galaxies. In Fontanot *et al.* (2018a), we build on the results of numerical simulations from Papadopoulos *et al.* (2011) to relate the position of the knee of the IMF with the cosmic ray density field ($U_{\rm CR}$, normalized to the Milky Way reference value $U_{\rm MW}$) associated with model galaxies (Fig. 1, *panel (b)*). In a variable IMF scenario, it is only possible to calibrate the models against photometric data; in detail we consider $z = 0$ luminosity functions (LF) in the $g-$, $r-$, $i-$, and $K-$band, and the redshift evolution of the K- and $V-$band LFs. In order to compute self-consistent synthetic photometry for our variable IMF realizations, we construct Single Stellar Populations libraries (for different ages and metallicities) corresponding to each assumed IMF shape, using an updated version of Bruzual & Charlot (2003) models. Synthetic photometry

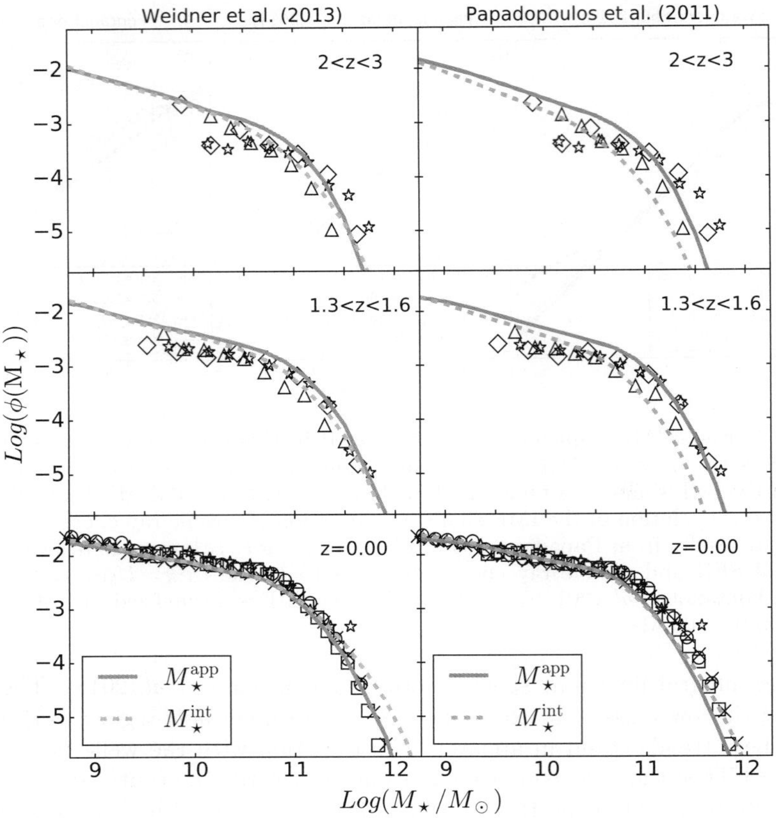

Figure 2. Evolution of the GSMF. In each panel, solid lines correspond to the GSMF as a function of the apparent photometrically-equivalent stellar mass $M_\star^{\rm app}$, while dot-dashed lines show the GSMF as a function of the intrinsic galaxy stellar mass $M_\star^{\rm int}$. Dark points show a collection of observational measurements as in Fontanot *et al.* (2009).

represents the only observable we can directly compare with observational constraints. Indeed, most of the estimates for physical properties of galaxies available in the literature are derived under the hypothesis of a universal IMF, and it could be problematic to compare them with the *intrinsic* properties of model galaxies, as predicted by our variable IMF runs. To better explain this issue, in the following we focus on galaxy stellar masses. We define a photometrically-equivalent *apparent* stellar mass ($M_\star^{\rm app}$) by applying to our synthetic photometry mass-to-light versus colour scalings typically used in photometric surveys (see e.g. Zibetti, Charlot & Rix 2009). In this way, we derive the $M_\star$ an observer would associate to our synthetic photometry assuming a universal IMF.

3. Results

Both our variable IMF realizations provide consistent results in terms of galaxy properties and their evolution. In particular, they are able to naturally solve the long-standing problem of α-enhancement in massive ETG (see e.g. Fig.B1 in Fontanot *et al.* 2018a). We compare the intrinsic stellar masses $M_\star^{\rm int}$ as predicted by the models with $M_\star^{\rm app}$ and we find that our realisations are able to reproduce the excess of dynamical mass (that closely correspond to $M_\star^{\rm int}$) with respect to the photometrically derived stellar mass (see e.g. Fig.B3 in Fontanot *et al.* 2018a). In the framework of photometric surveys, the most interesting results involve the study of galaxy evolution, as traced by statistical estimators such as the GSMF. In this case, the effect of assuming a variable IMF might be dramatic, as shown in Fig. 2. Solid lines represent the GSMF evolution derived using

$M_\star^{\rm app}$: in most cases it agrees quite well with the observational estimates (also obtained from photometric surveys assuming a universal IMF). On the other hand, dashed lines correspond to the intrinsic evolution of the GSMF in the same GAEA runs: it is clear from this comparison that the variable IMF hypothesis can change our understanding of galaxy evolution, as seen using photometric surveys. In particular, the real GSMF might differ significantly from estimates based on photometry, assuming a universal IMF.

It is worth stressing that all variable IMF approaches we consider so far do not predict a change in the low-mass end of the IMF. Therefore, they wouldn't be able to reproduce the spectral features observed in ETG spectra. Nonetheless, these models are able to reproduce the metal enrichment of ETGs better than universal IMF runs. In Fontanot *et al.* (2018b) we present a new derivation for the variable IMF, that combines the IGIMF theory with the numerical results of Papadopoulos *et al.* (2011). We assume the IMF shape depends on both the SFR and $U_{\rm CR}$. As a representative example, in Fig. 1, *panel (c)*, we show the evolution of the IMF shape as function of SFR, for a MW-like $U_{\rm CR}$. This scenario provides a possible explanation for the observed IMF and metal gradients in ETGs. Indeed, at increasing SFR the predicted IMF shape becomes *the same time* shallower at the high-mass end (which implies a larger number of massive stars) and steeper at the low-mass end (thus predicting an excess of low-mass stars).

References

Bruzual, G. & Charlot, S. 2003, *MNRAS*, 344, 1000

Cappellari, M., McDermid, R. M., Alatalo, K., Blitz, L., Bois, M., Bournaud, F., Bureau, M., Crocker, A. F., *et al.* 2012, *Nature*, 484, 485

Conroy, C., Dutton, A. A., Graves, G. J., Mendel, J. T., van Dokkum, P. G., *et al.* 2013, *ApJL*, 776, L26

Fontanot, F., De Lucia, G., Monaco, P., Somerville, R. S., Santini, P., *et al.* 2009, *MNRAS*, 397, 1776

Fontanot, F., De Lucia, G., Hirschmann, M., Bruzual, G., Charlot, S., Zibetti, S., *et al.* 2017, *MNRAS*, 464, 3812

Fontanot, F., Hirschmann, M., & De Lucia, G. 2017, *ApJL*, 842, 14

Fontanot, F., De Lucia, G., Xie, L., Hirschmann, M., Bruzual, G., Charlot, S., *et al.* 2018a, *MNRAS*, 475, 2936

Fontanot, F., La Barbera, F., De Lucia, G., Pasquali, A., Vazdekis, A., *et al.* 2018b, *MNRAS*, 479, 5678

Kroupa, P. 2001, *MNRAS*, 322, 231

Gunawardhana, M. L. P., Hopkins, A. M., Sharp, R. G., Brough, S., Taylor, E., Bland-Hawthorn, J., Maraston, C., Tuffs, R. J., Popescu, C. C., Wijesinghe, D., Jones, D. H., Croom, S., Sadler, E., Wilkins, S., Driver, S. P., Liske, J., Norberg, P., Baldry, I. K., Bamford, S. P., Loveday, J., Peacock, J. A., Robotham, A. S. G., Zucker, D. B., Parker, Q. A., Conselice, C. J., Cameron, E., Frenk, C. S., Hill, D. T., Kelvin, L. S., Kuijken, K., Madore, B. F., Nichol, B., Parkinson, H. R., Pimbblet, K. A., Prescott, M., Sutherland, W. J., Thomas, D., & van Kampen, E. 2011, *MNRAS*, 415, 1647

Hirschmann, M., De Lucia, G., & Fontanot, F. 2016, *MNRAS*, 461, 1760

Hoversten, E. A. & Glazebrook, K. 2008, *ApJ*, 675, 163

La Barbera, F., Ferreras, I., Vazdekis, A., de la Rosa, I. G., de Carvalho, R. R., Trevisan, M., Falcón-Barroso, J., Ricciardelli, E., *et al.* 2013, *MNRAS*, 433, 3017

Nanayakkara, Themiya; Glazebrook, Karl; Kacprzak, Glenn G; Yuan, Tiantian; Fisher, David; Tran, Kim-Vy; Kewley, Lisa J.; Spitler, Lee; Alcorn, Leo; Cowley, Michael; Labbe, Ivo; Straatman, Caroline & Tomczak, Adam, 2017, *MNRAS*, 468, 3071

Papadopoulos, P. P., Thi, W.-F., Miniati, F., Viti, S., *et al.* 2011, *MNRAS*, 414, 1705

Sarzi, M., Spiniello, C., La, Barbera F., Krajnović, D., van den Bosch, R., *et al.* 2018, *MNRAS*, 478, 4084

Weidner, C., Kroupa, P., Pflamm-Altenburg, J., Vazdekis, A., *et al.* 2013, *MNRAS*, 436, 3309
Zibetti, S., Charlot, S., & Rix, H. 2009, *MNRAS*, 400, 1181

Discussion

KENTARO NAGAMINE: Could you explain a little more on how you link SFR to the evolution of IMF? I.e. does a higher SFR imply a more top heavy IMF?

FABIO FONTANOT: In the IGIMF model, an higher SFR directly correspond to a top-heavy IMF. In the Papadopoulos *et al.* model I am using the SFR density as proxy for CR energy density. This implies that a higher SFR at fixed disc radius indeed provide a more top-heavy IMF. Several combination (of SFR and scale-radii) are possible in the GAEA framework, but in general higher SFRs tend to correspond to higher SFR densities (see e.g. Fig. 3 in Fontanot *et al.* 2018b).

GERGÖ POPPING: Do I understand correctly that the [α/Fe] trend comes from a top heavy IMF during bursts and the bottom heavy IMF from more passive star-formation?

FABIO FONTANOT: The [α/Fe] trends are mainly driven by the top-heavy IMF expected in the strong SFR events connected with the formation of more massive galaxies. The current model do not implement a bottom-heavy IMF yet, but we are working in this direction.

THEMIYA NANAYAKKARA: How can you link the models of IMF dependence in GAEA to reproduce high mass slope variations observed by Hoverste & Glazebrook (2008), Gunawardhana *et al.* (2011) at $z = 0$ and Nanayakkara *et al.* (2017) at $z = 2$?

FABIO FONTANOT: The observations of Gunawardhana *et al.* (2011) at $z = 2$ have been explicitly taken into account in the definition of the IGIMF model, so that they are reproduced by definition. We didn't check the other samples, but that would be an interesting test for our model.

Challenges in Panchromatic Modelling with Next Generation Facilities
Proceedings IAU Symposium No. 341, 2018
M. Boquien, E. Lusso, C. Gruppioni & P. Tissera, eds.

doi:10.1017/S1743921319004769

Modeling the panchromatic emission of galaxies with CIGALE

M. Boquien[1], D. Burgarella[2], Y. Roehlly[2], V. Buat[2], L. Ciesla[2], D. Corre[2], A. K. Inoue[3] and H. Salas[1]

[1]Centro de Astronomía (CITEVA),Universidad de Antofagasta, Avenida Angamos 601, Antofagasta, Chile
email: `mederic.boquien@uantof.cl`

[2]Aix-Marseille Université, CNRS, LAM (Laboratoire d'Astrophysique de Marseille) UMR 7326, 13388, Marseille, France

[3]Waseda University Department of Physics Waseda Research Institute for Science and Engineering Tōkyō, Japan

Abstract. Panchromatic modeling is one of the most powerful tools at our disposal to measure reliably the physical properties of galaxies across cosmic times. We present here an entirely new implementation in `python` of one such tool: `CIGALE`. Developed along three main design principles: simplicity, modularity, and efficiency, it has proven to be a versatile code that in addition to estimating the physical properties of galaxies (or regions within galaxies), can generate arbitrary sets of theoretical models or be used as a library to build other tools. Among its defining features, it is a truly panchromatic code ranging from the far-ultraviolet to the radio that takes into account numerous physical components (including active nuclei or synchrotron emission), that can fit non-photometric data, handle upper limits, determine photometric redshifts, and even build mock catalogs.

Keywords. galaxies: evolution, ultraviolet: galaxies, infrared: galaxies, methods: numerical, methods: statistical

1. Introduction

To undestand the formation and evolution of galaxies, it is critical that we are in a position to measure their physical properties reliably across cosmic times. Doing so is challenging. First, assembling the necessary observations is a considerable effort. Then, converting these observations into physical quantities such as the stellar mass, the attenuation, the star formation rate (SFR), etc. is not straightforward. Uncertainties and variations in the stellar populations (metallicity, rotation, binary evolution, etc.), the star formation history, the properties of the dust and its relative distribution to the stars, etc. all can affect tremendously the emerging radiation from galaxies.

While simple recipes can provide estimates of some physical properties from a single band (e.g., near-infrared to measure the stellar mass, or the far-infrared for the SFR), such recipes are inherently limited and rely on strong assumptions. To address some of the limitations, more advances recipes have been created combining two or more bands (e.g., near-infred and optical color for the stellar mass, far-infrared and far-ultraviolet or Hα for the SFR). Even though such recipes have the advantage of simplicity, they remain limited in their range of applicability. A more general solution that has been emerging over the past two decades is the modeling of the spectral energy distribution (SED) of galaxies. By combining observations over a broad range of wavelengths, it is not only possible to waive a certain number of assumptions made to conceive these simple recipes, but it also allows to break degeneracies plaguing these estimators (e.g., the combination of far-ultraviolet to far-infrared data permits to disentangle the age of stellar population from attenuation by dust). Such an approach taken by the `CIGALE` code (Burgarella *et al.* 2005; Noll *et al.* 2009), which has proven to be highly fruitful (e.g., Buat *et al.* 2011, 2012, 2014, 2018; Boquien *et al.* 2013, 2014, 2016; Ciesla *et al.* 2015, 2016, 2017; Salim *et al.* 2016, 2018). We present here an entirely new implementation in `python` of `CIGALE` (Boquien *et al.* 2018).

2. CIGALE: Code Investigating GALaxy Emission

In a nutshell `CIGALE` is a panchromatic (ultraviolet to radio) energy balance (the emission absorbed by dust in the ultraviolet-to-near-infrared) Bayesian (the physical properties are derived from the marginalized likelihood distributions) SED modeling code written in `python`. It is the successor of a code of the same name written in `FORTRAN` (Burgarella *et al.* 2005; Noll *et al.* 2009).

2.1. *Design principles*

It has been developed following three main design principle based on the experience from the `FORTRAN` version: simplicity, modularity, and efficiency.

2.1.1. Simplicity

The aim is for `CIGALE` 1) to be easy to run from the point of view of a user, 2) to be transparent as to what it does to avoid a "black box effect", and 3) to be easily adaptable to new use cases with simple modifications of the code.

To achieve these aims `CIGALE` has a dynamically generated configuration file that is abundantly documented and provides most of the information necessary to run the code. In addition `CIGALE` is written in a modern version of `python`, a language that is now very broadly used in astronomy, ensuring that a large part of the community can adapt the code if need be. To facilitate such modifications, the code follows a very modular structure (see Sect. 2.1.2). Finally to make the code accessible and to facilitate external contributions, `CIGALE` is developed in a fully open manner on a public git repository.

2.1.2. Modularity

The spectral models are built through a succession of modules, each of them dedicated to a specific component: SFH, stellar populations, nebular emission, dust absorption, dust emission, non-thermal radio emission, absorption from the intergalactic medium, redshifting. In a similar way the estimation of the physical properties is also a module in itself. Each module can be transparently substituted for another module of the same category (e.g., two different dust emission models). That is to say that modules are largely self-contained and independent from one another.

2.1.3. Efficiency

The code relies heavily on major `python` libraries such as Numpy, Scipy, or Astropy to carry out the heavy lifting of the computation. Where these algorithms proved to be bottlenecks, optimized routines have been implemented. Special care has been made to avoid computing identical quantities several times with a cache system. Finally, to exploit the architecture of modern computers, `CIGALE` has a high-level parallelism making use of multiple cores both to compute the models and estimate the physical properties (Fig. 1 left).

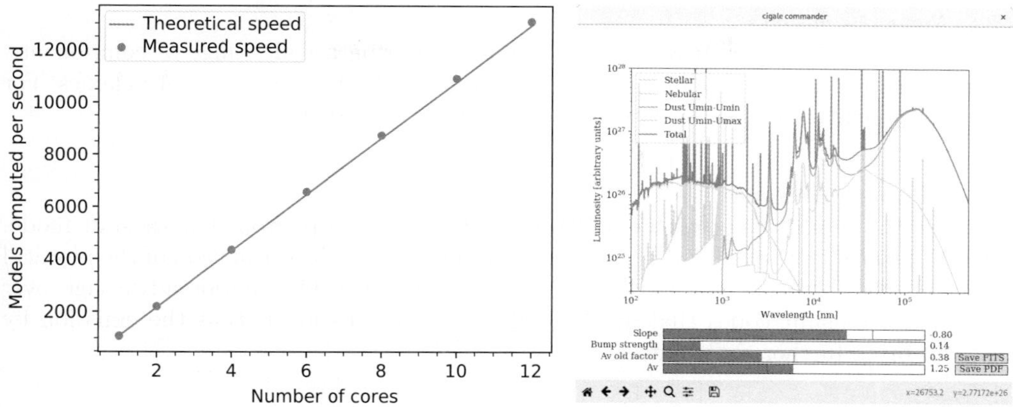

Figure 1. Left: Number of models computed per second as a function of the number of cores used. While the velocity depends somewhat on the actual parameter space explored, in this case computing a single model on a single core takes ~1 ms. Right: Example of a pedagogical tool built on `CIGALE` to explore the effect of the dust attenuation curve.

2.2. *A versatile code*

Following the aforementioned principles `CIGALE` proves to be a very versatile code. It can not only be used to estimate the physical properties of galaxies, its main goal, but it can also to be used to compute sets of theoretical models or even as a library to build other softwares. For instance a small pedagogical tool has been built to interactively explore the impact of various physical properties on the SED of galaxies (Fig. 1, right).

In addition to this, `CIGALE` is also a very feature-rich code. We describe hereafter some of its defining features in a non-exhaustive way.

2.2.1. Panchromatic code

Handling data ranging from the far-ultraviolet to the radio domain, `CIGALE` is a genuine panchromatic code. Such a broad wavelength coverage is essential to break degeneracies and measure the physical properties of galaxies as reliably as possible.

2.2.2. Bayesian code

The physical properties and the related uncertainties are estimated as the likelihood-weighted means and standard deviation of the model values on a grid, which can reach several hundred million models to sample the parameter space. This allows to take into account the fact that different models with different physical properties can be likely to reproduce the observations.

2.2.3. Fitting non-photometric data

Beyond the standard approach of fitting fluxes in passbands, `CIGALE` can also fit any spectral line flux, or intensive or extensive physical property computed by the code: ultraviolet slope β, D_n4000 index, equivalent width, total infrared luminosity, etc.

2.2.4. Upper limits

Astronomical observations are not always as deep as we would like. To take this reality into account, `CIGALE` takes into account the presence of upper limits following the approach of Sawicki (2012).

2.2.5. Numerous modules

Numerous modules are provided with CIGALE, covering a wide range of components, including active nuclei, nebular emission, and the synchrotron emission of galaxies. For some components, such as the dust, several modules are provided.

2.2.6. Photometric redshift

In case the redshift is unknown, CIGALE can be used in a photometric redshift mode, either to determine the redshift in isolation or to factor into the estimation of the physical properties the fact that the redshift is unknown a priori. One major advantage over photometric redshift codes that are limited to stellar emission is that the emission by dust can bring further contraints.

2.2.7. Mock catalogs

Mock catalogs are a useful tool to explore the self-consistency of the modeling and assess the reliability of the estimation of the physical properties. To this aim CIGALE can automatically compute mock-catalogs for any sample, allowing the comparison of known and estimated physical properties.

3. Conclusion

The CIGALE tool is pursuing further developments to fulfill the ever evolving needs in the field of SED modeling, including for instance full spectro-photometric fits, or X-ray and photo-dissociation region models. It is freely available for download at https://cigale.lam.fr.

References

Boquien, M., Boselli, A., Buat, V., Baes, M., Bendo, G., Boissier, S., Ciesla, L., Cooray, A., *et al.* 2013, *A&A*, 554, A14

Boquien, M., Buat, V., & Perret, V. 2014, *A&A*, 571, A72

Boquien, M., Burgarella, D., Roehlly, Y., Buat, V., Ciesla, L., Corre, D., Inoue, A. K., & Salas, H. 2018, *A&A*, 622, A103

Boquien, M., Kennicutt, R., Calzetti, D., Dale, D., Galametz, M., Sauvage, M., Croxall, K., Draine, B., *et al.* 2016, *A&A*, 591, A6

Buat, V., Boquien, M., Małek, K., Corre, D., Salas, H., Roehlly, Y., Shirley, R., & Efstathiou, A. 2018, *A&A*, 619, A135

Buat, V., Giovannoli, E., Heinis, S., Charmandaris, V., Coia, D., Daddi, E., Dickinson, M., Elbaz, D., *et al.* 2011, *A&A*, 533, A93+

Buat, V., Heinis, S., Boquien, M., Burgarella, D., Charmandaris, V., Boissier, S., Boselli, A., Le Borgne, D., & Morrison, G. 2014, *A&A*, 561, A39

Buat, V., Noll, S., Burgarella, D., Giovannoli, E., Charmandaris, V., Pannella, M., Hwang, H. S., Elbaz, D., *et al.* 2012, *A&A*, 545, A141

Burgarella, D., Buat, V., & Iglesias-Páramo, J. 2005, *MNRAS*, 360, 1413

Ciesla, L., Boselli, A., Elbaz, D., Boissier, S., Buat, V., Charmandaris, V., Schreiber, C., Béthermin, M., *et al.* 2016, *A&A*, 585, A43

Ciesla, L., Charmandaris, V., Georgakakis, A., Bernhard, E., Mitchell, P. D., Buat, V., Elbaz, D., LeFloc'h, E., *et al.* 2015, *A&A*, 576, A10

Ciesla, L., Elbaz, D., & Fensch, J. 2017, *A&A*, 608, A41

Mezcua, M., Civano, F., Marchesi, S., Suh, H., Fabbiano, G., & Volonteri, M. 2018, *MNRAS*, 478, 2576

Noll, S., Burgarella, D., Giovannoli, E., Buat, V., Marcillac, D., & Muñoz-Mateos, J. C. 2009, *A&A*, 507, 1793

Salim, S., Boquien, M., & Lee, J. C. 2018, *ApJ*, 859, 11

Salim, S., Lee, J. C., Janowiecki, S., da Cunha, E., Dickinson, M., Boquien, M., Burgarella, D., Salzer, J. J., & Charlot, S. 2016, *ApJS*, 227, 2
Sawicki, M. 2012, *PASP*, 124, 1208

Discussion

MIYAJI TAKAMITSU: Please explain the development of the X-ray implementation in a little more detail.

MÉDÉRIC BOQUIEN: This is an on-going effort led at the moment by Guang Yang. The main challenge is to relate the X-ray emission to the physical properties of the galaxy as X-ray binary evolution can be complex and depend on factors such as the star formation history. At the moment we rely on the results of Mezcua *et al.* (2018b). The first results appear promising but it requires further validation.

DAN TARANU: What is the fundamental problem with spectro-photometric modeling?

MÉDÉRIC BOQUIEN: In my view we face two main challenges. First, if we naively combine photometric with spectroscopic observations, the latter are so numerous relative to the former that they completely drive the likelihood. This strongly reduces the relative weight of panchromatic photometric data. The consequence is that bands bringing a lot of information (e.g., far-ultraviolet and far-infrared observations probe the emission massive stars and dust that by far unconstrained with optical spectra) end up having little bearing on the fit. The other issue, and it is what makes the first one problematic, is that fits are necessarily affected by systematic errors both on the models themselves (dust models or single stellar populations are not only sensitive to the choice of ingredients such as stellar atmosphere, rotation, or binary evolution, but also to the quality of the stellar spectra or models to build them) and on the observations (e.g., imperfect flat field or wavelength-dependent flux calibration). Even small errors will considerably lower the likelihood of models due to the sheer number of spectral data points relative to the photometric data points.

ADAM CARNALL: How do you include Bayesian priors in the grid-based approach?

MÉDÉRIC BOQUIEN: The priors on can be set by playing on the spacing between the input physical properties. For instance an evenly spaced set of parameters will be equivalent to a flat prior, whereas if a specific range of parameters has a shorter spacing, this will make this region denser relative to the rest, thus yielding a non-flat prior. The priors can either be given through an explict list of values or defined through any function provided in Numpy, such as `logspace` for instance.

Challenges in Panchromatic Modelling with Next Generation Facilities
Proceedings IAU Symposium No. 341, 2018
M. Boquien, E. Lusso, C. Gruppioni & P. Tissera, eds.

doi:10.1017/S1743921319004733

Going beyond galaxy ages with dense basis star formation history reconstruction

Kartheik G. Iyer[1,2] and Eric Gawiser[1]

[1]Department of Physics and Astronomy, Rutgers, the State University of New Jersey, 136 Frelinghuysen Road, Piscataway - 08854, New Jersey, USA

[2]Dunlap Institute for Astronomy and Astrophysics, University of Toronto, 50 St George St, Toronto, ON M5S 3H4, Canada
email: kartheik.iyer@dunlap.utoronto.ca

Abstract. Panchromatic SED fitting allows us to better resolve degeneracies between quantities like the star formation rate and dust. This in turn allows us to more robustly extract information about the different stellar populations that comprise a galaxy's Star Formation History (SFH). Using the Dense Basis SED fitting method (Iyer & Gawiser 2017), we reconstruct the SFHs with uncertainties for a large sample of galaxies using an atlas of SEDs corresponding to a physically motivated basis of SFHs. Using Gaussian Process Regression, we encode the parameters describing these SFHs in a functionally independent form. This give us more robust estimates for quantities like Stellar Masses and Star Formation Rates, that directly depend on the SFH. These SFHs can additionally be used to answer questions like the time at which a galaxy's star formation peaked, and how many major episodes of star formation occurred in a galaxy's past, allowing us to go beyond the traditionally estimated 'Galaxy Age', which is often poorly constrained. They also allow us to probe the high-redshift low-stellar mass regime of the SFR-M* correlation by constructing trajectories in SFR-M_* space for each galaxy.

Keywords. galaxies: star formation history, galaxies: evolution, techniques: photometric

1. Introduction

Galaxy surveys like CANDELS (Grogin *et al.* (2011); Koekemoer *et al.* (2011)) provide us with high-quality multiwavelength spectral energy distributions (SEDs) for $\mathcal{O}(10^5)$ galaxies, with this number rising to $\mathcal{O}(10^8)$ with upcoming surveys. Galaxy catalogs compiled by analyzing these SEDs often report quantities like stellar masses (M_*), star formation rates (SFR), and galaxy ages (Santini *et al.* (2015)). However, these quantities are summary statistics that depend on the star formation history (SFH): a record of when and how a galaxy formed its stars. Approximations for the SFH in SED fitting can thus cause biases in estimating these quantities, which could then propagate into more involved analyses of galaxy evolution such as mass functions (Baldry *et al.* (2008)), estimates of the cosmic star formation rate density (SFRD) (Madau & Dickinson (2014)), and scaling relations like the $SFR - M_*$ correlation (Speagle *et al.* (2014)) and the mass-metallicity correlation (Tremonti *et al.* (2004)).

Estimating physical quantities from broadband photometric SEDs has long been done using SFHs that follow a simple parametric form like exponentially declining (Skelton *et al.* (2014)) or lognormal SFHs (Gladders *et al.* (2013)), or are binned in time (Dye (2008)). In Iyer & Gawiser (2017) we quantified the bias and scatter in estimating different physical quantities due to different SFH parametrizations, while devising a flexible approach towards describing galaxy SFHs that encodes the maximal amount of SFH information in a minimal number of parameters, called the Dense Basis method. Recent improvements to the method (Iyer *et al.* (2019)) make the problem of estimating SFHs

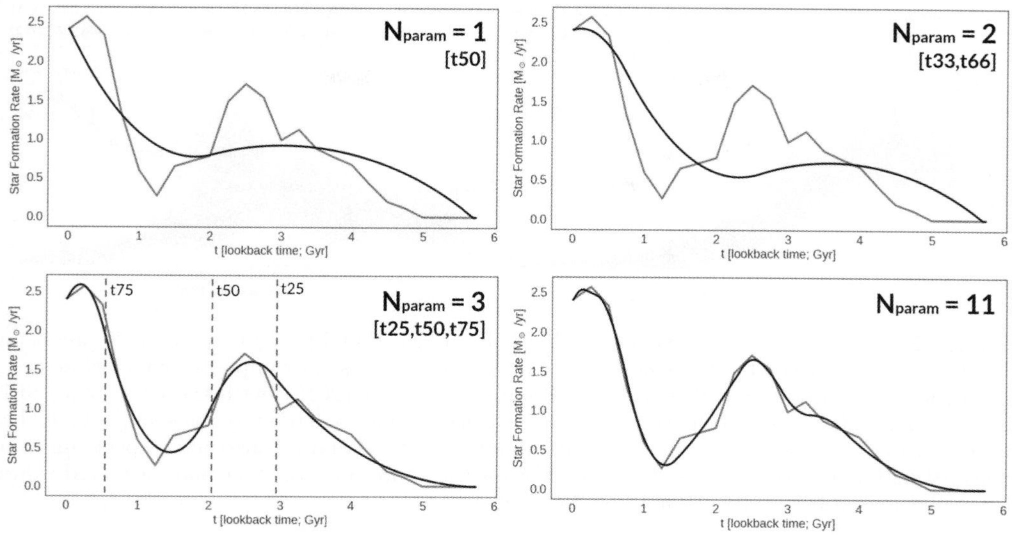

Figure 1. Constructing smooth, nonparametric star formation histories corresponding to an example galaxy in the Somerville, Popping & Trager (2015) semi-analytic model (blue line). Working here in SFH space, we estimate the SFH using varying numbers of parameters $\{t_x\}$, which denote the lookback times at which the galaxy forms various equally spaced quantiles of its stellar mass. The Dense Basis SFH constructed from these parameters is shown as a solid black line for various values of N_{param} given in the top-right corner of each plot. The accuracy of the GP-SFH approximation increases as we go to higher N_{param}, although we obtain a good approximation of the truth for as low as $N_{param} = 3$.

more well-conditioned, leading to novel methods of studying galaxy evolution that are now accessible with these smooth, nonparametric SFHs derived from the observations.

2. The Dense Basis method

The Dense Basis method provides a flexible, nonparametric description for galaxy SFHs that can be incorporated into any SED fitting code. In this framework, a star formation history is described using a tuple:

$$SFH \equiv (M_*, SFR, \{t_x\}) \tag{2.1}$$

where the set $\{t_x\}$ describes the lookback times at which a galaxy formed N uniformly spaced quantiles of its stellar mass, where N is a free parameter that can be informed by the data being fit. Figure 1 shows an example SFH from the Santa-Cruz semi-analytic model (Somerville, Popping & Trager (2015)), with the N = 3 panel showing the lookback times at which the galaxy formed 25%, 50% and 75% of its stellar mass at the time of observation, respectively. These parameters represent integral constraints that help specify the shape of the SFH. To construct smooth SFHs that satisfy these constraints, we use Gaussian Processes (Rasmussen & Williams (2006)), a machine learning technique for nonparametric regression that generalizes the gaussian probability distribution to the space of functions. In Figure 1 we calculate the $(M_*, SFR, \{t_x\})$ tuples for the example SFH with $N_{param} = 1, 2, 3$, and 11, and then use the GP-SFH routine to generate smooth, nonparametric SFHs corresponding to these N-tuples. While we chose an SFH that is not well described by any standard parametric form for this example, the results are generally applicable to any SFH, with the construction of the smooth SFH well approximating the true SFH shape with as little as 3 parameters for $\{t_x\}$. Extending this from working purely in SFH space to SED fitting is described in detail in Iyer *et al.* (2019). Examples of SFH reconstruction through SED fitting are shown in Figure 2. In this framework,

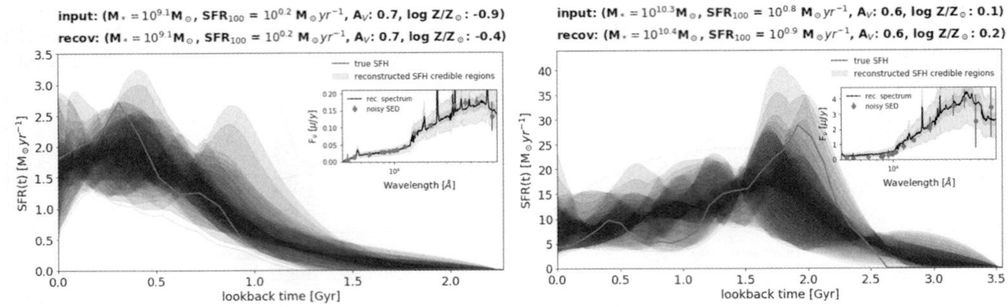

Figure 2. Reconstructing the SFHs of galaxies from broadband photometric SED fitting using the Dense Basis method. The panels show two randomly chosen examples of galaxies at $z \sim 1$ from the semi-analytic model Somerville, Popping & Trager (2015), with the true (blue) SFH and the reconstructed SFH posterior (black shaded regions - darker regions indicate higher probability). The inset panels show the simulated noisy SED and reconstructed spectrum from SED fitting for the two galaxies. Text above below the figures give the true and estimated values for stellar mass, SFR, dust attenuation and stellar metallicity.

the 'age' of a galaxy would be t_0, which formally has S/N $\rightarrow 0$ and is thus very poorly constrained by observations.

The method offers several advantages:

• The $\{t_x\}$ parameters are physically motivated, and directly trace galaxy growth.

• The parameters used to describe the SFHs are all continuous, have comparable S/N, and can be readily incorporated into any SED fitting code or bayesian analysis.

• Well motivated priors for each parameter can be designed using SFHs from both analytic models and cosmological simulations of galaxy formation and evolution.

3. Studying galaxy evolution with nonparametric SFHs

The smooth, nonparametric SFHs obtained through the Dense Basis method enable us to study galaxy evolution using approaches that were previously inaccessible through SED fitting. A flexible number of parameters allows us to specify the SFH to arbitrary precision, and in practice can be informed by the data that is being fit. Panchromatic SEDs bring in additional information that allows us to constrain more flexible dust models by balancing energy between dust absorption and re-emission in the far-IR (Smith & Hayward (2015)) and accounting for AGN contributions (Rivera *et al.* (2016)) using the full X-ray to radio SED. In turn, this allows us to better constrain galaxy SFHs and study the different modes by which galaxies in different demographics: quiescent, starburst, green valley, late bloomer, etc. form their stellar mass. While better constraining SFHs without parametric assumptions yields more robust estimates of stellar masses and star formation rates, the method can also be used to study:

• **The number of major episodes of star formation in a galaxy's past** (Iyer & Gawiser (2017), Iyer *et al.* (2019)). Galaxies with multiple strong episodes of star formation are particularly interesting because they bear imprints from processes that inefficiently quench the star formation such as AGN feedback or ram-pressure stripping, or are experiencing a rejuvenation event, where inflows or mergers trigger star formation.

• **The origin and evolution of scaling relations** like the SFR-M_* correlation (Iyer *et al.* (2018)). The SFHs of galaxies allow us to propagate galaxies backwards in time along their SFR-M_* trajectories to study the correlation at high redshifts and low masses, regions that are hard to probe with direct observations due to galaxies growing fainter. They allow us to test whether galaxies move along the evolving correlation as they grow, as opposed to the correlation resulting from the central limit theorem (Kelson+14).

• Studies have also considered using SFHs to estimate the cosmic SFRD (Leja *et al.* (2018)), and can be used to estimate the timescales on which galaxies become quiescent, building on the work by Carnall *et al.* (2017).

• In **comparison with distributions of SFHs from cosmological simulations of galaxy formation**, observational SFHs can provide useful constraints and help inform the next generation of simulations by testing mechanisms and implementations of preventive and ejective feedback chemical enrichment, and star formation.

References

Iyer, Kartheik & Eric Gawiser 2017, *ApJ*, 838.2 (2017): 127
Iyer, Kartheik, *et al.* 2018, *ApJ* 866.2 (2018): 120
Iyer, K., Gawiser, E., *et al.* 2019, arXiv preprint, https://iopscience.iop.org/article/10.3847/1538-4357/ab2052
Somerville, R. S., Popping, G. & Trager, S. C. 2015, *MNRAS*, 453(4), 4337–4367
Rasmussen, C. E., & Williams, C. K. 2006, *Gaussian process for machine learning. MIT press*
Pacifici, C. 2019, *in prep.*
Grogin, Norman A., *et al.* 2011, *ApJS*, 197.2 (2011): 35
Koekemoer, Anton M., *et al.* 2011, *ApJS*, 197.2 (2011): 36
Santini, P., *et al.* 2015, *ApJ*, 801.2 (2015): 97
Baldry, I. K., K. Glazebrook & S. P. Driver. 2008, *MNRAS*, 388.3 (2008): 945–959
Madau, P. & Dickinson, M. 2014, *ARAA*, 52, 415–486
Speagle, J. S., Steinhardt, C. L., Capak, P. L. & Silverman, J. D. (2014), *ApJS*, 214(2), 15
Tremonti, C. A., *et al.* 2004, *ApJ* 613.2 (2004): 898
Smith, D. J. B. & Hayward, C. C. (2015), *MNRAS* 453.2 (2015): 1597–1607
Rivera, G. C., *et al.* 2016, *ApJ*, 833(1), 98
Dye, S. 2008, *MNRAS*, 389(3), 1293–1305
Skelton, R. E., *et al.* 2014, *ApJS*, 214(2), 24
Gladders, M. D., *et al.* 2013, *ApJ*, 770(1), 64
Leja, J., *et al.* 2018, arXiv preprint, https://iopscience.iop.org/article/10.3847/1538-4357/ab133c
Carnall, A., *et al.* 2017, arXiv preprint, https://academic.oup.com/mnras/article/480/4/4379/5068189

Discussion

DENIS BURGARELLA: Can you talk more about results from the comparison between the different SED fitting codes?

KARTHEIK IYER: The code comparison exercise from Pacifici *et al.* (2019) uses estimates of stellar masses, star formation rates, and dust attenuation from 14 different SED fitting codes that make different model assumptions for dust, SPS models, and star formation histories to quantify the inter-code variability between different SED fitting methods, which contains information about the modeling uncertainties in addition to the formally estimated uncertainties on these parameters. By looking at the distribution in Stellar Mass, we find the inter-code variability to be ~ 0.16 dex, and for SFRs to be ~ 0.29 dex.

ADAM CARNALL: Will the code implementing the Gaussian Process Star Formation Histories be made public? I am interested in incorporating it in my SED fitting code.

KARTHEIK IYER: The Dense Basis code is available as a Github repository at http://github.com/kartheikiyer/dense_basis, and the documentation can be found at http://dense-basis.readthedocs.io. While our SED fitting code is highly optimized, capable of fitting $\sim 10,000$ galaxies per second per code, the Dense Basis SFH parametrization is general enough to be incorporated into any sophisticated SED fitting code.

Challenges in Panchromatic Modelling with Next Generation Facilities
Proceedings IAU Symposium No. 341, 2018
M. Boquien, E. Lusso, C. Gruppioni & P. Tissera, eds.

doi:10.1017/S1743921319002394

A hierarchical bayesian dust SED model and its application to the nearby universe

Frédéric Galliano

AIM, CEA, CNRS, Université Paris-Saclay, Université Paris Diderot, Sorbonne Paris Cité, F-91191 Gif-sur-Yvette, France
email: `frederic.galliano@cea.fr`

Abstract. In this paper, I review several dust evolution studies based on the DustPedia nearby galaxy sample. I first present the dust spectral energy distribution model, implementing a hierarchical Bayesian method, that we have developed. I then discuss the dust evolution trends we have derived among (integrated) and within (resolved) galaxies. In particular, we show that the trend of dust-to-gas ratio with metallicity is clearly non-linear, indicating the need for grain growth in the interstellar medium. Our trend is closer to the one derived with damped Lyα systems than what was suggested by previous studies. We finally demonstrate the universal processing of small amorphous carbon grains by stellar photons.

Keywords. methods: statistical, dust, ISM: evolution, galaxies: evolution, infrared: ISM

1. Introduction

The properties of interstellar dust (chemical composition, size distribution, *etc.*) depend on the local environmental conditions, and on the elemental enrichment history of the galaxy (*e.g.* Galliano *et al.* 2018). However, the individual processes controlling this evolution (dust production by stars, dust growth in the interstellar medium (ISM), dust destruction by shocks, *etc.*) are not known accurately enough to unambiguously model the evolution of galaxies. One of the ways to refine our knowledge of these evolutionary processes consists in studying the variations of local observed dust properties, measured via their spectral energy distribution (SED), as a function of the local physical conditions (ultraviolet (UV) field, metallicity, *etc.*). These trends between dust properties and physical conditions are valuable constraints on the nature and efficiency of the grain evolutionary processes. This leaves us with the task of properly deriving dust parameters and their uncertainties from broadband infrared (IR) observations.

2. A Hierarchical Bayesian Dust Spectral Energy Distribution Model

This task of properly deriving dust parameters is known to present several intricate degeneracies and noise-induced false correlations (*e.g.* Shetty *et al.* 2009). We have recently developed a model (HerBIE; Galliano 2018) to address this issue.

2.1. *The Microphysical Dust Model*

The microscopic framework of our code is the THEMIS dust model (Jones *et al.* 2017). This model is consistent with the extinction, depletion patterns and emission (including the *Planck* data) properties of the diffuse Galactic ISM. THEMIS assumes two grain types (cross-sections from laboratory measurements): (i) amorphous carbons, partially hydrogenated, noted a-C(:H), carrying the aromatic and aliphatic features; and (ii) a-C(:H)

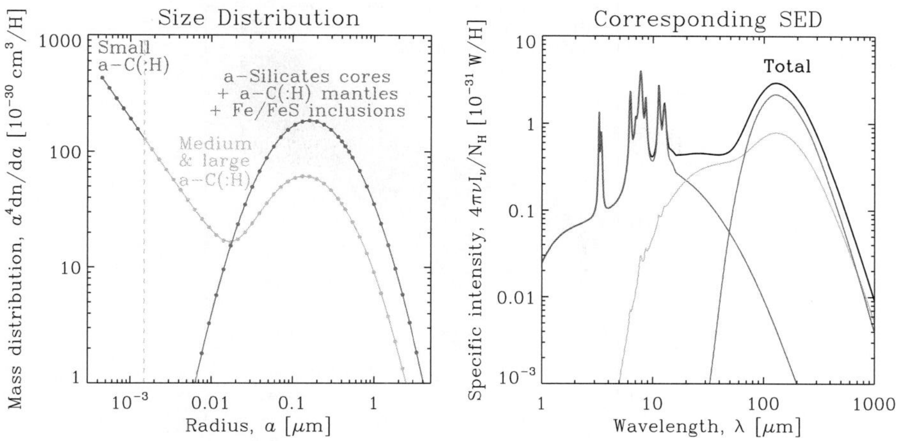

Figure 1. Parameterization of the THEMIS model implemented in HerBIE. The small (medium and large) a-C(:H) have a radius smaller (larger) than 1.5 nm. The color code of the three subcomponents is identical in the two panels.

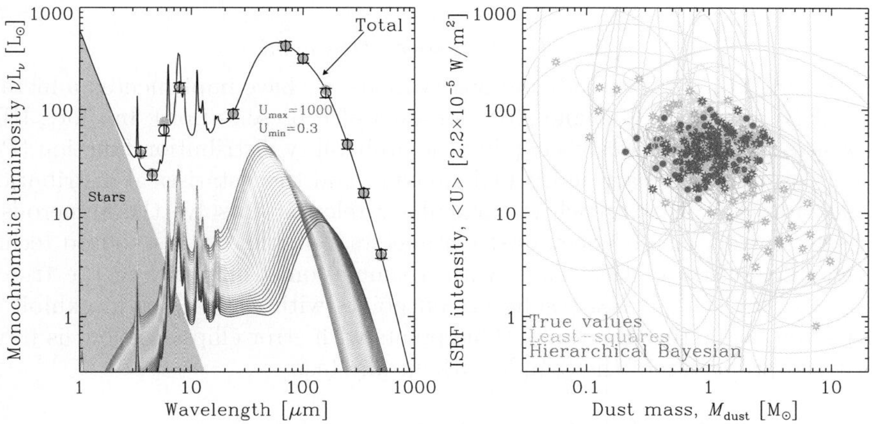

Figure 2. Left: distribution of starlight intensities (color curves) fit to an SED. Right: simulation to demonstrate the efficiency of different methods to fit dust SEDs.

coated amorphous silicates with iron inclusions. We parameterize the dust size distribution by scaling the fraction of small a-C(:H), the sub-component carrying the mid-IR features (Fig. 1).

2.2. *The Macroscopical Mixing of Physical Conditions*

When studying galaxies, we can not avoid the mixing of physical conditions within the observed region. To compensate our ignorance of the particular ISM structure of our sources, we adopt the empirical prescription of Dale *et al.* (2001), assuming that the distribution of starlight intensities heating the dust, U, follows a power-law: $dM_{\rm dust} \propto U^{-\alpha} dU$ for $U_{\rm min} < U < U_{\rm max}$. Thus, the shape of the observed far-IR peak constrains this distribution. This is shown in Fig. 2-left, where the sum of the colored curves is the total dust SED, shown with the black solid line. With this SED model, the main relevant parameters, which are weakly dependent on the assumptions we made, are: (i) the dust mass, $M_{\rm dust}$; (ii) the mean starlight intensity, $\langle U \rangle$; and (iii) the mass fraction of small a-C(:H), $q_{\rm AF}$.

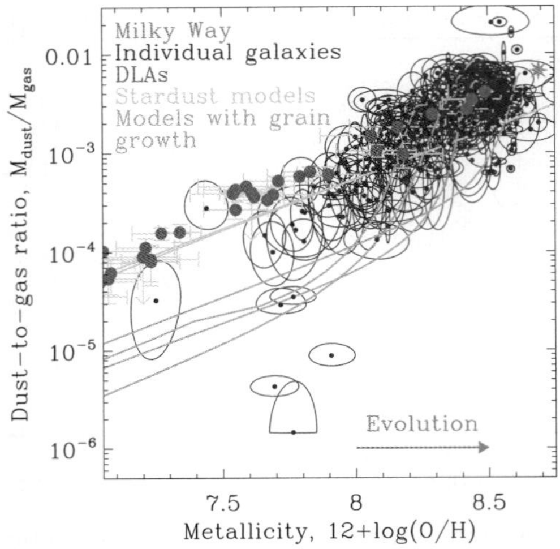

Figure 3. Derived evolution of the dust-to-gas ratio with metallicity for our nearby galaxy sample (blue) and for DLAs (red), compared to dust evolution models (lines).

2.3. *The Statistical Treatment*

To fit this SED model to broadband observations, we have implemented a hierarchical Bayesian (HB) approach (Galliano 2018, for a detailed presentation). In short, HB modelling consists in simultaneously sampling the probability distribution function (PDF) of the model parameters of each individual galaxies, and their statistical distribution over the whole sample. This approach is particularly relevant to solve the numerous noise-induced false correlations between dust parameters, obtained with standard techniques. Fig. 2-right demonstrates this point, with a simulation. Compared to the true values (red points), the least-squares results (green points with error ellipses) exhibit a false anticorrelation, while the HB results (blue points with error ellipses) allow us to recover the statistical properties of the sample without any bias.

3. The Studied Nearby Galaxy Sample

We have applied this model to the near-IR-selected, volume-limited, nearby galaxy sample, DustPedia (Davies *et al.* 2017). In order to improve our low-metallicity coverage, we included in our analysis the 34 sources of the dwarf galaxy sample (DGS; Madden *et al.* 2013; Rémy-Ruyer *et al.* 2013), that were not in DustPedia. In total, our sample contains 813 objects. The photometry comes from Clark *et al.* (2018) and the total gas mass and metallicities from De Vis *et al.* (2019).

4. Constraints on Dust Evolution

4.1. *Cosmic Evolution of the Dust Content*

The most important relation that we can derive from our modelling is the trend of dust-to-gas ratio as a function of metallicity. If we consider each galaxy as a snapshot of dust evolution at a given stage, then this relation shows how the dust content is built-up from the available heavy elements in the ISM. Fig. 3 shows that this relation from our nearby galaxy sample (blue points with error ellipses; Galliano *et al.*, *in prep.*) is non-linear. Compared to the dust evolution models of De Vis *et al.* (2017, yellow and orange lines), it indicates that dust growth in the ISM is a crucial ingredient. Indeed, in models where only stardust contributes to dust production, we are left with linear relation (yellow), inconsistent with the data. The particular care we gave to having an homogenized sample

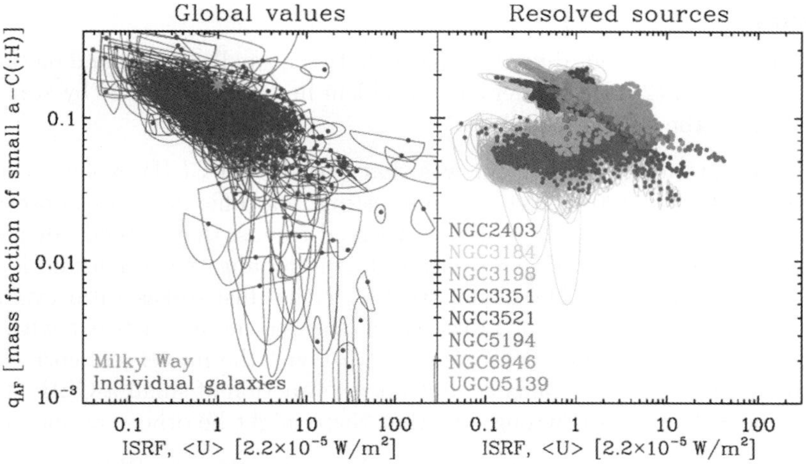

Figure 4. Processing of small a-C(:H) by intense ISRF, among and within galaxies.

and applying a rigorous statistical treatment to the fitting, allow us to provide the most accurate relation that has been published to date. Interestingly enough, this new trend is in much better agreement with the trend obtained from damped Lyα systems (DLA; De Cia *et al.* 2016), than the previous trends derived on nearby galaxies (Galliano *et al.* 2018, for a review).

4.2. *Evolution of the Aromatic Feature Carriers*

Fig. 4 shows the trend between the fraction of small a-C(:H) and the ISRF, among (left) and within (right) galaxies. It clearly confirms that these grains are destroyed by intense radiation fields, as expected.

References

Clark, C. J. R., Verstocken, S., Bianchi, S., Fritz, J., Viaene, S., Smith, M. W. L., Baes, M., Casasola, V., *et al.* 2018, *A&A*, 609, A37

Dale, D. A., Helou, G., Contursi, A., Silbermann, N. A., & Kolhatkar, S. 2001, *ApJ*, 549, 215

Davies, J. I., Baes, M., Bianchi, S., Jones, A., Madden, S., Xilouris, M., Bocchio, M., Casasola, V., *et al.* 2017, *PASP*, 129, 044102

De Cia, A., Ledoux, C., Mattsson, L., Petitjean, P., Srianand, R., Gavignaud, I. & Jenkins, E. B. 2016, *A&A*, 596, A97

De Vis, P., Gomez, H. L., Schofield, S. P., Maddox, S., Dunne, L., Baes, M., Cigan, P., Clark, C. J. R., *et al.* 2017, *MNRAS*, 471, 1743

De Vis, P., Jones, A., Viaene, S., Casasola, V., Clark, C. J. R., Baes, M., Bianchi, S., Cassara, L. P., *et al.* 2019, *A&A*, 623, A5

Draine & Li 2007, https://iopscience.iop.org/article/10.1086/511055

Galliano, F. 2018, *MNRAS*, 476, 1445

Galliano, F., De Vis, P., Galametz, M., *et al.* 2019, *in prep.*

Galliano, F., Galametz, M., & Jones, A. P. 2018, *ARA&A*, 56, 673

Jones, A. P., Köhler, M., Ysard, N., Bocchio, M., & Verstraete, L. 2017, *A&A*, 602, A46

Madden, S. C., Rémy-Ruyer, A., Galametz, M., Cormier, D., Lebouteiller, V., Galliano, F., Hony, S., Bendo, G. J., *et al.* 2013, *PASP*, 125, 600

Rémy-Ruyer, A., Madden, S. C., Galliano, F., Hony, S., Sauvage, M., Bendo, G. J., Roussel, H., Pohlen, M., *et al.* 2013, *A&A*, 557, A95

Seok *et al.* 2014, https://academic.oup.com/mnras/article/439/2/2186/1020880

Shetty, R., Kauffmann, J., Schnee, S., & Goodman, A. A. 2009, *ApJ*, 696, 676

Discussion

HIROYUKI HIRASHITA: Regarding the relation between small a-C(:H) and metallicity, you said that Seok *et al.* (2014)'s model has a problem in a scatter caused by star formation histories. Why is the scatter a problem?

FRÉDÉRIC GALLIANO: My argument was that the small a-C(:H) evolution mechanism proposed by Seok *et al.* (2014; S14) was interesting, but difficult to test, observationally. I tried overplotting the S14 tracks on my observed trends: they extend beyond the area covered by my data. This could be a mean of constraining the star formation histories of these galaxies. However, S14 model tracks for the dust-to-gas ratio evolution with metallicity, for a similar range of SFHs, cover an area more consistent with the data. It means that, for a given SF timescale (e.g. 0.5 Gyr), the model will correctly predict the dust-to-gas ratio of some galaxies, but not the fraction of small a-C(:H). It does not mean that the S14 process is wrong, but that there might be other parameters.

ADAM CARNALL: Are the "individual galaxies" fitted within the BMH?

FRÉDÉRIC GALLIANO: Yes, the results of the spatially resolved, individual galaxies are from a hierarchical run. In what I presented, there is an independent HB run for the integrated sources, and one independent run for each of the spatially resolved objects. All the physical parameters, as well as the dependencies (gas and stellar masses, metallicity, *etc.*) are in the hierarchical model.

ADAM CARNALL: How do you sample from the posterior distribution of the BMH?

FRÉDÉRIC GALLIANO: I performed Gibbs sampling on each of the parameters and hyperparameters. The sampler is a multithreaded Fortran code I developed. I ended up developing my own sampler, as all the publicly available samplers I tested were not adapted for high dimensionality. For instance, in some of the galaxies I presented, there are almost $100,000$ pixels. With a seven parameter model, I have to sample a posterior in a $\simeq 700,000$ dimension parameter space...

ADAM CARNALL: Can you construct conditional distributions for all your parameters?

FRÉDÉRIC GALLIANO: Yes, I can construct conditional posteriors for each of the model parameters and dependencies.

HELEN KIM: For your sample of 875 galaxies, what criteria did you use to determine they had sufficient IR data?

FRÉDÉRIC GALLIANO: Among the 3000, NIR and volume-limited selected, DustPedia galaxies, the 875 objects we modeled the dust SEDs were those which had been observed with *Herschel* and had a size larger than 1 arcminute.

FUMI EGUSA: How do you treat ionization state of dust grains in the model?

FRÉDÉRIC GALLIANO: There is currently no account of the effects of grain charging in the dust model I used (THEMIS; Jones *et al.* 2017). Other dust models, such as Draine & Li (2007), include a PAH component, constituted of neutral and charged molecules. The ratio of this mixture can be varied in order to empirically change the aromatic feature spectrum and obtain a better fit. However, there is, to my knowledge, no reliable constraints on the effect of charge of small a-C(:H), from laboratory data. In the modeling I presented, I am able to adapt the aromatic feature spectrum in order to fit the complexity of the data, by dividing the size distribution of small a-C(:H) into two bins, whose weight can be varied.

Challenges in Panchromatic Modelling with Next Generation Facilities
Proceedings IAU Symposium No. 341, 2018
M. Boquien, E. Lusso, C. Gruppioni & P. Tissera, eds.

doi:10.1017/S1743921319002746

Bayesian discrimination of the panchromatic spectral energy distribution modelings of galaxies

Yunkun Han[1,2,3], Zhanwen Han[1,2,3] and Lulu Fan[4]

[1]Yunnan Observatories, Chinese Academy of Sciences, 396 Yangfangwang, Guandu District, Kunming, 650216, P. R. China
email: hanyk@ynao.ac.cn

[2]Center for Astronomical Mega-Science, Chinese Academy of Sciences, 20A Datun Road, Chaoyang District, Beijing, 100012, P. R. China

[3]Key Laboratory for the Structure and Evolution of Celestial Objects, Chinese Academy of Sciences, 396 Yangfangwang, Guandu District, Kunming, 650216, P. R. China

[4]Institute of Space Science, Shandong University, Weihai, 264209, China

Abstract. Fitting the multi-wavelength spectral energy distributions (SEDs) of galaxies is a widely used technique to extract information about the physical properties of galaxies. However, a major difficulty lies in the numerous uncertainties regarding almost all ingredients of the SED modeling of galaxies. The Bayesian methods provide a consistent conceptual basis for dealing with the problem of inference with many uncertainties. While the Bayesian parameter estimation method have become quite popular in the field of SED fitting of galaxies, the Bayesian model comparison method, which is based on the same Bayes' rule, is still not widely used in this field. With the application of Bayesian model comparison method in a series of papers, we show that the results obtained with Bayesian model comparison are understandable in the context of stellar/galaxy physics. These results indicate that Bayesian model comparison is a reliable and very powerful method for the SED fitting of galaxies.

Keywords. galaxies: fundamental parameters, galaxies: stellar content, galaxies: statistics

---

1. Introduction

As shown in a series of papers with the same title by Conroy and collaborators (Conroy *et al.* 2009, 2010; Conroy & Gunn 2010), there are many important uncertainties in the stellar population synthesis modeling. To obtain the synthetic spectra of simple stellar populations (SSPs), we need to employ a library of stellar evolutionary tracks, a library of stellar spectra, and assumption about the stellar initial mass function (IMF), all of which suffer from some important uncertainties. It is still challenging to model the complex effects of convection, rotation and a binary companion in stellar evolution (Paxton *et al.* 2013). Some not well understood phases of stellar evolution, such as the thermally pulsating AGB stars (Marigo & Girardi 2007; Marigo *et al.* 2008; Rosenfield *et al.* 2016), horizontal branch stars (Heber 2009; Cassisi & Salaris 2013), and blue straggler stars (Boffin *et al.* 2014), have important contributions to the integrated light from a galaxy. On the other hand, the empirical libraries (e.g., Vazdekis *et al.* 2010) of stellar spectra are limited in the coverage in the parameter space, possible errors in the estimation of physical parameters, finite SNR, while the theoretical ones (e.g., Leitherer *et al.* 2010) suffer from shortcomings of atomic and molecular data, inappropriate abundance ratios, non-LTE, etc. Futhermore, the stellar initial mass functions (IMFs) may not be universal as traditionally assumed, but could be variable in the galaxy's formation history (Kroupa 2001; Bastian *et al.* 2010; Cappellari *et al.* 2012). Consequently, the numerous uncertainties in the ingredients of the modeling have led to the diversity of stellar population synthesis (SPS) models†. Except for the SSPs given by SPS modelings, we need models about the star formation history (SFH) and chemical evolution history of the interstellar medium, which are even more uncertain.

2. Bayesian model comparison

Given the numerous uncertainties regarding almost all ingredients of the SED modelings of galaxies, we need to find a method to allow a reasonable comparison of the many possible assumptions about them. While the Bayesian parameter estimation method has been widely used in the field of SED fitting galaxies, the Bayesian model comparaison method, which is based on the same Bayes' rule as Bayesian parameter estimation, is still rarely used in this field. A main reason for this situation is that the computation of Bayesian evidence, which is critical for Bayesian model comparison, is generally much more computationally expansive than the posterior sampling in Bayesian parameter estimation. However, with the advent of state-of-the-art Bayesian inference tool, such as MultiNest (Feroz *et al.* 2009), it has become possible to achieve efficient and robust computation of Bayesian evidence. The Bayesian evidence can be thought as a quantified Occam's razor such that a model with a better balance between its goodness-of-fit to the data and its complexity will be more favored (Gregory 2005).

A major difficulty for the application of Bayesian model comparison method to the comparison of different SED modelings lies in the fact that the computation of the SED modeling itself could be even more computationally expansive. In Han & Han (2012), we have employed the artificial neural network (ANN) technique to allow a rather rapid computation of model SEDs. With the building-up of a pre-released version of our BayeSED code, we have applied the Bayesian model comparison to the study of a sample of hyper-luminous infrared galaxies (HLIRGs) (Ruiz *et al.* 2010). We found that the more complicated Starburst+AGN model has larger Bayesian evidence than the simpler pure Starburst or pure AGN model. This result supports the idea that HLIRGs represent an important phase in the formation and evolution of galaxies where both the star formation and SMBH accretion are very active. In Fan *et al.* (2016a,b), we have applied the similar methods to the infrared SED decomposition of a sample of WISE-selected, hyperluminous hot dust-obscured galaxies. We found that the more complicated AGN torus+Greybody model has larger Bayesian evidence than the simpler pure AGN torus or pure Greybody model for these galaxies.

In Han & Han (2014), we have presented the first version of our BayeSED code. The performance of BayeSED for the estimation of physical parameters, such as stellar mass and star formation rate, has been extensively tested with a mock sample of galaxies and the comparaison with the results obtained with the FAST code (Kriek *et al.* 2009). With BayeSED V1.0, we have presented the first Bayesian comparison of SPS models. For a Ks-selected sample of 5467 low-z galaxies given by Muzzin *et al.* (2013), we found that the simpler BC03 model has larger Bayesian evidence than the M05 model. However, this comparison of SPS models has some important limitations. For example, we have only considered the most popular BC03 and M05 models, while there are many others. The SFH of galaxies is fixed to the commonly used Exp-dec form, and the dust attenuation law (DAL) is fixed to the Calzetti law. Since the SFH and DAL of different galaxies could be very different, the above assumption about them could introduce serious bias to the results of Bayesian SPS model comparaison.

† See http://www.sedfitting.org/Models.html for an incomplete list of these models.

In Han & Han (2019), we have presented a more comprehensive Bayesian Discrimination of SPS models, where 16 SSP models, 5 forms of SFH and 4 forms of DAL have been considered. We have introduced a method to define the Bayesian evidence for the SED modeling of a sample of galaxies with a universal and fixed SSP, but object-dependent and free SFH and DAL. Since the Bayesian evidence is defined for a sample of galaxies instead of an individual galaxy and the results are obtained with the marginalization over the different choices of SFH and DAL for different galaxies, we can achieve a much more robust comparison of SPS models. We found that the most widely used BC03 has much larger Bayesian evidence than many other more updated SPS models, including the CB07 model from the same authors. Our results indicate that the contribution of TP-AGB stars is still not properly considered in current SPS models. On the other hand, we found that the version of BPASS V2.0 model (Eldridge *et al.* 2008) without binaries has a slightly larger Bayesian evidence than the BC03 models. However, the version of BPASS V2.0 model with binaries actually has the lowest Bayesian evidence. For the two models with the consideration of binaries, the Yunnan-II model (Zhang *et al.* 2005) is more favored than the BPASS V2.0 model. Our results indicate that the effects of binaries are still not well considered in current SPS models, which is not surprising, given the uncertainties about the period and mass-ratio distributions of binary stars (Moe & Di Stefano 2017) and other not well understood complex physics in binary stellar evolution.

3. Conclusion

With the MultiNest-like Bayesian inference tools and methods for rapid generation of model SEDs, it has become possible to reliably and efficiently compute the Bayesian evidence for a large set of SED models. In a series of papers, we show that the results obtained with Bayesian model comparison are understandable in the context of stellar/galaxy physics. So, we believe that Bayesian model comparison is a reliable and very powerful method, and it should be more extensively used in the field of SED fitting and the study of galaxies. We have built the BayeSED code† for this kind of studies. The public version of the code is in a compiled binary format, which is generally faster than other interpreted language code (e.g. Python). The code is directly runable on MAC/LINUX platform and does not require users to resolve the often annoying library dependency problems. In the future, more and more sophisticated SED models will be added gradually to BayeSED. The suggestions about adding new SED models and other new features, and the reports about bugs are very welcome.

Acknowledgements

This work is supported by the National Natural Science Foundation of China (NSFC, Grant Nos. 11773063, 11521303, 11733008), Chinese Academy of Sciences (CAS, no. KJZD-EW-M06-01) and Yunnan Province (Grant No. 2017FB007, 2017HC018).

References

Bastian, N., Covey, K. R., & Meyer, M. R. 2010, *ARA&A*, 48, 339

Boffin, H., Carraro, G., & Beccari, G. 2014, *Ecology of Blue Straggler Stars* (Springer Berlin Heidelberg)

Cappellari, M., McDermid, R. M., Alatalo, K., Blitz, L., Bois, M., Bournaud, F., Bureau, M., Crocker, A. F., *et al.* 2012, *Nature*, 484, 485

Cassisi, S. & Salaris, M. 2013, *Old Stellar Populations: How to Study the Fossil Record of Galaxy Formation* (Wiley)

Conroy, C. & Gunn, J. E. 2010, *ApJ*, 712, 833

† https://bitbucket.org/hanyk/bayesed/

Conroy, C., Gunn, J. E., & White, M. 2009, *ApJ*, 699, 486
Conroy, C., White, M., & Gunn, J. E. 2010, *ApJ*, 708, 58
Eldridge, J. J., Izzard, R. G., & Tout, C. A. 2008, *MNRAS*, 384, 1109
Fan, L., Han, Y., Nikutta, R., Drouart, G., & Knudsen, K. K. 2016a, *ApJ*, 823, 107
Fan, L., Han, Y., Fang, G., Gao, Y., Zhang, D., Jiang, X., Wu, Q., Yang, J., *et al.* 2016b, *ApJ*, 822, L32
Feroz, F., Hobson, M. P., & Bridges, M. 2009, *MNRAS*, 398, 1601
Gregory, P. 2005, *Bayesian Logical Data Analysis for the Physical Sciences* (New York, NY, USA: Cambridge University Press)
Han, Y. & Han, Z. 2012, *ApJ*, 749, 123
—. 2014, *ApJS*, 215, 2
—. 2019, *ApJS*, 240, 3
Heber, U. 2009, *ARA&A*, 47, 211
Kriek, M., van Dokkum, P. G., Labbé, I., Franx, M., Illingworth, G. D., Marchesini, D., & Quadri, R. F. 2009, *ApJ*, 700, 221
Kroupa, P. 2001, *MNRAS*, 322, 231
Leitherer, C., Ortiz Otálvaro, P. A., Bresolin, F., Kudritzki, R.-P., Lo Faro, B., Pauldrach, A. W. A., Pettini, M., & Rix, S. A. 2010, *ApJS*, 189, 309
Marigo, P. & Girardi, L. 2007, *A&A*, 469, 239
Marigo, P., Girardi, L., Bressan, A., Groenewegen, M. A. T., Silva, L., & Granato, G. L. 2008, *A&A*, 482, 883
Moe, M. & Di Stefano, R. 2017, *ApJS*, 230, 15
Muzzin, A., Marchesini, D., Stefanon, M., Franx, M., Milvang-Jensen, B., Dunlop, J. S., Fynbo, J. P. U., Brammer, G., *et al.* 2013, *ApJS*, 206, 8
Paxton, B., Cantiello, M., Arras, P., Bildsten, L., Brown, E. F., Dotter, A., Mankovich, C., Montgomery, M. H., *et al.* 2013, *ApJS*, 208, 4
Rosenfield, P., Marigo, P., Girardi, L., Dalcanton, J. J., Bressan, A., Williams, B. F., & Dolphin, A. 2016, *ApJ*, 822, 73
Ruiz, A., Miniutti, G., Panessa, F., & Carrera, F. J. 2010, *A&A*, 515, A99+
Vazdekis, A., Sánchez-Blázquez, P., Falcón-Barroso, J., Cenarro, A. J., Beasley, M. A., Cardiel, N., Gorgas, J., & Peletier, R. F. 2010, *MNRAS*, 404, 1639
Zhang, F., Han, Z., Li, L., & Hurley, J. R. 2005, *MNRAS*, 357, 1088

Discussion

JOEL LEJA: Given the many uncertainties in SED modeling, can you be sure that a model comparison is uniquely showing that that parameter is favored?

YUNKUN HAN: No. When you want to answer questions about one specific parameter (e.g. SSP model) with Bayesian model comparison, other uncertainties (SFH, DAL) can be marginalized out to obtain robust answer. However, there are many possibilities for the other two. Indeed, to make the results of model comparison as robust as possible, we should make the selection of the other two as flexible as possible.

DAVID ROSARIO: Can you distinguish between the model parameter and systematic issues with the dataset?

YUNKUN HAN: Yes. Except for the normal physical parameters, we have added the parameter $err_{\rm sys}^{\rm obs}$ to take into account the possible systematic issues within the dataset.

Challenges in Panchromatic Modelling with Next Generation Facilities
Proceedings IAU Symposium No. 341, 2018
M. Boquien, E. Lusso, C. Gruppioni & P. Tissera, eds.

doi:10.1017/S1743921319002758

Stellar population synthesis of galaxies with chemical evolution model

Shiyin Shen[1,2] and Jun Yin[1]

[1]Key Laboratory for Research in Galaxies and Cosmology, Shanghai Astronomical Observatory, Chinese Academy of Sciences, 80 Nandan Road, Shanghai, 200030, China
[2]Key Lab for Astrophysics, Shanghai, 200234, China
email: ssy@shao.ac.cn

Abstract. The derivation of accurate stellar populations of galaxies is a non-trivial task because of the well-known age-metallicity degeneracy. We aim to break this degeneracy by invoking a chemical evolution model (CEM) for isolated disk galaxy, where its metallicity enrichment history (MEH) is modelled to be tightly linked to its star formation history (SFH). Our CEM has been successfully tested on several local group dwarf galaxies whose SFHs and MEHs have been both independently measured from deep colour-magnitude diagrams of individual stars. By introducing the CEM into the stellar population fitting algorithm as a prior, we expect that the SFH of galaxies could be better constrained.

Keywords. galaxies: stellar content, galaxies: evolution, galaxies: abundances

1. Introduction

Star formation history (SFH), i.e. the amount of stars formed in galaxies as function of time, is one of the elements of galaxy evolution. Stars are formed from cool phase gas and then evolved stars return metals into the interstellar medium (ISM). The ISM cools and new generation of stars form. Among this circle, the SFH is the key factor that links the gas cooling and metallicity enrichment history (MEH) of galaxies.

However, recovering the SFH of galaxies is difficult (Conroy, 2013). For very nearby galaxies, the deep colour-magnitude diagram (CMD) reaching the turn-off stars is known as the only direct and most reliable method that can recover both of their detailed SFHs and MEHs. For galaxies at larger distances, stellar population synthesis methods have been developed and used to fit their observed integrated spectral energy distributions (SEDs). In an idealized case, the SED of a galaxy can be viewed as a composition of single stellar populations (SSPs) with different ages and metallicities. However, because of the similarities of the old SSPs ($t > 1$ Gyr) and the age-metallicity degeneracy among SSPs, recovering the detailed weights of each SSPs, especially for those with $t > 1$ Gyr, is very difficult. In reality, the observables are further complicated by many other details, e.g. stellar kinematics, ionised gas emissions, central AGNs, dust attenuation and emission etc. Therefore, typically, only the first order description of the SFH, e.g. the average age or the fraction of young stellar populations, rather than its detailed shape, could be reliably derived from the observed SEDs of galaxies.

2. Recovering detailed SFH

In popular SED fitting algorithms (e.g. STARLIGHT, Cid Fernandes *et al.* 2005), the ages and metallicties of SSPs are considered as independent parameters. To recover the detailed SFH and MEH of a galaxy, a library of SSPs with different ages and metallicities

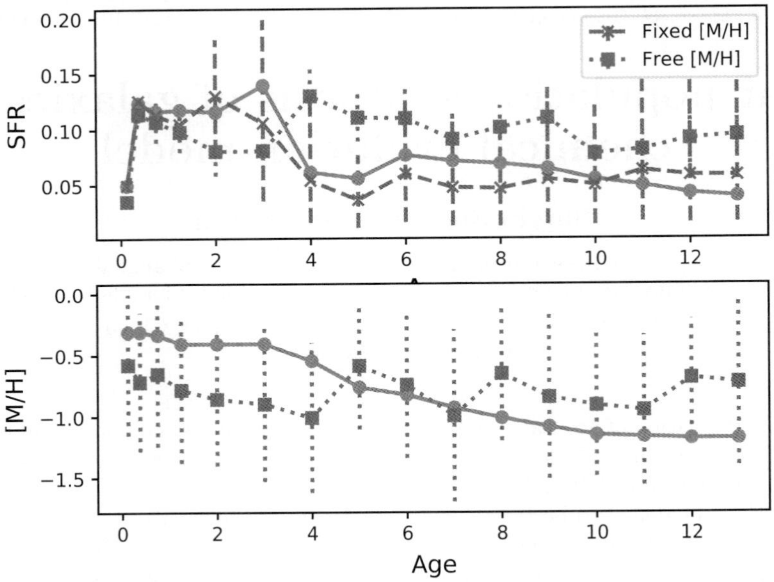

Figure 1. The SFH (top panel) and MEH (bottom panel) of a mock galaxy. The solid circles represent the input SFH and MEH, whereas the red squares show the SFH and MEH recovered from the mock spectrum with no prior information. The green crosses in the top panel show the recovered SFH when its MEH is assumed to be the blue circles in the bottom panel.

is first built, then the best combinations of these SSPs are searched in the library space. However, because of the age-metallicity degeneracy, the recovering of the SSP weights from SED is an ill-posed problem in mathematics.

To show this problem more intuitively, we make a mock galaxy spectrum and test how good a full spectrum fitting algorithm can recover its SFH and MEH. We take the SFH and MEH of the LMC bar region, which are derived from the deep CMD fitting (Ruiz-Lara *et al.* 2015) and are shown as the solid circles in Fig. 1. We use SSPs from MILES library (Falcón-Barroso *et al.*, 2011) to generate the mock spectrum in the wavelength range $3600 - 7500\mathring{A}$ with a resolution of $R = 2000$ and S/N=50 per pixel. For simplicity, we neither consider the kinematics of stellar populations nor consider the dust attenuation.

For this idealized case, can we recover the input SFH and MEH accurately using a full spectrum fitting? We take the same MILES SSP library and use a MCMC algorithm to probe the full age and metallicity space for the best solution. The resulting SFH and MEH, converted from the likelihood distributions of SSP weights, are shown as the red squares in each panel of Fig. 1. Except the weights of young stellar populations ($t < 1$ Gyr), the weights of the old stellar populations and the shape of the MEH are both not well constrained.

Is an accurate MEH knowledge helpful to recover the accurate SFH? To test this idea, we set the metallicities of the different age SSPs to follow the input MEH (the blue circles in the bottom panel of Fig. 1). We then make the full spectrum fitting again. Now, we only need to probe the weights of the SSPs in age space. The resulted SFH is shown as the green crosses in the top panel of Fig. 1. As can be seen, when MEH is known as a prior, the SFH is recovered with much better accuracy.

Fig. 1 shows that the prior information on the MEH, which breaks the age-metallicity degeneracy, is crucial for recovering the accurate SFH of galaxies. Actually, the importance of the prior information on SFH or MEH have already been explored by many

SED fitting codes. For an example, the code STECKMAP (Ocvirk *et al.* 2006) allows a penalization of the best fitting with an assumed MEH. However, a realistic MEH needs a physical justification. Indeed, that is the chemical evolution model we will discuss next.

3. Chemical evolution model

We consider galaxy as a pool of stars and gas. At any given time, there are both inflow and outflow of gas into the pool. The inflow gas is primordial, i.e. with zero metallicity. Stars formed from gas, and died stars return enriched gas into the pool. Some of the enriched gas makes up of the outflows.

We write the time dependent star formation rate (SFR), gas inflow and outflow rate as $\Psi(t)$, $f_{in}(t)$ and $f_{out}(t)$ respectively. Stars are born from gas pool following an initial mass function (IMF). We assume that the massive stars ($M > 1M_\odot$) die immediately and return gas into the pool, while the low mass stars ($M < 1M_\odot$) have infinite lifetime. We take the classical Salpter IMF, which has a gas return fraction $R \sim 0.3$. Thus, the evolution of gas in pool is

$$\frac{dM_{gas}(t)}{dt} = -(1-R)\Psi(t) + f_{in}(t) - f_{out}(t). \tag{3.1}$$

On the other hand, the surface star formation rate density of a galaxy is known to be tightly correlated with its surface gas density Σ_{gas}, i.e. the well-known Kennicutt-Schmidt relation, $\Psi_\Sigma \propto \Sigma_{gas}^{1.4}$ (Kennicutt, 1998). For gas outflow, we assume it is driven by supernova explosion so that it is proportional to SFR and inversely correlated with the potential well of the galaxy,

$$f_{out}(t) = \eta \left[0.5 + \left(\frac{v_{vir}}{70 \text{ km s}^{-1}}\right)^{-3}\right] \cdot \Psi(t) \tag{3.2}$$

where v_{vir} is the circular velocity of the galaxy halo, and η is the wind efficiency.

The evolution of metallicity $Z(t)$ is a balance between the star formation and metal outflow, which is written as

$$\frac{d(ZM_{gas})}{dt} = -Z(1-R)\Psi(t) + y(1-R)\Psi(t) - Zf_{out}, \tag{3.3}$$

where y is the yield and we take $y = 0.1$.

With above equations (3.1 to 3.3), once with structure parameters (to convert gas mass to surface density), for any given $\Psi(t)$ of a galaxy, we can predict both its $Z(t)$ and gas inflow/outflow histories.

4. Test on LCID galaxies

We test above CEM with LCID galaxies (Gallart, 2007), whose detailed SFHs and MEHs have been obtained using deep CMDs from HST. As an example, we take the SFHs of three different type LCID galaxies, Tucana (dSph), LGS-3 (dTran), IC 1613 (dIrr), and predict their $Z(t)$ from the above CEM. We calculate their average surface densities inside the half-light radii and use their stellar masses to estimate the circular velocities. The wind efficacy is set as $\eta = 0.4$ in all the cases. The results are shown in Fig. 2. For all three galaxies, our CEM reproduces their MEHs from SFHs reasonably well.

5. Conclusion

Encouraged by Fig. 2, we conclude that our CEM can be used to break the age-metallicity degeneracy in stellar population synthesis studies. Specifically, we may start from the SED fitting with both SFH and MEH free. With a preliminary SFH, a new MEH could be predicted from our CEM, which then could be used as a prior in the

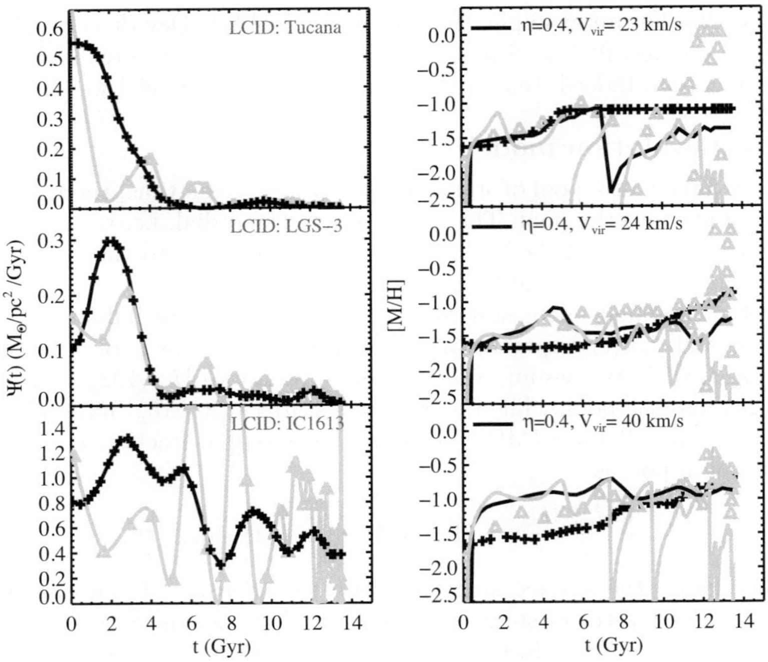

Figure 2. The SFH (left panels) and MEH (right panels) of 3 LCID galaxies. The black crosses show the best estimations of SFHs and MEHs from LCID project, whereas the yellow triangles show one of their Monte-Carlo realization (indicating the uncertainty, *Dan Weisz, private communication.*). The solid lines in the left panels are the continuous SFHs used as the input of CEM, while the model predicted MEHs are shown as the solid curves with corresponding colours in right panels.

second round SED fitting (as we done in Fig. 1). With several iterations, a self-consistent SFH and MEH could be finally obtained. We expect that this algorithm would reduce the uncertainties of the final SFH estimation, and is proper for isolated disk galaxies.

SS thanks supports from NSFC (No. 11573050 and 11433003). JY thanks support from Science Foundation of Shanghai (No. 16ZR1442100).

References

Cid Fernandes, R., Mateus A., Sodré L., Stasińska G., & Gomes J. M. 2005, *MNRAS*, 358, 363

Conroy, C. 2013, *ARA&A*, 51, 393

Falcón-Barroso, J., Sánchez-Blázquez, P., Vazdekis, A., Ricciardelli, E., Cardiel, N., Cenarro, A.J., Gorgas, J., & Peletier, R. F. 2011, *A&A*, 532, A95

Gallart, C. 2007, *Proceedings of the International Astronomical Union*, 2

Kennicutt, R. C. 1998, *ARA&A*, 36, 189

Ruiz-Lara, T., Pérez I., Gallart, C., Alloin, D., Monelli, M., Koleva, M., Pompei, E., Beasley, M., *et al.* 2015, *A&A*, 583, A60

Ocvirk, P., Pichon, C., Lançon, A., & Thiébaut, E. 2006, *MNRAS*, 365, 74

Discussion

THEMIYA NANAYAKKARA: Can your CEM be used to predict $[O/Fe]$ enhanced stellar populations as shown by Stedel et al. (2016) in $z \sim 2$ galaxies?

Shiyin Shen: No, currently, our model only considers global metallicity, and do not distinguish α and Fe metals. But this is a good idea, we will consider it in the future.

Themiya Nanayakkara: What about the role of mergers in the CEM?

Shiyin Shen: For the major merger events, the halo mass will change significantly. Our model can not deal with this situation yet.

Challenges in Panchromatic Modelling with Next Generation Facilities
Proceedings IAU Symposium No. 341, 2018
M. Boquien, E. Lusso, C. Gruppioni & P. Tissera, eds.

doi:10.1017/S1743921319004794

A new galaxy Spectral Energy Distribution model with the evolution of dust consistent with chemical evolution

Kazuki Y. Nishida[1], **Tsutomu T. Takeuchi**[1], **Takuma Nagata**[1], **Ryosuke S. Asano**[1] **and Akio K. Inoue**[2,3]

[1]Nagoya University, Japan
[2]Osaka Sangyo University, Japan
[3]Waseda University, Japan

Abstract. The spectral energy distribution (SED) model should treat the evolution of a galaxy from its birth. Dust in galaxies affects the formation and evolution of galaxies in various ways. For example, dust grains scatter and absorb stellar emitted ultraviolet (UV) photons and re-emit the radiation at infrared (IR) wavelengths. In this work, we construct a galaxy SED model based on our dust evolution model (Asano *et al.* 2013a,b, 2014) with a rigorous treatment of the chemical evolution. To reduce the computational cost, we adopt mega-grain approximation (MGA; Inoue, 2005). MGA regards a high density dusty region as a huge size (10 pc) dust grain for calculating dust scattering. In this approximation, we can solve the radiative transfer easily and provide SEDs and attenuation curves of galaxies. This model can be used to fit any galaxy in the wavelength range of 10 nm-3 mm.

Keywords. Panchromatic codes and modeling techniques

1. Introduction

Galaxy spectral energy distribution (SED) fitting is an important method which is frequently used for estimating several physical properties (e.g., star formation rate, stellar mass and dust mass). The main contributions of galaxy emission are separated into two components, stellar and dust. Dust absorbs UV light emitted by stars and re-emits as far-infrared (FIR) radiation. Therefore, dust in galaxies fundamentally determines the galaxy SED.

Some previous SED models (e.g., PEGASE; Fioc & Rocca-Volmerange, 1999) use an empirical dust model based on the properties of nearby galaxies. However, according to result from our new dust model (Asano model; Asano *et al.* 2013a,b, 2014), the dust-to-gas mass ratio suddenly rises up at a certain galaxy age. It implies that if we use such empirical dust mass distribution models to calculate distant galaxy SED, it might be inaccurate. Thus, we need to construct a new SED model based on Asano model.

2. Method

2.1. *Dust model*

We use a new dust evolution model proposed by Asano *et al.* It is a dust evolution model taking into account the chemical evolution and can predict the grain size distribution at each stage of the galaxy evolution. Figure 1 shows the most striking result which Asano model has shown. We clearly see a well-defined moment at which dust-to-gas mass ratio

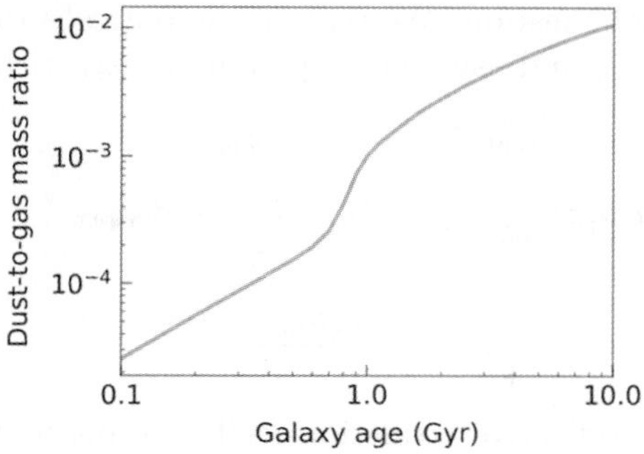

Figure 1. An example of the time dependence of the dust-to-gas mass ratio based on the Asano model. The dust-to-gas mass ratio suddenly rises up at early galaxy age. The age of the sudden increase depends on the supposed star formation history of a galaxy.

suddenly rises up. This age depends on star formation history. Figure 1 is a result for a Milky Way (MW)-like galaxy.

In this model, dust grains are treated as ten species (C, Si, Fe, FeS, Al_2O_3, $MgSiO_3$, Mg_2SiO_4, SiO_2, MgO, Fe_3O_4). However, since the detailed physical properties are not known for all the species, we divided them into two species for dust grains, carbon and silicate. We divide carbon grains into two subspecies, polycyclic aromatic hydrocarbons (PAH) and graphite, with a size boundary of 50 Å. PAHs are further subdivided into neutral and ionized PAHs. We regard the rest of the species as astronomical silicate. We use optical properties of PAHs, absorption coefficient, albedo, and asymmetry paramters, which were taken from Li & Draine (2001). Graphite and astronomical silicate grains properties were taken from Draine & Lee (1984a) and Laor & Draine (1993).

2.2. *Stellar emissions*

We use PEGASE for calculating stellar emissions except dust effects. PEGASE is an evolutionary SED model. We use the Salpeter (1955) initial mass function (IMF)

$$\phi(m) = Am^{-2.35}. \tag{2.1}$$

Where m is the stellar mass and A is a normalization constant defined from

$$\int_{0.1N_\odot}^{100M_\odot} m\phi(m)\mathrm{d}m = 1. \tag{2.2}$$

We use the Schmidt (1959) low to compute the star formation rate (SFR)

$$\psi(t) = \frac{M_{\mathrm{gas}}(t)}{\tau_{\mathrm{SF}}} \tag{2.3}$$

where $M_{\mathrm{gas}}(t)$ is the gas mass at time t and τ_{SF} is the star forming time scale.

2.3. *Mega-grain approximaion (MGA)*

We use two approximations to make the radiative transfer calculation faster. One is the mega-grain approximation (MGA; Inoue, 2005; Városi & Dwek 1999).

There are two thermally stable phases in the ISM, assuming thermal energy and chemical equilibria. One is the warm neutral medium (WNM) and the other is the cold neutral medium (CNM). We regard the WNM as a homogeneous inter-clump medium and CNM as a huge particle called a mega-grain. The size of mega-grain is set to be Jeans length $\sqrt{15p/4\pi G/\rho_{\mathrm{mg}}} = 10.4$ pc, where p is pressure of CNM, ρ_{mg} is mega-grain mass density and G is the gravitational constant. We replace the optical properties of normal grains (dust opacity $k_{\rm d}$, albedo $\omega_{\rm d}$ and asymmetry parameter $g_{\rm d}$) with the effective values

$$\kappa_{\rm eff} = \kappa_{\rm mg} + \kappa_{\rm icm} \tag{2.4}$$

$$\omega_{\rm cl} = \frac{\omega_{\rm mg}\kappa_{\rm mg} + \omega_{\rm d}\kappa_{\rm mg}}{\kappa_{\rm eff}} \tag{2.5}$$

$$g_{\rm cl} = \frac{g_{\rm mg}\kappa_{\rm mg} + g_{\rm d}\kappa_{\rm mg}}{\kappa_{\rm eff}} \tag{2.6}$$

where $\kappa_{\rm eff}, \omega_{\rm eff}, g_{\rm eff}$ are the effective extinction coefficient per unit length, effective albedo, effective asymmetry parameter, $\kappa_{\rm icm} = k_{\rm d} D \rho_{\rm icm}$ is the extinction coefficient per unit length of the interclump medium, $\omega_{\rm cl}$ and $g_{\rm cl}$ are the albedo and asymmetry parameter of a clump, respectively. We use MGA for solving the radiative transfer. However, in contrast we do not use MGA for treating dust re-emission, because the mega-grain has aout very large heat capacity, we use a normal dust grain for calculating dust temperature.

2.4. *One-dimensional galaxy approximation*

We suppose a one-dimensional plane-parallel geometry along the z-axis (Inoue, 2005). We put the gas-dust disk with a constant mean density at the mid-plane of the galaxy. This mid-plane contains mega-grains and young stars. Outside the gas-dust disk, we assume that there is no dust and only old stars diffusing from the mid-plane of the disk.

The optical depth is defined as $\mathrm{d}\tau = -\kappa_{\rm eff}\mathrm{d}z$ with $\tau = 0$ at $z = h_{\rm d}$. We assume effective optical parameters $\kappa_{\rm eff}, \omega_{\rm eff}, g_{\rm eff}$ are constant throughout the gas-dust disk. In this approximation, the radiative transfer equation is written by

$$\mu \frac{\mathrm{d}I(\tau,\mu)}{\mathrm{d}\tau} = S(\tau,\mu) - I(\tau,\mu) \tag{2.7}$$

where I is the intensity, μ is the cosine of the angle between the ray and the z-axis and S is the source function given by

$$S(\tau,\mu) = \eta_*(\tau)/\kappa_{\rm eff} + \omega_{\rm eff}\int_{-1}^{1} I(\tau,\mu')\Phi(g_{\rm eff},\mu,\mu')\mathrm{d}\mu' \tag{2.8}$$

where η_* is the stellar emissivity and Φ is the scattering phase function. We use the Henyey-Greenstein phase function (Henyey & Greenstein, 1941).

3. Result

Figure 2 shows one of the result of our SED model. The vertical axis represents galaxy luminosity times wavelength in units of solar luminosity per solar mass. Galaxy properties (e.g. total mass, star formation rate, disk radius and disk height) are fixed to the value corresponding to the face on ($\mu = 1$) MW-like galaxy at the age 10 Gyr.

4. Discussion

Figure 3 shows SED model at each galaxy age. 10 Gyr is same as figure 2. At short wavelengths, the emission of an old galaxy is more attenuated than for a young galaxy with increasing dust mass increasing. The FIR luminosity is very low at 100 Myr and becomes strong as time pass. However, we find the luminosity decreasing with elapsed time from 1 Gyr to 10 Gyr. It is affected by decreasing stellar luminosity which is the source of dust heating.

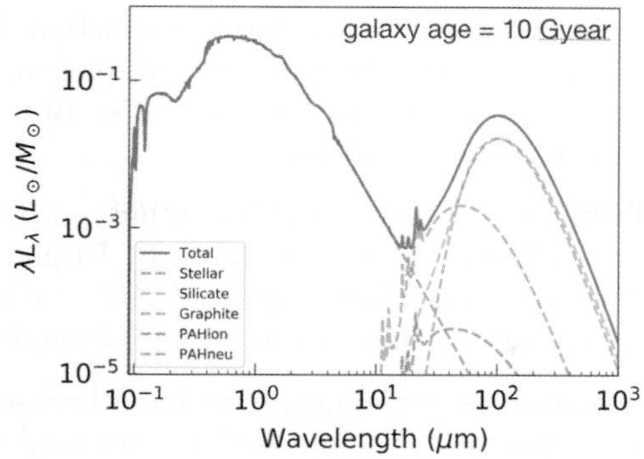

Figure 2. Example of our galaxy SED model. Parameters are fixed to MW ones. The black solid line represents the total galaxy emission, the blue dashed line is stellar emission, and others are the dust emission. This galaxy age is 10 Gyr.

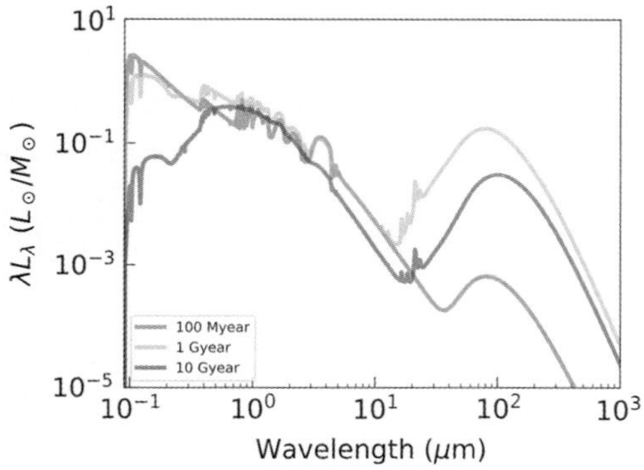

Figure 3. Example of SED models at each galaxy age. The blue, orange and green lines represent ages of 100 Myr, 1 Gyr and 10 Gyr.

References

Asano, S. R., Takeuchi, T. T., Hirashita, H., & Inoue, A. K. 2013a, *EP&S* 65, 213

Asano, R. S., Takeuchi, T. T., Hirashita, H., & Nozawa, T. 2013b, *MNRAS* 432, 637

Asano, R. S., Takeuchi, T. T., Hirashita, H., & Nozawa, T. 2014, *MNRAS* 440, 134

Da Cunha, E, Charlot, S, & Elbaz, D 2008, *MNRAS* 5, 81

Draine, T. B. & Lee, M. H. 1984, *ApJ* 285, 89

Draine, T. B. 2003, *ARA&A*, 41, 241

Fioc, M. & Rocca-Volmerange, B. 1999, *arXiv e-prints*

Henyey, C. L. & Greenstein, L. J 1941, *ApJ* 93, 70

Inoue, A. K. 2005, *MNRAS* 359, 171

Laor, A. & Draine, T. B. 1993, *ApJ* 402, 441

Li, A. & Draine, T. B. 2001, *ApJ* 554, 778

Salpeter, E. E 1955, *ApJ* 121, 161

Salpeter, M. 1959, *ApJ* 129, 243

Városi, F. & Dwek, E. 1999, *ApJ* 463, 681

Discussion

BUAT: On your resulting SEDs it seems that a UV absorption at 200 nm is still observed at 1 and 10 Gyr. Is the UV bump always present in your attenuation curves?

NISHIDA: The attenuation curve starts from a "gray" extinction, without a strong wavelength dependence at short wavelengths. Then it starts to steepen, and the bump emerges along with the increase of small dust grains. Later on, after 10 Gyrs, the curve becomes gradually flatter again and the bump weakens.

GALLIANO: Have you benchmarked your stochastic heating code? You could, for instance, compare your temperature distribution to these shown by Draine (2003). My question is motivated by the fact that your mid-infrared spectral energy distributions do not look right. It looks like your grains are not fluctuating to high enough temperatures.

NISHIDA: Indeed our mid-infrared spectral energy distributions seem to be weak. Grains are not heating to temperatures high enough. We have not made a benchmark test yet. We will do it and present tuned parameters for the Milky Way.

Challenges in Panchromatic Modelling with Next Generation Facilities
Proceedings IAU Symposium No. 341, 2018
M. Boquien, E. Lusso, C. Gruppioni & P. Tissera, eds.

doi:10.1017/S1743921319002412

The Spitzer Extragalactic Representative Volume Survey - measuring photometric redshifts for ~4 million galaxies - challenges and ways forward

Janine Pforr

Scientific Support Office, Directorate of Science and Robotic Exploration,
European Space Research and Technology Centre, (ESA/ESTEC), Keplerlaan 1,
2201 AZ Noordwijk, The Netherlands
email: janine.pforr@esa.int, Research Fellow

Abstract. We highlight the challenges as well as lessons learnt in the derivation of the photometric redshifts for ~4 million galaxies at $0 < z \lesssim 6$ contained in the Spitzer Extragalactic Representative Volume Survey (SERVS) and summarise the photometric redshift results recently published in Pforr *et al.* (2019). The inhomogeneous nature of the ancillary photometry for SERVS presents a similar situation to the one future, large, extragalactic surveys with e.g. LSST and JWST will face. We employ template SED-fitting to determine photometric redshifts. Our comparison of photometric redshifts to ~75.000 public, spectroscopic redshifts results in an average $\sigma_{\rm NMAD}$ of ~0.038 and outlier fraction of 3.7% for sources with the best photometric coverage. We find that photometric redshifts are determined most robustly when filter bands are numerous and cover a wide wavelength range. We highlight some possible improvements for the photometric redshifts in SERVS in the future.

Keywords. galaxies: general, galaxies: distances and redshifts, galaxies: fundamental parameters, surveys

1. The Spitzer Extragalactic Representative Volume Survey

Galaxy evolution studies have been driven by large galaxy surveys in the last decades. By now, some extragalactic fields are prime observation targets due to the large amount of existing ancillary data spanning a wide wavelength range. Particularly, for the robust derivation of photometric redshifts and other stellar population galaxy parameters such as age and stellar mass via spectral energy distribution (SED) fitting, a multi-wavelength approach is crucial. However, new facilities and technologies produce increasingly large data sets to process and analyse, providing a significant challenge for current tools and methods. One example large survey is the Spitzer Extragalactic Representative Volume Survey (SERVS; PI: M. Lacy; Mauduit *et al.* 2012) which covers five popular extragalactic fields in the 3.6 and 4.5 μm bands of the Spitzer Space Telescope Infrared Array Camera (IRAC, Fazio *et al.* 2004) carried out during the warm phase of the Spitzer Space Telescope mission. With a total of 18 deg^2 distributed over the Chandra Deep Field South (CDFS), ELAIS-South 1 (ES1), ELAIS-North 1 (EN1), Lockman Hole (LH) and XMM-LSS (XMM) fields SERVS is able to overcome cosmic variance. In this work we make use of the SERVS DR2 and data fusion containing photometry from the UV, optical, near- and mid-IR compiled by Vaccari (2015) (http://www.mattiavaccari.net/df) and carry out a comparison between photometric and spectroscopic redshifts for sources detected in both IRAC bands down to the coverage-weighted signal to noise CSNR = 3 in both bands. Details about the survey and data fusion can be found in Mauduit *et al.* (2012) and Vaccari (2015), respectively.

2. The SERVS photometric redshifts

We determine the SERVS photometric redshifts via template spectral energy distribution (SED) fitting using the HYPERZ code package (Bolzonella *et al.* 2000). Depending on the specific SERVS field, we are able to use a maximum of 12 filter bands for the SED fitting ranging from the observed-frame UV to the IR. In order to match the various magnitude measurements and point spread function (PSF) differences between ancillary surveys and SERVS, we use apertures in each band as close to 1.9" (the SERVS flux measurement aperture we adopt) as possible.

The fitting is then carried out using model template spectra. These template spectra are based on the Maraston (2005) stellar population models for a Salpeter (1955) initial mass function and 221 age steps, and have exponentially declining star formation histories (SFHs), so called τ-models with e-folding times of τ=0.1, 0.3, 1.0 and 2.0 Gyr. Our setup contains each τ-model in four metallicity flavours, i.e. 1/5, 1/2, 1, and 2 solar metallicity. We restrict the age of the model in the fitting to 0.1 Gyr as minimum and the age of the Universe at the probed redshift as the maximum. We further use a Calzetti *et al.* 2000 dust reddening law with values of A_V=0 to 3 in steps of 0.2. Each template is then redshifted from 0 to 6 in steps of 0.01. The reduced χ^2, χ_r^2, is calculated for each template for each object. The template with the lowest χ_r^2 is deemed the best fit for that object and provides its photometric redshift. This template setup was chosen according to the findings of Pforr *et al.* (2013) who found this setup to provide a good compromise between CPU runtime and photometric redshift accuracy as determined with the help of simulated galaxies from a galaxy formation model.

In order to evaluate the photometric redshifts for SERVS, we compare the photometric with $\sim$75000 publicly available spectroscopic redshifts and compute the normalised median absolute deviation $\sigma_{\rm NMAD} = 1.48 \times median[|\frac{(\Delta z - median(\Delta z))}{(1+z_{\rm spec})}|]$ (Huber 1981; Hoaglin *et al.* 1983) with $\Delta z = z_{\rm spec} - z_{\rm phot}$, and the percentage of catastrophic outliers (outlier fraction η) defined as $|\Delta z| > 0.15 \times (1 + z_{\rm spec})$. Both quantities are commonly used to evaluate the photometric redshift success (e.g., Brammer *et al.* 2009; Dahlen *et al.* 2013).

We investigate the dependency of $\sigma_{\rm NMAD}$ and outlier fraction η as a function of number of photometric filter bands in the fitting, for each SERVS fields, for each spectroscopic redshift source catalogue in the XMM field, and for SERVS as a whole, and as a function of i-band magnitude for each SERVS field. We find that overall, the photometric redshifts are determined best (smallest $\sigma_{\rm NMAD}$ and η) when the wavelength coverage is large, i.e. more filter bands are used in the fitting. We find a $\sigma_{\rm NMAD} = 0.038$ and an outlier fraction $\eta < 4\%$ for sources with the largest number of filter bands. We also find that photometric redshifts are determined best for the brightest objects. The majority of SERVS sources lie at $z < 1.5$. Details about the method and the dependency of the photometric redshift accuracy can be found in Pforr *et al.* (2019).

3. Challenges for the SERVS photometric redshifts

The determination of photometric redshifts for SERVS faces several, non-trivial challenges. The first challenge is presented by the inhomogeneity of the ancillary data set. SERVS itself covers a much larger area than most of the optical and near-IR surveys available in the SERVS fields. Unsurprisingly, the deepest data is only available in very small areas such as portions of the CDFS field. On the other hand, much shallower data exists for almost each of the fields yet often only in a few filter bands. While we select ancillary data in order to strike the best compromise between survey depth and area, differences between the surveys still arise. Differences are, e.g., the different PSF sizes between the optical, near-IR and Spitzer surveys which hamper source matching and deblending across different surveys, and different types of magnitude measurements such as AUTO vs corrected and uncorrected aperture magnitudes.

Another challenge arises from the spectroscopic surveys used to evaluate the reliability and robustness of the photometric redshifts. While there are plenty of spectroscopic redshifts available from a variety of surveys, they inherit their own selection function, often dictated by brightness limits. Consequently, often only the brightest objects with the best photometric wavelength coverage also have spectroscopic redshifts creating a potential bias for fainter, or more distant objects. Additionally, depending on the spectral resolution, the spectroscopic redshifts have different accuracies.

Finally, the number of sources to process provides a significant computing challenge in terms of CPU hours and disk storage space, a situation only expected to worsen with the advancement of new and larger facilities.

4. Possible improvements for the SERVS photometric redshifts

Given the challenges highlighted in Section 3 several methods could be used to improve future iterations of the SERVS photo-z and those of up-coming surveys. We highlight the following three examples.

Data homogenisation: In order to homogenise the data, recent surveys, e.g. CANDELS, have taken the approach of using so-called forced photometry for the creation of their multi-wavelength photometry which is subsequently used for the SED-fitting (Guo *et al.* 2013; Galametz *et al.* 2013). The goal of this technique is to improve the accuracy of the multi-band photometry for catalogues containing surveys with disparate angular resolutions. An example of an application of this type of technique to a large database is given in Lang (2016) which employed *The Tractor* image modelling code. *The Tractor* uses information on source positions, brightnesses, and shape parameters to model the source surface brightness profile at high resolution. This high resolution model is then convolved with the PSF of the other bands in the database with lower resolutions, thus ensuring accurate source cross-matching between bands, improved sensitivity to inherently faint sources, and deblending of sources in the lower resolution bands. For SERVS galaxies, this method gains fluxes for some bands that did not have an entry in the publicly available catalogues before (i.e. non- detections) and deals with blending issues in the IRAC photometry as shown by Nyland *et al.* (2017) for 1 deg^2 in the SERVS XMM field. Specifically, Nyland *et al.* (2017) were able to double the number of sources for which the maximum number of filter bands is available for the fitting which in turn greatly increases the number of accurate photometric redshifts compared to the position-matched catalogues. Additionally, in Nyland *et al.* (2017) the number of high-redshift sources (i.e. $z > 4$) increased due to the better photometry. However, for a survey of the size of SERVS the forced photometry approach results in a significant amount of reprocessing of ancillary data and presents a considerable computational challenge. Nevertheless, an effort to derive forced photometry for all SERVS sources is currently underway.

Computing-related issues: One major challenge for large surveys, existing and upcoming, relates to computing in terms of CPU runtime of any data analysis codes, and data storage of raw, reduced, and processed data. However, CPU runtime can be improved for example by parallelising and optimising existing codes. Alternatively, GPU processing offers new ways of fast computing. Data processing can then be carried out on large supercomputers to use many CPUs or GPUs simultaneously. Particularly for large surveys, standard data processing pipelines could be considered to offer easier access to processed and value-added data for large numbers of users.

Optimisation of the SED-fitting: Finally, the SED-fitting could be optimised itself, for example by using different sets of templates for specific objects such as AGN. Additionally, deeper photometric data could be used where it exists even if only for a comparatively small number of sources instead of the compromise we made by using shallower data that covers most of the fields. Finally, new fitting approaches such as those using neural networks, MCMC based codes and machine learning techniques could be employed (see, e.g., Acquaviva *et al.* 2011).

5. Summary

We provide a brief overview of the SERVS survey and the photometric redshifts (Pforr *et al.* 2019), highlighting the challenges of large data sets and inhomogeneous ancillary data in the photometric redshift determination. We further discuss improvements that can be made for SERVS and future, large surveys, such as data homogenisation via forced photometry, improvements in computing, and optimisation of the SED-fitting.

References

Acquaviva, V., Gawiser, E., & Guaita, L., 2011, *ApJ*, 737, 47

Bolzonella, M., Miralles, J.-M., & Pelló, R., 2000, *A&A*, 363, 476

Brammer, G. B., Whitaker, K. E., van Dokkum, P. G., Marchesini, D., Labbé, I., Franx, M., Kriek, M., Quadri, R. F., *et al.* 2009, *ApJL*, 706, 173

Calzetti, D., Armus, L., Bohlin, R. C., Kinney, A. L., Koornneef, J., & Storchi-Bergmann, T., 2000, *ApJ*, 533, 682

Dahlen, T., Mobasher, B., Faber, S. M., Ferguson, H. C., Barro, G., Finkelstein, S. L., Finlator, K., Fontana, A., *et al.* 2013, *ApJ*, 755, 93

Duncan, K. J., Brown, M. J. I., Williams, W. L., Best, P. N., Buat, V., Burgarella, D., Jarvis, M. J., Małek, K., *et al.* 2018, *MNRAS*, 473, 2655

Fazio, G. G., Hora, J. L., Allen, L. E., Ashby, M. L. N., Barmby, P., Deutsch, L. K., Huang, J.-S., Kleiner, S., *et al.* 2004, *ApJS*, 154, 10

Galametz, A., Grazian,, A., Fontana, A., Ferguson, H. C., Ashby, M. L. N., Barro, G., Castellano, M., Dahlen, T., *et al.* 2013, *ApJS*, 206, 10

Guo, Y., Ferguson, H. C., Giavalisco, M., Barro, G., Willner, S. P., Ashby, M. L. N., Dahlen, T., Donley, J. L., *et al.* 2013, *ApJS*, 207, 24

Hoaglin, D. C., Mosteller, F., & Tukey, J. W. 1983, *Wiley Series in Probability and Mathematical Statistics, New York: Wiley, 1983, edited by Hoaglin, David C.; Mosteller, Frederick; Tukey, John W.*

Huber, P. J., 1981 *Wiley Series in Probability and Mathematical Statistics, New York: Wiley, —c1981*

Lang, D., Hogg, D. W., & Schlegel, D. J., 2016, *AJ*, 151, 36

Maraston, C. 2005, *MNRAS*, 362, 799

Mauduit, J.-C., Lacy, M., Farrah, D., Surace, J. A., Jarvis, M., Oliver, S., Maraston, C., Vaccari, M., *et al.* 2012, *PASP*, 124, 714

Nyland, K., Lacy, M., Sajina, A., Pforr, J., Farrah, D., Wilson, G., Surace, J., Häußler, B., *et al.* 2017, *ApJS*, 230, 9

Pforr, J., Vaccari, M., Lacy, M., Maraston, C., Nyland, K., Marchetti, L., & Thomas, D., 2019, *MNRAS*, 483, 3168

Pforr, J., Maraston, C., & Tonini, C., 2013, *MNRAS*, 435, 1389

Salpeter, E. E. 1955, *ApJ*, 121, 161

Salvato, M., Hasinger, G., Ilbert, O., Zamorani, G., Brusa, M., Scoville, N. Z., Rau, A., Capak, P., *et al.* 2009, *ApJ*, 609, 1250

Vaccari, M. 2015, *"The Many Facets of Extragalactic Radio Surveys: Towards New Scientific Challenges", The Spitzer Data Fusion: Contents, Construction and Applications to Galaxy Evolution Studies*, p27

Discussion

David Rosario: Can you comment on the relative merits and performance of a model-based photo-z approach versus a pure template-base approach (often used for AGN, but also used for galaxies in XMM-XXL for instance)?

Janine Pforr: Pure template based approaches have produced very good photo-z results in the past, particularly for AGN (see Salvato *et al.* 2009). Important for photo-z derivation is that the SED shapes in the sample are adequately covered by the templates. However, care has to be take not to include too many options in order to avoid degeneracies which can result in catastrophic outliers. This might be easier to achieve with a pure template based approach. Overall, the last years have seen many excellent photo-z estimates using both approaches.

Véronique Buat: You mentioned that your data are similar (come from?) the data fusion results of the HELP project. Did you compare photo-z measures to those produced by HELP (Duncan *et al.* 2018 for the method) in the same fields?

Janine Pforr: Mattia Vaccari created the data fusion for SERVS and for the HELP project, so similar approaches were used for both. I have not yet compared the SERVS photo-z to other photo-z estimates for the same objects from other photo-z efforts, since the focus of the Pforr *et al.* (2019) paper was the comparison to spectroscopic redshifts and our overall photo-z robustness. We can certainly compare the different photo-z estimates in the future to learn about differences, similarities, and odd sources.

Denis Burgarella: We see bumps in the redshift distribution at, for instance $z = 3$, $z = 5$. Are they real?

Janine Pforr: Any structures in the high-z photo-z distribution need further investigation for several reasons. 1) Shown are only objects with more than 5 (8) filter bands available in the fitting, which biases the distribution particularly at high z where sources are more likely to be detected in fewer bands. I showed how this could be improved with forced photometry. 2) At high-z SERVS is only sensitive to the brightest and most massive galaxies. 3) We have very few z_{spec} available to test our photo-z at high redshift.

Challenges in Panchromatic Modelling with Next Generation Facilities
Proceedings IAU Symposium No. 341, 2018
M. Boquien, E. Lusso, C. Gruppioni & P. Tissera, eds.

doi:10.1017/S1743921319002539

Probing the building blocks of galaxies: Sub-galactic scaling relations between X-ray luminosity, SFR and stellar mass

K. Kouroumpatzakis[1,2], A. Zezas[1,2,3], P. H. Sell[1,2], P. Bonfini[1,2], M. L. N. Ashby[3] and S. P. Willner[3]

[1]Department of Physics, University of Crete, GR-70013 Heraklion, Greece
email: kkouroub@physics.uoc.gr

[2]Institute of Astrophysics, FORTH, GR-71110 Heraklion, Greece

[3]Center for Astrophysics | Harvard & Smithsonian, Cambridge, MA

Abstract. It is well known that X-ray luminosity (L_X) originating from high mass X-ray binaries (HMXBs) is tightly correlated with the host galaxy's star formation rate (SFR). We explore this connection using a sample representative of the star-formation activity in the local Universe (Star-Formation Reference Survey; SFRS) along with a comprehensive set of star-formation (radio, FIR, 24μm, 8 μm, Hα, UV, SED fitting) and stellar mass (K-band, 3.6 μm, SED fitting) indicators, and Chandra observations. We investigate the L_X – SFR and L_X– stellar mass ($M_\star$) scaling relations down to sub-galactic scales of $\sim 1\,\mathrm{kpc}^2$. This way we extend these relations to extremely low SFR ($\sim 10^{-6} M_\odot \cdot \mathrm{yr}^{-1}$) and $M_\star$ ($\sim 10^4 M_\odot$). We also quantify their scatter and their dependence on the age of the local stellar populations as inferred from the different age sensitive SFR indicators. These results are particularly important for setting the benchmark for the formation of X-ray binaries in vigorous, but low SFR objects such as galaxies in the early Universe.

Keywords. galaxies:star formation – X-rays: galaxies – X-rays:binaries

---

1. Introduction

Various star-formation rate (SFR) indicators (e.g. UV, ionization emission lines, polycyclic aromatic hydrocarbons (PAHs), IR, radio) are used in order to trace recent and on-going SF. These different SFR indicators are probing different mechanisms associated to young stars (e.g. UV emission from the photospheres of the massive young stars, emission from the ionized gas that surrounds them, emission from dust absorbed UV, that is thermally re-emitted in the IR light, etc.). As a result, the integrated emission they trace corresponds to different stellar populations and therefore each indicator probes different timescales of star-formation. Among the SFR tracers, it is the emission from the ionized gas that is the most sensitive to the youngest stellar populations, with little contamination from older stars (e.g. Murphy *et al.* 2011; Boquien *et al.* 2014; Cervino *et al.* 2016).

In the spectral range of the X-rays it is already known that emission generated by X-ray binaries is tightly correlated with SFR and stellar mass. The integrated X-ray luminosity of low mass X-ray binaries (LMXBs) is correlated with the stellar mass (e.g. Gilfanov, 2004) and that of high mass X-ray binaries (HMXBs) with SFR (e.g. Mineo *et al.* 2012). Theoretical models show that the bulk of the XRB emission is produced by the HMXBs in short timescales (∼10 Myr; e.g. Fragos *et al.* 2013), while there is almost constant X-ray output per unit stellar mass by the long-lived LMXBs that are active through longer times (several Gyr). Therefore correlating the HMXBs with the SFR indicators that probe stellar populations of similar ages (e.g. H_α) is important in order to constrain XRB population synthesis models, explore the cosmological evolution of compact objects, and investigate the effect of HMXBs in galaxy evolution through their effect in preheating the IGM.

In this work we address these correlations both in integrated galaxy emission (Sell *et al.* 2019) and also in sub-galactic scales (down to 1 kpc^2) in order to constrain the correlation in the regime of extremely low SFR ($\sim 10^{-6}\,\mathrm{M_\odot yr^{-1}}$), $\mathrm{M}_\star(\sim 10^4\,\mathrm{M}_\odot)$, and sSFR ($10^{-13}\,\mathrm{yr}^{-1}$).

2. Sample and results

Our sample is based on the Star Formation Reference Survey (SFRS; Ashby *et al.* 2011). SFRS is comprised by 369 galaxies, that represent all modes of SF in the local Universe. They cover the 3D space of three fundamental characteristics of galaxies: the SFR, indicated by the 60 μm luminosity; the specific SFR (sSFR), indicated by the $\mathrm{K_s} - \mathrm{F_{60}}$ (mag) color; and the dust temperature, indicated by the FIR (F_{100}/F_{60}) flux ratio. The SFRS also benefits from a panchromatic coverage of the electromagnetic spectrum, from radio wavelengths up to the X-rays, including spectroscopic emission line nuclear classification of the galaxies (Maragkoudakis *et al.* 2017) and H_α imaging.

The sample used for this work consists of 14 star-forming (non-AGN) SFRS galaxies (NGC5879, NGC3353, UGC5720, NGC4194, NGC5691, NGC6090, NGC5474, NGC5585, MCG632070, NGC3245, NGC5584, NGC3656, UGC9618, and NGC5204) that have *Chandra* data of adequate quality to study the x-ray emission and stellar populations in 1 kpc^2 scales. The integrated emission of the sample galaxies spans ∼4 dex in the SFR and ∼3 dex in sSFR. In sub-galactic scales these ranges become ∼6 dex and ∼5 dex in SFR and sSFR respectively.

The integrated X-ray luminosity has been measured through spectral fitting of the *Chandra* spectra with the use of the *CIAO* and *Sherpa* tools, while statistically accounting for background AGN contamination. In order to measure XRBs emission, and avoid any hot gas contamination, we account the emission attributed only to the power law component of the spectral fits. The details of the spectral analysis and corresponding results are presented in Sell *et al.* 2019. Fig. 1 show the $\mathrm{L_X} - \mathrm{SFR}$ correlations from this work based on different SFR indicators. In this correlation, as well as in the $\mathrm{L_X/SFR} -$ sSFR correlation, we find significant differences, both in slope and intercept, depending on the used SFR indicators. Similar differences are seen in the scatter of these correlations. We attribute these variations on the different age and attenuation sensitivity of each SFR indicator.

For the sub-galactic analysis we apply grids of 1×1, 2×2, and $3 \times 3\,\mathrm{kpc}^2$ on the IRAC 3.6 μm images (used to measure the stellar mass), the H_α, IRAC 8μm, MIPS 24μm (used to measure the SFR), and the *Chandra* images in the soft ($0.5 - 2.0$keV), the hard ($2.0 - 8.0$keV), and the total ($0.5 - 8.0$keV) bands. From these grids we generate SFR, sSFR, and X-ray luminosity maps (Fig. 2) which we use to correlate these parameters in each sub-galactic region.

In order to infer the X-ray luminosity of each tile we calculate the posterior probability distribution of the source counts using the BEHR tool (Park *et al.* 2006). We calculate the X-ray luminosity based on the integrated spectrum of each galaxy, and the hardness ratio of each region.

With this analysis we are capable of examining the $\mathrm{L_X} - \mathrm{SFR}$ correlation (Fig. 3; Left) down to extremely low SFRs ($\sim 10^{-6}\,\mathrm{M_\odot yr^{-1}}$). We see a good agreement with the established correlation (Mineo *et al.* (2014)) down to SFR $\sim 10^{-2}\,\mathrm{M_\odot yr^{-1}}$ but we

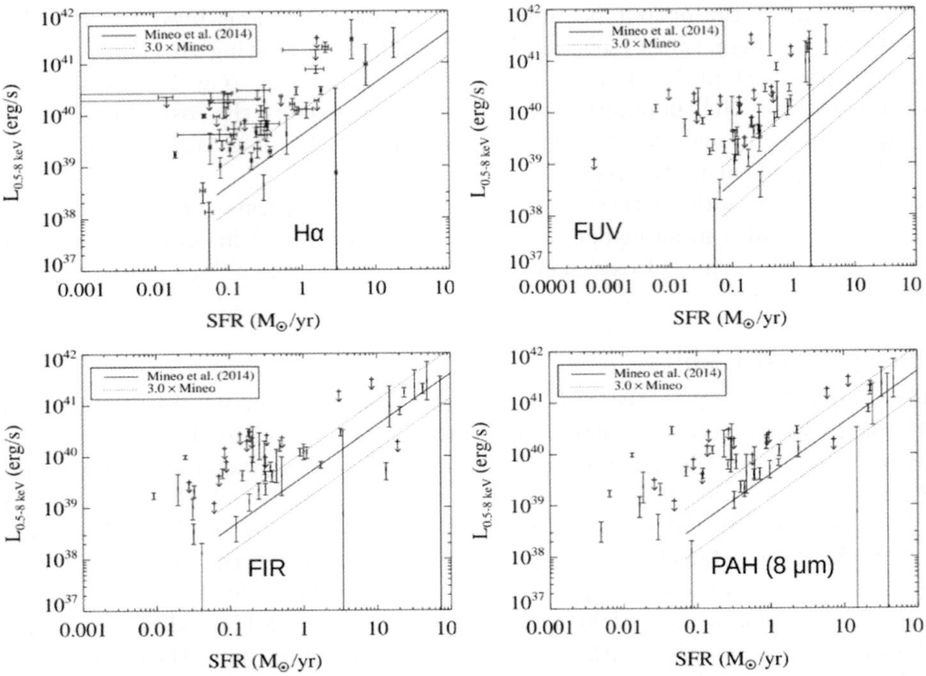

Figure 1. The $L_X - SFR$ correlations based on various SFR indicators. Clockwise from the top left we see H_α, *GALEX* FUV, PAH 8μm, and FIR indicators. The black points represent detections, blue points represent upper limits and the dashed line indicates the fit of Mineo *et al.* (2014). We see a systematic excess in the H_α and FUV correlations which we attribute to attenuation. We find a better agreement with the Mineo *et al.* (2014) relation with the IR indicators.

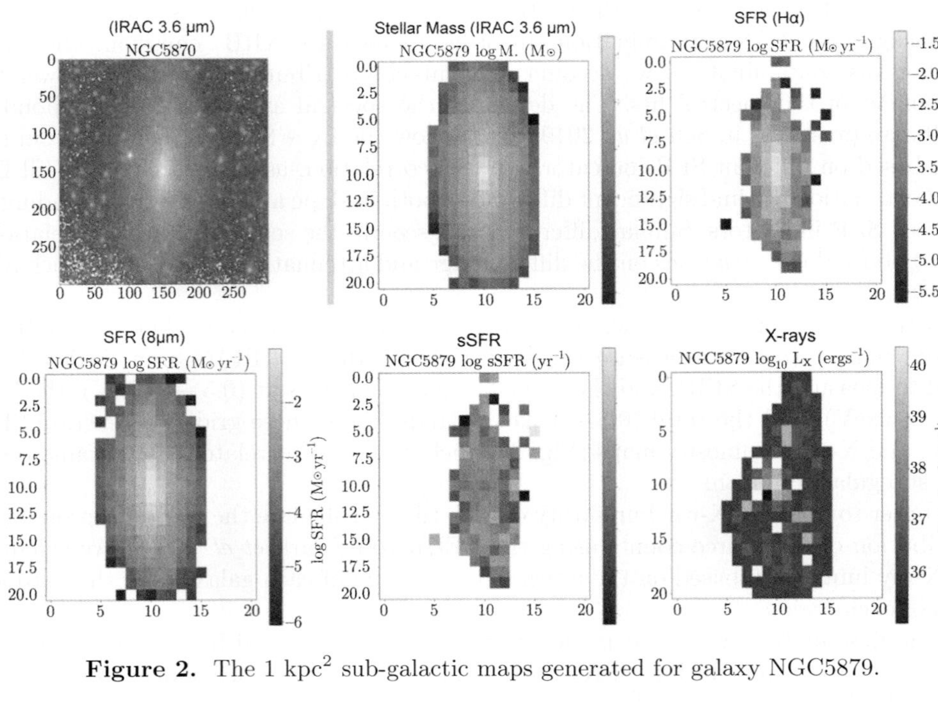

Figure 2. The 1 kpc^2 sub-galactic maps generated for galaxy NGC5879.

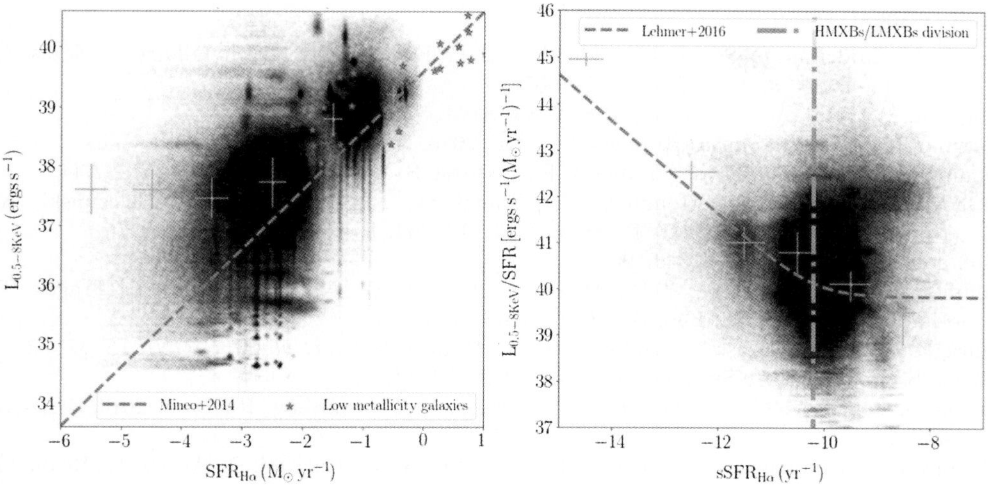

Figure 3. Left: The combined $L_X - SFR$ correlation in log space for all the tiles of the sample galaxies. Each tile is represented by sampling 1000 draws of the X-ray luminosity and SFR distributions, seen as the black density plot in the background. The dark orange points represent the modes and the 68% percentiles of the L_X and SFR of all the tiles in 1 dex SFR bins. The blue dashed line indicates the Mineo *et al.* (2014) fit, and the red stars represent the low metallicity galaxies from Douna *et al.* 2015. **Right:** Similarly, but showing the $L_X/SFR - SFR$ correlation. The dark orange points represent the modes and the 68% percentiles of the L_X SFR and sSFR of all the tiles in 1 dex sSFR bins. The blue dashed line indicates the Lehmer *et al.* (2016) fit and the red dashed-doted line indicates the division where the LMXB (left) and the HMXB (right) populations dominate.

find an excess of X-ray luminosity in lower SFRs. Similarly we examine the $L_X/SFR - sSFR$ correlation (Fig. 3; Right), where we find an exceptional agreement with the fit of Lehmer *et al.* (2016), while extending the correlation in lower values of sSFR. We also explore these correlations with other SFR indicators (PAH 8μm, 24μm continuum) and in different physical scales (1×1, 2×2, and $3 \times 3\,\mathrm{kpc}^2$) and we still find the same behavior in both correlations.

3. Conclusions

Our results are summarized in the following points:

• We find variations in the $L_X - SFR$ correlations depending on the different SFR indicator. We interpret this as the result of the different sensitivity of each indicator in stellar population age and attenuation.

• We find that the $L_X - SFR$ correlations hold down to SFR $\sim 10^{-2}\,M_\odot yr^{-1}$ and we find an excess of X-ray luminosity in the extremely low SFR regime.

• We find similar scatter in the $L_X - SFR$ correlation between the dwarf galaxies and the sub-galactic regions which is increasing in the low SFR regime.

• We find exceptional agreement on the $L_X/SFR - sSFR$ correlation with Lehmer *et al.* (2016), while extending this correlation to lower values of sSFR. This is also an indication that the X-ray luminosity excess in the low SFR regime is due to an underlying LMXBs emission.

• With this work we will be able to quantify the magnitude of the scatter with regard to the physical scales, and measure the stochasticity in the $L_X - SFR$ correlation introduced by the different SF histories of the individual galaxies.

References

Ashby, M., Mahajan, S., Smith, H. A., Willner, S. P., Fazio, G. G., Raychaudhury, S., Zezas, A., Barmby, P., *et al.* 2011, *PASP*, 123, 907
Boquien, M., Buat, V., Perret, V., *et al.* 2014, *A&A*, 571, A72
Cervino, M., Bongiovanni, A., Hidalgo, S., *et al.* 2016, *A&A*, 589, A108
Douna, V. M., Pellizza, L. J., Mirabel, I. F., Pedrosa, S. E., *et al.* 2015, *A&A*, 579, A44
Fragos, T., Lehmer, B., Tremmel, M., Tzanavaris, P., Basu-Zych, A., Belczynski, K., Hornschemeier, A., Jenkins, L., *et al.* 2013, *AJ*, 764, 1
Gilfanov, M. 2004, *MNRAS*, 349, 1
Lehmer, B. D., Basu-Zych, A. R., Mineo, S., Brandt, W. N., Eufrasio, R. T., Fragos, T., Hornschemeier, A. E., Luo, B., *et al.* 2016, *AJ*, 825, 1
Maragkoudakis, A., Zezas, A., Ashby, M. L. N., Willner, S. P., *et al.* 2017, *MNRAS*, 475, 2
Mineo, S., Gilfanov, M., Sunyaev, R., *et al.* 2012, *AJ*, 737, 2
Mineo, S., Gilfanov, M., Lehmer, B. D., Morrison, G. E., Sunyaev, R., *et al.* 2014, *MNRAS*, 437, 2
Murphy, E. J., Condon, J. J., Schinnerer, E., Kennicutt, R. C., Calzetti, D., Armus, L., Helou, G., Turner, J. L., *et al.* 2011, *MNRAS*, 419, 3
Park, T., Kashyap, V. L., Siemiginowska, A., van Dyk, D. A., Zezas, A., Heinke, C., Wargelin, B. J., *et al.* 2006, *AJ*, 652, 1
Sell, P., Zezas, A., Kouroumpatzakis, K., Bonfini P., Ashby, M., Willner S., *et al.* 2019, *In preparation*

Discussion

Daniel Schaerer: To include X-ray emission in evolutionary synthesis models and to test them we need observational constraints and the X-ray emissions of simple populations at different ages. Are there such constraints for your observations or others?

Konstantinos Kouroumpatzakis: Our analysis can help constrain the detected populations of LMXBs and HMXBs in different conditions of star-formation (low, high) and in regard with the time scales of the observed SFR in sub galactic regions of galaxies. Therefore we will be able to constrain what should be expected by the population synthesis models.

David Rosario: Looking at low sSFR systems where HMXBs are not important, should AGN boost the X-ray emission or contaminate that from LMXBs?

Konstantinos Kouroumpatzakis: We have tried to account for any possible contamination of AGN contribution in the L_x measured in our analysis. We statistically excluded it in the X-ray fits, and also excluded the sub-galactic nuclear regions for those that it is known (by spectral classification) that host an AGN.

Themiya Nanayannara: There is a known problem of lack of FUV photons in SSPs to explain observed spectra. This is related to He+ ionizing photons. Can we and how can HMXBs can be included to the SSPs to solve this problem?

Konstantinos Kouroumpatzakis: Constraining the SFR limit where the underlying LMXB L_x contribution is higher than the HMXB L_x, will help constrain the binary stellar evolutionary models and maybe indicate the lack of the UV photons in SSPs.

Tomo Goto: How strong is an SFR indicator in $L_x < 10^{38}$? Can you correlate $M_\star$ to do? In principal should be able to solve.

Konstantinos Kouroumpatzakis: When we observe high values of L_x, we know that the bulk of the emission is attributed to HMXBs. In the low SFR regime, due to the lack of recent star-formation, we attribute the L_x to LMXBs and therefore this L_x correlates with the $M_\star$ of the regions.

Challenges in Panchromatic Modelling with Next Generation Facilities
Proceedings IAU Symposium No. 341, 2018
M. Boquien, E. Lusso, C. Gruppioni & P. Tissera, eds.

doi:10.1017/S1743921319002771

Tips learned from panchromatic modeling of AGNs

Y. Sophia Dai(戴昱)

Chinese Academy of Sciences South America Center for Astronomy (CASSACA)/National Astronomical Observatories of China (NAOC), 20A Datun Road, Beijing 100101, China
email: daysophia@gmail.com

Abstract. I will review the tips learned from panchromatic modeling of active galactic nuclei (AGNs), based on our recent work to study the relationship between AGN and star formation (SF). Several AGN SED models are compared, and a significant AGN contribution is found in the IR luminosities and corresponding star formation rates (SFRs). I will review the AGN-SF relation and how different parameters and sample selections affect the observed correlation. I will then report on the constant ratio discovered between the SFR and the black hole mass accretion rate (BHAR), and their implications on the gas supply and galaxy formation history of these systems. Caveats and important questions to answer are summarized at the end.

Keywords. galaxies: active, galaxies: evolution, X-rays: galaxies, infrared: galaxies

1. Introduction

Twenty years have passed since the empirical connections were first found between the supermassive black hole (SMBH) mass ($M_\bullet$) and properties of their host galaxies, including the stellar velocity dispersion (M-σ), the bulge luminosity, and the bulge mass by e.g., Magorrian *et al.* (1998), Marconi & Hunt (2003). The stellar mass-$M_\bullet$ ratio shows a larger variation than the bulge mass-$M_\bullet$ ratios, which ranges from a few hundred to a few thousand. This possibly indicates intrinsically different relations for different galaxy types (Marconi & Hunt (2003), Kormendy & Ho (2013), Reines & Volonteri (2015)).

Given their significantly different sizes and masses, the finding of these correlations triggered the search for the intrinsic physical drive that connects SMBHs and galaxies. Various evolution models have been proposed, including the merger theory, e.g., Hopkins *et al.* (2006), the cold-flow scheme, e.g., Springel *et al.* (2005), or purely mathematical model, e.g., Peng (2007). In these models, AGN feedback is often needed to modulate the process, locally or globally, in forms of winds, jets or radiation perturbations, e.g., Fabian (2012).

In practice, statistical samples of AGNs or star forming galaxies are used to study the AGN-SF relation. Typical sample selection utilizes the fact the AGN and SF dominate different parts of the spectrum. AGNs are often selected in the X-ray, e.g., Lutz *et al.* (2010), Mullaney *et al.* (2012), Dai *et al.* (2018), or by highly ionized and often broad optical lines, e.g., Netzer (2009), Matsuoka & Woo (2015), Harris *et al.* (2016), while IR luminosity or certain lines (e.g Hα) is chosen to represent the SF. Since AGN selections in the X-ray and optical are not targeting the same nucleus regions, and optical lines and infrared luminosities are not tracing the same SF regions (gas *vs* cold dust), direct comparisons could be confusing. Recently, Dai *et al.* (2018) constructed a sample of IR-bright AGNs, with detections in both the X-ray and far-IR, aiming to focus on the phase where both BH accretion and SF are active.

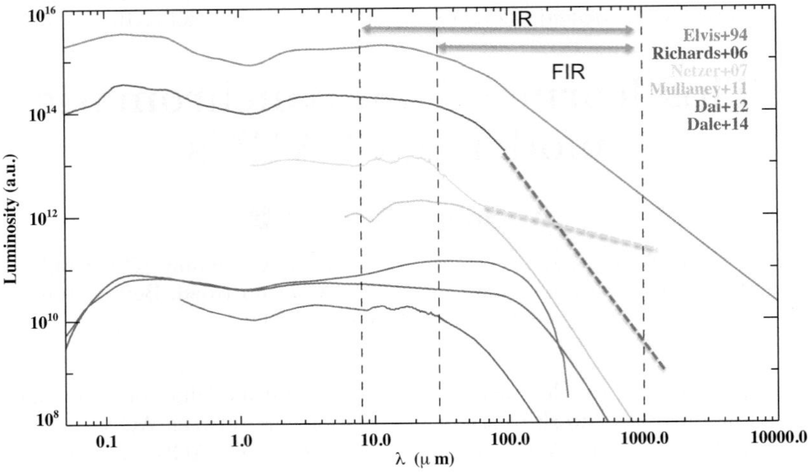

Figure 1. Comparison of AGN SEDs from the literature in arbitrary unit for display purpose. General consistency in the total IR luminosity and relatively larger variation in the far-IR luminosity are found (Sec 2).

2. SED Analysis

SED decomposition is one of the most common practices to derive the relative luminosities of AGN and SF components. This is typically achieved by fitting models, either using existing de-composition codes (e.g., CIGALE, MagPhys, GRASIL), or users' own decomposition templates, e.g., Rosario *et al.* (2018); Rivera *et al.* (2016). Despite large scatter, the AGN mean SEDs have shown surprising uniformity over z, luminosity, and Eddington ratios (e.g., Elvis *et al.* (1994); Richards *et al.* (2006); Dai *et al.* (2012)), with a big-blue-bump in the UV-optical, a near-IR bump, followed by a not-so-well-constrained far-IR decline due to limited observation data. With *Herschel*, a separation between the younger, far-IR bright population and the older, far-IR faint AGN population has been reported by e.g., Dai *et al.* (2012).

In Fig. 1, we compare the AGN IR SED models from Elvis *et al.* (1994); Richards *et al.* (2006); Mullaney *et al.* (2011); Dai *et al.* (2012); and Dale *et al.* (2014). A general consistency (~0.2 dex) is found for the integrated IR luminosities (8-1000μm, normalized at 6 μm) amongst different models. Counter-intuitively, after removing the near- to mid-IR, which is believed to be dominated by AGN thus affected by AGN variability, a larger discrepancy (~0.6 dex) is found in the far-IR (30–1000μm). The intrinsic variation of the AGN SED in the far-IR, and the interpolation for some of the templates, could both contribute to this far-IR inconsistency. The total variation, though, is still $\leqslant 1$ dex. To better use the known information in the X-ray, we developed a 3-step AGN IR decomposition method as described in Dai *et al.* (2018). Average AGN contributions of 23% and 11% are found for the total and far IR luminosities, respectively. We conclude that AGN removal is essential, but uncertain, in the IR.

3. AGN-SF correlation in two formats

3.1. $L_{\rm SFR}/L_{\rm AGN}$ *ratio*

Observationally, various, sometimes contradictory correlations have been reported between AGN and SF luminosities, be it suppressed SFR in luminous AGN host by e.g., Page *et al.* (2012); Barger *et al.* (2015); or flat or unrelated AGN-SF luminosities by e.g., Stanley *et al.* (2015)); or bi-modality and overall linear correlations, often with a

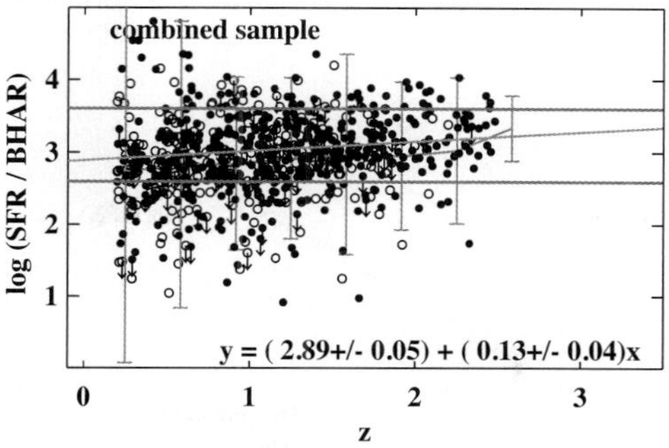

Figure 2. The SFR/BHAR ratio derived for the IR-bright AGNs from Dai *et al.* (2018). An overall flat trend is observed with a median and deviation of log(SFR/BHAR)= 2.9 ± 0.5. The two lines mark the $M_*/M_\bullet$ ratios of 2.6 and 3.6, as reported in Magorrian *et al.* (1998) and Reines & Volonteri (2015), respectively.

sample-dependent correlation coefficient, by e.g., Lutz *et al.* (2010); Harris *et al.* (2016), Pitchford *et al.* (2016); Shimizu *et al.* (2017); Dai *et al.* (2018). We note that AGN populations may be intrinsically different at the X-ray luminous and faint ends, as indicated by observations of X-ray bright sources with ALMA and SCUBA2, also indicated by the bi-modality mentioned above. Therefore, one has to be careful in interpreting the observed trend when stacking method is used in the X-ray or IR.

3.2. *SFR/BHAR ratio*

The radio between BHAR and SFR can be used to avoid a false correlation due to z effect. Although with a scatter of ~0.5 dex, we found a constant SFR/BHAR ratio over z, $M_\bullet$, and M_*, with the $\dot{M}$ ratio mostly in the log (SFR/BHAR)= 2.6-3.6 range (Fig. 2). This is consistent with the scenario that on an average basis, the galaxy and the black hole form at a fixed rate similar to the locally observed mass ratio. Though with large error bars, some recent studies have shown a M_* dependence of this $\dot{M}$ ratio, e.g., Yang *et al.* (2017); Cowley *et al.* (2018); while others find a constant ratio independent of M_*, e.g. Mullaney *et al.* (2012), Dai *et al.* (2018).

3.3. *Caveats of the correlation studies*

• Malmquist bias is prevalent, which could result in false increase at the luminous end.

• Sample-dependence is important. Fig. 3 illustrates a simplified picture assuming that the AGN-SF correlation only exists during the active phase, during which AGNs coexist with active star formation, resulting in the observed fixed-fraction of gas inflow. This explains the flattening slopes found in samples with higher fraction of obscured objects, e.g. recovered by stacks in the X-ray.

• The choice of binning could cause different correlation slopes, or the lack of any correlation. This can be due to 1) the different variability time scale for AGNs and SF; 2) purely mathematical effects due to choices of the dependent parameters (Reines & Volonteri (2015); Dai *et al.* (2018)).

4. Summary

Based on SED analysis of the IR-bright AGN sample in Dai *et al.* (2018), we found:

(*a*) $L_{\rm AGN}$-$L_{\rm IR,SF}$, and SFR-BHAR correlations are confirmed by various observational studies, consistent with the scenario of a common gas/mass supply for SMBH and the host galaxy.

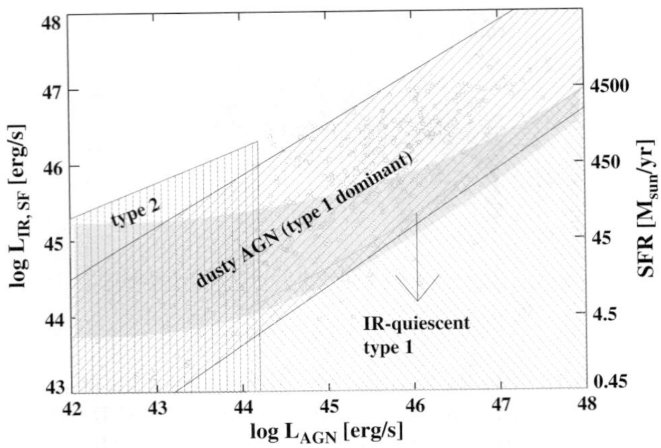

Figure 3. A toy-model illustration of the locations of different AGN populations in the $L_{\rm AGN}$-$L_{\rm IR,SF}$ plane. Sample selections can significantly affect the observed trend. This model assumes that the AGN-SF correlation are most significant during a phase where both AGN and SF are active (green region). Grey circles mark the Dai *et al.* (2018) sample. Stacking of the IR-undetected X-ray sources (blue region), or of the X-ray undetected IR sources (red region), could both flatten the observed relation.

(*b*) A nearly constant ratio of log(SFR/BHAR) $\sim$2.9 is observed, agreeing with the local stellar mass *vs* SMBH mass ratios, indicating homogeneous evolution across z. The effect of M_* is still debatable and under investigation.

(*c*) Several caveats can potentially mask out an intrinsic AGN-SF correlations, e.g. 1). intrinsic scatter and uncertainties ($\sim$0.5 dex) due to various SFR & BHAR estimators; 2). Selection bias and Malmquist bias; 3). Mixing of different evolutionary populations by stacking and sample selection; 4). Binning method (variability, choice of free parameter).

Future studies are needed to construct a clean AGN sample, of similar physical properties, at similar evolutionary stages, to study the intrinsic AGN SED, and to explain the flat SFR/BHAR ratios, their dependence, and their indication on the galaxy evolution history.

References

Barger, A. J., Cowie, L. L., Owen, F. N., Chen, C. C., Hasinger, G., Hsu, L. Y., Li, Y., *et al.* 2015, *ApJ*, 801, 87

Cowley, M. J., Spitler, L. R., Quadri, R. F., Goulding, A. D., Papovich, C., Tran, K. V. H., Labbé, I., Alcorn, L., *et al.* 2018, *MNRAS*, 473, 3710

Dai, Y. S., Wilkes, B. J., Bergeron, J., Kuraszkiewicz, J., Omont, A., Atanas, A., Teplitz, H. I., *et al.* 2018, *MNRAS*, 478, 4238

Dai, Y. S., Bergeron, J., Elvis, M., Omont, A., Huang, J.-S., Bock, J., Cooray, A., Fazio, G. G., *et al.* 2012, *ApJ*, 753, 33, D12

Dale, D. A., Helou, G., Magdis, G. E., Armus, L., Dìaz-Santos, T., Shi, Y., *et al.* 2014, *ApJ*, 784, 83

Elvis, M., Wilkes, B. J., McDowell, J. C., Green, R. F., Bechtold, J., Willner, S. P., Oey, M. S., Polomski, E., *et al.* 1994, *ApJS*, 95, 1

Fabian, A. C. 2012, *ARA&A*, 50, 455

Harris, K., Farrah, D., Schulz, B., Hatziminaoglou, E., Viero, M., Anderson, N., Béthermin, M., Chapman, S., *et al.* 2016, *MNRAS*, 457, 4179

Hopkins, P. F., Hernquist, L., Cox, T. J., Di Matteo, T., Robertson, B., Springel, V., *et al.* 2006, *ApJS*, 163, 1
Kormendy, J. & Ho, L. C. 2013, *ARA&A*, 51, 511
Lutz, D., Mainieri, V., Rafferty, D., Shao, L., Hasinger, G., Weiß, A., Walter, F., Smail, I., *et al.* 2010, *ApJ*, 712, 1287
Netzer, H. 2009, *MNRAS*, 399, 1907
Magorrian, J., Tremaine, S., Richstone, D., Bender, R., Bower, G., Dressler, A., Faber, S. M., Gebhardt, K., *et al.* 1998, *AJ*, 115, 2285
Marconi A. & Hunt L. K. 2003, ApJ, 589, L21
Matsuoka, K. & Woo, J.-H. 2015, *ApJ*, 807, 28
Mullaney, J. R., Daddi, E., Béthermin, M., Elbaz, D.; Juneau, S., Pannella, M., Sargent, M. T., Alexander, D. M., *et al.* 2012b, *ApJL*, 753, L30
Mullaney, J. R., Alexander, D. M., Goulding, A. D., Hickox, R. C., *et al.* 2011, *MNRAS*, 414, 1082
Page, M. J., Symeonidis, M., Vieira, J. D., Altieri, B., Amblard, A., Arumugam, V., Aussel, H., Babbedge, T., *et al.* 2012, *Nature*, 485, 213
Peng, C. Y. 2007, *ApJ*, 671, 1098
Pitchford, L. K., Hatziminaoglou, E., Feltre, A., Farrah, D., Clarke, C., Harris, K. A., Hurley, P., Oliver, S., *et al.* 2016, *MNRAS*, 462, 4067
Reines, A. E. & Volonteri, M. 2015, *ApJ*, 813, 82
Richards, G. T., Lacy, M., Storrie-Lombardi, L. J., Hall, P. B., Gallagher, S. C., Hines, D. C., Fan, X., Papovich, C., *et al.* 2006, *ApJS*, 166, 470
Rivera *et al.* 2016, *ApJ*, 833, 98C
Rosario, D. J., Santini, P., Lutz, D., Shao, L., Maiolino, R., Alexander, D. M., Altieri, B., Andreani, P., *et al.* 2012, *A&A*, 545, A45
Rosario, D. J., Burtscher, L., Davies, R. I., Koss, M., Ricci, C., Lutz, D., Riffel, R., Alexander, D. M., *et al.* 2018, *MNRAS*, 473, 5658
Shimizu, T. T., Mushotzky, R. F., Meléndez, M., Koss, Michael J., Barger, Amy J., Cowie, L. L., *et al.* 2017, *MNRAS*, 466, 3161
Springel, V., White, S. D. M., Jenkins, A., Frenk, Carlos S.; Yoshida, N., Gao, L., Navarro, J., Thacker, R., *et al.* 2005, *Nature*, 435, 629
Stanley, F., Harrison, C. M., Alexander, D. M., Swinbank, A. M., Aird, J. A., Del Moro, A., Hickox, R. C., Mullaney, J. R., *et al.* 2015, *MNRAS*, 453, 591
Yang, G., Chen, C.-T. J., Vito, F., Brandt, W. N., Alexander, D. M., Luo, B., Sun, M. Y., Xue, Y. Q., *et al.* 2017, *ApJ*, 842, 72

Discussion

David Rosario: Existing work on panchromatic modelling (e.g., Rosario+2018) do include x-ray constraints. An important bias is that luminous AGN tend to be in more massive hosts, which, regardless of AGN, do show higher SFRs. How does this affect $L_{\rm AGN} - L_{\rm SFR}$ relationships?

Dai Y. Sophia: Good point. The higher AGN fraction at the luminous massive end is likely to contribute to the increase in SFR in general. Thus it is helpful to make the comparison in a mass normalized fashion, i.e. $L_{\rm AGN}/M_{*} vs$ SFR, or $L_{\rm AGN}/M_{\rm BH}$ vs SFR/M_{*}, or $L_{\rm AGN}/M_{\rm BH}$ vs SFR. It would be interesting to see how these parameters look in large samples. A possible bias would be the uncertainties in the mass estimates.

Challenges in Panchromatic Modelling with Next Generation Facilities
Proceedings IAU Symposium No. 341, 2018
M. Boquien, E. Lusso, C. Gruppioni & P. Tissera, eds.

doi:10.1017/S1743921319002540

X-ray - Infrared relation of AGNs and search for highly obscured accretion in the *AKARI* NEP Field

Takamitsu Miyaji and *AKARI* NEP Survey Team

Instituto de Astronomía, Universidad Nacional Autónoma de México, Km. 103 Carret. Tijuana-Ensendada, Ensenada, 22860 Mexico
email: miyaji@astro.unam.mx

Abstract. The infrared Astronomical Satellite *AKARI* conducted deep (∼0.4 deg^2) and wide (∼5.4 deg^2) surveys around the North Ecliptic Pole (NEP) with its InfraRed Camera (IRC) with nine filters continuously covering the 2–25 μm range. These photometric bands include three filters that fill the "*Spitzer* gap" between the wavelength coverages of IRAC and MIPS. This unique feature has enabled us to make sensitive mid-infrared detection of AGN candidates at z∼1-2, based on the Spectral Energy Distribution (SED) fitting including hot dust emission in the AGN torus. This enables us to compare X-rays and the AGN torus component of the infrared emission to help us identify highly absorbed AGNs, including Compton-thick ones. We report our results of the *Chandra* observation of the *AKARI* NEP Deep Field and discuss the prospects for upcoming *Spectrum-RG* (eROSITA+ART-XC) on the *AKARI* Wide field.

Keywords. galaxies: active – infrared: galaxies – X-rays: galaxies

1. Introduction

There are three major bumps in the electromagnetic continuum in a typical AGN: 1) optically-thick thermal emission from the accretion disk in the ultraviolet, 2) the primary X-ray continuum by the inverse-Compton scattering of the disk ultraviolet photons by hot coronae, and 3) the infrared emission from hot dust in the circum-nuclear torus. Additionally, reprocessed primary X-ray continuum by reflections at surrounding material produces a characteristic "Comption hump" in hard X-rays. Observational determination of how these three components are inter-related is a key to panchromatic modelings of the AGN component in galaxies.

In multiwavelength extragalactic surveys, X-rays are considered the most efficient marker of the AGN activity. However it is subject to absorption by intervening gas by photo-electric absorption, which occurs mainly at photon energies $E < 10$ keV. If the line of sight column density becomes $N_{\rm H} \gtrsim 10^{24}{\rm cm}^{-2}$, the effects Compton scattering further attenuates X-ray emission even at $E \gtrsim 10$ keV.

While the infrared emission from hot dust heated by AGNs is less subject to absorption by intervening material, it has to be separated from stars and warm dust from star-formation regions by an SED decomposition to measure the IR luminosity of the AGN component alone.

The continuous 9-band 2–25 μm photometric data available from *AKARI* NEP Survey are ideal for such SED decomposition of galaxies. The availability of three photometric bands in the 11–20 μm range, which is the wavelength gap between IRAC and MIPS instruments of *Spitzer*, makes the *AKARI* IRC dataset a very strong tool for

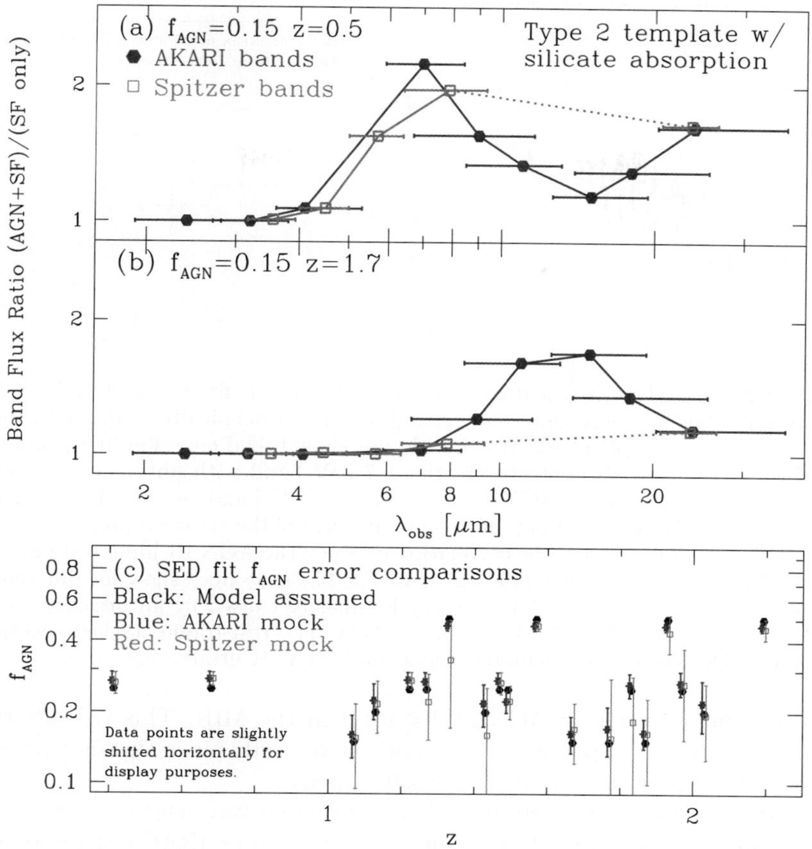

Figure 1. *(a) & (b):* Excess MIR emission ([AGN+Starburst]/Starburst), due to the presence of a type 2 AGN (with the 9.7 μm silicate absorption feature) observed through the *AKARI* IRC (blue/filled hexagon) and *Spitzer* IRAC+MIPS (red/open square) filter sets, for (a) $z = 0.5$ and (b) $z = 1.7$ objects. An AGN fraction $f_{\rm AGN} = 0.15$ of the total IR luminosity is assumed. *(c):* Comparisons of expected $f_{\rm AGN}$ errors between SED fits to a number of simulated SEDs (black/filled symbol without error bars) through *AKARI+Herschel* (blue/filled symbol with error bars) and *Spitzer+Herschel* (red/open square with error bars). In calculating the errors, imaging depths of ANEPD and COSMOS are assumed for the former and the latter respectively. The calculations are made by the Cigale software using the model in Ciesla *et al.* (2015).

separating AGN hot dust emission from stellar+star-formation components at $1 \lesssim z \leqslant 2$ (the "Cosmic Noon").

In this proceedings article, we summarize the results and prospects of our X-ray followup studies of the *AKARI* NEP field with emphasis on X-ray data and the connection between X-rays and infrared. These include our *Chandra* followup studies of the *AKARI* NEP Deep Field and prospects of upcoming *Spectrum-RG* (eROSITA+ART-XC) observations.

2. The *AKARI* NEP Survey

AKARI was a Japanese IR astronomical satellite (Matsuhara *et al.* 2006; see also Goto *et al.*, this symposium), which was operated from 2006 to 2011. One of the unique features of the observatory was the availability of continuous wavelength coverage over 2–25 μm in 9 filters with its IR Camera (IRC), including three filters in the 11–19 μm range. These filters fill the 9–20 μm gap between the *Spitzer* IRAC and MIPS instruments and

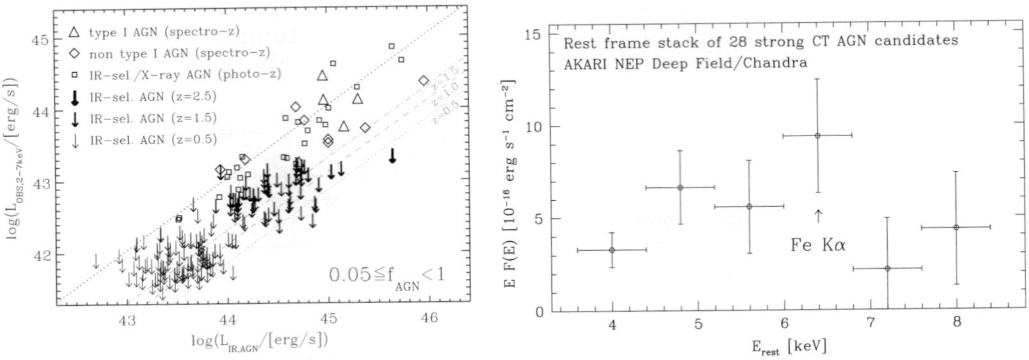

Figure 2. *Left:* Infrared AGN rest-frame luminosity from SED fits vs. hard (2–7 keV) observed X-ray luminosity. The symbols represent: triangles – spectroscopically-confirmed type I AGNs, diamonds – objects with optical spectra that do not show broad emission lines, small squares – IR-selected AGN that are also detected in the 2–7 keV band with photometric redshifts only, downward arrows – IR-selected AGN undetected 2–7 keV band, where the vertical position corresponds to the 2–7 keV flux upper limit. The thickness of the arrows represents the redshift of the source. The dotted line shows the correlation between the AGN IR luminosity and the X-ray luminosity ($\langle L_X/L_{IR}\rangle = 0.11$) for unabsorbed AGNs. The gray lines show for different redshifts the expected attenuation of the 2–7 keV X-ray luminosity caused by an intrinsic absorption of $N_H = 10^{24}$ cm^{-2}. Adopted from K15 (Fig. 18). *Right:* The rest-frame stacked spectrum of the 28 strong CT AGN candidates calculated using the CSTACK utility.

allows an efficient selection of AGN at $1 \lesssim z \lesssim 2$ in the MIR. This redshift range (the so-called Cosmic Noon) is particularly important to study since that is where both the X-ray detected number and luminosity densities peak.

Figure 1 demonstrates the advantage of the continuous wavelength coverage of *AKARI* 9-band photometry (including 10–19 μm) over the *Spizer* IRAC+MIPS photomtetric bands in detecting the AGN torus component. (See caption.)

As a legacy program of *AKARI*, a major portion of all pointed observations during its liquid helium phase was devoted to surveys around North Ecliptic Pole (NEP), namely the AKARI NEP Deep (~0.4 deg^2; hereafter ANEPD) and Wide (~5.4 deg^2; ANEPW) field surveys with IRC. At ANEPD, the *AKARI* data have 5σ sensitivity limits of, e.g., ~9 and 87 μJy at 3.2 and 15 μm respectively (Murata *et al.* 2013), while the limiting fluxes are approximately twice as large in NEPW (Kim *et al.* 2012).

3. Chandra Observations of the *AKARI* NEP Deep Field

We were granted a 250 ks dense tiling of a 3 × 4 *Chandra* ACIS-I mosaic survey in the ANEPD field (Krumpe *et al.* 2015; hereafter K15). Including additional ~50 ks of data from archive, we cover the field with exposure times ranging from ~40–80 ks. The source catalog and the first analysis of 457 X-ray sources is presented in K15.

In K15, (see Fig. 2 *left*), we compare the observed X-ray (2–7 keV) luminosity $L_{\rm OBS,2-7keV}$, i.e. luminosity calculated without K-correction or absorption correction) and the AGN component of the IR luminosity $L_{\rm IR,AGN}$ obtained from SED fitting using an updated version of the Hanami *et al.* (2012)'s procedure. By this comparison, we selected 28 strong Compton-thick AGN (CTAGN) candidates, i.e., AGNs with a line of sight column density of $N_H > 10^{24}$[cm^{-2}].

The CTAGNs are characterized by a large equivalent width of the Fe Kα line ($EW \gtrsim 1$ keV). Using the CSTACK utility†, we have made a rest-frame stacking of the X-ray spectra of the 28 strong CT AGN candidates (Fig. 2 *right*), obtaining an equivalent width

† http://lambic.astrosen.unam.mx/cstack

of the Fe Kα of $EW = 1.0 \pm 0.6$ keV. This suggests that a majority of these candidates are CTAGNs with possible contamination from non-CT AGNs.

4. Prospects for eROSITA/ART-XC observations

The spacecraft *Spectrum-RG*, which is planned to be launched in the near future (Merloni, this symposium), carries two instruments, eROSITA (sensitive in 0.2–10 keV) and ART-XC (5–30 keV). The plan of the mission is to perform 8 all-sky surveys over four years and the every great circle scan will go through the Ecliptic pole regions. *Spectrum-RG* will provide with us the best-matched dataset for our studies of highly obscured AGNs in the ~5.4 deg^2 ANEPW field, which is centered at the exact NEP. The combination of the ANEPW survey with the eROSITA dataset, we expect to probe a high luminosity - highly absorbed AGN population.

Using the current determination of the X-ray luminosity function of AGNs at $0.4 < z < 2$ (Miyaji *et al.* 2015), $N_{\rm H}$ function (Ueda *et al.* 2014), and the expected flux limit of eROSITA in the 2–10 keV band (Merloni *et al.* 2012), we expect to identify ~270 CTAGN/semi-CTAGN candidates using the method described above. The ART-XC instrument is sensitive in 5–30 keV, we expect to improve the distinction between non-CT AGNs and CT-AGNs by adding the ART-XC dataset.

5. Concluding Remarks

We are in the process of improving the investigation. The improvements include:

• Spectroscopic followup of ANEPD X-ray sources, IR-selected AGNs and and CTAGN candidates, using large telescopes including Gran Telescopio Canarias, Keck, and Large Binocular Telescope.

• We will use updated models for the AGN torus. These include the Clumpy Torus model (Nenkova *et al.* 2008) on the infrared side. An X-ray torus clumpy model that has the consistent geometrical configurations/parameter definitions as the Nenkova *et al.* (2008) model is being developed by a group at Kyoto University (Tanimoto *et al.*, Ogawa *et al.* this symposium). We will use these models for a consistent IR-X-ray modeling in selecting CT-AGNs.

TM is supported by UNAM-DGAPA PAPIIT (IN104216, IN111319) and CONACyT (252531). He thanks the conference organizers/NAOJ for a travel support.

References

Ciesla, L., Charmandaris, V., Georgakakis, A., Bernhard, E., Mitchell, P. D., Buat, V., Elbaz, D., LeFloc'h, E., *et al.* 2015, *A&A*, 576, A10

Hanami, H., Ishigaki, T., Fujishiro, N., Nakanishi, K., Miyaji, T., Krumpe, M., Umetsu, K., Ohyama, Y., *et al.* 2012, *PASJ*, 64, 70

Kim, S. J., Lee, H. M., Matsuhara, H., Wada, T., Oyabu, S., Im, M., Jeon, Y., Kang, E., *et al.* 2012, *A&A*, 548, A29

Krumpe, M., Miyaji, T., Brunner, H., Hanami, H., Ishigaki, T., Takagi, T., Markowitz, A. G., Goto, T., *et al.* 2015, *MNRAS*, 446, 911

Matsuhara, H., Wada, T. Matsuura, S., Nakagawa, T., Kawada, M., Ohyama, Y., Pearson, C. P., Oyabu, S., *et al.* 2006, *PASJ*, 58, 673

Merloni, A., Predehl, P., Becker, W., Böhringer, H., Boller, T., Brunner, H., Brusa, M., Dennerl, K., *et al.*, 2012, *eROSITA Science Book: Mapping the Structure of the Energetic Universe*, arXiv:1209.3114

Miyaji, T., Hasinger, G., Salvato, M., Brusa, M. Cappelluti, N., Civano, F., Puccetti, S., Elvis, M., *et al.* 2015, *ApJ*, 804, 104

Murata, K., Matsuhara, H., Wada, T., Arimatsu, K., Oi, N., Takagi, T., Oyabu, S., Goto, T., *et al.* 2013, *A&A*, 559, A132

Nenkova, M., Sirocky, M. M., Ivezić, Ž., & Elitzur, M. 2008, *ApJ*, 685, 147

Ueda, Y., Akiyama, M., Hasinger, G., Miyaji, T., & Watson, M. G. 2014, *ApJ*, 786, 104

Discussion

TRIANI: Is there any effect of inclination angle in the AGN SED in IR?

MIYAJI: Yes, there is. Especially, there is silicate absorption in the case of an edge-on view of the torus. We are investigating in detail the IR SED of tori including the angle dependence of the Nenkova "Clumpy Torus" model. Also on the X-ray side, we will revise the model using the new "X-clumpy" torus model, which has matching geometries and parameter definitions of the Nenkova model. This is under development by a group at Kyoto University and some results of its preliminary version of are presented in this conference (posters P36 by Tanimoto and P22 by Ogawa).

Challenges in Panchromatic Modelling with Next Generation Facilities
Proceedings IAU Symposium No. 341, 2018
M. Boquien, E. Lusso, C. Gruppioni & P. Tissera, eds.

doi:10.1017/S1743921319002898

The diagnostic power of radio spectra from star-forming galaxies

Eric J. Murphy

National Radio Astronomy Observatory, 520 Edgemont Road,
Charlottesville, VA 22901, U.S.A.
email: emurphy@nrao.edu

Abstract. Radio continuum emission from galaxies is powered by a combination of distinct physical processes, each providing unique diagnostic information. Over frequencies spanning ~1–120 GHz, radio spectra of star-forming galaxies are primarily comprised of: (1) non-thermal synchrotron emission powered by accelerated cosmic-ray electrons/positrons; (2) free-free emission from young massive star-forming (HII) regions; (3) anomalous microwave emission, which is a dominant, but completely unconstrained, foreground in cosmic microwave background experiments; and (4) cold, thermal dust emission that accounts for most of the dust and total mass content in the interstellar medium in galaxies. In this proceeding, we discuss these key energetic processes that contribute to the radio emission from star-forming galaxies, with an emphasis on frequencies $\gtrsim 30$ GHz, where current investigations of star formation within nearby galaxies show that the free-free emission begins to dominate over non-thermal synchrotron emission. We also discuss how planned radio facilities that will access these frequencies, such as a next-generation Very Large Array (ngVLA), will be transformative to our understanding of the star formation process in galaxies.

Keywords. Star Formation, Radio Continuum

1. Introduction

Radio continuum observations have proven to be a workhorse in our understanding of the star formation process (i.e., stellar birth and death) from galaxies both in the nearby universe and out to the highest redshifts. Despite the fact that radio continuum emission is energetically weak with respect to a galaxy's bolometric luminosity, it provides critical information on the process of massive star formation, as well as a mechanism to access the relativistic [magnetic field + cosmic rays (CRs)] component in the interstellar medium (ISM) of galaxies. Stars more massive than $\sim 8\,M_\odot$ end their lives as core-collapse supernovae, whose remnants are thought to be the primary accelerators of CR electrons (e.g., Koyama *et al.* 1995) giving rise to the diffuse synchrotron emission observed from star-forming galaxies (Condon 1992). These same massive stars are also responsible for the creation of HII regions that produce radio free-free emission, whose strength is directly proportional to the production rate of ionizing (Lyman continuum) photons.

Radio frequencies spanning ~1–120 GHz, which are observable from the ground, are particularly useful in probing such processes. The non-thermal emission component typically has a steep spectrum ($S_\nu \propto \nu^\alpha$, where $\alpha \sim -0.8$; e.g., Condon 1992), while the thermal (free-free) component is relatively flat ($\alpha \sim -0.1$). Accordingly, for globally integrated measurements of star-forming galaxies, lower frequencies (e.g., 1.4 GHz) are generally dominated by non-thermal emission, while the observed thermal fraction of the emission increases with frequency, eventually being dominated by free-free emission once beyond ~30 GHz (Condon & Yin 1990). On ~100 pc scales, the thermal fraction at

33 GHz for typical star-forming regions can be considerably higher, being $\gtrsim$80% (Murphy *et al.* 2011; Linden *et al.* 2020). Thus, observations at such frequencies, which are largely unbiased by dust, provide an excellent diagnostic for the current star formation rate (SFR) of galaxies.

It is worth noting that the presence of an anomalous microwave emission (AME) component in excess of free-free emission between $\sim$10 and 90 GHz, generally attributed to electric dipole rotational emission from ultrasmall ($a \lesssim 1$ nm) grains (Erickson 1957; Draine & Lazarian 1998; Planck Collaboration *et al.* 2011; Hensley & Draine 2017) or magnetic dipole emission from thermal fluctuations in the magnetization of interstellar dust grains (Draine & Lazarian 1999; Hensley *et al.* 2016), may complicate this picture. For a single outer-disk star-forming region in NGC 6946, Murphy *et al.* (2010) reported an excess of 33 GHz emission relative to what is expected given existing lower frequency radio data. This result has been interpreted as the first detection of so-called "anomalous" dust emission outside of the Milky Way. More recently, by surveying a sample of 112 star-forming complexes at 33 GHz in 56 nearby galaxies with the VLA at $\sim 2''$ angular resolution as part of the Star Formation in Radio Survey (SFRS; Murphy *et al.* 2018a), resulting in the detection of 225 discrete sources, a second detection of AME was found in a compact, optically-faint region of NGC 4725 (Murphy *et al.* 2018b). Since this initial discovery in NGC 6946, a number of searches for extragalactic AME have been undertaken with WMAP and *Planck* data (e.g., the Magellanic Clouds, Bot *et al.* 2010; NGC 4945, Peel *et al.* 2011), all of which have been inconclusive, suggesting that this emission component is most likely sub-dominant for globally integrated measurements.

Due to the faintness of galaxies at high (i.e., $\gtrsim$15 GHz) radio frequencies, existing work has been restricted to the brightest objects, and small sample sizes. For example, past studies demonstrating the link between high-frequency free-free emission and massive star formation include investigations of Galactic star-forming regions (Mezger & Henderson 1967), nearby dwarf irregular galaxies (Klein & Graeve 1986), galaxy nuclei (Turner & Ho 1983, 1994), nearby starbursts (Klein *et al.* 1988; Turner & Ho 1985), and super star clusters within nearby blue compact dwarfs (Turner *et al.* 1998; Kobulnicky & Johnson 1999). And while these studies focus on the free-free emission from galaxies, each was conducted at frequencies $\lesssim$30 GHz. With recent improvements to the backends of existing radio telescopes, such as the Caltech Continuum Backend (CCB) on the Robert C. Byrd Green Bank Telescope (GBT) and the Wideband Interferometric Digital ARchitecture (WIDAR) correlator on the Karl G. Jansky Very Large Array (VLA), the availability of increased bandwidth is making it possible to conduct investigations for large samples of objects at frequencies $\sim$30 GHz. In this proceeding, we discuss recent results that have taken advantage of these significant electronic upgrades and describe how next-generation radio facilities operating in the $\sim$1–120 GHz frequency range will provide significantly improved access to the diagnostic information encoded in the radio spectra of galaxies, and in turn transform our understanding of star formation both within and among galaxies.

2. The Energetic Processes Powering Radio Continuum Emission

Radio continuum emission from galaxies covering $\sim$1–120 GHz is powered by an eclectic mix of physical emission processes, each providing completely independent information on the star formation and ISM properties of galaxies (see Figure 1). These processes include non-thermal synchrotron, free-free (thermal bremsstrahlung), anomalous microwave, and thermal dust emission that are directly related to the various phases of the ISM and provide a comprehensive picture of how galaxies convert their gas into stars. Each of these emission components, described in detail below, are of low surface

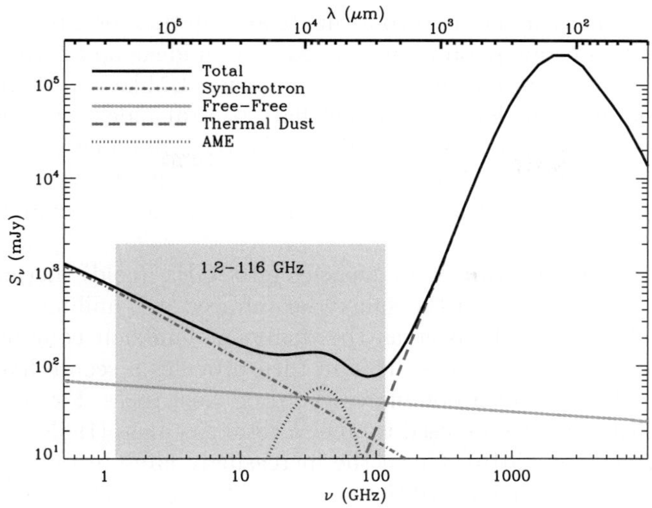

Figure 1. A model spectrum illustrating the various emission processes (non-thermal synchrotron, free-free, AME, thermal dust) that contribute to the observed radio frequency range to be covered by the ngVLA. Only in the proposed ngVLA frequency range (1.2−116 GHz, highlighted) do all major continuum emission mechanisms contribute at similar levels, making this range uniquely well-suited to next-generation continuum studies.

brightness in the ∼30−120 GHz frequency range, and therefore difficult to map in a spatially resolved manner at ∼10−100 pc scales in the general ISM of nearby galaxies using existing facilities. Consequently, our current knowledge about the emissions processes over this frequency range is limited to the brightest star-forming regions/nuclei in the most nearby sources (Leroy *et al.* 2011; Clemens *et al.* 2008, 2010; Murphy 2013; Murphy *et al.* 2015; Barcos-Muñoz *et al.* 2015), providing no information on how the situation may change in drastically different ISM conditions (Tabatabaei *et al.* 2018) that may be more representative of those in high-redshift galaxies, where we have to rely on globally integrated measurements.

The key energetic processes powering radio spectra from star-forming galaxies over the ∼1−120 GHz frequency range include:

• **Non-Thermal Synchrotron Emission:** At ∼GHz frequencies, radio emission from galaxies is dominated by non-thermal synchrotron emission resulting, indirectly, from star formation. Stars more massive than $\sim 8\,M_\odot$ end their lives as core-collapse supernovae, whose remnants are thought to be the primary accelerators CR electrons, giving rise to the diffuse synchrotron emission observed from star-forming galaxies. Thus, the synchrotron emission observed from galaxies provides a direct probe of the still barely understood relativistic (magnetic field + CRs) component of the ISM. As illustrated in Figure 1, the synchrotron component has a steep spectral index, typically scaling as $S_\nu \propto \nu^{-0.83}$ with a measured rms scatter of 0.13 (Niklas *et al.* 1997). Over the frequency range spanning 1−120 GHz, radio observations will be sensitive to CR electrons spanning more than an order of magnitude in energy (i.e., ∼ 1−50 GeV) for a typical range of ISM magnetic field strengths, including the population that may drive a dynamically-important CR-pressure term in galaxies (Socrates *et al.* 2008).

• **Free-Free Emission:** The same massive stars whose supernovae are directly tied to the production of synchrotron emission in star-forming galaxy disks are also responsible (during their lifetime) for the creation of HII regions. The ionized gas produces free-free emission, which is directly proportional to the production rate of ionizing (Lyman continuum) photons and optically-thin at radio frequencies. In contrast to optical recombination line emission, no hard-to-estimate attenuation term is required to link the free-free emission to ionizing photon rates, making it an ideal, and perhaps the most reliable, measure of the current star formation in galaxies. Unlike non-thermal synchrotron emission, free-free emission has a relatively flat spectral index, scaling as $S_\nu \propto \nu^{-0.1}$. Globally, free-free emission begins to dominate the total radio emission in normal star-forming galaxies at $\gtrsim$30 GHz (Condon 1992), exactly the frequency range for which the ngVLA will be delivering an order of magnitude improvement compared to any current or planned facility.

• **Thermal Dust Emission:** At frequencies $\gtrsim$100 GHz, (cold) thermal dust emission on the Rayleigh-Jeans portion of the galaxy far-infrared/sub-millimeter spectral energy distribution can begin to take over as the dominant emission component for regions within normal star-forming galaxies. This in turn provides a secure handle on the cold dust content in galaxies, which dominates the total dust mass. For a fixed gas-to-dust ratio, the total dust mass can be used to infer a total gas mass (Heiles *et al.* 1988; Dame *et al.* 2001; Scoville *et al.* 2016). Given the increasingly large instantaneous bandwidth offered by current and planned facilities (e.g., ALMA, ngVLA), such observations will simultaneously provide access to the $J = 1 \rightarrow 0$ line of CO revealing the molecular gas fraction for entire disks of nearby galaxies. Alternatively, combining HI observations with $J = 1 \rightarrow 0$ CO maps, one can instead use the thermal dust emission to measure the spatially varying gas-to-dust ratio directly.

• **Anomalous Microwave Emission (AME):** In addition to the standard Galactic foreground components described above, an unknown component has been found to dominate over these at microwave frequencies between ~10–90 GHz, and is seemingly correlated with 100 μm thermal dust emission. Cosmic microwave background (CMB) experiments were the first to discover the presence of AME (Kogut *et al.* (1996); Leitch *et al.* 1997), whose origin still remains unknown (for a review, see Dickinson *et al.* 2018). Its presence as a foreground is problematic as the precise characterization and separation of foregrounds remains a major challenge for current and future CMB experiments (BICEP2/Keck Collaboration *et al.* 2015; Planck Collaboration *et al.* 2016). At present, the most widely accepted explanation for AME is the spinning dust model (Erickson 1957; Draine & Lazarian 1998; Planck Collaboration *et al.* 2011; Hensley & Draine 2017) in which rapidly rotating very small grains, having a nonzero electric dipole moment, produce the observed microwave emission. Increasing the mapping speed at these frequencies with future instruments will lead to unprecedented investigations into the origin and prominence of this emission component both within our own galaxy and others, ultimately helping to improve upon the precision of future CMB experiments.

3. Recent Results

Murphy *et al.* (2012) presented 33 GHz photometry of 103 galaxy nuclei and extranuclear star-forming regions taken with the CCB on the GBT as part of the Star Formation in Radio Survey (SFRS). This study was limited to the 25″ resolution of the GBT beam at 33 GHz, which corresponds to a linear scale of ~1 kpc given the typical distance of the SFRS galaxies. Among the sources without evidence for an active galactic nucleus, and also having lower frequency radio data available, the median thermal fraction at 33 GHz was ≈76% with a dispersion of ≈24%. For all sources resolved on scales $\lesssim$0.5 kpc, the thermal fraction was found to be even larger at $\gtrsim$90% suggesting that radio emission at 33 GHz provides a sensitive measure of the ionizing photon rate from young star-forming regions. The 33 GHz to total IR flux ratio was found to increase as the radio spectral index (measured between 1.7 and 33 GHz) flattens independent of the GBT projected beam area. Consequently, the ratio of non-thermal to total IR emission appears relatively constant on $\gtrsim$1 kpc scales, suggesting only moderate variations in the cosmic-ray electron injection spectrum and ratio of synchrotron to total cooling processes among star-forming complexes. Assuming that this trend solely arises from an increase in the thermal fraction sets a maximum on the scatter of the non-thermal spectral indices among the star-forming regions of $\sigma_{\alpha^{\rm NT}} \lesssim 0.13$, consistent with Niklas *et al.* (1997).

This work was been recently advanced though new 33 GHz imaging for the SFRS using the VLA, mapping 112 pointings on $\gtrsim 2''$ scales, compared to the $\approx 25''$ single-beam GBT photometry. These galaxies, which are included in the *Spitzer* Infrared Nearby Galaxies Survey (SINGS; Kennicutt *et al.* 2003) and Key Insights on Nearby Galaxies: a Far-Infrared Survey with *Herschel* (KINGFISH; Kennicutt *et al.* 2011) legacy programs, are well studied and have a wealth of ancillary data available. A comparison with 33 GHz GBT single-dish observations indicates that the interferometric VLA observations recover 78% $\pm$ 4% of the total flux density over 25$''$ regions ($\approx$kpc scales) among all fields. On these scales, the emission being resolved out is most likely diffuse non-thermal synchrotron emission. Consequently, on the $\approx$30–300 pc scales sampled by VLA observations, the bulk of the 33 GHz emission is recovered and primarily powered by free-free emission from discrete H II regions, making it an excellent tracer of massive star formation. Of the 225 discrete regions used for aperture photometry, 162 are extranuclear (i.e., having galactocentric radii $r_{\rm G} \gtrsim 250$ pc) and detected at $> 3\sigma$ significance at 33 GHz and in Hα. Assuming a typical 33 GHz thermal fraction of 90% based on previous GBT results, the ratio of optically-thin 33 GHz to uncorrected Hα SFRs indicates a median extinction value on $\approx$30–300 pc scales of $A_{\rm H\alpha} \approx 1.26 \pm 0.09$ mag, with an associated median absolute deviation of 0.87 mag. Roughly 10% of these sources are highly embedded (i.e., $A_{\rm H\alpha} \approx\gtrsim$ 3.3 mag), suggesting that on average, H II regions remain embedded for $\lesssim$1 Myr.

New complementary interferometric observations at approximately matched resolution in the S- (2–4 GHz) and Ku- (12–18 GHz) bands (VLA/13B-215; PI. Murphy) has extended this analysis by allowing for the construction of spectral index maps for all regions which in turn results in a much more robust thermal/non-thermal decompositions (Linden *et al.* 2020). In the left panel of Figure 2, spectral index distributions are shown for frequencies between 3 and 15 GHz, 15 and 33 GHz, as well as measured via a ordinary least squares fit across all three frequencies for those sources detected in at least two of the three radio maps. Median and standard deviation (rms) values are given for each distribution, which are fairly consistent suggesting that the spectrum does not exhibit much curvature. Some spectral indices appear to be flatter than what is expected from pure free-free emission, and even positive in some cases. Such instances may be the result of a chance GHz peaking background galaxy, or perhaps due to AME (e.g., Murphy *et al.* 2010, 2018b), and will be followed up.

For 329 sources detected in at least two of the three radio maps, the typical aperture diameter used for the photometry corresponds to a median linear scale of $\approx$150 pc. Using the spectral index information over this lever-arm (that is an order-of-magnitude in frequency), we can estimate the thermal emission fraction at 33 GHz if we assume a fixed non-thermal spectral index. A constant non-thermal radio spectral index of $\alpha^{\rm NT} = 0.83$ is assumed given that this is the value found by Niklas *et al.* (1997; i.e., $\alpha^{\rm NT} = 0.83$ with a scatter of $\sigma_{\alpha^{\rm NT}} = 0.13$) and consistent to the average non-thermal spectral index found among the 10 star-forming regions studied in NGC 6946 by Murphy *et al.* (2011). For sources with measured spectral indices steeper than -0.83, that value less by -0.1 is used. In the right panel of Figure 2 the resulting distribution of the 33 GHz thermal fractions is shown, having median thermal fraction at 33 GHz is $\approx$93% with an rms of $\approx$24%. It is possible that circum-nuclear sources in the sample may be contaminated by energetics not associated with star formation. Limiting the analysis to only the 291

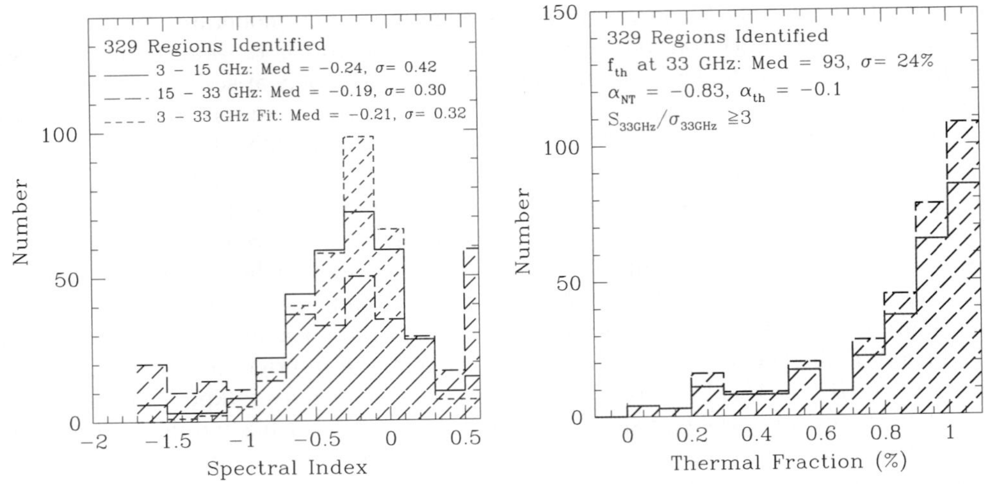

Figure 2. Adapted from Linden *et al.* (2020), which we refer the reader to for final plots and results. *Left:* The spectral index distributions measured between 3 and 15 GHz, 15 and 33 GHz, and measured via a ordinary least squares fit across all three frequencies measured at a common (i.e., the lowest) resolution of the native of the radio maps. Median and standard deviations (rms) values are given for each distribution. For the 329 sources detected in at least two of the three radio maps, this corresponds to a median linear scale of ≈150 pc *Right:* The corresponding distribution of thermal fractions measured at 33 GHz for all 329 sources assuming a non-thermal spectral index of −0.83 (Niklas *et al.* 1997; Murphy *et al.* 2011). With a median value of ≈93% and a standard deviation (rms) of ≈24%, the 33 GHz flux density appears to be a very sensitive measure of the ionizing flux of these star-forming regions.

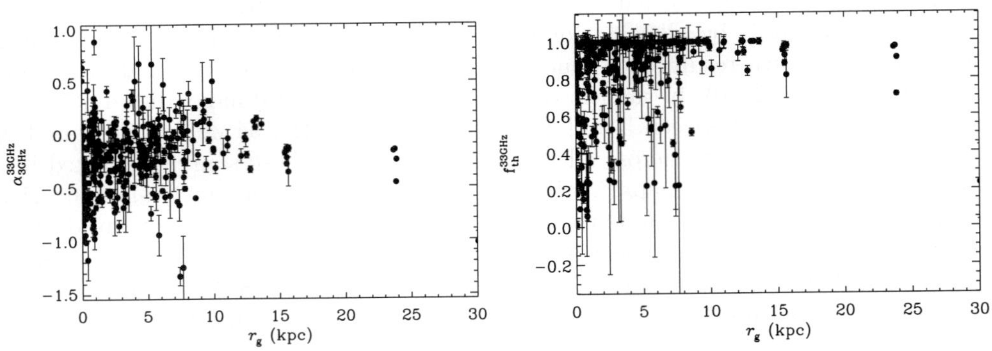

Figure 3. Adapted from Linden *et al.* (2020), which we refer the reader to for final plots and results. *Left:* Spectral index plotted against galactodentric radius. *Right:* Thermal fraction at 33 GHz plotted agains galactocentric radius. The scatter in spectral index (and thermal fraction) appears to decrease with increasing galactocentric radius.

sources with $r_G \gtrsim 250$ pc results in a similar value for the median 33 GHz thermal fraction, being ≈96% with an rms of ≈23%. Consequently, on the scales of $\lesssim$ a few hundred pc, the 33 GHz flux density appears to be dominated by free-free emission and therefore a sensitive measure of the ionizing flux of newly formed massive stars.

What is also found is that the scatter in spectral indices and thermal fractions decreases with increasing r_G (Figure 3). This result most likely arises from different star formation histories in the centers of galaxies versus further out in the disk, with the galaxy centers have longer periods of ongoing star formation. This in turn results in a larger spread of ages and radio spectral indices, as well as populations of CR electrons/positrons to be (re-)accelerated by passing shocks from next-generation star formation events.

4. Prospects with Future Facilities – A next-generation VLA

In the 2020's there will be a powerful suite of next-generation radio/mm facilities available that will transform our understanding in many areas of astrophysics. As currently envisioned, phase one of the Square Kilometre Array (SKA1-MID) will be the premier radio interferometer at decimetric wavelengths, consisting of up to 133 15 m (+ 64 13.5 m MeerKAT) dishes with a maximum baseline of up to ~150 km and eventually covering a frequency range spanning 350 MHz (85 cm) to 14 GHz (2 cm). ALMA, on the other hand, is a radio interferometer optimized to take advantage of the sub-mm (THz) windows that are only accessible at high, dry sites such as the Chajnantor plateau, which is at ~5000 m altitude. The currently envisioned next-generation Very Large Array (ngVLA) will be optimized for observations at frequencies (i.e., 1.2–116 GHz) between the exquisite performance of ALMA at submm wavelengths, and the future SKA-1 at decimeter to meter wavelengths, thus lending itself to be highly complementary with these facilities.

The ngVLA will revolutionize our understanding of the emission mechanisms that power the radio continuum emission in and around galaxies by enabling the routine construction of ~1.2–116 GHz radio spectral maps. Coupled with its nearly order of magnitude increased sensitivity compared to the current VLA, the ngVLA will make it possible to use this frequency window to investigate the distinct physical processes linked to stellar birth and death for large, heterogeneous samples of galaxies. Delivering such a finely sampled spectrum over this entire frequency range with a single instrument will allow robust separation of the various emission components, which is currently the main challenge for multi-frequency radio studies. Each observation will provide enough sensitivity and spectral coverage to robustly decompose and accurately quantify the individual energetic components powering the radio continuum, thus providing unique information on the non-thermal plasma, ionized gas, and cold dust content in the disks and halos of galaxies.

4.1. *Mapping Star Formation within Nearby Galaxies on a Range of Physical Scales*

For a proper decomposition of the radio continuum emission into its component parts, one needs to have spectral coverage at frequencies low enough (i.e., < 10 GHz) to be dominated by the non-thermal, steep (i.e., $\alpha \approx -0.8$) spectrum component, and at frequencies high enough (i.e., > 50 GHz) where the emission becomes completely dominated by the flat (i.e., $\alpha \approx -0.1$) spectrum free-free emission. Given the potential for a significant contribution from AME (Murphy *et al.* 2010, 2018b; Scaife *et al.* 2010; Hensley *et al.* 2015), peaking at frequencies ~ 20–40 GHz, coarse coverage spanning that spectral region is critical to account for such features. To date, the shape of the AME feature is largely unconstrained, and has not been carefully measured in the ISM of extragalactic sources. This is largely due to the insensitivity of current facilities to conduct a proper search for AME in nearby galaxies and map the feature with enough frequency resolution to provide useful constraints on its shape.

Broadband imaging spanning a frequency range of 1 – 120 GHz (i.e., largely covered by the ngVLA) will therefore be extremely powerful to properly decompose radio continuum emission from galaxies into its constituent parts. Additionally having frequency coverage below ~10 GHz provides sensitivity to free-free absorption, which is common in nearby luminous infrared galaxies (Condon *et al.* 1991; Clemens *et al.* 2010; Barcos-Muñoz *et al.* 2015). For individual (ultra compact) H II regions, the turnover frequency can be as high as ≈20 GHz (Murphy *et al.* 2010). A fundamental goal of the ngVLA will be to produce star formation maps for a large, heterogenous samples of nearby galaxies at $\approx 1''$ resolution. This will deliver Hα-like images that optical astronomers have relied on heavily for decades without having the additional complications of extinction and contamination by nearby [N II] emission, which make such images challenging to interpret.

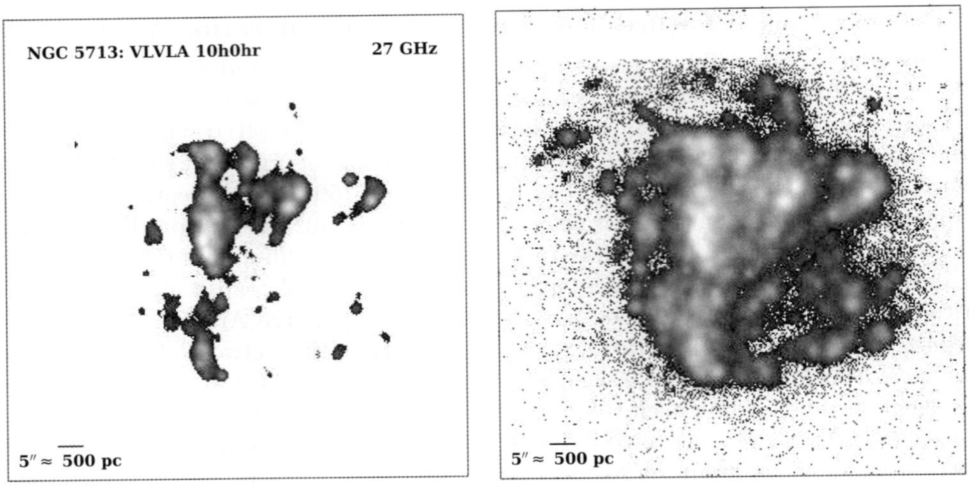

Figure 4. Both panels show a model 27 GHz free-free emission image of NGC 5713 ($d_L \approx$ 21.4 Mpc, SFR $\approx 4\,M_\odot\,\mathrm{yr}^{-1}$) based on an existing H$\alpha$ narrow band image at its native ($\approx 1''$) angular resolution taken from SINGS (Kennicutt *et al.* 2003). The left and right panels indicate the emission that would be detected at the 3σ level after a 10 hr on-source integration time using the current VLA in C-configuration ($1\sigma \approx 1.5\,\mu\mathrm{Jy\,bm}^{-1}$) and the ngVLA with a $1''$ sculpted beam ($1\sigma \approx 0.17\,\mu\mathrm{Jy\,bm}^{-1}$). In a relatively modest integration, the ngVLA is able to recover a significant fraction of star formation activity that is completely missed by the VLA.

By achieving arcsecond-like resolution that is commensurate with ground-based optical facilities, the ngVLA will be able to probe $\approx$100 pc scales out to the distance of Virgo (the nearest massive cluster at $d \approx 16.6$ Mpc), which are the typical sizes of giant molecular clouds (GMCs) and giant HII regions. At an rms sensitivity of $0.15\,\mu\mathrm{Jy\,bm}^{-1}$ at 27 GHz, such radio maps will reach a sensitivity expressed in terms of SFR density of $\approx 0.005\,M_\odot\,\mathrm{yr}^{-1}\mathrm{kpc}^{-2}$, matching the sensitivity of extremely deep Hα images such as those included in the Local Volume Legacy survey (Kennicutt *et al.* 2008). An example of this is illustrated in Figure 4 where an existing Hα narrow band image taken from SINGS (Kennicutt *et al.* 2003) was used to create a model 27 GHz free-free emission map at $1''$ resolution for the nearby star-forming galaxy NGC 5713. With a 10 hr integration the ngVLA will be able to map the entire disk of NGC 5713 down to an rms of $\approx 0.15\,\mu\mathrm{Jy\,bm}^{-1}$ ($\approx$35 mK). A comparison of what can currently be delivered with the VLA for the same integration time is also shown, indicating that only the brightest star-forming peaks are able to be detected. To make a map to the same depth using the current VLA would take $\gtrsim$800 hr! This is the same amount of time it would take to roughly survey $\gtrsim$80 galaxies with the ngVLA.

Using the same data, but applying a different imaging weighting scheme to create finer resolution maps (i.e., $0\overset{''}{.}1$, or even higher for brighter systems), similar multi-frequency radio continuum analyses can be performed for discrete HII regions and supernova remnants (SNRs) to complement high-resolution, space-based optical/NIR observations (e.g., *HST*, *JWST*, *WFIRST*, etc.). At an angular resolution of $0\overset{''}{.}1$, the data would sample $\approx$10 pc scales in galaxies out to the distance of Virgo to resolve and characterize (e.g., size, spectral shape, density, etc.) discrete HII regions and SNRs with a sensitivity to diffuse free-free emission corresponding to a SFR density of $\approx 0.5\,M_\odot\,\mathrm{yr}^{-1}\mathrm{kpc}^{-2}$.

Consequently, the ngVLA will provide a transformational step for studies of star formation in the local universe covering a large, heterogeneous set of astrophysical conditions. This statement is independent of the fact that with such observations using the ngVLA, having its wide-bandwidth, a number of RRL's will come for free. The detection of such lines (individually or through stacking), coupled with the observed continuum emission, can be used to quantify physical conditions for the HII regions such as electron temperature. It is without question that the ngVLA will make radio observations a critical component for investigations carried out by the entire astronomical community studying star formation and the ISM of nearby galaxies.

5. Summary

Radio spectra are powered by a combination of energetic processes, each of which contains a wealth of diagnostic information. By taking advantage of new wide-frequency backend capabilities of current radio facilities, observations of faint, high-frequency (i.e., $\gtrsim 30$ GHz) radio emission has become possible for samples of relatively bright regions within nearby galaxies. Such studies have shown that emission at these frequencies is dominated by free-free emission, being $\gtrsim$90% on $\lesssim$ a few hundred pc scales, making it a highly sensitive measurement of the ionizing photon rate of massive stars. To substantially advance these studies will take next-generation facilities covering frequencies spanning $\sim$1−120 GHz, such as the ngVLA. With an order-of-magnitude improvement in sensitivity over current and planned facilities at these frequencies, the ngVLA will enable transformational advancements in studies of star formation in the local universe and complement studies carried out with contemporaneous facilities like the SKA1 and ALMA, creating the ultimate suite of radio telescopes.

Acknowledgement

EJM thanks S. Linden for helping prepare figures from our currently unpublished work.

References

Barcos-Muñoz, L., Leroy, A. K., Evans, A. S., *et al.* 2015, *ApJ*, 799, 10
Barger, A., Kohno, K., Murphy, E. J., *et al.* 2018, ASP Monograph 7, Title of the Book, ed. E. Murphy (San Francisco, CA: ASP)
Balser, D. S., Anderson, L. D., Bania, T. M., *et al.* 2018, ASP Monograph 7, Title of the Book, ed. E. Murphy (San Francisco, CA: ASP)
BICEP2/Keck Collaboration, Planck Collaboration, Ade, P. A. R., *et al.* 2015, Physical Review Letters, 114, 101301
Bot, C., Ysard, N., Paradis, D., *et al.* 2010, *A&A*, 523, A20
Clemens, M. S., Vega, O., Bressan, A., *et al.* 2008, *A&A*, 477, 95
Clemens, M. S., Scaife, A., Vega, O., & Bressan, A. 2010, *MNRAS*, 405, 887
Condon, J. J. & Yin, Q. F. 1990, *ApJ*, 357, 97
Condon, J. J., Huang, Z.-P., Yin, Q. F., & Thuan, T. X. 1991, *ApJ*, 378, 65
Condon, J. J. 1992, *ARA&A*, 30, 575
Dame, T. M., Hartmann, D., & Thaddeus, P. 2001, *ApJ*, 547, 792
Dickinson, C., Ali-Haïmoud, Y., Barr, A., *et al.* 2018, NewAR, 80, 1
Draine, B. T. & Lazarian, A. 1998, *ApJL*, 494, L19
—. 1999, *ApJ*, 512, 740
Erickson, W. C. 1957, *ApJ*, 126, 480
Heiles, C., Reach, W. T., & Koo, B.-C. 1988, *ApJ*, 332, 313
Hensley, B., Murphy, E., & Staguhn, J. 2015, *MNRAS*, 449, 809
Hensley, B. S., Draine, B. T., & Meisner, A. M. 2016, *ApJ*, 827, 45
Hensley, B. S. & Draine, B. T. 2017, *ApJ*, 836, 179
Klein, U. & Graeve, R. 1986, *A&A*, 161, 155
Klein, U., Wielebinski, R., & Morsi, H. W. 1988, *A&A*, 190, 41
Kennicutt, Jr., R. C., Armus, L., Bendo, G., *et al.* 2003, *PASP*, 115, 928
Kennicutt, R. C., Jr., Lee, J. C., Funes, J. G., *et al.* 2008, *ApJS*, 178, 247–279
Kennicutt, R. C., Calzetti, D., Aniano, G., *et al.* 2011, *PASP*, 123, 1347
Kobulnicky, H. A. & Johnson, K. E. 1999, *ApJ*, 527, 154
Kogut, A., Banday, A. J., Bennett, C. L., *et al.* 1996, *ApJ*, 460, 1

Koyama, K., Petre, R., Gotthelf, E. V., *et al.* 1995, *Nature*, 378, 255
Leroy, A. K., Evans, A. S., Momjian, E., *et al.* 2011, *ApJL*, 739, L25
Leitch, E. M., Readhead, A. C. S., Pearson, T. J., & Myers, S. T. 1997, *ApJL*, 486, L23
Linden, S. T., Murphy, E. J., Dong, D., *et al.* 2020, *ApJS*, submitted
Mezger, P. G. & Henderson, A. P. 1967, *ApJ*, 147, 471
Murphy, E. J., Helou, G., Condon, J. J., *et al.* 2010a, *ApJL*, 709, L108
Murphy, E. J., Condon, J. J., Schinnerer, E., *et al.* 2011, *ApJ*, 737, 67
Murphy, E. J., Bremseth, J., Mason, B. S., *et al.* 2012, *ApJ*, 761, 97
Murphy, E. J. 2013, *ApJ*, 777, 58
Murphy, E. J., Dong, D., Leroy, A. K., *et al.* 2015, *ApJ*, 813, 118
Murphy, E. J., Momjian, E., Condon, J. J., *et al.* 2017, *ApJ*, 839, 35
Murphy, E. J., Dong, D., Momjian, E., *et al.* 2018a, *ApJS*, 234, 24
Murphy, E. J., Linden, S. T., Dong, D., *et al.* 2018b, *ApJ*, 862, 20
Murphy, T., Cohen, M., Ekers, R. D., *et al.* 2010b, *MNRAS*, 405, 1560
Niklas, S., Klein, U., & Wielebinski, R. 1997, *A&A*, 322, 19
Peel, M. W., Dickinson, C., Davies, R. D., Clements, D. L., & Beswick, R. J. 2011, *MNRAS*, 416, L99
Planck Collaboration, Ade, P. A. R., Aghanim, N., *et al.* 2011, *A&A*, 536, A20
Planck Collaboration, Adam, R., Ade, P. A. R., *et al.* 2016, *A&A*, 586, A133
Scaife, A. M. M., Nikolic, B., Green, D. A., *et al.* 2010, *MNRAS*, 406, L45
Scoville, N., Sheth, K., Aussel, H., *et al.* 2016, *ApJ*, 820, 83
Selina, R. and Murphy, E. J., 2017, ngVLA Memo #17
Socrates, A., Davis, S. W., & Ramirez-Ruiz, E. 2008, *ApJ*, 687, 202–215
Tabatabaei, F. S., Minguez, P., Prieto, M. A., & Fernández-Ontiveros, J. A. 2018, *Nature Astronomy*, 2, 83
Turner, J. L. & Ho, P. T. P. 1983, *ApJL*, 268, L79
—. 1985, *ApJL*, 299, L77
—. 1994, *ApJ*, 421, 122
Turner, J. L., Ho, P. T. P., & Beck, S. C. 1998, *AJ*, 116, 1212

Discussion

ROSARIO: Given the range in radio spectral shapes and thermal fractions in individual SF regions, can you reflect on how much uncertainty one must keep in mind when using radio measurements as SF tracers in higher redshift surveys?

MURPHY: This question really requires 2 answers. In the case where the IMF is known and the galaxy being studied is known to be star-forming (not an AGN), the uncertainty on the radio derived SFR is quite good, and will be more reliable for higher frequency observations (say $\sigma_{\rm SFR} < 2\times$, and perhaps as good as 20-50% uncertainty). If one folds in the uncertainty in the IMF and/or the contribution from an AGN, the uncertainty is $> 2\times$, just like any other SFR diagnostic.

SCHAERER: When going to higher redshifts, we are interested in compact SFGs, maybe resembling low-z dwarfs. How do these fit into the observation and trends you have shown?

MURPHY: Such galaxies should be detectable in the radio. Depending on how porous these galaxies are to cosmic rays, they may only be detectable in their free-free emission.

NAGAO: Are radio spectra of ULIRGs free from possible AGN contributions? Removing very low-luminosity AGNs from the ULRIG sample may be difficult.

MURPHY: It is almost certain that ULRIGs will contain some fraction of radio emission that is due to AGN. However, determining the fraction (or dominance) can be carried out using other diagnostics. In the case of what was presented, PAH equivalent widths were used, which I believe has reliably determined those sources denoted by AGN.

Challenges in Panchromatic Modelling with Next Generation Facilities
Proceedings IAU Symposium No. 341, 2018
M. Boquien, E. Lusso, C. Gruppioni & P. Tissera, eds.

doi:10.1017/S1743921319002862

Challenges in modelling the rest-frame ultraviolet/optical spectra of galaxies at the high-redshift frontier

Erik Zackrisson and Anton Vikaeus

Observational Astrophysics, Department of Physics and Astronomy,
Uppsala University, Box 516, SE-751 20 Uppsala, Sweden
email: erik.zackrisson@physics.uu.se

Abstract. New challenges in the modelling of galaxy spectra are bound to emerge as upcoming telescopes like the James Webb Space Telescope will allow us to detect galaxies at fainter flux levels and higher redshifts than ever before. Here, we highlight three modelling problems that may become relevant for upcoming observations in the rest-frame ultraviolet/optical at the high-redshift frontier: stellar initial mass function sampling effects, Population III signatures and the leakage of ionizing radiation into the intergalactic medium.

Keywords. Galaxies: high-redshift, galaxies: stellar content, ISM: lines and bands

1. Introduction

Spectral synthesis models predict the spectra or spectral energy distributions (SEDs) of astronomical objects as a function of astrophysical parameters, either for the purpose of analyzing data, forecasting future observations or providing input for numerical simulations or other calculations. Current observations of the rest-frame ultraviolet(UV)/optical properties of galaxies at both low and high redshift have already revealed a number of challenges that need to be addressed in order for such codes to provide robust interpretations of galaxies at the high-redshift frontier. This includes problems in interpreting the rest-frame UV lines of the brightest Lyα-emitters at $z \gtrsim 6$ (e.g. Stark 2018; Mainali *et al.* 2018), problems in explaining HeII emission properties at low metallicities (e.g. Senchyna *et al.* 2017; Berg *et al.* 2018), and mismatches between the metallicities of stars and gas in intermediate-redshift galaxies (Steidel *et al.* 2016). As new observational facilities like the *James Webb Space Telescope (JWST), the Wide Field Infrared Survey Telescope (WFIRST), the Giant Magellan Telescope (GMT), the Thirty Meter Telescope (TMT),* and *the Extremely Large Telescope (ELT)* bring galaxies of unprecedented faintness at the high-redshift frontier within reach, one can also expect that the spectral synthesis community will soon have to face other modelling challenges. Here, we will highlight three such situations that we predict will become increasingly relevant for the high-redshift galaxy community in coming years: the modelling of Population III signatures, stellar initial mass function (IMF) sampling effects and Lyman continuum (LyC) leakage. Throughout this paper, we focus our discussion exclusively on galaxies at $z \gtrsim 6$, even though some of these issues are relevant at lower redshifts as well.

2. Population III galaxies

The detection of signatures of chemically pristine stars, the so-far elusive Population III (hereafter Pop III), is sometimes referred to as the holy grail in the study of the high-redshifts Universe. Due to the cooling and fragmentation properties of zero-metallicity

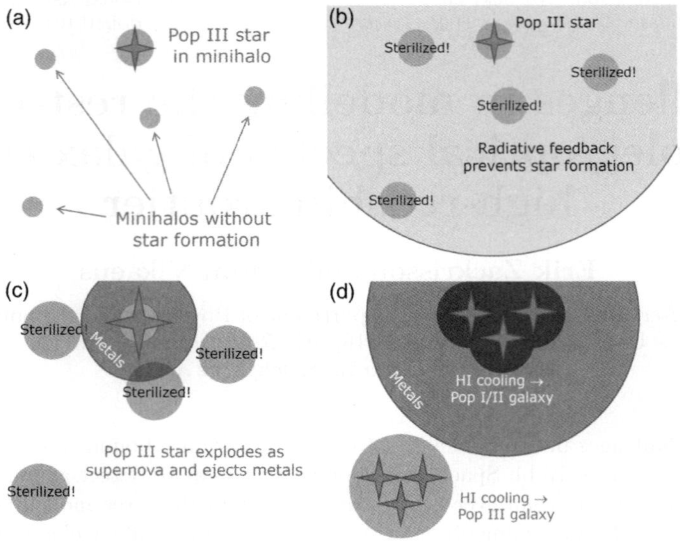

Figure 1. Schematic illustration of one potential scenario that leads to the formation of a Pop III galaxy. **a)** A minihalo (a halo in the sub-HI cooling mass range) forms a single pop III star (or alternatively a small number of such stars), while adjacent minihalos at somewhat lower masses have yet to experience any star formation. **b)** The Pop III star emits Lyman-Werner radiation (blue region) that inhibits star formation in the adjacent minihalos. **c)** After having expired its lifetime, the pop III star goes supernova and ejects metals (red region) into the surrounding medium, while the nearby minihalos – still incapable of forming stars – continue to accrete material and increase in mass. **d)** Several of the now metal-enriched minihalos merge with the post-Pop III minihalo and reach the HI cooling limit required for prolonged star formation. This object then forms a small galaxy, dominated by Pop I/II stars. At the same time, a rare minihalo that has remained chemically pristine due to its distance from the Pop III supernova blast site also reaches the HI cooling limit, but instead forms a Pop III galaxy.

gas, the stellar initial mass function of Pop III stars is expected to be top-heavy, with a characteristic mass somewhere in the 1–1000 $M_\odot$ range (e.g. Stacy *et al.* 2016; Hosokawa *et al.* 2016).

While individual Pop III stars at high redshift may be beyond the reach of upcoming telescopes (unless subject to extremely high magnification due to gravitational lensing; Rydberg *et al.* 2013; Windhorst *et al.* 2018; Diego 2018), there may be other viable avenues for studying the properties of such stars – including Pop III supernovae (and their ejecta), Pop III gamma-ray bursts – and Pop III galaxies. The latter are small systems in which large numbers of Pop III stars manage to form within a limited time window, thereby potentially boosting the integrated luminosity of the Pop III stars to detectable levels.

Pure Pop III galaxies may form from pristine gas in dark matter halos that have managed to reach the mass regime for HI cooling (total mass $\gtrsim 10^7$–10^8 $M_\odot$, depending on redshift) without experiencing prior star formation or being polluted by metals from nearby objects. Figure 1 features a schematic illustration of a possible scenario of this type. While such systems are expected to be quite numerous (Stiavelli & Trenti 2010), most are expected to attain total stellar masses $\lesssim 10^4 M_\odot$ and hence lie well below the detection thresholds of currently planned telescopes (unless subject to very high magnifications along fortuitous sightlines; Zackrisson *et al.* 2015). However, the stir in the community caused by the potential Pop III nature of the very bright Lyα-emitter CR7 (Sobral *et al.* 2015) – albeit no longer believed to be a Pop III galaxy (e.g. Sobral *et al.* 2019) – pushed the community to think hard about whether more

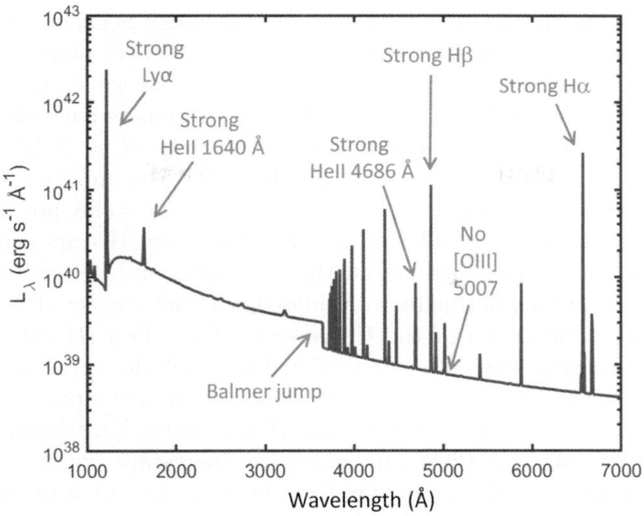

Figure 2. Some of the rest-frame UV/optical Pop III spectral signatures discussed in the literature. Please note that Lyα (and the continuum at wavelengths directly shortward of this line) may be absorbed by the neutral intergalactic medium at $z > 6$. This particular SED has been generated using the Yggdrasil (Zackrisson *et al.* 2011) model, under the assumption of a Pop III.1 stellar IMF (characteristic mass $\sim 100\ M_\odot$), an age of 1 Myr, a total stellar mass of $10^6\ M_\odot$ and negligible leakage of ionizing photons.

massive systems could be created in rare situations. For instance, Visbal *et al.* (2017); Yajima & Khochfar (2017); Inayoshi *et al.* (2018) all argue that Pop III objects with total stellar mass $\sim 10^6\ M_\odot$ could potentially form. If the Pop III stars within one of these objects were to form within a time window of $\approx 10^7$ yr, with a stellar initial mass function centered around $\sim 10\ M_\odot$, the apparent rest-frame UV magnitude of such a galaxy would be $m_{\rm AB} \approx 30$–31 AB at $z \approx 6$–10. Due to the low probabilities of such relatively massive Pop III galaxies, large survey areas combined with a magnification boost due to strong lensing would likely be required to find them, and this may have to await the launch of WFIRST. However, Johnson & Aykutalp (2018) have outlined a scenario in which star formation could be prevented in a metal-free halo close to a powerful quasar until the halo reaches a mass of $\sim 10^{10}\ M_\odot$. As such halos have the potential to form very massive Pop III galaxies (possibly with $m_{\rm AB} < 30$ mag), this motivates a search for such objects around known high-redshift quasars, which could be realized already with JWST.

In coming years, new Pop III galaxy candidates are likely to be discovered, and developing SED models that can help determine the nature of such objects is an important task. Several spectral synthesis models geared to Pop III systems are already available, including those by Schaerer (2002, 2003); Inoue (2011); Raiter *et al.* (2010); Zackrisson *et al.* (2011); Mas-Ribas *et al.* (2016), and several potential spectral diagnostics in the rest-frame UV/optical for such objects are known – some of which are summarized in Figure 2. However, there are a number of predicted features of such systems have not yet made their way into the existing spectral synthesis models.

Rotation creates increased mixing within Pop III stars and could, for sufficiently high rotation rates, lead to chemically homogenous evolution. In general, rotation acts to make Pop III stars live longer, attain higher luminosities and higher surface temperatures, which leads to boosted ionizing fluxes. As shown by Yoon *et al.* (2012), the hydrogen-ionizing fluxes may be boosted by factors of several for Pop III stars in the ≈ 20–$300\ M_\odot$ range, provided that the rotational rate is sufficiently high, and the effects for He^{+}-ionizing radiation can be even more dramatic. Such effects have not yet been incorporated into spectral synthesis models for Pop III galaxies, although Schaerer (2003) does explore the consequences of a rotationally-induced Pop III Wolf Rayet phase.

Because of recent interest in the potential gravitational wave signatures of merging black holes left behind from Pop III stars, a number of papers have attempted to model or simulate the Pop III binary population (e.g Kinugawa *et al.* 2014; Belczynski *et al.* 2017; Inayoshi *et al.* 2017; Hirano *et al.* 2018). However, the impact that Pop III binary evolution may have on the spectra of Pop III galaxies have so far not been explored.

Some fraction of the ionizing radiation produced by Pop III stars could evade absorption in the dense interstellar medium in these objects, and instead contribute to the ionization of the intergalactic medium. While this is not expected to be a significant contributor to the reionization of the Universe (because Pop III galaxies are so rare), this would still alter the strength of emission lines and the ratio between direct star light and nebular emission in these objects. While some attempts to model a few different scenarios of this type have been made (Inoue 2011; Zackrisson *et al.* 2011), the simulations by Johnson *et al.* (2009) indicate that the escape could be highly selective, with hydrogen-ionizing radiation escaping more easily than He^+-ionizing radiation, with significantly altered emission-line ratios as a result.

Once a candidate Pop III galaxy has been identified, possibly based on some combination of the spectral signature shown in Figure 2, there are primarily two questions that observers will attempt to answers: 1) Is this really a Pop III system; and 2) Is the stellar IMF top-heavy, as predicted by theory? While the solution to the first problem would be to establish that the system has a Pop III-like metallicity, detection thresholds for faint, high-redshift systems may make this far from trivial. Since Lyα may be absorbed in the neutral IGM at $z > 6$, Hα and [OIII]5007 are usually expected to be the strongest rest-frame UV/optical emission lines for reionization-epoch galaxies (Shimizu *et al.* 2016). However, Hα redshifts out of the reach of JWST/NIRSpec range at $z > 7$, which may make Hβ the strongest hydrogen recombination line that can be detected at the high-redshift frontier in the foresable future. If Hβ is detected, but not [OIII], then this would be a strong indication of a very low metallicity, but how low? The exact value of the ratio [OIII]/Hβ would be very hard to determine observationally for [OIII]/H$\beta < 0.1$, which corresponds to [M/H]$\lesssim -3$ (Inoue 2011). However, this would still be consistent with a very metal-poor, yet definitely metal-enriched galaxy that has already started forming low-mass stars (e.g. Schneider *et al.* 2012). If no signatures of a top-heavy IMF like (a very strong HeII line (e.g Raiter *et al.* 2010) is found in an object of this type, it would be hard to draw any conclusion on the Pop III IMF from this.

Another problem lies in the fact that there may be a significant population of "hybrid" Pop III/II/I galaxies, i.e. galaxies that contains both metal-free and metal-enriched stars, as for instance predicted by the simulations of Sarmento *et al.* (2018). Another possibility is highlighted by the simulations by Xu *et al.* (2016), where a small halo forming Pop III stars photoionize gas that has been metal-enriched by other nearby halos. While the Pop III halo studied by Xu *et al.* (2016) does not lie in the mass range where it would be readily detectable, one could envision situations like this giving rise to a Pop III galaxy that only contains Pop III stars (and, as a result, a very hard ionizing continuum) nonetheless displays metal emission lines. Problems of this type are already acute at low redshifts, for which there are claims that certain very metal-poor galaxies (but at a metallicity above the Pop III regime) display Pop III signatures within specific regions, primarily based on the strength of HeII emission lines (e.g. Kehrig *et al.* 2015, 2016; Ucci *et al.* 2019; Kehrig *et al.* 2018). It is not immediately clear, however, that the discrepancy between the observed HeII fluxes and those predicted by models can be completely detached from the more general problem in explaining the strength of HeII emission in low-metallicity galaxies (e.g. Senchyna *et al.* 2017; Berg *et al.* 2018).

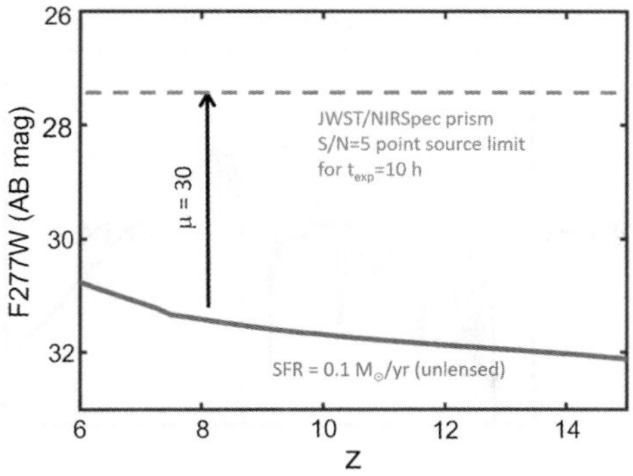

Figure 3. The redshift dependence of the JWST/NIRCam F277W flux (measured in AB magnitudes) for a $Z = 0.008$ galaxy that has experienced a constant SFR $= 0.1\ M_\odot\ \mathrm{yr}^{-1}$ for 100 Myr, compared to the JWST/NIRSpec spectroscopic $R = 100$ point-source $S/N = 5$ detection limit (dashed horizontal line) for an exposure time of 10 h. While the intrinsic flux of this galaxy (considered to be at the bright limit of the regime where IMF sampling effects may become important) lies comfortably below the JWST spectroscopic detection limit, strong lensing with magnification $\mu = 30$ (indicated by arrow) or more by a foreground galaxy cluster may well lift it into the detectable range. Hence, galaxies for which the SED is significantly affected by IMF sampling effects may turn up in deep spectroscopic studies of cluster fields with JWST.

3. Stochastic IMF sampling

A standard assumption when modelling the SED of galaxies is that the stellar initial mass distribution can be treated as continuous, i.e. that the IMF is fully sampled at all points in time when new stars are formed. This assumption may break down in low-mass or low-SFR systems (for a review, see Cerviño 2013), with the result that two objects with identical ages, metallicities and star formation histories may exhibit different SEDs due to differences in the number of massive stars present at a given time. This problem is widely recognized in the study of local, low-mass systems like unresolved star clusters (e.g. Krumholz *et al.* 2019), but has been mostly ignored by the high-redshift community since the masses and SFRs of systems studied tend to be much too large for such effects to have any significant impact. However, as new telescopes push the detection thresholds to intrinsically fainter galaxies, IMF stochasticity may become an increasingly important problem in modelling and interpreting galaxy properties at high redshifts as well. Forero-Romero & Dijkstra (2012) argue that this may have a noticeable effect on studies of Lyα equivalent widths and Lyman continuum leakage even for galaxies with SFRs as high as $\sim 1\ M_\odot\ \mathrm{yr}^{-1}$, and Paalvast & Brinchmann (2017) predict non-negligible effects on the ratios of various strong optical emission lines for $< 0.1\ M_\odot\ \mathrm{yr}^{-1}$. It has also been argued that IMF sampling effects may also affect the impact of feedback recipes in simulations of low-mass galaxies (stellar mass $\lesssim 10^6\ M_\odot$) at $z \approx 0$ (Su *et al.* 2018) and galaxies close to the HI cooling limit ($\sim 10^8\ M_\odot$) at $z \gtrsim 6$ (Applebaum *et al.* 2018).

If we take SFR $\lesssim 0.1\ M_\odot\ \mathrm{yr}^{-1}$ as the limit where IMF stochasticity may start to have non-negligible effects on galaxy SEDs (e.g. Paalvast & Brinchmann 2017), at what apparent magnitudes do we expect the affected systems to turn up at $z \gtrsim 6$? In Figure 3, we plot the JWST/NIRCam F277W AB magnitudes for Yggdrasil (Zackrisson *et al.* 2011) $Z = 0.008$ galaxies at $z = 6$–15 that have experienced a constant SFR $= 0.1\ M_\odot\ \mathrm{yr}^{-1}$ for 100 Myr. As seen, the intrinsic AB magnitudes are in the range $m_{277} \approx 31$–32 mag†, which – in the absence of gravitational lensing – is below the photometric detection threshold of even the longest JWST exposure times, and UV dust attenuation would

† These galaxies are modelled assuming a fully sampled IMF, and IMF sampling effects would introduce slight galaxy-to-galaxy variations in their broadband fluxes.

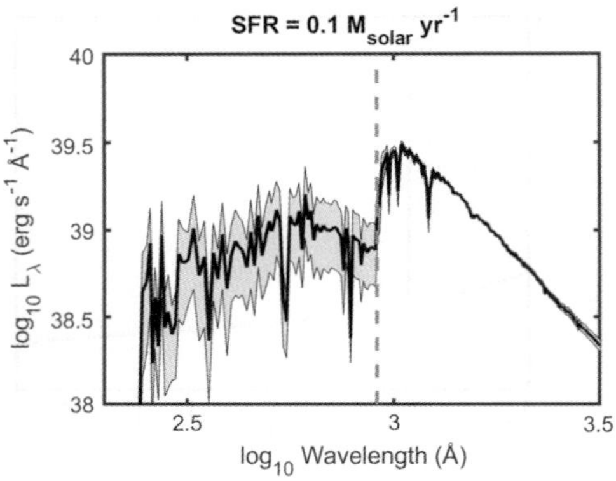

Figure 4. The mean stellar population SED (black line), predicted by SLUG (da Silva *et al.* 2011) at rest-frame UV wavelengths for a $Z = 0.15\ Z_\odot$ galaxy that has maintained a constant star formation rate at a level of $\mathrm{SFR} = 0.1\ M_\odot\ \mathrm{yr}^{-1}$ for 100 Myr (giving a total stellar population mass of $10^7\ M_\odot$) in the case where are stars are born in stellar clusters. The coloured patch represents the galaxy-to-galaxy standard deviation of flux levels around around the mean SED after a normalization to the average galaxy flux at 1500 Å, indicating significant (factor of ≈ 2.5) variations in ionizing flux (at wavelengths shortward of the Lyman limit at 912 Å, marked by the vertical, red dashed line) despite potentially having similar 1500 Å fluxes.

make them even fainter still. However, such objects may still appear well above JWST detection thresholds in strongly lensed fields. For example, a magnification of $\mu > 30$ could potentially push such systems above the JWST spectroscopic detection threshold ($m_{\mathrm{AB}} \approx 27.5$ mag for $S/N = 5$ with JWST/NIRSpec prism after a 10 h exposure time). Hence, spectroscopic studies of $z \gtrsim 6$ galaxies in cluster lensing fields may well be probing galaxies for which sampling effects are significant.

What effect is this IMF stochasticity expected to have on such systems? In Figure 4, we show the variations predicted in the stellar component of the rest-frame ultraviolet and optical spectra predicted by the SLUG code (da Silva *et al.* 2011) for an $\mathrm{SFR} = 0.1\ M_\odot$ yr^{-1}, $Z = 0.15\ Z_\odot$ galaxy at an age of 100 Myr in the case where all stars are born in star clusters (thereby maximizing the impact of sampling effects). As seen, there are in this case highly significant (factor of ≈ 2.5) galaxy-to-galaxy variations in the hydrogen-ionizing UV flux due to variations in the number of the most massive stars compared to the non-ionizing UV flux (which are produced by somewhat older stars of lower masses). This will result in variations in line equivalent widths (especially at wavelengths where the spectrum is not dominated by nebular continuum) that JWST will be able to study spectroscopically. Hence, the use of models that assume a stable star formation rate and a fully sampled IMF may lead to incorrect conclusions when applied to the analysis some of the faint, lensed galaxies that JWST will allow us to study in the years to come.

4. Lyman-continuum leakage

If galaxies are to provide the ionizing photons for the reionization of the Universe, then some fraction of the LyC radiation produced by young, hot stars in these galaxies must be able to escape the interstellar medium and make it into the intergalactic medium. The LyC escape fraction f_{esc} required for cosmic reionization at $z > 6$ is typically estimated at $f_{\mathrm{esc}} \approx 0.1$, but simulations often predict substantial temporal variations in this quantity, sometimes with escape fractions of $f_{\mathrm{esc}} > 0.5$ attained for limited periods of time (e.g. Trebitsch *et al.* 2017).

Since LyC leakage into the intergalactic medium (or the low-density circumgalactic medium) means that fewer ionizing photons will contribute to photoionization in the interstellar medium (where densities are high and recombination time scales short), LyC leakage lowers the relative contribution of nebular emission (in the form of emission lines and nebular free-bound/free-free continuum) to the observed galaxy SED. Depending on the escape mechanism, emission line ratios may also be affected. There are basically two different scenarios that can give rise to LyC escape (e.g. Zackrisson *et al.* 2013): an ionization-bounded nebula with holes (also known as the "picket fence" model) and a density-bounded nebula. To first order, the first mechanism will simply act to reduce all nebular contributions by a factor $1 - f_{\rm esc}$. If the escape fraction is low ($f_{\rm esc} \lesssim 0.1$), this corresponds to a very small effect ($\lesssim 10\%$) on emission line fluxes and an even smaller effect on emission line equivalent widths (because part of the continuum may be nebular), which is often well within the observational error bars. Only at very high escape fractions ($f_{\rm esc} > 0.5$) may the effects be sufficiently pronounced to give rise to telltale signatures. Density-bounded leakage, on the other hand, can give rise to detectable effects already at small $f_{\rm esc}$. For instance, Nakajima & Ouchi (2014) demonstrate how the [OIII]5007/[OII]3727 line ratio can change significantly even when going from $f_{\rm esc} = 0$ to 0.1. In the context of classical HII regions, this happens because part of the [OII] zone of the nebula is truncated, whereas the [OIII] zone further in remains intact.

While LyC leakage has so far not been considered an important ingredient in spectral modelling of galaxies, a failure to recognize cases of significant LyC leakage may lead to incorrect inference on star formation activity and nebular conditions for such objects. Accurately diagnosing LyC leakage is admittedly very challenging amidst the many other effects that can give rise to similar spectral signatures, especially given the small number of emission lines detectable at the high-redshift frontier. Changes in $f_{\rm esc}$ from ionization-bounded leakage with holes are for instance degenerate with rapid star formation fluctuations (Binggeli *et al.* 2018), which cause fluctuations in the ratio between the ionizing and the non-ionizing UV flux (e.g. Boquien *et al.* 2014), and hence in emission line strengths. The [OIII]/[OII] line ratios relevant for density-bounded leakage are also affected by the ionization parameter, the metallicity and the presence of shocks (e.g Allen *et al.* 2008; Nakajima & Ouchi 2014). Another problem is that some fraction of the ionizing photons may be absorbed by dust rather than gas, a scenario that can only be constrained through rest-frame far-infrared data (e.g. Inoue *et al.* 2001; Hirashita *et al.* 2003), and even then only with great difficulty.

References

Allen, M. G., Groves, B. A., Dopita, M. A., Sutherland, R. S., & Kewley, L. J. 2008, *ApJS*, 178, 20

Applebaum, E., Brooks, A. M., Quinn, T. R., & Christensen, C. R. 2018, arXiv:1811.00022

Belczynski, K., Ryu, T., Perna, R., Berti, E., Tanaka, T. L., & Bulik, T. 2017, *MNRAS*, 471, 4702

Berg, D. A., Erb, D. K., Auger, M. W., Pettini, M., & Brammer, G. B. 2018, *ApJ* 859, 164

Binggeli, C., Zackrisson, E., Pelckmans, K., Cubo, R., Jensen, H., & Shimizu, I. 2018, *MNRAS*, 479, 368

Boquien, M., Buat, V., & Perret, V. 2014, *A&A*, 571, 72

Cerviño, M. 2013, *New Astron. Revs* 57, 123

da Silva, R. L., Fumagalli, M., & Krumholz, M. 2011, *ApJ*, 745, 145

Diego, J. M. 2018, arXiv:1806.04668

Forero-Romero, J. E. & Dijkstra, M. 2012, *MNRAS*, 428, 2163

Hirano, S., Yoshida, N., Sakurai, Y., & Fujii, M. S. 2018, *ApJ*, 855, 17

Hirashita, H., Buat, V., & Inoue, A. K. 2003, *A&A*, 410, 83

Hosokawa, T., Hirano, S., Kuiper, R., Yorke, H. W., Omukai, K., & Yoshida, N. 2016, *ApJ*, 824, 119

Inayoshi, K., Hirai, R., Kinugawa, T., & Hotokezaka, K. 2017. *MNRAS*, 468, 5020

Inayoshi, K., Li, M., & Haiman, Z. 2018, *MNRAS*, 479, 4017
Inoue, A. K., Hirashita, H., & Kamaya, H. 2001, *ApJ*, 555, 613
Inoue, A. K. 2011, *MNRAS*, 415, 2920
Johnson, J. L., Greif, T. H., Bromm, V., Klessen, R. S., & Ippolito, J. 2009, *MNRAS*, 399, 37
Johnson, J. L. & Aykutalp, A., 2018, arXiv:1806.07901
Kehrig, C., Vílchez, J. M., Pérez-Montero, E., Iglesias-P'aramo, J., Brinchmann, J., Kunth, D., Durret, F., & Bayo, F. M. 2015, *ApJ*, 801, L28
Kehrig, C., Vílchez, J. M., Pérez-Montero, E., Iglesias-P'aramo, J., Hernández-Fernández, J. D., Duarte Puertas, S., Brinchmann, J., Durret, F., *et al.* 2016, *MNRAS*, 459, 2992
Kehrig, C., Vílchez, J. M., Guerrero, M. A., Iglesias-Páramo, J., Hunt, L. K., Duarte-Puertas, S., & Ramos-Larios, G. 2018, *MNRAS* 480, 1081
Kinugawa, T., Inayoshi, K., Hotokezaka, K., Nakauchi, D., & Nakamura, T. 2014, *MNRAS*, 442, 2963
Krumholz, M. R., Adamo, A., Fumagalli, M., & Calzetti, D. 2019, *MNRAS*, 482, 3550
Mainali, R., Zitrin, A., Stark, D. P., Ellis, R. S., Richard, J., Tang, M., Laporte, N., Oesch, P., *et al.* 2018, *MNRAS*, 479, 1180
Mas-Ribas, L., Dijkstra, M., & Forero-Romero, J. E. 2016, *ApJ*, 833, 65
Nakajima, K. & Ouchi, M. 2014, *MNRAS*, 442, 900
Paalvast, M. & Brinchmann, J. 2017, *MNRAS* 470, 1612
Raiter, A., Schaerer, D., & Fosbury, R. A.E. 2010, *A&A* 523, 64
Rydberg, C.-E., Zackrisson, E., Lundqvist, P., & Scott, P. 2013, *MNRAS* 429, 3658
Sarmento, R., Scannapieco, E., & Cohen, S. 2018, *ApJ*, 854, 75
Schaerer, D. 2002, *A&A*, 382, 28
Schaerer, D. 2003, *A&A*, 397, 527
Schneider, R., Omukai, K., Bianchi, S., & Valiante, R. 2012, *MNRAS*, 419, 1566
Senchyna, P., Stark, D. P., Vidal-Garcia, A., Chevallard, J., Charlot, S., Mainali, R., Jones, T., Wofford, A., *et al.* 2017, *MNRAS*, 472, 2608
Shimizu, I., Inoue, A. K., Okamoto, T., & Yoshida, N. 2016, *MNRAS*, 461, 3563
Sobral, D., Matthee, J., Darvish, B., Schaerer, D., Mobasher, B., Röttgering, H. J.A., Santos, S., & Hemmati, S. 2015, *ApJ*, 808, 139
Sobral, D., Matthee, J., Brammer, G., Ferrara, A., Alegre, L., Röttgering, H., Schaerer, D., Mobasher, B., *et al.* 2019, *MNRAS*, 482, 2422
Stacy, A., Bromm, V., & Lee, A. T. 2016, *MNRAS*, 462, 1307
Stark, D. P. 2018, *ARA&A*, 54, 761
Steidel, C. C., Strom, A. L., Pettini, M., Rudie, G. C., Reddy, N. A., & Trainor, R. F. 2016, *ApJ*, 826, 159
Stiavelli, M. & Trenti, M. 2010, *ApJ*, 716, L190
Su, K.-Y., Hopkins, P. F., Hayward, C. C., Ma, X., Boylan-Kolchin, M., Kasen, D., Keres, D., Faucher-Giguére, C.-A., *et al.* 2018, *MNRAS*, 480, 1666
Trebitsch, M., Blaizot, J., Rosdahl, J., Devriendt, J., & Slyz, A. 2017, *MNRAS*, 470, 224
Ucci, G., Ferrara, A., Gallerani, S., Pallottini, A., Cresci, G., Kehrig, C., Hunt, L. K., Vilchez, J. M., *et al.* 2019, *MNRAS*, 483, 1295
Visbal, E., Bryan, G. L., & Haiman, Z. 2017, *MNRAS*, 469, 1456
Windhorst, R. A., Timmes, F. X., Wyithe, J., Stuart B., Alpaslan, M., Andrews, S. K., Coe, D., Diego, J. M., *et al.* 2018, *ApJS*, 234, 41
Xu, H., Norman, M. L., O'Shea, B. W., & Wise, J. H. 2016, *ApJ*, 823, 140
Yajima, H. & Khochfar, S. 2017, *MNRAS*, 467, L51
Yoon, S. C., Dierks, A., & Langer, N. 2012, *A&A*, 542, 113
Zackrisson, E., Rydberg, C.-E., Schaerer, D., Östlin, G., & Tuli, M. 2011, *ApJ*, 740, 13
Zackrisson, E., Inoue, A. K., & Jensen, H. 2013, *ApJ*, 777, 39
Zackrisson, E., González, J., Eriksson, S., Asadi, S., Safranek-Shrader, C., Trenti, M., & Inoue, A. K. 2015, *MNRAS* 449, 3057

Discussion

KON CHEN: How do you tell the difference between the Pop III galaxies and first galaxies?

ERIK ZACKRISSON: A Pop III galaxy is usually defined as a galaxy either consisting exclusively of Pop III stars or dominated by such stars. Chronologically, they are not expected to be the very first galaxies. An explanation of why this is may be found in Stiavelli & Trenti (2010).

HIROYUKI HIRASHITA: Do you expect any progress for the escape fractions of Ly continuum by JWST?

ERIK ZACKRISSON: The spectral features in the rest-frame UV/optical that are affected by Lyman-continuum leakage are definitely observable with JWST for large numbers of $z > 6$ galaxies, but there are many modelling degeneracies that may complicate the analysis as we attempt to apply the indirect techniques (which attempt to estimate $f_{\rm esc}$ without actually observing the Lyman-continuum flux) to galaxies in this redshift range. The main problem is that many of the highest-redshift galaxies display properties that are difficult to reproduce with current SED models and galaxy formation simulations. The first few years of JWST observations will hopefully tell us what's going on in these objects, which should allow for more reliable estimates of the escape fractions.

YUICHI MATSUDA: How big is the (opposite) Balmer Break of Pop III galaxies? Is this detectable with JWST?

ERIK ZACKRISSON: In f_λ units, the flux may drop by factors of a few when going from the shortward to longward side of this discontinuity. This is readily observable through low resolution spectroscopy with JWST up to $z > 10$, but please note that this is a signature of dominant nebular emission (youth of the stellar population) and not necessarily of Population III.

DANIEL SCHAERER: For HeII, I'd be more optimistic for the Pop III detection. The typical strength of nebular HeII is EW(1640) ~ 2 Å, whereas Pop III could produce a line $\sim$10 times stronger. This should be distinguishable from "non-Pop III".

ERIK ZACKRISSON: I agree that for a pure Pop III Galaxy, there may indeed be clear-cut cases, but for all kinds of hybrids (Pop II/I and Pop III in the same object), the interpretation may be far from straightforward (as in the case of all current claims of Pop III signature detections based on HeII).

FABIO FONTANOT: How does the typical dark matter halo mass hosting a Pop III galaxy compare with the characteristic mass of the minimum dark matter hosting SF (Okamoto 2008)?

ERIK ZACKRISSON: In the early universe, the first galaxies are expected to form in halos with virial temperatures $\sim 10^4$ K (capable of HI cooling; $M_{\rm halo} \sim 10^7$–$10^8\ M_\odot$). This holds both for pristine and metal enriched halos – pop III galaxies would simply form in the small subset of such halos and have managed to reach the HI cooling mass without prior star formation (internal enrichment) or external chemical pollution from nearby halos.

Challenges in Panchromatic Modelling with Next Generation Facilities
Proceedings IAU Symposium No. 341, 2018
M. Boquien, E. Lusso, C. Gruppioni & P. Tissera, eds.

doi:10.1017/S1743921319002801

Systematic errors in dust mass fits: The role of dust opacity

Lapo Fanciullo[1], Francisca Kemper[1], Sundar Srinivasan[2], Peter Scicluna[1] and James M. Simpson[1]

[1]Academia Sinica Institute of Astronomy and Astrophysics, 11F of AS/NTU Astronomy-Mathematics Building, No.1, Sec. 4, Roosevelt Rd, Taipei 10617, Taiwan, R.O.C.
email: lfanciullo@asiaa.sinica.edu.tw

[2]Instituto de Radioastronomía y Astrofísica, Universidad Nacional Autónoma de México, Antigua Carretera a Pátzcuaro # 8701 Ex-Hda. San José de la Huerta, Morelia, Michoacán. México. C.P.58089

Abstract. The estimation of interstellar dust masses is an important pursuit in our understanding of both local and early Universe – see e.g. the "dust budget crisis". One of the most used methods of estimating dust masses – dust emission fitting – requires an estimate of the dust opacity at far-infrared and submillimeter wavelengths, but in most models this quantity is based on extrapolation rather than on actual measurements. It is becoming more and more evident that the opacity in typical dust models differs from that of dust analogs measured in the lab, meaning that astronomical dust mass estimations may need to be revised. To estimate the systematic errors introduced by this mismatch, we calculated dust emission for a model where dust far-infrared opacity is the same as that measured in lab samples, then we fit the synthetic emission with a typical (modified blackbody) dust model. Our results show that, if interstellar dust is indeed similar to the lab dust analogs, most fits may overestimate dust masses by as much as an order of magnitude.

Keywords. methods: laboratory, methods: numerical, ISM: dust, infrared: ISM, submillimeter

1. Context and motivation

Interstellar dust is an important contributor to the spectral energy distribution (SED) of galaxies, which it dominates – through thermal emission – at far infrared (FIR) and submillimeter (submm) wavelengths. Determining interstellar dust masses provides important constraints on galactic and cosmic evolution, as exemplified by the "dust budget crisis": too much dust is observed at high redshift to have formed in the wind of evolved stars, which are the main contributors to dust in the local Universe; dust from supernovae or grain growth in the interstellar medium are usually invoked to solve the discrepancy (see Rowlands *et al.* (2014); Watson *et al.* (2015) and refs. therein). Similar issues are encountered in nearby galaxies, such as the Magellanic Clouds (e.g., Srinivasan *et al.* 2016).

One of the most common methods for estimating dust masses is to fit the dust FIR/submm SED – normally in the form of multi-wavelength photometry – with a modified blackbody (MBB) model (Casey 2012). For an optically thin, pointlike (unresolved) source, the SED follows

$$I_\nu(\lambda) \propto M_d \; \kappa(\lambda) \; B_\nu(T_d), \tag{1.1}$$

where M_d is the dust mass, T_d the temperature, and the opacity $\kappa(\lambda)$ is in units of $cm^2\,g^{-1}$. The opacity is usually approximated as a power law $\kappa(\lambda) = \kappa_0 \; (\lambda/\lambda_0)^\beta$ with a value of β variable between 1.0 and 2.5, but often ~1.5–2 (e.g., Clements *et al.* 2018). Eq. 1.1 is usually valid for $\lambda > 50$ μm in the rest frame.

As is evident from eq. 1.1, there is a degeneracy between the fit-derived dust mass and the opacity $\kappa(\lambda)$. Dust mass estimations therefore require the dust FIR/submm in input. Many dust models use lab measurements on dust analogs (carbon, silicates) for the optical properties in the ultraviolet to mid-infrared range, but due to the sparsity of lab measurements for $\lambda > 100$ μm, FIR/submm opacity has traditionally been extrapolated from the optical properties at shorter wavelength. As FIR/submm lab measurements become more widespread, however, it is becoming clear that the long-wavelength opacity of dust analogs is more complex than previously thought. More specifically (e.g., Coupeaud *et al.* 2011; Demyk *et al.* 2017a,b):

- $\kappa(\lambda)$ often cannot be expressed as a simple power law;
- The $\kappa(\lambda)$ measured in the lab are generally higher – even by an order of magnitude – than the ones in typical dust models;
- For amorphous materials, $\kappa(\lambda)$ also depends on temperature and tends to increase with higher T.

If the optical properties of interstellar dust are close to those measured in the lab, there could be systematic errors in the dust masses derived from MBB fits. To test for this possibility we build synthetic observations of dust emission and fit them with a typical MBB model, then control whether the fit recovers the true value of the model input parameters.

2. Methods

Our synthetic emission is calculated from eq. 1.1, using $\kappa(\lambda)$ values from lab measurements. We made a bibliographic search for lab-measured, temperature-dependent dust analog opacities in the FIR/submm. From Mennella *et al.* (1998); Coupeaud *et al.* (2011); Demyk *et al.* (2017a,b) we obtained the $\kappa(\lambda)$ for various types of amorphous carbon and amorphous silicates. The materials vary considerably in composition, grain structure and – for silicates – oxidation state of metals.

The final purpose of the project is to construct a dust model grid spanning a wide variety of dust compositions, temperature distributions and redshifts, to see how the fit results depend on these factors. In the present proceedings we limit ourselves to dust composed of 30% in mass amorphous carbon (material BE from Mennella *et al.* 1998) and 70% amorphous $Mg_{0.7}Fe_{0.3}SiO_3$ silicate (material E30R from Demyk *et al.* 2017b), close to the average composition for Milky Way dust (e.g. Compiègne *et al.* 2011). The grid spans six redshift values (0.00, 0.25, 0.50, 0.75, 1.00 and 1.50) and four dust temperatures (20, 30, 40 and 50 K). After creating the synthetic SEDs we convolved them with the SPIRE 250, 350 and 500 μm and the JCMT SCUBA2 850 μm filter profiles, to produce four-band photometry which we then fit with a MBB model.

The MBB fit used an opacity of the form $\kappa_0\ (\lambda/\lambda_0)^\beta$ and left mass, temperature and β as free parameters. Following James *et al.* (2002) we used a value of $\kappa_0 = 0.7$ cm^2 g^{-1} at $\lambda_0 = 850$ μm. In the following, we call the fit-derived mass and temperature $M_{\rm fit}$ and $T_{\rm fit}$ respectively, to distinguish them from the values $M_{\rm d}$ and $T_{\rm d}$ used in the creation of the synthetic SED. The purpose of our analysis is to see whether $M_{\rm fit}$ ($T_{\rm fit}$) differs significantly from $M_{\rm d}$ ($T_{\rm d}$).

3. Fit results

The fit results are shown in Fig. 1. The best fit results are those for the temperature, $T_{\rm fit}$. The values of β show a large dispersion and are relatively low compared to the typical ISM values of ~1.5–2, but they remain within the physically plausible range of

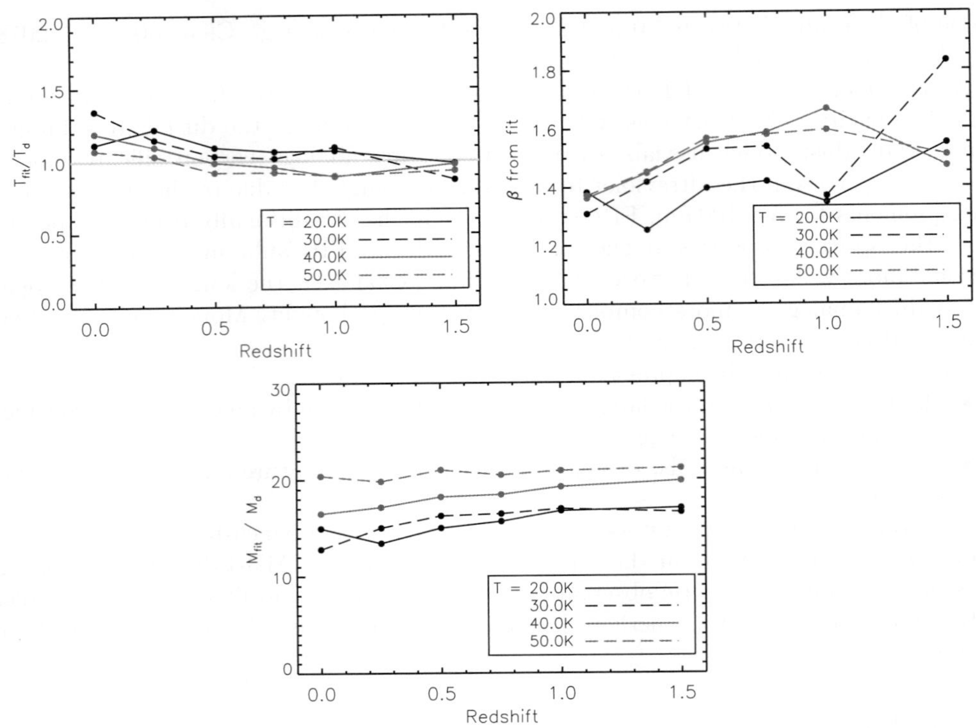

Figure 1. Parameter results from the MBB fit to synthetic photometry. Top left: T_{fit}/T_d, with a grey horizontal line showing $T_{fit}/T_d = 1$. Top right: β. Bottom: M_{fit}/M_d.

1.0–2.5. The values of M_{fit} are by far the most problematic, since they overestimate the dust mass by a factor ~15–20. The large value of M_{fit}/M_d is due mainly to the higher opacity of laboratory materials compared to the κ_0 values used in dust models. The ratio M_{fit}/M_d also depends on temperature, but in the range explored (20 – 50 K) the effect is minor compared to the baseline M_{fit}/M_d value.

We hypothesized that the low β might be due to our using the Mennella *et al.* (1998) carbon, which has a very shallow dependence of opacity on wavelength. To test this hypothesis we repeated the photometry synthesis and fit with a "carbon poor" model using the same materials, but now containing only 20% carbon. The results, shown in fig. 2, support our hypothesis since the carbon poor model has a higher β, comparable to the values typical of the ISM. Since carbon has a higher opacity than silicates, decreasing its opacity also has the effect of reducing M_{fit}/M_d; nonetheless, M_{fit} remains one order of magnitude larger than M_d, so our previous remarks remain qualitatively valid.

4. Astrophysical implications

If the composition of interstellar dust is close to the materials studied in the lab the implications of the high M_{fit}/M_d ratio are far-reaching, and dust mass estimates from SED fits may have to be reduced by an order of magnitude. This result alone, if confirmed, would solve the dust budget crisis. Since this is a preliminary study, there are some issues to be addressed before this finding's full significance can be assessed.

In our dust models, we used the $\kappa(\lambda)$ from the lab as-is, but dust opacity is influenced by the size and structure of grains. Lab samples are often in the form of porous aggregates, which have a higher FIR/submm opacity than an equivalent mass of compact grains (see e.g. discussion in Demyk *et al.* 2017b); the $\kappa(\lambda)$ measured in the lab may therefore be

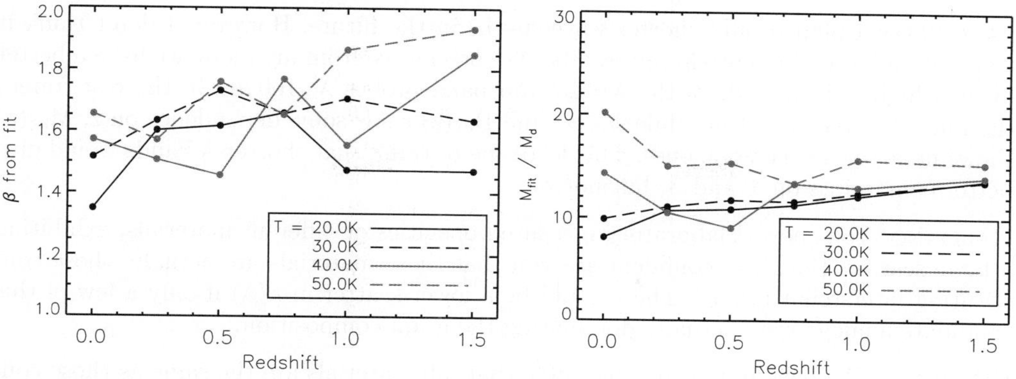

Figure 2. Fit results for the "carbon poor" model (see text). Left: β. Right: M_{fit}/M_d.

an overestimate. However, models show that aggregation tends to increase opacity by a factor ~2 (Köhler *et al.* 2015; Demyk *et al.* 2017a,b), which falls short of explaining the difference between M_{fit} and M_d.

Another point to consider is that SED fitting is not the only dust mass tracer. Revised estimates of dust masses from FIR/submm fits will have to be checked for consistency against other tracers, such as elemental depletion in the gas phase (e.g., James *et al.* 2002).

References

Casey, C. M. 2012, *MNRAS*, 425, 3094

Clements, D. L., Pearson, C., Farrah, D., Greenslade, J., Bernard-Salas, J., González-Alfonso, E., Afonso, J., Efstathiou, A., Rigopoulou, D., Lebouteiller, V., Hurley, P. D., & Spoon, H. 2018, *MNRAS*, 475, 2097

Compiègne, M., Verstraete, L., Jones, A., Bernard, J.-P., Boulanger, F., Flagey, N., Le Bourlot, J., Paradis, D., & Ysard, N. 2011, *A&A*, 525, A103

Coupeaud, A., Demyk, K., Meny, C., Nayral, C., Delpech, F., Leroux, H., Depecker, C., Creff, G., Brubach, J.-B., & Roy, P. 2011, *A&A*, 535, A124

Demyk, K., Meny, C., Lu, X.-H., Papatheodorou, G., Toplis, M. J., Leroux, H., Depecker, C., Brubach, J.-B., Roy, P., Nayral, C., Ojo, W.-S., Delpech, F., Paradis, D., & Gromov, V. 2017, *A&A*, 600, A123

Demyk, K., Meny, C., Leroux, H., Depecker, C., Brubach, J.-B., Roy, P., Nayral, C., Ojo, W.-S., & Delpech, F. 2017, *A&A*, 606, A50

James, A., Dunne, L., Eales, S., & Edmunds, M. G. 2002, *MNRAS*, 335, 753

Köhler, M., Ysard, N., & Jones, A. P. 2015, *A&A*, 579, A15

Mennella, V., Brucato, J. R., Colangeli, L., Palumbo, P., & Bussoletti, E. 1998, *ApJ*, 496, 1058

Rowlands, K., Gomez, H. L., Dunne, L., Aragón-Salamanca, A., Dye, S., Maddox, S., da Cunha, E., & van der Werf, P. 2014, *MNRAS*, 441, 1040

Srinivasan, S., Boyer, M. L., Kemper, F., Meixner, M., Sargent, B. A., & Riebel, D. 2016, *MNRAS*, 457, 2814

Darach Watson, Lise Christensen, Kirsten Kraiberg Knudsen, Johan Richard, Anna Gallazzi, & Michał Jerzy Michałowski 2015, *Nature*, 519, 327

Discussion

Goto: Would adding shorter wavelengths ($\lambda < 250\,\mu m$) help because the wavelength dependence is different?

FANCIULLO: I plan to add shorter wavelengths in the future. However, I don't think it will make a large difference in my results. As you can see from my plots, we have a better fit at a high redshift (where the $\lambda = 250\,\mu$m band probes $\lambda \sim 100\,\mu$m in the rest frame) than at a low redshift. But while $T_{fit}(z)$ and $B_{fit}(z)$ show some dependence on z, $M_{fit}(z)$ shows none (or a very weak one). This leads me to think that shorter-λ bands would give better constraints on T and β, but not M.

GALLIANO: You showed laboratory measured opacities of different materials, exhibiting a large scatter. But, how confident are you that these materials are actually those constituting interstellar grains? There could be a lower scatter in $\kappa(\lambda)$ if only a few of this candidate analogs were actually dominating the grain composition.

FANCIULLO: We have indeed no guarantee that lab materials are the same as those constituting interstellar dust. However, no dust analog ever studied has shown FIR/submm opacity as low as those used in interstellar dust models. If we solve the mass fit problem by assuming that interstellar material are not the same as in the lab, we now have the issue of an unknown interstellar material with the same stoichiometry as silicates and carbon (as revealed by elemental depletions) but much lower (FIR/submm) opacity.

Challenges in Panchromatic Modelling with Next Generation Facilities
Proceedings IAU Symposium No. 341, 2018
M. Boquien, E. Lusso, C. Gruppioni & P. Tissera, eds.

doi:10.1017/S174392131900276X

Properties of LBGs with [OIII] detection at z > 3: The importance of including nebular emission data in SED fitting

Fang-Ting Yuan[1], Denis Burgarella[2], David Corre[2], Veronique Buat[2], Médéric Boquien[3] and Shiyin Shen[1]

[1]Shanghai Astronomical Observatory, CAS, Nandan Road 80, 200030, Shanghai, China
email: yuanft@shao.ac.cn

[2]Aix Marseille Université, CNRS, CNES, LAM (Laboratoire d'astrophysique de Marseille)
UMR 7326, 13388, Marseille, France

[3]Centro de Astronomía (CITEVA), Universidad de Antofagasta,
Avenida Angamos 601, Antofagasta, Chile

Abstract. At high redshift, the contribution of strong emission lines to the broadband photometry can cause large uncertainties when estimating galaxy physical properties. To examine this effect, we investigate a sample of 54 LBGs at $3 < z_{\rm spec} < 3.8$ with detected [OIII] line emissions. We use CIGALE to fit simultaneously the rest-frame UV-to-NIR SEDs of these galaxies and their emission line data. By comparing the results with and without emission line data, we show that spectroscopic data are necessary to constrain the nebular model. We examine the K-band excess, which is usually used to estimate the emissions of [OIII]+Hβ lines when there is no spectral data, and find that the difference between the estimation and observation can reach up to > 1 dex for some galaxies, showing the importance of obtaining spectroscopic measurements of these lines. We also estimate the equivalent width of the Hβ absorption and find it negligible compared to the Hβ emission.

Keywords. galaxies: high-redshift, galaxies: formation, galaxies: emission lines

1. Introduction

Lyman Break Galaxies (LBGs) are star-forming galaxies with relatively low dust attenuation. These galaxies are important tracers for cosmic star formation rate at high redshift. However, there are many uncertainties in deriving the properties of LBGs from the spectral energy distribution (SED) fitting. At high redshift, an important uncertainty is the contamination to broadband photometry from the strong nebular emission lines (de Barros *et al.* 2014). Stark *et al.* (2013) have shown that [OIII]5007, Hβ and $H\alpha$ can contaminate K, *Spitzer* IRAC1 (3.6μm) or IRAC2 (4.5μm) bands when redshift > 3. On the other hand, information from nebular emission may give additional constraints on properties such as star formation rate (SFR), stellar mass and dust attenuation.

However, detecting emission lines is non-trivial. Deep near-infrared spectroscopy with sufficient spectral resolution is needed. Due to the lack of spectroscopic data, many studies assume the observed flux excess compared to the SED model flux in K/IRAC1/IRAC2 band as the line emission redshifted to that band. This method may provide only a rough estimation of the emission lines, thus its accuracy should be examined more carefully.

To better understand the properties of high-z LBGs, we selected a sample of LBGs with detected emission lines ([OII], [NeIII], Hβ, [OIII]4959,5007) at $3 < z_{\rm spec} < 3.8$. We analyze the properties of these galaxies (e.g., SFR, stellar mass, dust attenuation) using the SED fitting that considers both broadband photometry and emission line data.

2. Data

Our sources are taken from the samples published by Schenker *et al.* (2013); Troncoso *et al.* (2014) and Holden *et al.* (2016). The observations are taken with the Keck I MOSFIRE or VLT SINFONI spectrographs. The detected lines include [OII], [NeIII], Hβ and [OIII]4959, 5007. The photometric data are collected from 3D-HST, GOODS-MUSIC and also from the work of Magdis *et al.* (2010), and cover the wavelength range from U-band to Mid-infrared ($\sim$850Å to 2μm at restframe). In total, we obtain photometric and emission line data for 54 galaxies.

Since our data are selected from different works, we need to control the data quality. We divided our sample into Sample-A and Sample-B. Galaxies in Sample-A should satisfy the following criteria: 1) The galaxy is a secure match between the line and photometric data; 2) At least five bands have a detection better than 3σ; 3) At least one of the four IRAC bands has a detection better than 3σ; and 4) At least one emission line has a detection better than 3σ. In total, we have 26 out of the 54 galaxies in Sample-A. The others are in Sample-B. Although Sample-B has a poorer quality than Sample-A, galaxies in Sample-B are detected in at least three bands. Upper limits are included in the SED fitting using the method presented by Sawicki (2012) (see also Boquien *et al.* 2018).

3. Method

The SED fitting is performed with CIGALE (Code Investigating GALaxy Emission, Burgarella *et al.* 2005; Noll *et al.* 2009; Boquien *et al.* 2018). This code adopts the energy balance strategy and is able to fit the star and dust emission simultaneously. It applies the Bayesian analysis to estimate the output parameters. In this work, we use a new version of CIGALE which enables fitting the photometric and the nebular emission data at the same time.

We use the Salpeter IMF (Salpeter 1955) and the stellar populations from Bruzual & Charlot (2003). The star formation history is assumed to follow the delayed form: $SFR \propto t\exp(-t/\tau)$. We assume that the dust attenuation for the stellar part obeys the Calzetti Law (Calzetti *et al.* 2000) with a modification on the slope (Noll *et al.* 2009).

The nebular emission is modelled based on the templates of Inoue (2011) using the CLOUDY code. The input parameters include the metallicity, the radiation strength U, the dust attenuation and the escape fraction of the Lyman continuum photons, $f_{\rm esc}$. The metallicity is assumed to be the same as that of stars (i.e. $0.02Z_\odot$ to $Z_\odot$). The radiation strength U is fixed to $\log U = -2.5$. We have tested different $\log U$ values following Inoue (2011), and find that the value of -2.5 gives minimum χ^2 for our fitting. The dust attenuation for the nebular emission is assumed to follow the Milky Way extinction curve, in which $R_V = 3.1$. The $f_{\rm esc}$ is fixed to 0.2 (e.g. Stark *et al.* 2013).

The nebular emission and stellar emission are connected through $f_{\rm esc}$ and $f_{\rm ebv}$. The parameter $f_{\rm ebv}$ is the ratio of the $E(B-V)_{\rm star}$ to the $E(B-V)_{\rm neb}$. Usually, the value is taken as 0.44. Many recent studies, however, have found that this factor can vary up to 1 (e.g., Puglisi *et al.* 2016; Yuan *et al.* 2018; Buat *et al.* 2018). In this work, we take this parameter as free (in the range 0.1–1.0) and find that the fitting is improved compared to that by using a fixed value.

We examine the degeneracies between different parameters using mock galaxies and the probability density function (Yuan *et al.* in preparation). We find that in the 'Withline' run, the output parameters (SFR, M_*, A_V and age) are better constrained than in the 'Noline' and 'Nonebular' runs.

Table 1. Models and Data used for the 'Withline', 'Noline' and 'Nonebular' runs of CIGALE.

Name	Models	Data
Withline	stellar+nebular	photometric + line
Noline	stellar+nebular	photometric
Nonebular	stellar	photometric

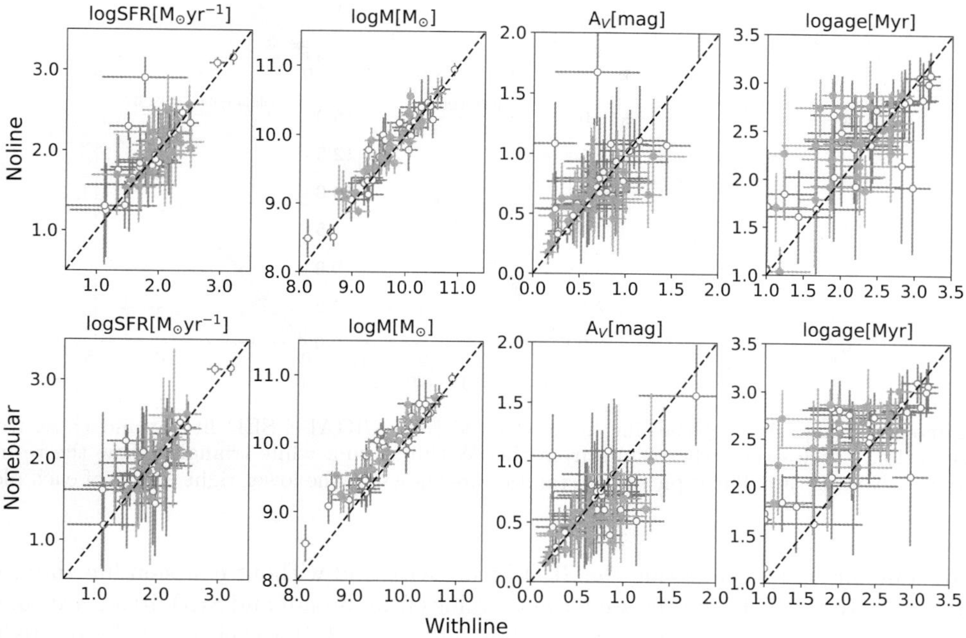

Figure 1. *Top*: Comparison of the results of the 'Withline' and 'Noline' runs. *Bottom*: Comparison of the results of the 'Withline' and 'Nonebular' runs. Orange dots are galaxies in Sample-A. Blue circles are galaxies in Sample-B.

4. Result

We investigate the influence of including nebular emission model and data in the fitting by comparing the results of three runs as listed in Table 1. First, we compare the results with and without line data (the 'Withline' and 'Noline' runs) and find that the SFR and stellar mass derived show no significant difference (Figure 1, *Top*). However, with no emission line data, the SED fitting is not able to predict line emission correctly (Figure 2).

If no nebular model is applied to the photometric data, the resulting stellar masses are systematically overestimated by 0.23 dex (Figure 1, *Bottom*). The fit results in a larger amount of aged stellar populations than those with nebular models. This is consistent with the fact that the K-band fluxes are boosted by the [OIII] emission.

The line emission can be estimated from the photometric data using SED fitting. The difference between the model predicted K-band flux and the observed one is attributed to the fluxes of the [OIII]+Hβ emission. We compared the results from this method with the observation. The average equivalent width estimated from photometric data is 503Å, while the observed value is 487Å, both with a large dispersion of 672Å, implying that although the estimation is systematically unbiased, it is not valid for an individual galaxy.

We calculate the Hβ absorption from the best model and find that the Balmer absorption is not significant for these galaxies. The equivalent width for this absorption is $\sim$1Å at rest-frame on average, comparing to the $\sim$20Å for the emission.

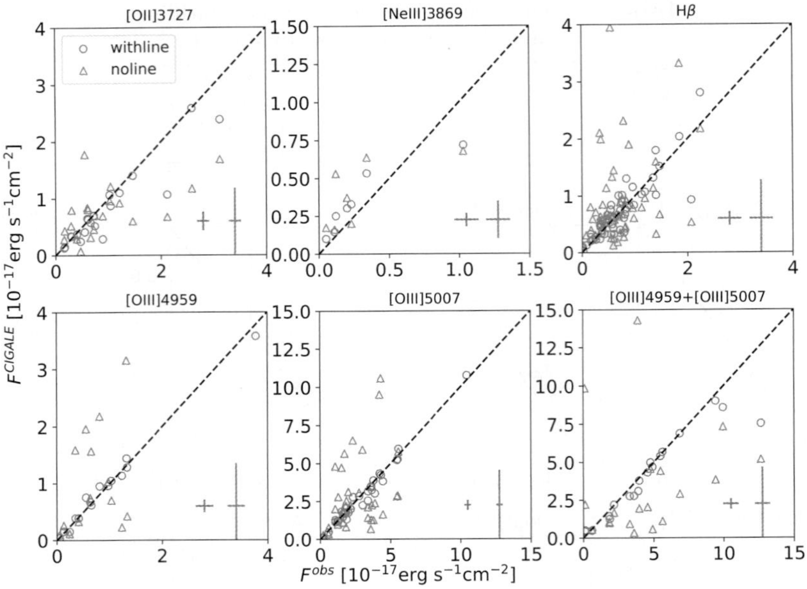

Figure 2. Comparison of the line fluxes derived from CIGALE SED fitting and those from observation. Circles show the results from the 'Withline' run, while triangles show the results from the 'Noline' run. The typical uncertainties are shown at the lower right corner of each plot.

To summarize, by comparing the SED fitting with and without emission line data, we find that both the SFR and the stellar mass values from broadband SED fitting are barely affected by the contamination from emission lines at $z \sim 3$. However, it is risky to predict line fluxes using photometric data alone. We also find that Balmer absorption plays an insignificant role in measuring the line fluxes, due to the fact that the stellar populations are not old enough to produce deep absorption features. Our UV-to-NIR SED and the emission line data together still cannot constrain the dust attenuation law and the star formation history very well. The Hα emission line data from JWST may help anchor the dust attenuation and thus better constrain the properties of these galaxies.

References

Boquien, M., Burgarella, D., Roehlly, Y., Buat, V., Ciesla, L., Corre, D., Inoue, A. K., Salas, H., *et al.* 2018, arXiv:1811.03094

Bruzual, G. & Charlot, S. 2003, *MNRAS*, 344, 1000

Buat, V., Boquien, M., Małek, K., Corre, D., Salas, H., Roehlly, Y., Shirley, R., Efstathiou, A., *et al.* 2018, *A&A*, 619, A135

Burgarella, D., Buat, V., Iglesias-Páramo, J., *et al.* 2005, *MNRAS*, 360, 1413

Calzetti, D. 1997, *AJ*, 113, 162

Calzetti, D., Armus, L., Bohlin, R. C., Kinney, A. L., Koornneef, J., Storchi-Bergmann, T., *et al.* 2000, *ApJ*, 533, 682

de Barros, S., Schaerer, D., Stark, D. P., *et al.* 2014, *A&A*, 563, A81

Holden, B. P., Oesch, P. A., González, V. G., Illingworth, G. D., Labbé, I., Bouwens, R., Franx, M., van Dokkum, P., *et al.* 2016, *ApJ*, 820, 73

Inoue, A. K. 2011, *MNRAS*, 415, 2920

Magdis, G. E., Elbaz, D., Hwang, H. S., Daddi, E., Rigopoulou, D., Altieri, B., Andreani, P., Aussel, H., *et al.* 2010, *ApJ*, 720, L185

Noll, S., Burgarella, D., Giovannoli, E., Buat, V., Marcillac, D., Muñoz-Mateos, J. C., *et al.* 2009, *A&A*, 507, 1793

Puglisi, A., Rodighiero, G., Franceschini, A., Talia, M., Cimatti, A., Baronchelli, I., Daddi, E., Renzini, A., *et al.* 2016, *A&A*, 586, A83
Salpeter, E. E. 1955, *ApJ*, 121, 161
Sawicki, M. 2012, *PASP*, 124, 1208
Schenker, M. A., Ellis, R. S., Konidaris, N. P., Stark, D. P., *et al.* 2013, *ApJ*, 777, 67
Stark, D. P., Schenker, M. A., Ellis, R., Robertson, B., McLure, R., Dunlop, J., *et al.* 2013, *ApJ*, 763, 129
Troncoso, P., Maiolino, R., Sommariva, V., Cresci, G., Mannucci, F., Marconi, A., Meneghetti, M., Grazian, A., *et al.* 2014, *A&A*, 563, A58
Yuan, F.-T., Argudo-Fernández, M., Shen, S., Hao, L., Jiang, C., Yin, J., Boquien, M., Lin, L., *et al.* 2018, *A&A*, 613, A13

Discussion

Kentaro Motohara: Why does the [OII]3427 estimate change according to model's $\log U$?

Yuan: Because in the fitting, the line ratios depend on $\log U$.

Tomo Goto: Without line data (photometry only), how much does it improve by including line model to the no line model?

Yuan: With the nebular model, the offset to the reference will decrease about 0.15 dex for stellar mass estimation. However, the scatters are not improved. Both are 0.20 dex compared to the reference run.

Helen Kim: Do your galaxies have any FIR detections? (e.g. Herschel)

Yuan: No, we do not have any restframe far-infrared data in our sample. Some of these galaxies have been observed by Spitzer MIPS, but only upper limits are obtained, because these galaxies are faint at FIR.

Daniel Ceverino: Did you try other SFR histories in your SED fit? Simulations of galaxies at these redshifts support increasing SFR with time.

Yuan: In the current stage, we only test the delayed and delay+burst star formation histories. From the fit assuming the delayed SFH, we find the shape close to an increasing SFH.

Challenges in Panchromatic Modelling with Next Generation Facilities
Proceedings IAU Symposium No. 341, 2018
M. Boquien, E. Lusso, C. Gruppioni & P. Tissera, eds.

doi:10.1017/S1743921319002618

Census of Lyα, [OIII]5007, Hα, and [CII]158μm line emission with ~1000 galaxies at $z = 4.9 - 7.0$ revealed with Subaru/HSC, Spitzer, and ALMA

Yuichi Harikane

Institute for Cosmic Ray Research, The University of Tokyo, 5-1-5 Kashiwanoha, Kashiwa, Chiba 277-8582, Japan

Department of Physics, Graduate School of Science, The University of Tokyo, 7-3-1 Hongo, Bunkyo, Tokyo, 113-0033, Japan
email: hari@icrr.u-tokyo.ac.jp

Abstract. We investigate rest-frame UV to far-infrared emission lines and SEDs from 1124 galaxies at $z = 4.9-7.0$. Our sample is composed of 1092 Lyα emitters (LAEs) at $z = 4.9-7.0$ identified by Subaru/Hyper Suprime-Cam (HSC) narrowband surveys and 34 galaxies at $z = 5.148-7.508$ with deep [CII]158μm ALMA data. The SEDs clearly show flux excesses in the Spitzer/IRAC 3.6 and 4.5μm bands, suggesting strong rest-frame optical emission lines of [OIII] and/or Hα. We model the galaxy SEDs with a flexible code combining stellar population and photo-ionization models (BEAGLE; Chevallard & Charlot 2016), and investigate relations between the emission lines of Lyα, [OIII], Hα, and [CII]. We find 1) a positive correlation between the rest-frame Hα equivalent width (EW) and the Lyα EW, $EW^0_{\mathrm{Ly}\alpha}$, 2) an interesting turn-over trend that the [OIII]/Hα flux ratio increases in $EW^0_{\mathrm{Ly}\alpha} \simeq 0-30$ Å, and then decreases out to $EW^0_{\mathrm{Ly}\alpha} \simeq 130$ Å, and 3) a $> 99\%$ anti-correlation between a [CII] luminosity to star-formation rate ratio $L_{[\mathrm{CII}]}/SFR$) and $EW^0_{\mathrm{Ly}\alpha}$. Modeling with BEAGLE also suggests that a simple anti-correlation between $EW^0_{\mathrm{Ly}\alpha}$ and metallicity explains self-consistently all of the relations of Lyα, Hα, [OIII]/Hα, and [CII] in our study, indicative of detections of very metal-poor ($\sim 0.03 Z_\odot$) galaxies with $EW^0_{\mathrm{Ly}\alpha} \sim 200$ Å.

Keywords. galaxy evolution, galaxy formation, high-redshift galaxy

1. Introduction

Probing physical conditions of the inter-stellar medium (ISM) is fundamental in understanding star formation and gas reprocessing in galaxies across cosmic time. Early ALMA observations found surprisingly weak [CII]158μm emission in Lyα emitters (LAEs) at $z \sim 6-7$ ([CII] deficit; e.g., Ouchi *et al.* 2013; Schaerer *et al.* 2015; Maiolino *et al.* 2015). A theoretical study discusses that the [CII] deficit can be explained by very low metallicity (0.05 $Z_\odot$) in the ISM (Vallini *et al.* 2015; Olsen *et al.* 2017). Thus estimating metallicities of the high-redshift galaxies is crucial to our understanding of the origin of the [CII] deficit.

The ISM property is also important for cosmic reionization. Observations by the Planck satellite and high redshift UV luminosity functions (LFs) suggest that faint and abundant star-forming galaxies dominate the reionization process (e.g., Robertson *et al.* 2015), and understanding ionizing properties of such star forming galaxies is important. Various studies constrain ionizing photon production efficiencies of star forming galaxies to be $\log\xi_{\mathrm{ion}}/[\mathrm{Hz\ erg}^{-1}] = 24.8-25.3$ at $z \sim 0-2$ (e.g., Matthee *et al.* 2017; Shivaei *et al.*

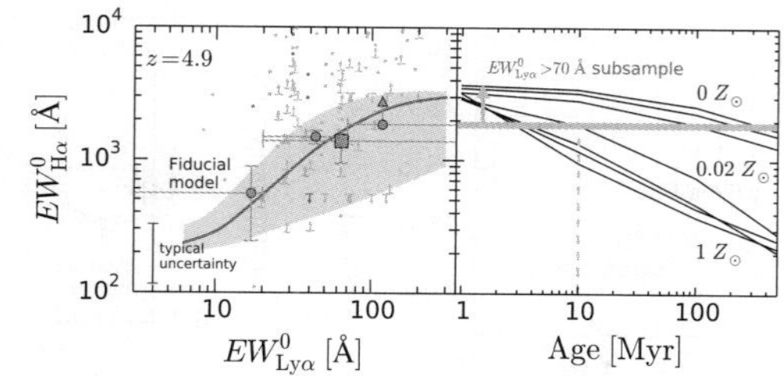

Figure 1. Left panel: Hα EWs as a function of Lyα EWs at $z = 4.9$. The red square and circles are the results from the stacked images of the subsamples, and the gray dots show the EWs of the individual objects detected in the [3.6] and/or [4.5] bands. The upward and downward arrows represent the 2σ lower and upper limits, respectively. The dark gray curve and the shaded region show the prediction from the fiducial model. **Right panel:** Inferred stellar age and metallicity from the constrained $EW^0_{\mathrm{H}\alpha}$. The red solid line shows the lower limit of $EW^0_{\mathrm{H}\alpha} \gtrsim 2000$ Å in the 70 Å $< EW_{\mathrm{Ly}\alpha} < 1000$ Å subsample at $z = 4.9$. The black curves represent $EW^0_{\mathrm{H}\alpha}$ calculated in Inoue (2011) with metallicities of $Z = 0$, 5×10^{-6}, 5×10^{-4}, 0.02, 0.2, 0.4, and 1 $Z_\odot$.

2017). Since the faint star-forming galaxies are expected to be strong line emitters, it is important to estimate ξ_{ion} of LAEs at higher redshift, as their ISM properties are likely more similar to the ionizing sources.

2. Data & Analysis

We use LAE samples at $z = 4.9$, 5.7, 6.6, and 7.0 selected with the NB filters of $NB718$, $NB816$, $NB921$, and $NB973$, respectively (Shibuya *et al.* 2018; Itoh *et al.* 2018 Zhang *et al.* in prep.), obtained in our Subaru/Hyper Suprime-Cam (HSC) survey (Aihara *et al.* 2018b). We divide our LAE samples into subsamples by the Lyα equivalent width (EW) bins. We cut out 12" × 12" images of the LAEs in HSC $grizyNB718NB816NB921NB973$ ($grizyNB816NB921$), VIRCAM JHK_s (WFCAM JHK), and IRAC [3.6][4.5] bands in the UD-COSMOS (UD-SXDS) field. Then we generate median-stacked images of the subsamples in each band.

We generate the model SEDs at $z = 4.9$, 5.7, 6.6, and 7.0 using BEAGLE (Chevallard & Charlot 2016). We estimate rest-frame optical emission line fluxes by comparing the stacked SEDs with the model SEDs. We calculate the flux differences between the stacked SEDs and the model SEDs in the [3.6] band at $z = 4.9$, and [3.6] and [4.5] bands at $z = 5.7$, 6.6, and 7.0. The flux differences are corrected for dust extinction with the τ_{V} values in the models, assuming the Calzetti *et al.* (2000) extinction curve. We estimate the Hα, Hβ, and [OIII]λ5007 line fluxes from these flux differences.

In addition to our HSC LAE samples, we compile previous ALMA and PdBI observations targeting [CII]158μm in galaxies at $z > 5$. We use results of 34 galaxies from the literature (see Table 3 in Harikane *et al.* 2018).

3. Results

The left panel in Figure 1 shows rest-frame Hα EWs ($EW^0_{\mathrm{H}\alpha}$) as a function of Lyα EWs at $z = 4.9$. The Hα EW increases from $\sim$600 Å to > 1900 Å with increasing Lyα EW. This high $EW^0_{\mathrm{H}\alpha}$ value indicates very young stellar age of < 10 Myr or very low metallicity of < 0.02 $Z_\odot$ (the right panel in Figure 1).

We estimate the ionizing photon production efficiencies of the $z = 4.9$ LAEs from their Hα fluxes and UV luminosities. The left panel in Figure 2 shows estimated ξ_{ion} values as a function of UV magnitude. We calculate the values of the ξ_{ion} in two cases; $f^{\mathrm{ion}}_{\mathrm{esc}} = 0$ and $f^{\mathrm{ion}}_{\mathrm{esc}} = 0.1$. The ionizing photon production efficiency is estimated to be logξ_{ion}/

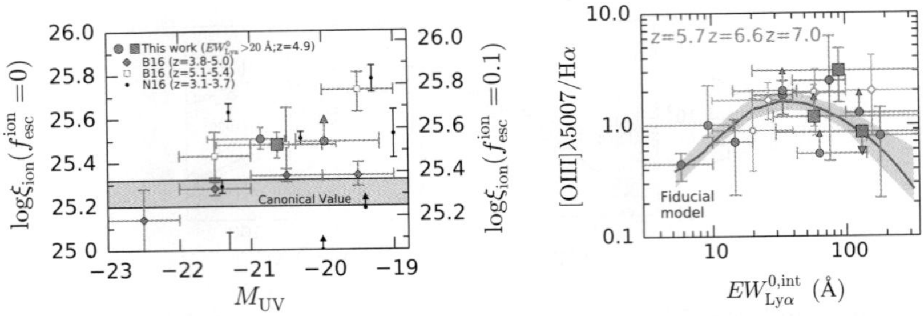

Figure 2. Left panel: Inferred ionizing photon production efficiencies of the LAEs at $z = 4.9$ as a function of UV magnitude. The left and right axes represent the efficiencies with the ionizing photon escape fractions of 0 and 10%, respectively. The red circles and square show the results of the subsamples divided by $EW^0_{\mathrm{Ly}\alpha}$, and the upward arrow represents the 2σ lower limit. The ξ_{ion} values of LBGs at $z = 3.8$–5.0 ($z = 5.1$–5.4) in Bouwens *et al.* (2016) are represented as the blue diamonds (green open squares). We plot the ξ_{ion} values of LAEs at $z = 3.1$–3.7 (Nakajima *et al.* 2016) with the black circles. The gray shaded region indicates typically assumed ξ_{ion} (see Table 2 Bouwens *et al.* 2016). **Right panel:** [OIII]/Hα ratio as a function of the Lyα EW. The blue, red, and green circles and squares are the [OIII]λ5007/Hα flux ratios at $z = 5.7$, 6.6, and 7.0, respectively. The open gray diamonds and circles are the ratios of $z = 2.5$ and 0.3 galaxies (Trainor *et al.* 2016; Cowie *et al.* 2011), respectively. We also plot the fitting result of the fiducial model with the dark gray curve with the shaded region representing the 1σ uncertainty.

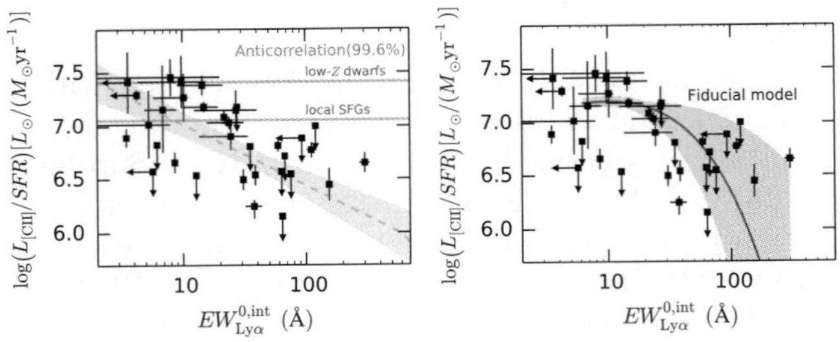

Figure 3. Left panel: Ratio of the [CII] luminosity to the SFR as a function of rest-frame Lyα EW. We find the anti-correlation in the $L_{[\mathrm{CII}]}/SFR - EW^{0,\mathrm{int}}_{\mathrm{Ly}\alpha}$ plane at the 99.6% confidence level. The green horizontal lines show the $L_{[\mathrm{CII}]}/SFR$ ratios for low-metallicity dwarf galaxies and local star-forming galaxies in De Looze *et al.* (2014) for $SFR = 10\ \mathrm{M_\odot yr^{-1}}$. The red-dashed line and the shaded region denote the best-fit $L_{[\mathrm{CII}]}/SFR - EW^{0,\mathrm{int}}_{\mathrm{Ly}\alpha}$ relation. **Right panel:** Same as the left panel but with the prediction from the fiducial model.

[Hz erg^{-1}] $= 25.48^{+0.06}_{-0.06}$ for the $EW^0_{\mathrm{Ly}\alpha} > 20$ Å subsample with $f^{\mathrm{ion}}_{\mathrm{esc}} = 0$. This value is systematically higher than those of LBGs at the similar redshift and UV magnitude ($\log\xi_{\mathrm{ion}}$/[Hz erg^{-1}] $\simeq 25.3$; Bouwens *et al.* 2016) by 60–100%.

The [OIII]/Hα ratios of the $z = 5.7$, 6.6, and 7.0 LAEs are presented in the right panel in Figure 2. We find that the ratio increases with increasing $EW^0_{\mathrm{Ly}\alpha}$ from 7 Å to 20 Å, then decreases to ∼130 Å, showing the turn-over trend at the 2.3σ confidence level.

In the left panel in Figure 3, we plot the ratios of the [CII] luminosity to SFR, $L_{[\mathrm{CII}]}/SFR$ as a function of Lyα EW corrected for the IGM absorption, $EW^{0,\mathrm{int}}_{\mathrm{Ly}\alpha}$. We find an anti-correlation in the $L_{[\mathrm{CII}]}/SFR - EW^{0,\mathrm{int}}_{\mathrm{Ly}\alpha}$ plane at the 99.6% confidence level.

4. Discussion

We investigate physical quantities explaining our observed [OIII]/Hα ratios as a function of Lyα EW. We simply parameterize the metallicity, Z_{neb}, the ionization parameter,

$U_{\rm ion}$, and the stellar age with the Lyα EW in units of Å. We fit our observational results of the [OIII]/Hα ratios with this model, and the best-fit relations with 1σ errors are

$$\log Z_{\rm neb} = -0.33^{+0.16}_{-0.11}\left(\log EW^{0,{\rm int}}_{{\rm Ly}\alpha}\right)^2 + 0.35^{+0.10}_{-0.18}, \tag{4.1}$$

$$\log U_{\rm ion} = -0.09^{+0.66}_{-0.22}\log EW^{0,{\rm int}}_{{\rm Ly}\alpha} - 2.58^{+0.33}_{-0.81}, \tag{4.2}$$

$$\log {\rm Age} = -0.29^{+1.66}_{-0.77}\log EW^{0,{\rm int}}_{{\rm Ly}\alpha} + 8.90^{+0.47}_{-3.21}. \tag{4.3}$$

The result suggests an anti-correlation between the metallicity and the Lyα EW, implying the very metal-poor ISM ($\sim$0.03 $Z_\odot$) in the galaxies with $EW^{0,{\rm int}}_{{\rm Ly}\alpha} \sim 200$ Å (see also; Nagao *et al.* 2007). Hereafter, we call this model "the fiducial model". We find that this fiducial model agrees well with the observed $EW^0_{{\rm H}\alpha} - EW^0_{{\rm Ly}\alpha}$ relation (the left panel in Figure 1). The fiducial model can also nicely explains the $L_{\rm [CII]}/SFR - EW^{0,{\rm int}}_{{\rm Ly}\alpha}$ anti-correlation (the right panel in Figure 3), indicating that the [CII] deficit in high Lyα EW galaxies may be due to the low metallicity.

References

Aihara, H., Arimoto, N., Armstrong, R., Arnouts, S., Bahcall, A., Bickerton, S., Bosch, J., Bundy, K., *et al.* 2018b, *PASJ*, 70, S4

Bouwens, R. J., Smit, R., Labbé, I., Franx, M., Caruana, J., Oesch, P., Stefanon, M., & Rasappu, N. 2016, *ApJ*, 831, 176

Calzetti, D., Armus, L., Bohlin, R. C., Kinney, L., Koornneef, J., & Storchi-Bergmann, T. 2000, *ApJ*, 533, 682

Chevallard, J. & Charlot, S. 2016, *MNRAS*, 462, 1415

Cowie, L. L., Barger, A. J., & Hu, E. M. 2011, *ApJ*, 738, 136

De Looze, I., Cormier, D., Lebouteiller, V., Madden, S., Baes, M., Bendo, J., Boquien, M., Boselli, A., *et al.* 2014, *A&A*, 568, A62

Harikane, Y., Ouchi, M., Shibuya, T., Kojima, T., Zhang, H., Itoh, R., Ono, Y., Higuchi, R., *et al.* 2018, *ApJ*, 859, 84

Inoue, A. K. 2011, *MNRAS*, 415, 2920

Itoh, R., Ouchi, M., Zhang, H., Inoue, A., Mawatari, K., Shibuya, T., Harikane, Y., Ono, Y., *et al.* 2018, ArXiv e-prints, arXiv:1805.05944

Maiolino, R., Carniani, S., Fontana, A., Vallini, L., Pentericci, L., Ferrara, A., Vanzella, E., Grazian, A., *et al.* 2015, *MNRAS*, 452, 54

Matthee, J., Sobral, D., Best, P., Khostvan, A., Oteo, I., Bouwens, R., & Rottgering, H. 2017, *MNRAS*, 465, 3637

Nagao, T., Murayama, T., Maiolino, R., Marconi, A., Kashikawa, N., Ajiki, M., Hattori, T., Ly, C., *et al.* 2007, *A&A*, 468, 877

Nakajima, K., Ellis, R. S., Iwata, I., Inoue, A., Kusakabe, H., Ouchi, M., & Robertson, B. 2016, *ApJ*, 831, L9

Olsen, K., Greve, T. R., Narayanan, D., Thompson, R., Dave, R., Niebla Rios, L., & Stawinski, S. 2017, *ApJ*, 846, 105

Ouchi, M., Ellis, R., Ono, Y., Nakanishi, K., Kohno, K., Momose, R., Kurono, Y., Ashby, M., *et al.* 2013, *ApJ*, 778, 102

Robertson, B. E., Ellis, R. S., Furlanetto, S. R., & Dunlop, J. S. 2015, *ApJ*l, 802, L19

Schaerer, D., Boone, F., Zamojski, M., Staguhn, J., Dessauges-Zavadsky, M., Finkelstein, S., & Combes, F. 2015, *A&A*, 574, A19

Shibuya, T., Ouchi, M., Konno, A., Higuchi, R., Harikane, Y., Ono, Y., Shimasaku, K., Taniguchi, Y., *et al.* 2018, *PASJ*, 70, S14

Shivaei, I., Reddy, N. A., Siana, B., Shapley, A., Kriek, M., Mobasher, B., Freeman, W., Sanders, R., *et al.* 2017, ArXiv e-prints, arXiv:1711.00013

Trainor, R. F., Strom, A. L., Steidel, C. C., & Rudie, G. C. 2016, *ApJ*, 832, 171

Vallini, L., Gallerani, S., Ferrara, A., Pallottini, A., & Yue, B. 2015, *ApJ*, 813, 36

Discussion

Tomo Goto: The OIII/Hα turn over seems constant across a range of redshifts. But abundance might decline at higher redshifts?

Yuichi Harikane: We do not detect any significant evolution with redshift. We may be able to improve measurements with JWST.

Hiroyuki Hirashita: For the metallicity-Lyα equivalent width relation, does dust play a role?

Yuichi Harikane: We do not think so. The spectral energy distributions of Lyα emitters show a very low dust content. The dust does not significantly affect the results.

Daniel Schaerer: Whats is the lowest metallicity you obtain?

Yuichi Harikane: 2–3% solar.

Adam Carnall: Is the metallicity-Lyα equivalent width correlation being driven by the data or is it a feature of the models you are fitting?

Yuichi Harikane: From the data, the anti-correlation comes from the turnover trend of OIII/Hα.

Challenges in Panchromatic Modelling with Next Generation Facilities
Proceedings IAU Symposium No. 341, 2018
M. Boquien, E. Lusso, C. Gruppioni & P. Tissera, eds.

doi:10.1017/S1743921319002436

Detections of far-infrared [OIII] and dust emission in a galaxy at $z = 8.312$: Early metal enrichment in the heart of the reionization era

Y. Tamura[1], K. Mawatari[2], T. Hashimoto[3,4], A. K. Inoue[3], E. Zackrissonm[5], L. Christensen[6], C. Binggeli[5], Y. Matsuda[4], H. Matsuo[4], T. T. Takeuchi[1], R. S. Asano[1], K. Sunaga[1], I. Shimizu[7], T. Okamoto[8], N. Yoshida[9,10], M. Lee[1,4], T. Shibuya[11], Y. Taniguchi[12], H. Umehata[13], B. Hatsukade[14], K. Kohno[14] and K. Ota[15]

[1]Department of Physics, Nagoya University, Japan
email: ytamura@nagoya-u.jp
[2]Institute for Cosmic Ray Research, The University of Tokyo, Japan
[3]Department of Environmental Science and Technology, Osaka Sangyo University, Japan
[4]National Astronomical Observatory of Japan, Japan
[5]Department of Physics and Astronomy, Uppsala University, Sweden
[6]Dark Cosmology Centre, Niels Bohr Institute, University of Copenhagen, Denmark
[7]Department of Earth & Space Science, Osaka University, Japan
[8]Department of Cosmosciences, Hokkaido University, Japan
[9]Department of Physics, The University of Tokyo, Japan
[10]Kavli Institute for the Physics and Mathematics of the Universe, The University of Tokyo, Japan
[11]Department of Computer Science, Kitami Institute of Technology, Japan
[12]The Open University of Japan, Japan
[13]RIKEN Cluster for Pioneering Research, Japan
[14]Institute of Astronomy, The University of Tokyo, Japan
[15]Kyoto University Research Administration Office, Japan

Abstract. We present ALMA detection of the [O III] 88 μm line and 850 μm dust continuum emission in a Y-dropout Lyman break galaxy, MACS0416_Y1. The [O III] detection confirms the object with a spectroscopic redshift to be $z = 8.3118 \pm 0.0003$. The 850 μm continuum intensity (0.14 mJy) implies a large dust mass on the order of $4 \times 10^6 M_\odot$. The ultraviolet-to-far infrared spectral energy distribution modeling, where the [O III] emissivity model is incorporated, suggests the presence of a young ($\tau_{\rm age} \approx 4$ Myr), star-forming (SFR $\approx 60\ M_\odot\ {\rm yr}^{-1}$), and moderately metal-polluted ($Z \approx 0.2 Z_\odot$) stellar component with a stellar mass of $3 \times 10^8 M_\odot$. An analytic dust mass evolution model with a single episode of star formation does not reproduce the metallicity and dust mass in ≈ 4 Myr, suggesting an underlying evolved stellar component as the origin of the dust mass.

Keywords. galaxies: formation, galaxies: high-redshift, galaxies: ISM

1. Introduction

How and when metal enrichment happened in the epoch of reionization (EoR) is one of the most fundamental questions in modern astronomy. Recent *Planck* results suggest an instantaneous reionization redshift of $z_{\rm re} = 7.68 \pm 0.79$ Planck Collaboration (2018), and the latest *Hubble Space Telescope* (*HST*) surveys have revealed more than a hundred of candidate $z \gtrsim 8$ Lyman break galaxies (LBGs). Furthermore, based on the samples of $z \gtrsim 8$ LBGs, Oesch *et al.* (2018) reported strong evolution of the ultraviolet (UV) luminosity function from $z \sim 10$ to $\sim$8, implying a rapid increase of the cosmic star-formation rate density by an order of magnitude within a very short time-scale ($\lesssim 200$ Myr). Hence, characterizing star formation and interstellar medium (ISM) of galaxies at the pre-reionization era ($z \gtrsim 8$) is an important next step to our understanding of the earliest evolution of galaxies.

Inoue *et al.* (2014) predicted that the [O III] 88 μm line, which is often observed as the brightest far-infrared (FIR) line in local H II regions, is detectable with ALMA and can be used as an instantaneous tracer of massive star formation at $z \sim 8$. Since our first ALMA detection of the [O III] line in a $z = 7.212$ Lyα emitter, SXDF-NB1006-2 (Inoue *et al.* 2016), there have been mounting detections of the [O III] line in the EoR (e.g., Hashimoto *et al.* 2018; Sunaga *et al.* 2018). Furthermore, the [O III] line plays an important role in characterizing physical properties of $z \gtrsim 8$ galaxies because it is extinction-free and sensitive to SFR, stellar age, and gas-phase oxygen abundance.

In this paper, we report ALMA detections of [O III] and dust emission in a Y-dropout galaxy MACS0416_Y1 (Tamura *et al.* 2018), which is a robust candidate for a $z \sim 8$ LBG found behind the Frontier Field cluster, MACS J0416.1−2403 (Laporte *et al.* 2015).

2. ALMA Observations and Results

In our ALMA Cycle 4 observations of MACS0416_Y1, four different tunings were assigned to cover a frequency range between 340.0 and 366.4 GHz. The total on-source time was 7.3 hr. The data sets are combined to make the continuum image and a spectral cube with a frequency resolution of 31.25 MHz (≈ 26 km s^{-1}) to search for the [O III] line.

We detect 850-μm continuum emission at the position of MACS0416_Y1 as shown in Fig. 1a. The emission is spatially resolved ($\approx 0\farcs3 \times 0\farcs1$) and is similar in spatial distribution on a $\sim$1 kpc scale to that of the rest-frame UV emission. The bulk of dust emission is likely to be associated with the eastern 'E' knot, **while the 'C' and 'W' knots also have the dust emission** (Fig. 1c). The observed flux density of $S_{850\,\mu{\rm m}} = 137 \pm 26$ μJy corresponds to a de-lensed total IR luminosity of $L_{\rm IR} = (1.7 \pm 0.3) \times 10^{11} L_\odot$ and a dust mass of $M_{\rm dust} = (3.6 \pm 0.7) \times 10^6 M_\odot$ if assuming a dust temperature of $T_{\rm dust} = 50$ K, a dust emissivity index of $\beta = 1.5$, a dust emissivity of $\kappa_{\rm d}(850\ \mu{\rm m}) = 0.15$ m^2 kg^{-1} and the magnification factor of $\mu_{\rm g} = 1.43 \pm 0.04$.

At the position of the dust emission, we detect an emission line feature at 364.377 ± 0.012 GHz, strongly suggesting the [O III] 88 μm emission line at $z = 8.3118 \pm 0.0003$ (Fig. 1b). The apparent line flux is $F_{\rm [O\,III]} = 0.66 \pm 0.16$ Jy km s^{-1}, corresponding to the delensed line luminosity of $L_{\rm [O\,III]} = (1.2 \pm 0.3) \times 10^9 L_\odot$.

3. UV-to-FIR SED Modeling

This dust continuum detection is somewhat surprising because the UV slope is blue ($\beta_{\rm UV} \approx -2$, **where $\beta_{\rm UV}$ is defined as a power-law index of the UV flux density $F_\lambda \propto \lambda^{\beta_{\rm UV}}$**). Furthermore, the *Spitzer*/IRAC photometry shows a red color in the rest-frame optical ($[3.6]-[4.5] > 0.38$). These spectral features are similar to those found in lower-z low mass star-forming galaxies, while the full SED should be assessed for the better understanding of the physical property of MACS0416_Y1.

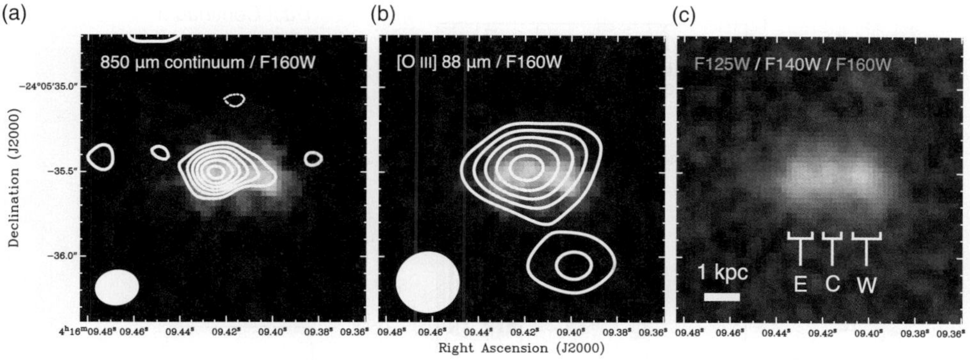

Figure 1. (a) The ALMA 850 μm continuum image of MACS0416_Y1 (contours) overlaid on the *HST*/WFC3 near-infrared image in the F160W band. The contours are drawn at -2σ, 2σ, $3\sigma, \ldots, 7\sigma$, where $\sigma = 10.9$ μJy beam^{-1}. The synthesized beam size is indicated at the bottom-left corner. (b) The ALMA [O III] 88 μm integrated intensity image (contours) overlaid on the *HST*/F160W image. The contours are drawn in the same manner as (a), but $\sigma = 55$ mJy beam^{-1} km s^{-1}. The image is optimally tapered with a $0.''35$ Gaussian kernel to maximize the signal-to-noise ratio. (c) The *HST*/WFC3 image taken with F160W (red), F140W (green) and F125W (blue) bands. The letters 'E', 'C' and 'W' denote the positions of the eastern, central, and western clumps seen in the rest-frame UV, respectively. The physical scale of 1 kpc on the image plane is indicated by the bar at the bottom-left corner.

We characterize the SED to investigate the physical properties of MACS0416_Y1 by template fits. The model is based on the prescription presented by Mawatari *et al.* (2018)†, where the emission components of a stellar continuum (Bruzual & Charlot 2003), the rest-frame UV-to-optical nebular lines and continuum (Inoue 2011), and the dust continuum (Rieke *et al.* 2009) are accounted for, **with the assumption that the absorbed UV-to-optical emission is reprocessed as the FIR emission.** In addition, we take into account the [O III] 88 μm emissivity model (Inoue *et al.* 2014). We use the photometric data of the 850 μm continuum and [O III] line in addition to the rest-frame UV-to-optical bands to model the SED of MACS0416_Y1.

The results are shown in Fig. 2. One of the important outcomes is that there exist solutions which reasonably explain the large amount of dust coexisting with the young stellar components. Regardless of standard extinction laws, the SED fits favor a young, high-SFR solution, where large equivalent widths of the enhanced [O III] $\lambda\lambda4959, 5007$ Å and Hβ lines contribute to the $[3.6]-[4.5]$ color. The SFR and age are estimated to be ≈ 60 $M_\odot$ yr^{-1} and ≈ 4 Myr, respectively, suggesting that MACS0416_Y1 is at the onset of a starburst phase. The stellar mass visible in the UV-to-optical is estimated to be $M_{\rm star} \approx 3 \times 10^8 M_\odot$. The best-fitting metallicity already reaches $Z \approx 0.2 Z_\odot$ at $z = 8.3$ despite a large uncertainty, suggesting rapid enrichment of heavy elements in the middle of the reionization era.

4. Evolved Stellar Component as the Origin of Dust

The SED analysis does not, however, explain how the galaxy has *obtained* the large amount of dust, even though **the SED model treats the energy balance self-consistently** if assuming that the dust *preexists*. The inferred dust-to-stellar mass ratio ($\sim 10^{-2}$) is two orders of magnitude higher than those found in dusty star-forming galaxies, suggesting the presence of a passively-evolved stellar component which does not

† Panchromatic Analysis for Nature of HIgh-z galaxies Tool (PANHIT), which is available for download at http://www.icrr.u-tokyo.ac.jp/%7Emawatari/PANHIT/PANHIT.html.

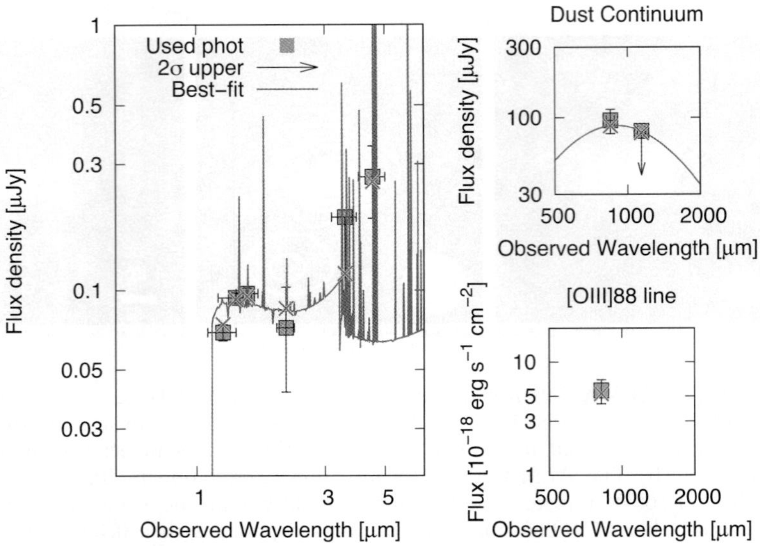

Figure 2. The best-fitting spectral energy distribution (SED, solid curve) obtained for the Calzetti *et al.* (2000) dust extinction law. The filled squares represent the observed photometric data points. The ALMA constraints are shown in the small panels. The crosses are flux densities (or flux for the [O III] 88 μm line) predicted from the model.

dominate the UV-to-optical light. Furthermore, a dust formation model (e.g., Asano *et al.* 2013) assuming a single episode of star formation does not reproduce the dust mass and metallicity in the galaxy age of ≈ 4 Myr. This strongly suggests that the ISM was already metal-polluted before the onset of the current episode of star formation.

Obviously, the disagreement should be mitigated if assuming the presence of an underlying evolved stellar component assembled in a past star-formation activity. We find not only that a 0.3 Gyr-old stellar component can reproduce the dust mass and metallicity, but also that the addition of the evolved stellar component does not substantially change the stellar SED. Therefore, it is likely that the mature (with an age of $\sim$0.3 Gyr) stellar population with no or little ongoing star formation may be the origin of the very early enrichment of metal and dust.

References

Asano, R., Takeuchi, T., Hirashita, H., & Inoue, A. K. 2013, *Earth, Planet. & Space*, 65, 213

Bruzual, G. & Charlot, S. 2003, *MNRAS*, 344, 1000

Calzetti, D., Armus, L., Bohlin, R. C., Kinney, A. L., Koornneef, J., & Storchi-Bergmann, T. 2000, *ApJ*, 533, 682

Hashimoto, T., Laporte, N., Mawatari, K., Ellis, R. S., Inoue, A. K., Zackrisson, E., Roberts-Borsani, G., Zheng, W., *et al.* 2018, *Nat.*, 557, 392

Inoue, A. K. 2011, *MNRAS*, 415, 2920

Inoue, A. K., Shimizu, I., Tamura, Y., Matsuo, H., Okamoto, T. & Yoshida, N. 2014, *MNRAS*, 442, 1805

Inoue, A. K., Tamura, Y., Matsuo, H., Mawatari, K., Shimizu, I., Shibuya, T., Ota, K., Yoshida, N., *et al.* 2016, *Science*, 352, 1559

Laporte, N., Streblyanska, A., Kim, S., Pelló, R., Bauer, F. E., Bina, D., Brammer, G., De Leo, M. A., *et al.* 2015, *A&A*, 575, A92

Mawatari, K., *et al.* 2018, *Proc. IAU Symp. No. 341*, in this volume

Oesch, P. A., Bouwens, R. J., Illingworth, G. D., Labbé, I., & Stefanon, M. 2018, *ApJ*, 855, 105

Planck Collaboration 2018, arXiv:1807.06209

Rieke, G. H., Alonso-Herrero, A., Weiner, B. J., *et al.* 2009, *ApJ*, 692, 556
Shimizu, I., Inoue, A. K., Okamoto, T., & Yoshida, N. 2014, *MNRAS*, 440, 731
Sunaga, K. *et al.* 2018, *Proc. IAU Symp. No. 341*, in this volume
Tamura, Y., Mawatari, K., Hashimoto, T., Inoue, A. K., Zackrisson, E., Christensen, L., Binggeli, C., Matsuda, Y., *et al.* 2019, *ApJ*, 874, 27

Discussion

TOMO GOTO: [O III] and 850 μm might have a spatial offset. If separated, both components might be on the local relation on the $L_{[\mathrm{OIII}]}/L_{\mathrm{TIR}}$ versus L_{TIR} plot?

YOICHI TAMURA: Yes, I believe so. The current spatial resolution for [O III] is not enough to resolve the galaxy, but if we spatially resolve it — actually we are going to do so in the current cycle of ALMA — the three components might follow the local relation.

FABIO FONTANOT: I am wondering if a top-heavy initial mass function can loosen the tensions you find in the dust content?

YOICHI TAMURA: As you pointed out, thermally pulsating asymptotic giant branch stars will supply more dust, which is great, but I think more frequent supernovae will destroy dust grains, which might suppress the increase in dust mass. I could not tell which effect is more efficient, but we can test how the change in the initial mass function would affect the dust mass by modifying the dust model.

DAVID ROSARIO: In low-mass galaxies with high specific star formation rates, simulations show a substantial loss of metal-enriched gas in feedback-driven outflows. If this also applies to MACS0416_Y1, how is the star formation history required to produce the altered dust? Is this effect included in the dust evolution models used?

YOICHI TAMURA: I do not think I can answer to your question quantitatively since our dust evolution model is a "closed box". However, I would say qualitatively that it would be necessary to put a star formation activity with longer star formation histories and lower star formation rates into the model. In this case, the formation epoch dates back to $z \sim 15$ or even higher, but a long duration of star formation can trigger non-linear grain growth in the interstellar medium, which facilities reproducing the observed dust mass.

Challenges in Panchromatic Modelling with
Next Generation Facilities
Proceedings IAU Symposium No. 341, 2018
M. Boquien, E. Lusso, C. Gruppioni & P. Tissera, eds.

doi:10.1017/S1743921319002096

Dust mass and dust production efficiencies on the redshift frontier

Hiroyuki Hirashita[1], Denis Burgarella[2] and Rychard J. Bouwens[3]

[1]Institute of Astronomy and Astrophysics, Academia Sinica, Astronomy-Mathematics Building, AS/NTU No. 1, Sec. 4, Roosevelt Road, Taipei 10617, Taiwan
email: hirashita@asiaa.sinica.edu.tw

[2]Aix-Marseille Université, CNRS, LAM, UMR 7326, 13388 Marseille, France

[3]Leiden Observatory, Leiden University, NL-2300 RA Leiden, Netherlands

Abstract. In order to clarify the dust production in the early Universe, we constrain the dust mass in high-redshift ($z \gtrsim 5$) galaxies using the upper limits obtained by ALMA. We perform fitting to the rest-frame UV–far-infrared spectral energy distribution (SED) of a giant Lyα emitter, Himiko, at $z = 6.6$ and a composite SED of $z > 5$ Lyman break galaxies (LBGs). For Himiko, we obtain a high dust temperature > 70 K. This high dust temperature puts a strong upper limit on the total dust mass $M_{\rm d} \lesssim 2 \times 10^6$ M$_\odot$, and the dust mass produced per supernova (SN) $m_{\rm d,SN} \lesssim 0.1$ M$_\odot$. Such a low $m_{\rm d,SN}$ suggests significant loss of dust by reverse shock destruction or outflow. For the LBG sample, we only obtain an upper limit for $m_{\rm d,SN}$ as $\sim$2 M$_\odot$. This clarifies the importance of observing UV-bright objects (like Himiko) to constrain the dust production by SNe.

Keywords. dust, dust extinction, galaxy evolution, high redshift, ISM, far infrared

1. Introduction

Dust plays an important role in various aspects of galaxy evolution. In particular, dust absorbs and scatters stellar light and reemits it into far-infrared (FIR) wavelengths. Therefore, dust dramatically modifies the observed spectral energy distributions (SEDs) of galaxies (e.g., Takeuchi *et al.* 2005). Dust extinction could be estimated to match the SED at ultraviolet (UV) and optical wavelengths with a given stellar intrinsic SED. Because the stellar radiation energy absorbed by dust is emitted in the FIR, the total FIR emission constrains the total dust extinction as well. This energy balance between dust absorption and emission is the key to understand the effect of dust extinction on the SED shape ranging from UV to FIR.

Recently, it has become possible to investigate the dust production and evolution in high-z galaxies. The most sensitive dust search at high z is possible by the Atacama Large Millimeter/submillimeter Array (ALMA). The highest-z galaxies for which the dust emission is detected by ALMA are located at $z > 7$ (e.g., Watson *et al.* 2015). The high sensitivity of ALMA enables us to constrain the dust enrichment processes in those galaxies (Mancini *et al.* 2015; Wang *et al.* 2017). However, dust continuum has not been detected for most Lyman break galaxies (LBGs) at $z \gtrsim 6$ (e.g., Bouwens *et al.* 2016, hereafter B16).

Ouchi *et al.* (2013) observed a Lyα-emitting gas blob "Himiko" at $z = 6.6$ using ALMA. Hirashita *et al.* (2014) developed a method to constrain the dust mass formed per SN based on Ouchi *et al.* (2013)'s result. However, they treated the UV and FIR SEDs separately. As mentioned above, the SEDs in those two wavelength ranges are tightly

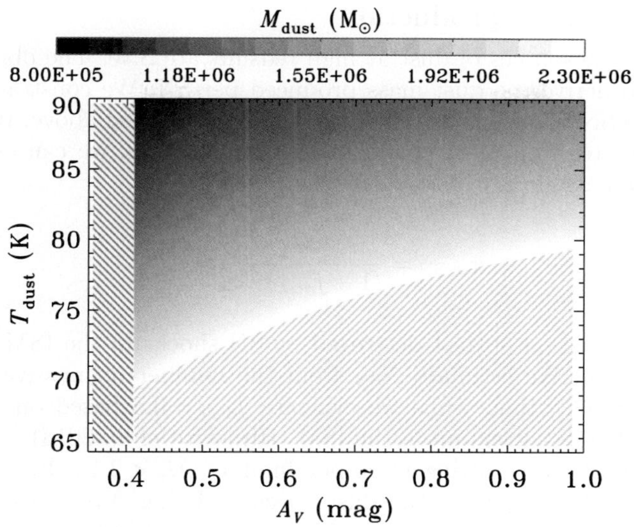

Figure 1. Dust mass derived from parameter set (A_V, T_d) for Himiko. The dust mass corresponding to (A_V, T_d) is shown by the gray scale, and the level of dust mass is shown in the bar on the top. The dust mass is only shown in the area where the SED fitting is successful. The yellow shaded area is the region where the mm flux exceeds the ALMA upper limit, while the red shaded area is the region where we do not obtain a satisfactory fit to the rest UV–optical data (i.e. reduced $\chi^2 \geqslant 3$).

related through dust absorption and reemission. It would be interesting to reexamine the above constraint on the SN dust production by treating those two wavelength ranges consistently with a single SED model.

2. SED fitting

We adopted Himiko at $z = 6.6$ to constrain the dust mass. We also composed a stacked SED using the 78 LBGs at $z \geqslant 5$ in B16. For the SED fitting, we use CIGALE (Noll *et al.* 2009; Boquien *et al.* 2019). The details of the method including the adopted parameters for the SED fitting are described in Hirashita *et al.* (2017). The varied parameters are the extinction in the V band (A_V), the dust temperature (T_d), the extinction of the old population relative to that of the young population (η) and the attenuation curve slope (δ). We minimize χ^2 to obtain the stellar mass (M_*). We only extract the solution which satisfies reduced $\chi^2 < 3$ and the ALMA upper limit. We adopt roughly a constant star formation history with a mean stellar age of 200 Myr. The fitting results below broadly hold as long as we have the present star formation activity. From the FIR dust emission, we obtain the total dust mass, M_d. The obtained dust mass depends on the assumed mass absorption coefficient; thus, we considered some variations in the dust species (unless otherwise states, we show the results for silicate).

First, we perform the SED fitting to Himiko with $\eta = 0$ and $\delta = -1.1$ (these parameters give the most conservative constraint for T_d). For a given set of parameters (A_V, T_d), we obtain the dust mass. In Fig. 1, we show the obtained dust mass at each point on the (A_V, T_d) plane by the gray scale. We only show the dust mass in the area of (A_V, T_d) where the fitting is satisfactory.

Next, we apply the same fitting procedure as above to the B16 LBG sample. Because the ALMA upper limit flux relative to the UV flux is higher than that of Himiko, the constraint on the parameters is weaker.

3. Constraint on dust production

The most probable sources of dust at high redshift are SNe. The obtained dust mass above are used to derive the dust mass produced per SN. We constrain the dust mass formed in a single SN by using the upper limit of $M_{\rm d}$ obtained above. If we assume that all the dust originates from dust condensation in SN ejecta, we can estimate the dust mass ejected from a single SN, $m_{\rm d,SN}$ as

$$m_{\rm d,SN} = \frac{M_{\rm d}}{(1 - f_{\rm dest})N_{\rm SN}}, \tag{3.1}$$

where $f_{\rm dest}$ is the fraction of dust destroyed by SN shocks in the ISM, and $N_{\rm SN}$ is the total number of SNe. We estimate $N_{\rm SN}$ from the number of massive ($> 8\ {\rm M}_{\odot}$) stars obtained by the SED fitting, while we evaluate $f_{\rm dest} = 0.5$ based on a dust evolution model including shock destruction in the ISM (Hirashita *et al.* 2014).

For Himiko, we obtain a stringent upper limit of $M_{\rm d} < 2.1 \times 10^6\ {\rm M}_{\odot}$ (for silicate; 0.97–5 $\times 10^6\ {\rm M}_{\odot}$, depending on the dust species). Using $N_{\rm SN}$ obtained by the SED fitting, we obtain $m_{\rm d,SN} < 0.067\ {\rm M}_{\odot}$ (for silicate; 0.031–0.16 ${\rm M}_{\odot}$ depending on the dust species). For the LBG sample, we obtain $m_{\rm d,SN} < 2.3\ {\rm M}_{\odot}$. The upper limit of $m_{\rm d,SN}$ obtained for the LBGs is too large to put a useful constraint on the SN dust production theory. Thus, we concentrate on the upper limit obtained for Himiko.

4. Discussion

In the above, we suggested a high dust temperature for Himiko. Indeed, some studies have suggested that the dust temperatures in high-redshift star-forming galaxies are high (e.g., Ouchi *et al.* 1999). Our recent cosmological simulation results (Aoyama *et al.* 2019) also show that the dust temperatures become higher at higher redshifts. The high dust temperatures are most probably due to high interstellar radiation field: since galaxies become more dense and compact at higher redshift, the mean distance between dust and stars becomes shorter. Other theoretical estimates also show high dust temperatures at high redshift (Ferrara *et al.* 2017; Narayanan *et al.* 2018).

The upper limit for $m_{\rm d,SN}$ obtained for Himiko indicates either that reverse shock destruction in SNe is efficient because of high ambient medium density ($\gtrsim 10\ {\rm cm}^{-3}$) if we compare it with dust condensation calculations by Nozawa *et al.* (2007) (see also Bianchi & Schneider 2007). Alternatively, the dust may be lost by galactic winds, etc. This possibility is shown to occurs by Hou *et al.* (2017), based on a numerical simulation developed by Aoyama *et al.* (2017). They showed that dust can be transported into the circum-galactic space by SN feedback.

We assumed $\delta = -1.1$ for the steepness of the attenuation curve. This is as steep as the Small Magellanic Cloud extinction curve. However, as shown by Asano *et al.* (2014), the extinction curve is expected to be flat in the early phase of galaxy evolution when SNe are the dominant source of dust (see also Hou, Hirashita, & Michałowski 2016). If we adopt a flat attenuation curve, we need more dust extinction to explain the rest-frame UV SED, leading to a higher dust temperature. In this case, we expect that $m_{\rm d,SN}$ is even lower than that obtained above. Alternatively, we can consider a scenario in which SN dust production is not dominant any more, but dust growth in the ISM is already efficient. In this case, $m_{\rm d,SN}$ should be much smaller than the above, since only a limited fraction of dust originates from SNe. In summary, regardless of whether SNe are the dominant source of dust or not, the dust mass per SN is expected to be smaller than 0.1 ${\rm M}_{\odot}$, supporting efficient dust destruction within SNe through the so-called reverse-shock destruction.

5. Conclusion

We investigate the possibility of constraining the dust mass in high-redshift ($z > 5$) galaxies by applying SED fitting (CIGALE) to rest UV–optical photometric data and the ALMA upper limits. For Himiko, we obtain $T_d > 70$ K and $M_d < 2 \times 10^6$ $M_\odot$. Based on this value, the dust mass produced per SN is estimated as $\lesssim 0.1$ $M_\odot$. This low value indicates that dust once condensed is destroyed in the shocked region associated with the SN, or that dust is lost out of the main body of the galaxy.

Acknowledgement

HH is supported by the Ministry of Science and Technology grant, MOST 105-2112-M-001-027-MY3 and MOST 107-2923-M-001-003-MY3 (RFBR 18-52-52-006).

References

Aoyama, S., Hirashita, H., Lim, C.-F., Chang, Y.-Y., Wang, W.-H., Nagamine, K., Hou, K.-C., Shimizu, I., *et al.* 2019, *MNRAS*, in press (arXiv:1809.10416)
Aoyama, S., Hou, K.-C., Shimizu, I., Hirashita, H., Todoroki, K., Choi, J.-H., & Nagamine, K. 2017, *MNRAS*, 466, 105
Asano, R. S., Takeuchi, T. T., Hirashita, H., & Nozawa, T. 2014, *MNRAS*, 440, 134
Bianchi, R. & Schneider, R. 2007, *MNRAS*, 378, 973
Boquien, M., Burgarella, D., Roehlly, Y., Buat, V., Ciesla, L., Corre, D., Inoue, A. K., & Salas, H. 2019, *A&A*, 622, 603
Bouwens, R. J., Aravena, M., Decarli R. Walter, F., da Cunha, E., Labbé, I., Bauer, F. E., Bertoldi, F., *et al.* 2016, *ApJ*, 793, 115 (B16)
Ferrara, A., Hirashita, H., Ouchi, M., Fujimoto, S., *et al.* 2017, *MNRAS*, 471, 5018
Hirashita, H., Burgarella, D., & Bouwens, R.J. 2017, *MNRAS*, 472, 4587
Hirashita, H., Ferrara, A., Dayal, P., & Ouchi, M. 2014, *MNRAS*, 443, 1704
Hou, K.-C., Hirashita, H., & Michałowski, M. J. 2016, *PASJ*, 68, 94
Hou, K.-C., Hirashita, H., Nagamine, K., Aoyama, S., & Shimizu, I. 2017, *MNRAS*, 469, 870
Mancini, M., Schneider, R., Graziani, L., Valiante, R., Dayal, P., Maio, U., Ciardi, B., & Hunt, L. K. 2015, *MNRAS*, 451, L70
Narayanan, D., Davé, R., Johnson, B., Thompson, R., Conroy, C., & Geach, J. E. 2018, *MNRAS*, 474, 1718
Noll, S., Burgarella, D., Giovannoli, E., Buat, V., Marcillac, D., & Muñoz-Mateos, J. C. 2009, *A&A*, 507, 1793
Nozawa, T., Kozasa, T., Habe, A., Dwek, E., Umeda, H., Tominaga, N., Maeda, K., & Nomoto, K. 2007, *ApJ*, 666, 955
Ouchi, M., Ellis, R. Ono, Y., Nakanishi, K., Kohno, K., Momose, R., Kurono, Y., Ashby, M. L. N., *et al.* 2013, *ApJ*, 778, 102
Ouchi, M., Yamada, T., Kawai, H., & Ohta, K. 1999, *ApJ*, 517, L19
Takeuchi, T. T., Ishii, T. T., Nozawa T., Kozasa, T., & Hirashita, H. 2005, *MNRAS*, 362, 592
Wang, W.-C., Hirashita, H., & Hou, K.-C. 2017, *MNRAS*, 465, 3475
Watson, D., Christensen, L., Knudsen, K. K., Richard, J., Gallazzi, A., & Michałowski, M. J. 2015, *Nature*, 519, 327

Discussion

TOMO GOTO: Declining star formation histories might increase $M_{\rm dust}$?

HIROYUKI HIRASHITA: Declining star formation histories put a loose constraint on the dust mass, but as long as there is recent star formation activity as indicated by the strong Lyα emission, the resulting constraint on $M_{\rm dust}$ does not change much.

AKIO INOUE: The spectral energy distribution fit for Himiko shows the best solution with the SMC law. However the grain size distribution model shows that in the early phase, the grain size tends to be larger, suggesting a flatter extinction curve. Are these consistent?

HIROYUKI HIRASHITA: There is a tension between these two are separate results. The assumed steep extinction curve gives the most conservative constraint on the dust mass. This is why we assumed the SMC-like law in the spectral energy distribution fitting.

KE-JUNG CHEN: How do you know the number of supernovas and the destruction fraction of the dust?

HIROYUKI HIRASHITA: We estimated the number of supernovae based on the spectral energy distribution fitting since the UV luminosity is proportional to the number of massive stars. The destruction fraction is derived from our dust evolution model separately.

HIROSHI SHIBAI: What happens if you do not assume uniformity?

HIROYUKI HIRASHITA: I expect only a little change in the obtained dust mass constraint because we still need significant extinction. However further tests would be useful for the geometry effect.

Challenges in Panchromatic Modelling with Next Generation Facilities
Proceedings IAU Symposium No. 341, 2018
M. Boquien, E. Lusso, C. Gruppioni & P. Tissera, eds.

doi:10.1017/S1743921319002886

The onset of star formation 250 million years after the Big Bang

Takuya Hashimoto[1,2]

[1]Department of Environmental Science and Technology, Faculty of Design Technology, Osaka Sangyo University, 3-1-1, Nagaito, Daito, Osaka 574-8530, Japan

[2]National Astronomical Observatory of Japan, 2-21-1 Osawa, Mitaka, Tokyo 181-8588, Japan
email: thashimoto@est.osaka-sandai.ac.jp

Abstract. In this IAU symposium, we present results of our recent paper, Hashimoto *et al.* (2018a) focusing on its spectral energy distribution modeling. We present spectroscopic observations of MACS1149-JD1, a gravitationally lensed galaxy originally discovered by Zheng *et al.* (2012) via the dropout technique. Using the Atacama Large Millimeter/submillimeter Array (ALMA), we detect an emission line of doubly ionized oxygen, [OIII] 88 μm, at a redshift of 9.1096±0.0006. This precisely determined redshift indicates that the red rest-frame optical colour observed with the Spitzer Space Telescope arises from a dominant stellar component that formed about 250 million years after the Big Bang, corresponding to a redshift of about 15. MACS1149-JD1 clearly demonstrates the importance and power of spectral energy distribution modeling to understand the earliest star formation history of the Universe.

Keywords. galaxies: evolution, galaxies: formation, galaxies: high-redshift, galaxies: individual (MACS1149-JD1)

1. Introduction

With the advent of the Atacama Large Millimeter/Submillimeter Array (ALMA) telescope, recent studies demonstrate how the far-infrared (FIR) fine structure line of doubly ionized oxygen ([OIII] 88 μm) is powerful to identify galaxies in the reionization epoch, redshift (z) beyond 6. In two years after the first detection of [OIII] in the reionization epoch by Inoue *et al.* (2016) at $z = 7.21$, twelve [OIII] detections at $z > 6$ are reported in the literature (Inoue *et al.* 2016; Carniani *et al.* 2017; Laporte *et al.* 2017; Marrone *et al.* 2018; Hashimoto *et al.* 2018a,b,c; Tamura *et al.* 2018; Walter *et al.* 2018). Among these, Laporte *et al.* (2017) and Tamura *et al.* (2018) have detected [OIII] in A2744_YD4 at $z = 8.38$ and MACS0416_Y1 at $z = 8.31$, respectively. Interestingly, these two $z \approx 8$ galaxies also show dust continuum detections, indicating fast growth of dust grains in the early Universe. Remarkably, Hashimoto *et al.* (2018a) have detected [OIII] at the significance level of 7.4σ in a gravitationally-lensed galaxy, MACS1149-JD1 (Zheng *et al.* 2012), at $z = 9.1096 \pm 0.0006$ using ALMA. The dust continuum emission is undetected in MACS1149-JD1. Based on VLT/X-Shooter observations, Lyα is also detected at 4.0σ. MACS1149-JD1 is the most distant object whose emission lines have been identified.

2. Spectral Energy Distribution Analyses

MACS1149-JD1 is currently the only one galaxy which has a rich data set from rest-frame UV, optical, to far-infrared at $z > 9$. A striking feature of the target is its red

Spitzer/IRAC colour. Previous to Hashimoto *et al.* (2018a), the origin of the colour was not uniquely determined: Without its precise spectroscopic redshift, the colour can be reproduced by both (1) contamination from strong nebular emission lines by young metal-poor stellar populations and (2) contamination from the Balmer break indicative of old stellar populations (Hoag et al. 2018). Understanding the IRAC colour is a key to unveil properties of MACS1149-JD1.

To build the SED of our target, we use data from the three deep *HST* images (Zheng *et al.* 2017), the deep K_s image obtained with VLT/HAWK-I (Hoag et al. 2018) as well as the deep *Spitzer*/IRAC 3.6 and 4.5 μm data (Zheng *et al.* 2017). In addition to these, we use our deep upper limit on dust continuum emission as well as our measurement of the [OIII] flux (see Table 1 of Hashimoto *et al.* 2018a). We use our own SED fitting code named PANHIT (Mawatari *et al.* 2016, Mawatari *et al.* in preparation), which is also adopted in Hashimoto *et al.* (2018b) and Tamura *et al.* (2018). For the details, we refer the readers to the website, http://www.icrr.u-tokyo.ac.jp/ mawatari/PANHIT/PANHIT.html.

We first adopted single stellar population models. We find that, with an exponentially declining star formation history (SFH), the [OIII] line flux cannot be reproduced while the HST and IRAC data points are relatively well reproduced: Under the assumption of the declining SFH, the star formation rate (SFR) at $z = 9.11$ becomes too small for the observed [OIII] flux. In the case of the constant SFR, we obtain a very large reduced χ^2 (χ^2_ν) value of $\simeq 11.4$, implying that the models are inappropriate. What is worse, we obtain the stellar age of $\simeq 500$ Myr, similar to the age of the Universe at this redshift, which we consider physically unacceptable. In the constant SFR case, one might think that the IRAC colour could be reproduced by young metal-poor models with strong nebular emission lines. Indeed, the Hβ plus [OIII] 4959 Å lines can contaminate the IRAC channel 2 at $z = 9.11$. Our nebular emission line model of Inoue (2011) accounts for extremely strong emission line cases with rest-frame equivalent widths up to EW(Hβ+[OIII] 4959 Å) ~ 1550 Å for young metal-poor stellar populations (see Figures 9 and 10 of Inoue 2011). Based on our detailed analyses, we find that the IRAC channel 2 is indeed boosted due to strong Hβ plus [OIII] 4959 Å lines in young metal-poor cases. However, the IRAC channel 1 is also boosted by nebular continuum and [OII] 3727 Å line, counterbalancing the IRAC colour. To summarize, single stellar population models cannot reproduce the observed SED.

We next considered a two stellar component SED comprising a young starburst and an old population. In order to reproduce the observed SED with as small a number of parameters as possible, we assumed constant SFRs for both components. Such assumptions are also adopted in Hashimoto *et al.* (2018b). We adopted several star formation durations for the old component, namely $\tau =$ 10, 100, 200, 300, and 400 Myr. In these two component models, we can reproduce the observations with $\chi^2_\nu \sim 2.8 - 3.0$. Models with $\tau = 300$ and 400 Myr lead to a stellar age $>$ 400 Myr at $z = 9.1$, similar to the age of the Universe at this redshift, which we consider physically unacceptable. The other three models give a similar quality fit and ages for the old stellar population ranging from 230 to 320 Myr. Due to the similarity of the physical and statistical results produced by the three remaining models, we cannot easily discriminate one over the other. For this reason, we select the $\tau =$ 100 Myr model as our fiducial case but include uncertainties which reflect the standard deviation caused by the choice of a $\tau =$ 10 or 200 Myr model.

Figure 1 shows our fiducial SED models, showing that all the data are well reproduced. The top, middle, and bottom panel of Figure 2 shows the SFH, mass assembly history, and the Balmer break growth, respectively, for the models with $\tau =$ 10, 100, and 200

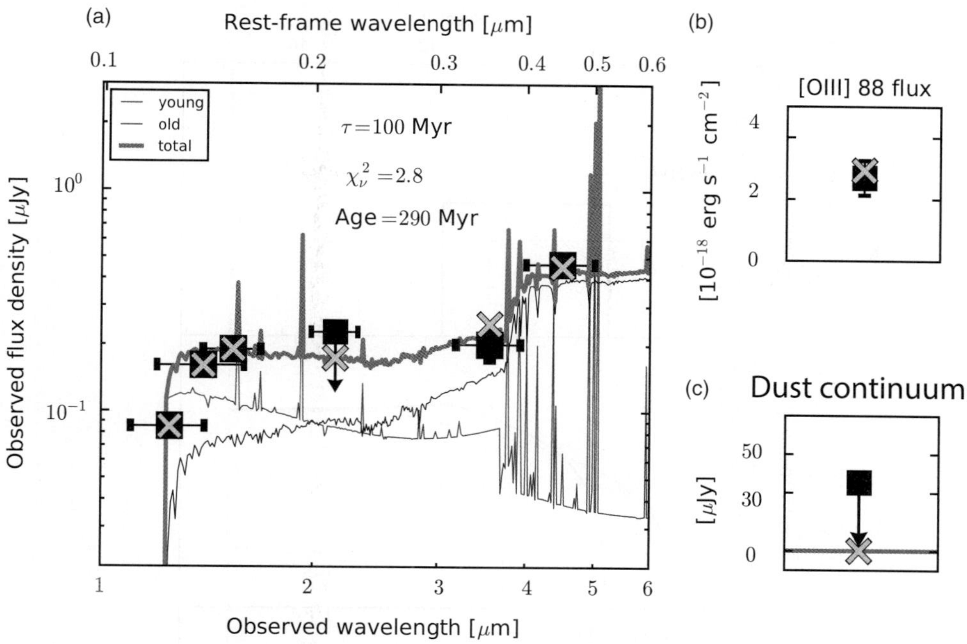

Figure 1. The SED of MACS1149-JD1 in the case of our fiducial model with $\tau = 100$ Myr (see the text for the details). In panel (a), black squares show the observations: F125W, F140W, and F160W data from *HST*, a 2σ upper limit for the K_s−band from VLT/HAWK-I, and 3.6μm and 4.5μm fluxes from *Spitzer*/IRAC. Horizontal and vertical error bars show the wavelength range of the filters and 1σ measurement uncertainties, respectively. The red solid line indicates the SED model and the corresponding magnitudes are shown via yellow crosses. Blue and black lines represent the contributions from the young and old component, respectively. In panel (b), the black square is the observed [OIII] emission line flux and its 1σ uncertainty, while the yellow cross indicates the model prediction. In panel (c), the black square shows the 2σ upper limit for the dust continuum flux density, and the yellow cross the model prediction.

Myr, demonstrating that MACS1149-JD1 has already started its first star formation as early as 250 million yeas after the Big Bang ($z \approx 15$).

3. Discussion and Summary

Our results clearly highlight the importance and power of multi-wavelength SED fitting to trace the earliest star formation history of the Universe. Very recently, Rocca-Volmerange *et al.* (2018) have analyzed the SED of three spectroscopically identified galaxies at $z \approx 8$ with Pegase.3. The authors have shown that the IRAC excess in these three galaxies are also due to the Balmer breaks. Is it possible that current bright $z \approx 8-9$ galaxies are biased against rare matured objects? Obviously, we need to increase the sample of spectroscopically identified galaxies at this epoch. *JWST* will help to investigate detailed properties of fainter galaxies at $z \approx 8-9$. It would be also interesting to examine how such Balmer breaks are common in the early Universe in simulations.

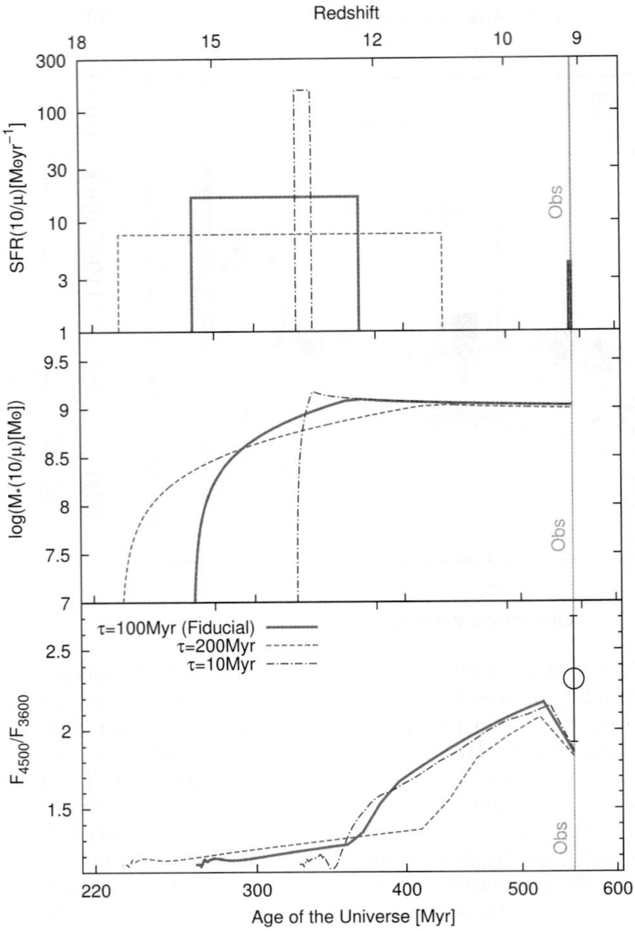

Figure 2. The star formation rate (a), stellar mass assembly (b) and Balmer break (c) as a function of redshift for three possible models. In each panel, the blue dot-dashed, green dashed, and red solid lines correspond to the models with star formation bursts of duration 10, 100, and 200 Myr, respectively. Each is capable of reproducing the Balmer break observed at $z = 9.1$ (shown by the black data point). The full SED fits shown in Figure 1 corresponds to our fiducial model corresponding to a burst of duration 100 Myr. The bulk of the stellar mass in MACS1149-JD1 observed at $z = 9.1$ was produced within a short period corresponding to the redshift interval $12 < z < 16$.

References

Hashimoto, T., Laporte, N., Mawatari, K., Ellis, R. S., Inoue, A. K., Zackrisson, E., Roberts-Borsani, G., Zheng, W., *et al.* 2018a, *Nature*, 557, 392

Zheng, W., Postman, M., Zitrin, A., Moustakas, J., Shu, X., Jouvel, S., Host, O., Molino, A., *et al.* 2012, *Nature*, 489, 406

Inoue, A. K., Tamura,Y., Matsuo, H., Mawatari, K., Shimizu, I., Shibuya, T., Ota, K., Yoshida, N., *et al.* 2016, *Science*, 352, 1559

Carniani, S., Maiolino, R., Pallottini, A., Vallini, L., Pentericci, L., Ferrara, A., Castellano, M., Vanzella, E., *et al.* 2017, *A&A*, 605, A42

Laporte, N., Ellis, R. S., Boone, F., Bauer, F., Quénard, D., Roberts-Borsani, G., Pelló, R., Pérez-Fournon, I., *et al.* 2017, *ApJL*, 837, L21

Marrone, D. P., Spilker, J. S., Hayward, C. C., Vieira, J. D., Aravena, M., Ashby, M. L. N., Bayliss, M. B., Béthermin, M., *et al.* 2018, *Nature*, 553, 51

Hashimoto, T., Inoue, A. K., Mawatari, K., Tamura, Y., Matsuo, H., Furusawa, H., Harikane, Y., Shibuya, T., *et al.* 2018b, submitted to *PASJ*, arXiv:1806.00486

Hashimoto, T., Inoue, A, K., Tamura, Y., Matsuo, H., Mawatari, K., & Yamaguchi Y. 2018c, submitted to *PASJ*, arXiv:1811.00030

Tamura, Y., Mawatari, K., Hashimoto, T., Inoue, A. K., Zackrisson, E., Christensen, L., Binggeli, C., Matsuda, Y., *et al.* 2018, submitted to *ApJ*, arXiv:1806.04132

Walter, F., Riechers, D., Novak, M., Decarli, R., Ferkinhoff, C., Venemans, B., Bañados, E., Bertoldi, F., *et al.* 2018, *ApJ*, 869L, 22

Hoag, A., Bradač, M., Brammer, G., Huang, K.-H., Treu, T., Mason, C. A., Castellano, M., Di Criscienzo, M., *et al.* 2018, *ApJ*, 854, 39

Zheng, W., Zitrin, A., Infante, L., Laporte, N., Huang, X., Moustakas, J., Ford, H. C., Shu, X., *et al.* 2017, *ApJ*, 836, 210

Mawatari, K., Yamada, T., Fazio, G. G., Huang, J.-S., & Ashby, M. L. N. 2016, *PASJ*, 68, 46

Inoue, A. K. 2011, *MNRAS*, 415, 2920

Rocca-Volmerange, B., Hilberer, A., & Fioc, M., 2018, submitted to *A&A* arXiv:1812.04283

4. Q&A

SPEAKER NAME: Takuya Hashimoto

QUESTIONER: Kentaro Motohara

QUESTION: What about the exponentially declining star formation histories to reproduce the observed spectral energy distribution?

ANSWER: We have also performed spectral energy distribution fitting with the exponentially declining star formation histories. While the Balmer break can be reproduced, [OIII] cannot be reproduced.

Challenges in Panchromatic Modelling with Next Generation Facilities
Proceedings IAU Symposium No. 341, 2018
M. Boquien, E. Lusso, C. Gruppioni & P. Tissera, eds.

doi:10.1017/S1743921319002527

Extreme variations in star formation activity in the first galaxies

Christian Binggeli[1], Erik Zackrisson[1], Xiangcheng Ma[2], Akio K. Inoue[3], Anton Vikaeus[1], Takuya Hashimoto[3,4], Ken Mawatari[3,5], Ikkoh Shimizu[6] and Philip F. Hopkins[7]

[1]Department of Physics and Astronomy, Uppsala University, Box 516, SE-751 20 Uppsala, Sweden
email: christian.binggeli@physics.uu.se

[2]Department of Astronomy, 501 Campbell Hall 3411, University of California, Berkeley, CA 94720-3411, USA

[3]Department of Environmental Science and Technology, Faculty of Design Technology, Osaka Sangyo University, 3-1-1, Nagaito, Daito, Osaka 574-8530, Japan

[4]National Astronomical Observatory of Japan, 2-21-1 Osawa, Mitaka, Tokyo 181-8588, Japan

[5]Institute for Cosmic Ray Research, The University of Tokyo, Kashiwa, Chiba 277-8582, Japan

[6]Theoretical Astrophysics, Department of Earth & Space Science, Osaka University, 1-1 Machikaneyama, Toyonaka, Osaka 560-0043, Japan

[7]TAPIR, MC 350-17, California Institute of Technology, Pasadena, CA 91125, USA

Abstract. Recently, spectroscopic detections of O[III] 88 μm and Ly-α emission lines from the $z \approx 9.1$ galaxy MACS1149-JD1 have been presented, and with these, some interesting properties of this galaxy were uncovered. One such property is that MACS1149-JD1 exhibits a significant Balmer break at around rest-frame 4000 Å, which may indicate that the galaxy has experienced large variations in star formation rate prior to $z \sim 9$, with a rather long period of low star formation activity. While some simulations predict large variations in star formation activity in high-redshift galaxies, it is unclear whether the simulations can reproduce the kind of variations seen in MACS1149-JD1. Here, we utilize synthetic spectra of simulated galaxies from two simulation suites in order to study to what extent these can accurately reproduce the spectral features (specifically the Balmer break) observed in MACS1149-JD1. We show that while the simulations used in this study produce galaxies with varying star formation histories, galaxies such as MACS1149-JD1 would be very rare in the simulations. In principle, future observations with the *James Webb Space Telescope* may tell us if MACS1149-JD1 represents something rare, or if such galaxies are more common than predicted by current simulations.

Keywords. galaxies: high-redshift, early universe

1. Introduction

While understanding galaxy evolution at high redshifts and the formation of the first stars is important to create a complete picture of the evolution of the universe, these processes are still largely shrouded in mystery. Much of our understanding regarding how stars are formed in the earliest galaxies comes from theory and numerical simulations. At the same time, the lack of observations at the highest redshifts make it difficult to verify predictions by simulations. Recent (and future) observations of galaxies at high redshifts could, however, provide us with a simple test to determine how well simulations are able to reproduce properties of the first galaxies. Until recently, the only spectroscopically confirmed galaxy at $z \gtrsim 9$ was GN-z11 at a redshift of $z \sim 11$ (Oesch *et al.* 2016). Using observations of the [OIII] 88μm and Ly-α emission lines from the galaxy MACS1149-JD1 made it possible to constrain the redshift of this object to $z \approx 9.1$, adding a second object to the list of spectroscopically confirmed galaxies at $z \gtrsim 9$ (Hashimoto *et al.* 2018).

One interesting feature of JD1 is that *Spitzer*/IRAC observations of the object indicate that it exhibits a strong Balmer break. A possible explanation for this feature is that the galaxy has experienced large variations in the star formation rate (SFR), with a significant period of low star formation activity prior to $z \sim 9$. While certain simulations predict large variations in the SFRs of high-redshift galaxies (e.g. Kimm *et al.* 2015; Trebitsch *et al.* 2017; Hopkins *et al.* 2018; Ma *et al.* 2018) it is unclear whether galaxies such as JD1 are commonly seen in simulations, or if JD1 represents an outlier. Here, we use simulated galaxies from two independent cosmological simulations and generate synthetic spectra for these in order to study the distributions in Balmer break strengths exhibited by galaxies in the simulations. Through this, we are able to discuss how likely we are to find objects like JD1 among the simulated galaxies. We also show simulated distributions of Balmer break strengths that could be compared to future observations with the *James Webb Space Telescope* (JWST) to test if simulations are able to reproduce what is seen in observations, or if simulations may be missing some key ingredient.

2. Model

In this study, we use simulated galaxies from the FIRE-2 simulations (Ma *et al.* 2018; Hopkins *et al.* 2018) and the simulations by Shimizu *et al.* (2016, hereafter S16). Star formation histories and metallicity distributions of stars in simulated galaxies with $M_\star > 10^8 M_\odot$ are extracted in order to obtain galaxies with masses comparable to JD1, which has a stellar mass of $M_\star = 1.08^{+0.53}_{-0.18} \times 10^9 M_\odot$ assuming the fiducial magnification of $\mu = 10$ (Hashimoto *et al.* 2018). We select galaxies in the redshift interval $z \sim 9.5 - 8.5$ for the FIRE-2 galaxies and $z = 9$ for the S16 galaxies, which, in total, provides us with 150 galaxies from the FIRE-2 simulations and 182 galaxies from the S16 simulations. In order to generate synthetic spectra for these, we use synthetic single stellar population spectra from Yggdrasil (Zackrisson *et al.* 2011) consisting of Starburst99 Padova-AGB models (Leitherer *et al.* 1999; Vázquez & Leitherer 2005) for metallicities $Z = 0.0004 - 0.050$ and Raiter *et al.* (2010) models for $Z = 10^{-7} - 10^{-5}$. In order to account for nebular emission, the CLOUDY photo-ionization code (Ferland *et al.* 2013) is used. In order to model the distributions of Balmer breaks, we consider effects of dust reddening on the spectra of simulated galaxies. The S16 simulations provide us with predictions of the galaxy-wide dust extinction at 1500 Å. To account for dust-reddening effects in these galaxies, we utilize the Pei (1992) extinction curve for the Small Magellanic Cloud (SMC). For the FIRE-2 simulations we apply the SMC extinction curve while assuming a fixed extinction in the V-band of $A_V = 0.5$ for all galaxies. This crude recipe provides us with a test of the effects of dust extinction on the overall distribution of Balmer breaks.

However, we also utilize a set of the FIRE-2 galaxies that have undergone post-processing with the SKIRT dust radiative transfer code (Camps & Baes 2015) in order to account for more realistic dust reddening. The radiative transfer code is run while assuming SMC-type grain distributions (Weingartner & Draine 2001), the dust-to-metal ratio is assumed to be 0.4 and all gas hotter than 10^6 K is assumed to be dust-free. Further details about this version of FIRE-2 simulations and effects of the radiative transfer calculations are presented in Ma *et al.* (2019). For this process, the single stellar populations consist only of Starburst99 Padova-AGB models, and only nebular emission from hydrogen is considered. The difference in the distributions of Balmer break is, however, small, and does not significantly change our results. From the synthetic spectra,

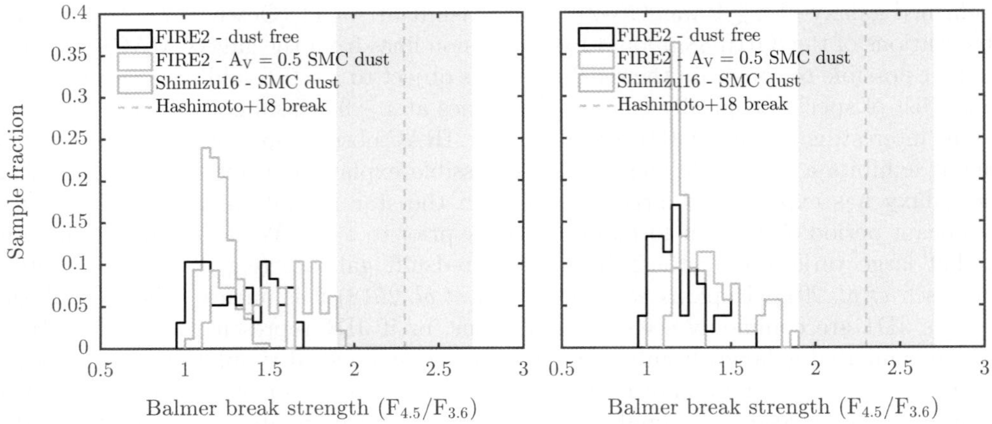

Figure 1. Balmer break strengths as measured by the *Spitzer*/IRAC 4.5/3.6 μm flux ratio for simulated galaxies with $10^8 \leqslant M_\star \leqslant 5 \times 10^8$ (left) and $M_\star \geqslant 5 \times 10^8$ (right) compared to the observed Balmer break in JD1. The yellow area shows the photometric error of the Balmer break observed in JD1.

we also generate fluxes in *Spitzer*/IRAC channel 1 and 2 (3.6 and 4.5 μm), assuming a redshift of $z = 9.1$ in order to compare results to JD1.

3. Results

Figure 1 shows the distributions of Balmer breaks in the simulated galaxies as observed by *Spitzer*/IRAC at $z = 9.1$ for stellar masses $10^8 M_\odot \leqslant M_\star \leqslant 5 \times 10^8 M_\odot$ (left panel) and $M_\star \geqslant 5 \times 10^8 M_\odot$ (right panel). Galaxies from the FIRE-2 simulations exhibit larger Balmer breaks than those seen in the S16 simulation, likely owing to the more stochastic star formation histories seen in the former. While the effect of dust on the synthetic spectra is to shift the distribution toward higher Balmer breaks, $A_V = 0.5$ represents a quite extreme case and leads to average UV slopes that are significantly redder than those observed in the high-redshift universe (e.g. Bouwens *et al.* 2014). However, even in the FIRE-2 galaxies with this extreme dust scenario, JD1 would be a rare type of object.

Figure 2 shows the $F_\nu(4200\text{Å})/F_\nu(3500\text{Å})$ distribution of galaxies in the simulations, where $F_\nu(4200\text{Å})$ and $F_\nu(3500\text{Å})$ are flux densities at rest-frame 4200 and 3500 Å, respectively. Continuum fluxes at these wavelengths should be observable with *JWST*/NIRSpec even at the lowest resolution setting without emission/absorption lines blending into the relevant spectral bins. The right panel in figure 2 show the FIRE-2 galaxies that have been post-processed using SKIRT for five different viewing-angles without dust and the same five viewing-angles when dust is taken into account. The effect of dust on the distributions of Balmer breaks is in this case insignificant. While dust is suggested as a possible explanation for the large Balmer break observed in JD1 by Katz *et al.* (2018), none of the post-processed FIRE-2 galaxies exhibit a large difference on the Balmer break as an effect of dust reddening. This could, of course, be an effect of differences in the exact treatment of dust extinction.

4. Summary

We have presented distributions of Balmer breaks in simulated galaxies from two independent simulations. While there is a span in observed Balmer breaks exhibited by the simulated galaxies, objects such as MACS1149-JD1 represents a type of galaxy that seems to be extremely rare in the simulations. Other studies have suggested that effects of dust reddening could explain the Balmer break observed in JD1 without the need of

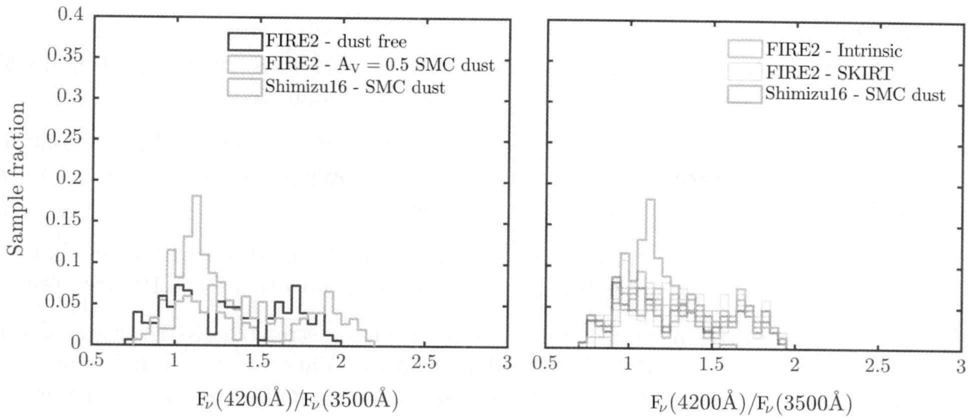

Figure 2. $F_\nu(4200Å)/F_\nu(3500Å)$ distributions in the simulated galaxies, where $F_\nu(4200Å)$ and $F_\nu(3500Å)$ are the flux densities at rest-frame 4200 and 3500 Å, respectively. Left panel: The S16 (blue) simulations, the dust-free FIRE-2 galaxies (black) and the FIRE-2 galaxies with the simplified dust treatment ($A_V = 0.5$, orange). Right panel: S16 galaxies (blue) and the FIRE-2 galaxies that have been post processed in SKIRT for five viewing-angles without dust (gray) and with dust taken into account (pale orange).

extreme variations in star formation activity. We do not see such an effect in the galaxies used here. We argue that future JWST observations of Balmer breaks in high-redshift galaxies may reveal if JD1 is just an extremely rare type of galaxy, or if models may be missing some key physics required to explain such large Balmer breaks.

References

Bouwens, R. J., Illingworth, G. D., Oesch, P. A., Labbé, I., van Dokkum, P. G., Trenti, M., Franx, M., Smit, R., *et al.* 2014, *ApJ*, 793, 115

Camps, P. & Baes, M. 2015, Astronomy and Computing, 9, 20

Ferland, G. J., Porter, R. L., van Hoof, P. A. M., Williams, R. J. R., Abel, N. P., Lykins, M. L., Shaw, G., Henney, W. J., *et al.* 2013, *Rev. Mexicana Astron. Astrofis.*, 49, 137

Hashimoto, T., Laporte, N., Mawatari, K., Ellis, R. S., Inoue, A. K., Zackrisson, E., Roberts-Borsani, G., Zheng, W., *et al.* 2018, Nature, 557, 392

Hopkins, P. F., Wetzel, A., Kereš, D., Faucher-Giguère, C.-A., Quataert, E., Boylan-Kolchin, M., Murray, N., Hayward, C. C., *et al.* 2018, *MNRAS*, 480, 800

Katz, H., Laporte, N., Ellis, R. S., Devriendt, J., & Slyz, A. 2018, ArXiv e-prints, 1809, arXiv:1809.07210

Kimm, T., Cen, R., Devriendt, J., Dubois, Y., & Slyz, A. 2015, *MNRAS*, 451, 2900

Leitherer, C., Schaerer, D., Goldader, J. D., Delgado, R. M. G., Robert, C., Kune, D. F., de Mello, D. F., Devost, D., *et al.* 1999, *ApJS*, 123, 3

Ma, X., Hopkins, P. F., Garrison-Kimmel, S., Faucher-Giguère, C.-A., Quataert, E., Boylan-Kolchin, M., Hayward, C. C., Feldmann, R., *et al.* 2018, *MNRAS*, 478, 1694

Ma *et al.* 2019, *MNRAS*, 487, 1864

Oesch, P. A., Brammer, G., van Dokkum, P. G., Illingworth, G. D., Bouwens, R. J., Labbé, I., Franx, M., Momcheva, I., *et al.* 2016, *ApJ*, 819, 129

Pei, Y. C. 1992, *ApJ*, 395, 130

Raiter, A., Schaerer, D., & Fosbury, R. A. E. 2010, *A&A*, 523, A64

Shimizu, I., Inoue, A. K., Okamoto, T., & Yoshida, N. 2016, *MNRAS*, 461, 3563

Trebitsch, M., Blaizot, J., Rosdahl, J., Devriendt, J., & Slyz, A. 2017, *MNRAS*, 470, 224

Vázquez, G. A. & Leitherer, C. 2005, *ApJ*, 621, 695

Weingartner, J. C. & Draine, B. T. 2001, *ApJ*, 548, 296

Zackrisson, E., Rydberg, C.-E., Schaerer, D., Östlin, G., & Tuli, M. 2011, *ApJ*, 740, 13

Discussion

Shohei Aoyama: Recently, high-z galaxies such as JD1 have been detected. In this sense, do simulation need to be improved in order to explain these objects?

Christian Binggeli: If future observations reveal that Balmer breaks like the one seen in JD1 are common and we find that here are very rare in simulations, this may indicate that simulations are missing some crucial ingredient.

Themiya Nanayakkara: What effect would you think stellar libraries have on reproducing the observed Balmer break? Going from Padova to FPSP, BPASS, etc.

Christian Binggeli: I have calculated Balmer break strength distributions for the simulated galaxies using the BPASSv2 stellar libraries. While there is a difference in the shape of the Balmer break strength distributions, they are still inconsistent with what is seen in JD1.

Daniel Ceverino: You may need to increase your sample of simulated galaxies if you want to find an analogue of the JD1 galaxy (Hashimoto *et al.* 2018). This is because of the large diversity of star formation histories of galaxies at high z.

Christian Binggeli: Yes, if we want to find an analogue or estimate a more exact probability of finding a galaxy like JD1, a larger sample may be required. However, we may still be able to determine if Balmer breaks like the ones seen in JD1 are common in simulations using a more limited sample.

Challenges in Panchromatic Modelling with Next Generation Facilities
Proceedings IAU Symposium No. 341, 2018
M. Boquien, E. Lusso, C. Gruppioni & P. Tissera, eds.

doi:10.1017/S1743921319002795

Subaru/HSC identifications of protocluster candidates at $z \sim 6-7$: Implications for cosmic reionization

Ryo Higuchi[1,2], Masami Ouchi[1,3], Yoshiaki Ono[1], Takatoshi Shibuya[1], Jun Toshikawa[1], Yuichi Harikane[1,2], Takashi Kojima[1,2], Yi-Kuan Chiang[4], Eiichi Egami[5], Nobunari Kashikawa[6,7], Roderik Overzier[8], Akira Konno[1,9], Akio K. Inoue[10], Kenji Hasegawa[11], Seiji Fujimoto[1,9], Tomotsugu Goto[12], Shogo Ishikawa[13,14], Kei Ito[7], Yutaka Komiyama[6,7] and Masayuki Tanaka[6]

[1]Institute for Cosmic Ray Research, The University of Tokyo, 5-1-5 Kashiwanoha, Kashiwa, Chiba 277-8582, Japan

[2]Department of Physics, Graduate School of Science, The University of Tokyo, 7-3-1 Hongo, Bunkyo-ku, Tokyo 113-0033, Japan

[3]Kavli Institute for the Physics and Mathematics of the Universe (Kavli IPMU, WPI), The University of Tokyo, 5-1-5 Kashiwanoha, Kashiwa, Chiba 277-8583, Japan

[4]Department of Physics & Astronomy, Johns Hopkins University, 3400 N. Charles Street, Baltimore, MD 21218, USA

[5]Steward Observatory, University of Arizona, 933 North Cherry Avenue, Tucson, AZ 85721, USA

[6]Optical and Infrared Astronomy Division, National Astronomical Observatory, Mitaka, Tokyo 181-8588, Japan

[7]Department of Astronomical Science, Graduate University for Advanced Studies (SOKENDAI), Mitaka, Tokyo 181-8588, Japan

[8]Observatório Nacional, Rua José Cristino, 77. CEP 20921-400, São Cristóvão, Rio de Janeiro-RJ, Brazil

[9]Department of Astronomy, Graduate School of Science, The University of Tokyo, 7-3-1 Hongo, Bunkyo-ku, Tokyo 113-0033, Japan

[10]Department of Environmental Science and Technology, Faculty of Design Technology, Osaka Sangyo University, 3-1-1, Nakagaito, Daito 574-8530 Osaka, Japan

[11]Department of Physics and Astrophysics, Nagoya University Furo-cho, Chikusa-ku, Nagoya, Aichi 464-8602, Japan

[12]Institute of Astronomy, National Tsing Hua University, No. 101,Section 2, Kuang-Fu Road, Hsinchu, Taiwan

[13]Center for Computational Astrophysics, National Astronomical Observatory of Japan, Mitaka, Tokyo 181-8588, Japan

[14]Department of Science, Faculty of Science and Engineering, Kindai University, Higashi-Osaka, Osaka 577-8502, Japan

email: `rhiguchi@icrr.u-tokyo.ac.jp`

Abstract. We report fourteen and twenty-eight protocluster candidates at $z = 5.7$ and 6.6 over 14 and 19 deg^2 areas, respectively, selected from 2,230 Lyα emitters (LAEs) photometrically identified with Subaru/Hyper Suprime-Cam (HSC) deep images. Six out of the 42 protocluster candidates include at least 1 spectroscopically confirmed LAEs at redshifts up to $z = 6.574$. By the comparisons with the cosmological Lyα radiative transfer (RT) model reproducing LAEs with the reionization effects, we find that more than a half of these protocluster candidates might be progenitors of the present-day clusters with a mass of $\gtrsim 10^{14} M_\odot$. We also investigate the correlation between LAE overdensity and Lya rest-frame equivalent width (EW), because the cosmological Lyα RT model suggests that a slope of EW-overdensity relation is steepened towards the epoch of cosmic reionization (EoR), due to the existence of the ionized bubbles around galaxy overdensities easing the escape of Lyα emission from the partly neutral intergalactic medium. The available HSC data suggest that the slope of the EW-overdensity correlation does not evolve from the post-reionization epoch $z=5.7$ to the EoR $z=6.6$ beyond the moderately large statistical errors.

Keywords. galaxies: high-redshift - galaxies: evolution - galaxies: formation

---

1. Introduction

It is important to study the physical process of cosmic reionization in astronomy today. In theoretical models, it is predicted that star-forming galaxies make ionized regions in the IGM around galaxies, called ionized bubbles. Large ionized bubbles are expected to form in galaxy high-density regions, and it is suggested that the cosmic reionization proceeds from high- to low-density regions (Overzier 2016). This process is called 'inside-out scenario'. The observation of galaxy high-density regions and identification of signatures of ionized bubbles are keys to testing the inside-out scenario of cosmic reionization. Observations of galaxy high-density regions near the epoch of cosmic reionization (EoR) are also important for a study of the early galaxy formation. In standard structure formation models, it is predicted that a large fraction of high-z galaxy high-density regions evolve into massive galaxy clusters at $z=0$. These galaxy high-density regions are called protoclusters. A protocluster is usually defined as a structure expected to evolve into a galaxy cluster with a halo mass $M_{\rm h} > 10^{14}\,{\rm M}_{\odot}$ (Chiang *et al.* 2013; Overzier 2016). Protoclusters at the EoR would be important examples of the early galaxy cluster formation (e.g. Ishigaki *et al.* 2016). Although the importances of high-z galaxy high-density regions are well recognized, there are only a few protoclusters at $z>6$ reported (ouchi *et al.* 2005; Utsumi *et al.* 2010; Toshikawa *et al.* 2012, 2014; Chanchaiworawit *et al.* 2017). To enlarge samples of protocluters at $z>6$, we need large field survey of galaxy high-density regions. In this study, we conduct protocluster survey at $z=5.7$ and 6.6 based on the samples of Lyα emitters (LAEs) obtained with Subaru/Hyper Suprime-Cam (HSC).

2. Data

HSC LAE Sample. We use LAE samples of HSC SSP data to calculate galaxy overdensity and identify protocluster candidates (see also Shibuya *et al.* 2018a). Shibuya *et al.* (2018a) select LAEs based on the HSC datasets. The color selection criteria are defined as

$$i - NB816 \geqslant 1.2 \text{ and } g > g_{3\sigma} \text{ and } [(r \leqslant r_{3\sigma} \text{ and } r - i \geqslant 1.0) \text{ or } (r > r_{3\sigma})] \tag{2.1}$$

and

$$\begin{gathered} z - NB921 \geqslant 1.0 \text{ and } g > g_{3\sigma} \text{ and } r > r_{3\sigma} \text{ and} \\ [(z \leqslant z_{3\sigma} \text{ and } i - z \geqslant 1.0) \text{ or } (z > z_{3\sigma})] \end{gathered} \tag{2.2}$$

for $z=5.7$ and 6.6 LAEs, respectively. We find 1,077 (1,153) LAEs at $z=5.7$ (6.6).

SC LAE Sample. In addition to the HSC LAE samples, we use photometric samples of Ouchi *et al.* (2008) (Ouchi *et al.* 2010) to select the spectroscopic targets of $z=5.7$ (6.6) LAEs. Ouchi *et al.* (2008) and Ouchi *et al.* (2010) find 401 and 207 LAEs at $z=5.7$ and 6.6, respectively.

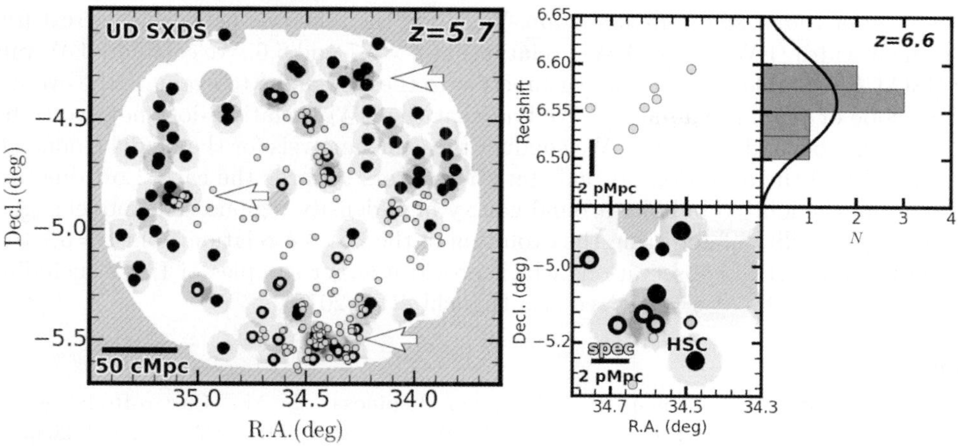

Figure 1. (Left) Example of the sky distribution of the LAEs with δ contours (gray colors) at $z = 5.7$. Black filled circles indicate HSC LAEs used for δ calculation. The gray filled circles show spec-LAEs. Masked regions and shallow regions are shown with gray regions. White quivers show the position of protocluster candidates. (Right) Example of our protocluster candidates at $z = 6.6$. The bottom panel is same as the left figure, but for the example of protocluster candidates at $z = 6.6$. The top-left panel presents the distribution of the spec-LAEs on the plane of R.A. vs. redshift directions. The top-right panel shows the redshift distribution of the spec-LAEs with the mean expected number of LAEs (black line).

Spectroscopic Sample. We carry out spectroscopic observations for our LAE samples. The details of spectroscopic observations for the HSC (SC) samples are shown in Shibuya *et al.* (2018b) (Higuchi *et al.* 2018). In addition to the spectroscopic sample of Shibuya *et al.* (2018b) and Higuchi *et al.* (2018), we refer other redshift catalogues of confirmed LAEs at $z = 5.7$ (6.6) taken from Ouchi *et al.* (2005), Ouchi *et al.* (2008) and Mallery *et al.* (2012) (Chanchaiworawit *et al.* 2017, and Guzmán *et al.* 2017). We make unified catalogs of ~ 200 spectroscopically confirmed LAEs (spec-LAEs) at $z = 5.7$ and 6.6.

3. Results and Discussions

Overdensity Measurements. We calculate LAE overdensities with the HSC LAE samples. The LAE overdensity δ is defined as $\delta = \frac{n - \overline{n}}{\overline{n}}$, where n ($\overline{n}$) is the total (average) number of LAEs found in a circle for the δ measurements. We use a circle with a radius of 0.07 deg (10 cMpc), which would be a typical size of protoclusters at $z \sim 6$ (Chiang *et al.* 2013). We show an example of the HSC LAE sky distribution and the δ map at $z = 5.7$ in Figure 1. We find some regions where δ values significantly exceed beyond those expected by random distribution. We call these regions as high-density regions (HDRs). We define a HDR as a region which has at least 4 LAEs in a radius of 0.07 deg. We identify 14 (28) $z = 5.7$ (6.6) HDRs in total.

Halo Mass Estimates. We estimate the probability of HDRs evolving into massive galaxy clusters at $z = 0$. From the theoretical model of Inoue *et al.* (2018), we derive a relation between the halo mass and δ ($M_{\rm h}$ - δ relation). We calculate the present-day halo masses of the haloes at $z = 5.7$ and 6.6, using the $M_{\rm h}$ - δ relation and extended Press-Schechter model of Hamana *et al.* (2006). We find that $\sim$60% of the haloes in the HDRs are expected to evolve into haloes with a mass of $> 10^{14}\,{\rm M}_{\odot}$ by $z = 0$. Because more than a half of the haloes in the HDRs are supposed to be progenitors of the present-day clusters, these HDRs can be regarded as protocluster candidates (the properties of protocluster candidates are listed in Higuchi *et al.* 2018).

Implications for Csomic Reionization. We study the relations between Lyα rest-frame equivalent width (EW) and δ (EW-δ relation) at $z = 5.7$ and 6.6. We calculate EW values for HSC LAE samples and fit a linear function to the EW and δ to evaluate the evolution of the slope of the linear function. We find that the EW-δ relation does not evolve from $z = 5.7$ to 6.6 beyond the errors. We conduct the same analysis for the model (Inoue *et al.* 2018), and find the evolution beyond statistical errors towards the early EoR due to the existence of the ionized bubbles around galaxy high-density regions. The model suggests there is a possibility of detecting the evolution of the EW - δ relation from $z = 5.7$ to 7.3 by the upcoming HSC observations which provides larger samples of LAEs including a new sample of LAEs at $z = 7.3$ (see also Higuchi *et al.* 2018).

References

Chanchaiworawit, K., Guzmán, R., Rodríguez Espinosa, J. M., Castro-Rodríguez, N., Salvador-Solé, E., Calvi, R., Gallego, J., Herrero, A., *et al.* 2017, *MNRAS*, 469, 2646

Chiang, Y.-K., Overzier, R., Gebhardt, K., *et al.* 2013, *ApJ*, 779, 127

Guzmán, R., Chanchaiworawit,, K., Rodríguez-Espinosa, J. M., Calvi, R. and Salvador-Solé, E., Manrique, A., Marín-Franch, A., Gallego, J., *et al.* 2017, in Early stages of Galaxy Cluster Formation, 12

Hamana, T., Yamada, T., Ouchi, M., Iwata, I., Kodama, T., *et al.* 2006, *MNRAS*, 369, 1929

Higuchi, R., Ouchi, M., Ono, Y., Shibuya, T., Toshikawa, J., Harikane, Y., Kojima, T., Chiang, Y.-K., *et al.* 2018, ArXiv e-prints, arXiv:1801.00531

Inoue, A. K., Hasegawa, K., Ishiyama, T., Yajima, H., Shimizu, I., Umemura, M., Konno, A., Harikane, Y., *et al.* 2018, *PASJ*, 70, 55

Ishigaki, M., Ouchi, M., Harikane, Y., *et al.* 2016, *ApJ*, 822, 5

Mallery, R. P., Mobasher, B., Capak, P., Kakazu, Y., Masters, D., Ilbert, O., Hemmati, S., Scarlata, C., *et al.* 2012, *ApJ*, 760, 128

Ono, Y., Ouchi, M., Harikane, Y., *et al.* 2018, *PASJ*, 70, S10

Ouchi, M., Shimasaku, K., Akiyama, M., Sekiguchi, K., Furusawa, H., Okamura, S., Kashikawa, N., Iye, M., *et al.* 2005, *ApJ*, 620, L1

Ouchi, M., Shimasaku, K., Akiyama, M., Simpson, C., Saito, T., Ueda, Y., Furusawa, H., Sekiguchi, K., *et al.* 2008, *ApJS*, 176, 301

Ouchi, M., Shimasaku, K., Furusawa, H., Saito, T., Yoshida, M., Akiyama, M., Ono, Y., Yamada, T., *et al.* 2010, *ApJ*, 723, 869

Overzier, R. A. 2016, *A&A Rev.*, 24, 14

Pavesi, R., Riechers, D. A., Capak, P. L., *et al.* 2016, *ApJ*, 832, 151

Shibuya, T., Ouchi, M., Konno, A., Higuchi, R., Harikane, Y., Ono, Y., Shimasaku, K., Taniguchi, Y., *et al.* 2018a, *PASJ*, 70, S14

Shibuya, T., Ouchi, M., Harikane, Y., Rauch, M., Ono, Y., Mukae, S., Higuchi, R., Kojima, T., *et al.* 2018b, *PASJ*, 70, S15

Toshikawa, J., Kashikawa, N., Ota, K., Morokuma, T., Shibuya, T., Hayashi, M., Nagao, T., Jiang, L., *et al.* 2012, *ApJ*, 750, 137

Toshikawa, J., Kashikawa, N., Overzier, R., Shibuya, T., Ishikawa, S., Ota, K., Shimasaku, K., Tanaka, M. , *et al.* 2014, *ApJ*, 792, 15

Utsumi, Y., Goto, T., Kashikawa, N., Miyazaki, S., Komiyama, Y., Furusawa, H., & Overzier, R., *et al.* 2010, *ApJ*, 721, 1680

Discussion

T. Goto: Do lyman break galaxies cluster around PCCs?

R. Higuchi: We have not checked lyman break galaxies around PCCs because we do not have GOLDRUSH sample at $z \sim 7$ (see Ono *et al.* 2018). I remember Pavesi *et al.* (2016) referred to our study and suggested that a dusty, starbursting galaxy at $z = 5.7$ exists around our PCCs.

Challenges in Panchromatic Modelling with Next Generation Facilities
Proceedings IAU Symposium No. 341, 2018
M. Boquien, E. Lusso, C. Gruppioni & P. Tissera, eds.

doi:10.1017/S1743921319002552

Probing the ISM of He II λ1640 emitters at $z = 2 - 4$ via MUSE

Themiya Nanayakkara[1], **Jarle Brinchmann**[1,2] **and The MUSE Collaboration**

[1]Leiden Observatory, Leiden University, PO Box 9513, 2300 RA Leiden, The Netherlands
email: nanayakkara@strw.leidenuniv.nl

[2]Instituto de Astrofisica e Ciencias do Espaco, Universidade do Porto, CAUP, Rua das Estrelas, 4150-762 Porto, Portugal.
email: jarle@astro.up.pt

Abstract. He II λ1640 emission in the absence of other metal lines is the most sought-after emission line to detect and characterize metal free stellar populations. However, even recent stellar population models with sophisticated treatment of stellar evolution also lack sufficient He^+ ionising photons to reproduce observed He II fluxes. We use VLT/MUSE GTO observations to compile a catalogue of 15 $z \sim 2 - 4$ He II λ1640 emitters from $\sim 10 - 30$ hour pointings. We show that both He II λ1640 detections and non-detections occupy similar distribution in UV absolute magnitudes. Rest-UV emission line analysis of our sample shows that the emission lines of our He II λ1640 emitters are driven by star-formation in solar to moderately sub-solar (∼1/20th) metallicity conditions. However, we find that even after considering effects from binary stars, we are unable to reproduce the He II λ1640 equivalent widths. Alternative mechanisms are necessary to compensate for the missing He^+ ionising photons.

Keywords. galaxies: high-redshift, galaxies: ISM, ultraviolet: ISM

1. Introduction

Several works have tried to obtain observational signatures for the existence of metal-free (pop-III) stars in the early Universe via current ground and space based telescopes (e.g., Cassata *et al.* 2013; Sobral *et al.* 2015). The failure to obtain observational evidence for these first generation of stars has been attributed to reasons such as the short life-time (∼ 1Myr) of pop-III systems, photometric calibration, presence of active galactic nuclei (AGN), pristine cold mode gas accretion to galaxies, and limited understanding of high-redshift stellar populations and the inter-stellar-medium (ISM). All these effects have contributed in varying degrees to the complexity of detecting and identifying pop-III host systems (Fardal *et al.* 2001; Yang *et al.* 2006; Sobral *et al.* 2015; Agarwal *et al.* 2016; Bowler *et al.* 2017; Matthee *et al.* 2017; Shibuya *et al.* 2017; Sobral *et al.* 2018).

With large samples of high-z galaxies, candidates for galaxies containing a significant population of pop-III stars can be selected due to the presence of strong Lyα and He II emission lines in the absence of other prominent emission features. These lines can be interpreted as existence of pristine metal poor stellar populations (Tumlinson *et al.* 2003; Raiter *et al.* 2010; Inoue *et al.* 2011; Sobral *et al.* 2015). This interpretation is however challenging in the face of other processes that can produce He^+ ionising photons (E> 54.4 eV, $\lambda <$ 228 Å). Therefore, to make compelling constraints of stellar populations in the presence of strong He II emission and link with pop-III hosts, a comprehensive understanding of He II emission mechanisms is required.

Stellar populations in a variety of ages and physical/chemical conditions undergo various mechanisms that may contribute to He II emission, such as young O/B type stars (e.g., Shirazi & Brinchmann 2012), hydrogen-stripped massive evolved Wolf-Rayet stars (e.g., Shirazi & Brinchmann 2012), very massive low-Z WNh stars (hydrogen rich WN stars; Gräfener & Vink 2015), post-asymptomatic giant branch stars (e.g., Binette *et al.* 1994), X-ray binary stars (e.g., Casares *et al.* 2017), radiative shocks (e.g., Izotov *et al.* 2012), and AGN (e.g., Shirazi & Brinchmann 2012). Binary interactions and stellar rotation prolong the lifetime of young O/B stars extending the total amount of He^+ photons present at a given star-formation history (e.g., Eldridge *et al.* 2017; Götberg *et al.* 2017). Even with a variety of such mechanisms, we still lack E> 54.4 eV photons in stellar population models to produce observed He II λ1640 line profiles consistently with other rest-UV emission lines (e.g., Shirazi & Brinchmann 2012; Senchyna *et al.* 2017; Berg *et al.* 2018).

2. Data & Analysis

The new generation of sensitive multiplexed optical instruments in 8-10m class telescopes such as the The Multi Unit Spectroscopic Explorer (MUSE; Bacon *et al.* 2010) has enabled astronomers to obtain spatially-resolved spectroscopy of galaxies throughout cosmic time in unprecedented numbers (e.g., Inami *et al.* 2017). Here, we present a sample of 15 He II λ1640 detections (including 3 AGN) obtained from deep $\sim 10-30$ hour pointings as a part of multiple MUSE guaranteed time observation programs (Bacon *et al.* 2015, 2017; Epinat *et al.* 2018; Marino *et al.* 2018) between $z = 1.93 - 4.67$. The Universe at $z \sim 2-4$ was reaching the peak of the cosmic star-formation rate density (Madau & Dickinson 2014), with galaxies in a diverse range of physical and chemical properties (e.g., Kacprzak *et al.* 2016; Kewley *et al.* 2016; Steidel *et al.* 2016; Nanayakkara *et al.* 2016, 2017; Strom *et al.* 2017). With MUSE we are able to obtain rest-UV spectroscopy of young, low-metallicity, highly star-forming systems which may give rise to a diverse range of exotic phenomena capable of producing high-energy ionizing photons.

We remove AGN from our sample and use multiple emission line diagnostics from Gutkin *et al.* (2016) and Xiao *et al.* (2018) to explore the ISM conditions of the He II λ1640 emission in the non-active galaxy sample. Additional details on sample selection are given in Nanayakkara *et al.*, (submitted). In Figure 1 (left panel), we show the C III]/He II λ1640 vs O III]/He II λ1640 line ratio diagrams for single-star stellar population models from Gutkin *et al.* (2016). Our values agree with literature data of high-z sources (Patrício *et al.* 2016; Berg *et al.* 2018) and have considerably lower metallicities compared to $z = 0$ sources from Senchyna *et al.* (2017). In this line ratio diagram, our galaxies occupy a region, that can be described by star-forming galaxies with solar to $\sim 1/20$th solar metallicities. We note that, when effects of binary stellar evolution are considered, the line-ratio diagnostics become more degenerate (also see Xiao *et al.* 2018). However, the line-ratios of our He II λ1640 emitters fall within the range powered by star-formation.

The main discrepancy between model and data arise only once line EWs are compared. As shown in the right panel of Figure 1, Xiao *et al.* (2018) binary models are able to reproduce the observed C III] EWs but lacks sufficient He^+ ionising photons to reproduce the observed He II λ1640 EWs, which is expected to be driven by the lack of photons below $\lambda < 228$ Å in BPASS models (e.g., Berg *et al.* 2018). By matching observed C III] luminosities to model C III] luminosities, we find that only extreme sub-solar metallicities ($\sim 1/200$th) are able to accurately predict the observed He^+ ionising photons in BPASS models, which is strongly in contrast with predictions from line-ratio diagnostics.

3. Conclusions & Future directions

We used deep optical spectroscopy from MUSE to obtain a sample of He II λ1640 detections at $z \sim 2-4$ to study their ISM conditions using state-of-the-art stellar-population/

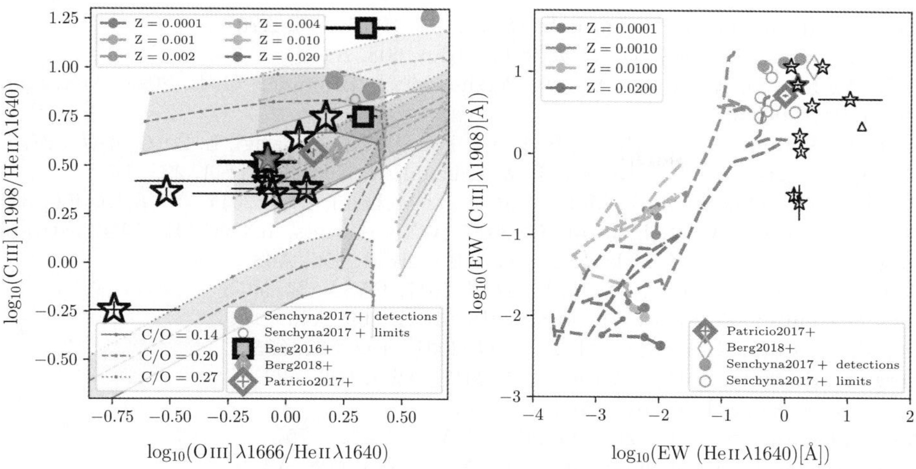

Figure 1. Left: Rest-frame C III]/He II λ1640vs O III]/He II λ1640 emission line ratios of the MUSE He II λ1640 sample. Only galaxies with SNR>3 for all four emission lines are shown here. The tracks are from Gutkin *et al.* (2016) models. Each set of tracks with similar colour resemble three C/O ratios and the region between the minimum and maximum C/O tracks are shaded by the same colour. From top to bottom the ionization parameter increases. Line ratios from Patrício *et al.* (2016), Senchyna *et al.* (2017), and Berg *et al.* (2016, 2018) are shown for comparison. MUSE line ratios of the Lyman continuum emitter from Naidu *et al.* (2017) is shown by the filled star. **Right:** C III] vs He II λ1640 equivalent widths of Xiao *et al.* (2018) models for a star-burst stellar population. Model parameters are similar to the left panel.

photo-ionization models. Emission line ratios of our He II λ1640 emitters could mostly be explained by Z$\sim 0.05 - 1.0$ $Z_\odot$ photo-ionisation models, but, even the BPASS binary models lack sufficient ionising photons to re-produce observed He II λ1640 EWs. In order to reproduce the observed He II λ1640 luminosities, BPASS stellar population models with extreme sub-solar metallicities ($\sim 1/200$th) are required. Such low metallicities are in contradiction with our line-ratio diagnostics and stellar populations models can suffer large uncertainties due to lack of empirical calibrations in this regime. Extra contribution to the number of ionizing photons from X-Ray binaries, sub-dominant AGN, or effects related to stellar rotations at high metallicities might be necessary to alleviate the tension between models and observations. In addition, top-heavy initial-mass-functions in star-forming galaxies (e.g., see Nanayakkara *et al.* 2017), will contribute to higher levels of ionising photons, which could increase the He^+ ionising photon budget.

Future deep surveys such as the MUSE extreme deep field survey (PI R. Bacon), a single 160 hour pointing by MUSE, along with several deep pointings in *Hubble* frontier field parallels (PI L. Wisotzki) will provide extremely high signal-to-noise rest-UV spectra at $z = 2 - 4$ to perform spectro-photometric analysis by simultaneous combination of nebular emission features with weaker ISM and photospheric emission and absorption features. By constraining the stellar population properties to finer detail within this epoch, we will be able to make accurate predictions for future surveys in the era of *James Webb Space Telescope*. Given that individual detections of pop-III stars will be unlikely until proposed future space telescopes such as LUVOIR, we should push the current instruments to their maximum potential to constrain the stellar population properties of galaxies leading to the buildup of the peak of the cosmic star-formation rate density.

References

Agarwal, B., Johnson, J. L., Zackrisson, E., *et al.* 2016, *MNRAS*, 460, 4003

Bacon, R., Accardo, M., Adjali, L., *et al.* 2010, in *Proc. SPIE*, Vol. 7735, Ground-based and Airborne Instrumentation for Astronomy III, 773508

Bacon, R., Brinchmann, J., Richard, J., *et al.* 2015, *A&A*, 575, A75
Bacon, R., Conseil, S., Mary, D., *et al.* 2017, *A&A*, 608, A1
Berg, D. A., Erb, D. K., Auger, M. W., Pettini, M., & Brammer, G. B. 2018, ArXiv e-prints, arXiv:1803.02340
Berg, D. A., Skillman, E. D., Henry, R. B. C., Erb, D. K., & Carigi, L. 2016, *ApJ*, 827, 126
Binette, L., Magris, C. G., Stasińska, G., & Bruzual, A. G. 1994, *A&A*, 292, 13
Bowler, R. A. A., Dunlop, J. S., McLure, R. J., & McLeod, D. J. 2017, *MNRAS*, 466, 3612
Casares, J., Jonker, P. G., & Israelian, G. 2017, ArXiv e-prints, arXiv:1701.07450 [astro-ph.HE]
Cassata, P., Le Fèvre, O., Charlot, S., *et al.* 2013, *A&A*, 556, A68
Eldridge, J. J., Stanway, E. R., Xiao, L., *et al.* 2017, PASA, 34, e058
Epinat, B., Contini, T., Finley, H., *et al.* 2018, *A&A*, 609, A40
Fardal, M. A., Katz, N., Gardner, J. P., *et al.* 2001, *ApJ*, 562, 605
Götberg, Y., de Mink, S. E., & Groh, J. H. 2017, *A&A*, 608, A11
Gräfener, G., & Vink, J. S. 2015, *A&A*, 578, L2
Gutkin, J., Charlot, S., & Bruzual, G. 2016, *MNRAS*, 462, 1757
Inami, H., Bacon, R., Brinchmann, J., *et al.* 2017, *A&A*, 608, A2
Inoue, A. K., Kousai, K., Iwata, I., *et al.* 2011, *MNRAS*, 411, 2336
Izotov, Y. I., Thuan, T. X., & Privon, G. 2012, *MNRAS*, 427, 1229
Kacprzak, G. G., van de Voort, F., Glazebrook, K., *et al.* 2016, *ApJL*, 826, L11
Kewley, L. J., Yuan, T., Nanayakkara, T., *et al.* 2016, *ApJ*, 819, 100
Madau, P., & Dickinson, M. 2014, *ARA&A*, 52, 415
Marino, R. A., Cantalupo, S., Lilly, S. J., *et al.* 2018, *ApJ*, 859, 53
Matthee, J., Sobral, D., Boone, F., *et al.* 2017, *ApJ*, 851, 145
Naidu, R. P., Oesch, P. A., Reddy, N., *et al.* 2017, *ApJ*, 847, 12
Nanayakkara, T., Brinchmann, J., & The MUSE Collaboration. 2018, arXiv e-prints, arXiv:1809.10970
Nanayakkara, T., Glazebrook, K., Kacprzak, G. G., *et al.* 2016, *ApJ*, 828, 21
—. 2017, *MNRAS*, 468, 3071
Patrício, V., Richard, J., Verhamme, A., *et al.* 2016, *MNRAS*, 456, 4191
Raiter, A., Schaerer, D., & Fosbury, R. A. E. 2010, *A&A*, 523, A64
Senchyna, P., Stark, D. P., Vidal-García, A., *et al.* 2017, ArXiv e-prints, arXiv:1706.00881
Shibuya, T., Ouchi, M., Harikane, Y., *et al.* 2017, ArXiv e-prints, arXiv:1705.00733
Shirazi, M., & Brinchmann, J. 2012, *MNRAS*, 421, 1043
Sobral, D., Matthee, J., Darvish, B., *et al.* 2015, *ApJ*, 808, 139
Sobral, D., Matthee, J., Brammer, G., *et al.* 2018, *MNRAS*, 2683
Steidel, C. C., Strom, A. L., Pettini, M., *et al.* 2016, *ApJ*, 826, 159
Strom, A. L., Steidel, C. C., Rudie, G. C., *et al.* 2017, *ApJ*, 836, 164
Tumlinson, J., Shull, J. M., & Venkatesan, A. 2003, *ApJ*, 584, 608
Xiao, L., Stanway, E. R., & Eldridge, J. J. 2018, *MNRAS*, 477, 904
Yang, Y., Zabludoff, A. I., Davé, R., *et al.* 2006, *ApJ*, 640, 539

Discussion

YUICHI HARIKANE: Do you think the strong He II λ1640 emission in Berg *et al.* (2018) can be explained if you assume ISM metallicity to be different to that of the stars?

THEMIYA NANAYAKKARA: He II λ1640 emission strength does show a dependence on stellar metallicity with BPASS models predicting a higher amount of He^+ ionising photons at low metallicities (also see Figure 3 of Nanayakkara *et al.* 2018). So if the stellar metallicity is low compared to the interstellar medium (as also argued by Steidel *et al.* 2016) it is possible that we might be under-predicting the amount of He^+ ionising photons produced by the stars in our systems. However, as I showed during my talk a $\sim 1/200\ Z_\odot$ metallicities are required to alleviate the tensions between the models and data, which might be somewhat unrealistic for UV bright systems at $z \simeq 1.6$. So additional changes in stellar populations might also be necessary.

TOMO GOTO: How do you explain the existence of the galaxy at the extreme lower left in the line ratio plot (Figure 1 left panel)?

NANAYAKKARA: The models by Gutkin *et al.* (2016) have quite a few free parameters, which can be altered to reproduce this line ratio. Most ISM and stellar properties at high-z are not well constrained. Simultaneous analysis of rest-frame UV/optical emission and absorption is required to constrain these properties, thus data from very deep surveys such as the MUSE extreme deep field and future *James Webb Space Telescope* observations will be crucial to properly understand these systems.

Challenges in Panchromatic Modelling with Next Generation Facilities
Proceedings IAU Symposium No. 341, 2018
M. Boquien, E. Lusso, C. Gruppioni & P. Tissera, eds.

doi:10.1017/S1743921319004757

Panchromatic study of the first galaxies in cosmological simulations

Hidenobu Yajima[1], **Shohei Arata**[2], **Makito Abe**[1] **and Kentaro Nagamine**[2,3,4]

[1]Center for Computational Sciences, University of Tsukuba, 1-1-1 Tennodai, Tsukuba, Ibaraki 305-8577, Japan
email: yajima@ccs.tsukuba.ac.jp

[2]Department of Earth and Space Science, Graduate School of Science, Osaka University, Toyonaka, Osaka 560-0043, Japan

[3]Department of Physics & Astronomy, University of Nevada, Las Vegas, 4505 S. Maryland Pkwy, Las Vegas, NV 89154-4002, USA

[4]Kavli IPMU (WPI), The University of Tokyo, 5-1-5 Kashiwanoha, Kashiwa, Chiba, 277-8583, Japan

Abstract. Recent discoveries of high-redshift galaxies have revealed the diversity of their physical properties, from normal star-forming galaxies to starburst galaxies. To understand the properties of these observed galaxies, it is crucial to understand the star formation (SF) history, and the radiation properties associated with the SF activity. Here we present the results of cosmological hydrodynamic simulations with zoom-in initial conditions, and show the formation of the first galaxies and their evolution towards observable galaxies at $z = 6$. In addition, we show their multi-wavelength radiative properties. We find that star formation occurs intermittently due to supernova (SN) feedback at $z > 10$, and their radiation properties rapidly change with time. We suggest that the first galaxies are bright at UV wavelengths just after the starburst phase, and become extended Lyman-alpha sources. We also show that massive galaxies cause dusty starburst and become bright at infrared wavelengths.

Keywords. galaxies: formation, radiative transfer, methods: numerical

1. Introduction

Understanding galaxy formation is one of the main goals of today's astronomy. Recent observations have allowed us to detect distant galaxies in the early Universe (e.g., Finkelstein *et al.* 2013). The most distant galaxy confirmed by the observation of line emission has reached $z = 9.1$ (Hashimoto *et al.* 2018), and recent deep surveys are providing large samples of galaxies in the era of cosmic reionization at $z \gtrsim 6$ (e.g., Ouchi *et al.* 2018). In the next decade, the next generation telescopes, e.g., James Webb Space Telescope (JWST), Thirty Meter Telescope (TMT) and European Extremely Large Telescope (E-ELT) are going to study star formation in the first galaxies at $z \gtrsim 10$ focusing on UV continuum, Lyman-α line, and Hα line. Thus, it is vital to model the formation of galaxies in the current theoretical framework before the next generation telescopes will start to work. In addition, the first galaxies can be considered to be main ionizing sources for cosmic reionization (e.g., Yajima *et al.* 2011). Therefore, radiation properties of the first galaxies would also provide indispensable information for the cosmic reionization history.

2. Method

We use the smoothed particle hydrodynamics (SPH) code GADGET-3 (Springel 2005) with sub-grid models developed in the *First Billion Year* (FiBY) project (e.g., Khochfar & Dalla Vecchia *et al.* in prep.). We here use the star formation model based on the Kennicutt-Schmidt law (Schaye & Dalla Vecchia 2008), and the stochastic thermal feedback model for supernovae (Dalla Vecchia & Schaye 2012).

We make three zoom-in initial conditions that produce halos with $M_{\rm h} = 2.4 \times 10^{10}\, h^{-1}{\rm M}_{\odot}$ (Halo-10), $1.6 \times 10^{11}\, h^{-1} M_{\odot}$ (Halo-11) and $7.5 \times 10^{11}\, h^{-1}{\rm M}_{\odot}$ (Halo-12) at $z = 6$. The entire simulation box size is $(20\, h^{-1}{\rm Mpc})^3$ for Halo-10 and Halo-11, and $(100\, h^{-1}{\rm Mpc})^3$ for Halo-12 in a comoving units. In order to investigate the impacts of star formation and stellar feedback, using the initial condition of Halo-11, we also carry out other simulations with the lower star formation efficiency (Halo-11lowSF) and without stellar feedback (Halo-11noSN). In this work, we set the constant gravitational softening length of 200 h^{-1} pc in comoving scale and the minimum smoothing length of 20 h^{-1} pc.

As post-processing, we carry out the radiative transfer calculations for modeled galaxies by using the radiative transfer code, All-wavelength Radiative Transfer with Adaptive Refinement Tree ($\rm ART^2$:Yajima *et al.* 2012). $\rm ART^2$ is based on a Monte Carlo technique. We assume the Chabrier IMF with the mass range from 0.1 to 100 $\rm M_{\odot}$. $\rm ART^2$ uses an adaptive refinement grid structure. We set the maximum refinement level to 12 which achieves the resolution of 2.7 h^{-1}pc for Halo-11 at $z = 6$. First, we calculate the ionization structure of inter-stellar medium in galaxies. Then Lyα photon transfer is calculated with the emissivity based on the pre-calculated ionization structure. Finally we calculate the radiative transfer of stellar continuum and dust thermal emission. The dust temperature is estimated under the assumption of radiative equilibrium state. Thus our radiation transfer simulations show the multi-wavelength radiative properties of galaxies.

3. Results & Discussion

Figure 1 presents the maps of projected gas density for the Halo-11, Halo-11lowSF and Halo-11noSN at redshift $z = 7$. In the case of Halo-11 in our fiducial model, the gas structure shows highly inhomogeneous and clumpy features. No extended galactic gas disk is seen within the central ~ 2 kpc of Halo-11. On the other hand, disk-like structures are seen in Halo-11-noSN and Halo-11-lowSF models. This suggests the SF efficiency or the SN feedback have a large impact on the galactic morphology. The star formation is rather concentrated in the central nuclear region for the Halo-11-noSN, but the stellar distribution is more extended than the other runs due to larger amount of stars produced by $z = 7$. As a result, both SF efficiency and SN feedback affect the stellar distribution as shown in the lower panels of the figure.

Star formation histories are presented in Figure 2. At $z \gtrsim 10$, star formation is almost completely quenched due to stellar feedback after experiencing a star formation event with SFR $\sim 0.1\ {\rm M_{\odot}\ s^{-1}}$ in Halo-11. As the SFR decreases, SN feedback ceases to work. This allows the gas to fall down towards the galactic center, which leads to the next starburst. Repeating this cycle causes the oscillation of SFR. The peak and the minimum of SFR is different by a factor of $\gtrsim 2$ dex at $z \gtrsim 10$.

We find that a more massive halo (Halo-12) shows a rapid variation in SFR, similarly to Halo-11 at $z \gtrsim 10$. After the star formation at $z = 12.2$, the SN feedback evacuates the gas and quenches the star formation for a while until $z = 10.5$. As halos become more massive towards lower redshift, the deeper gravitational potential well can hold the

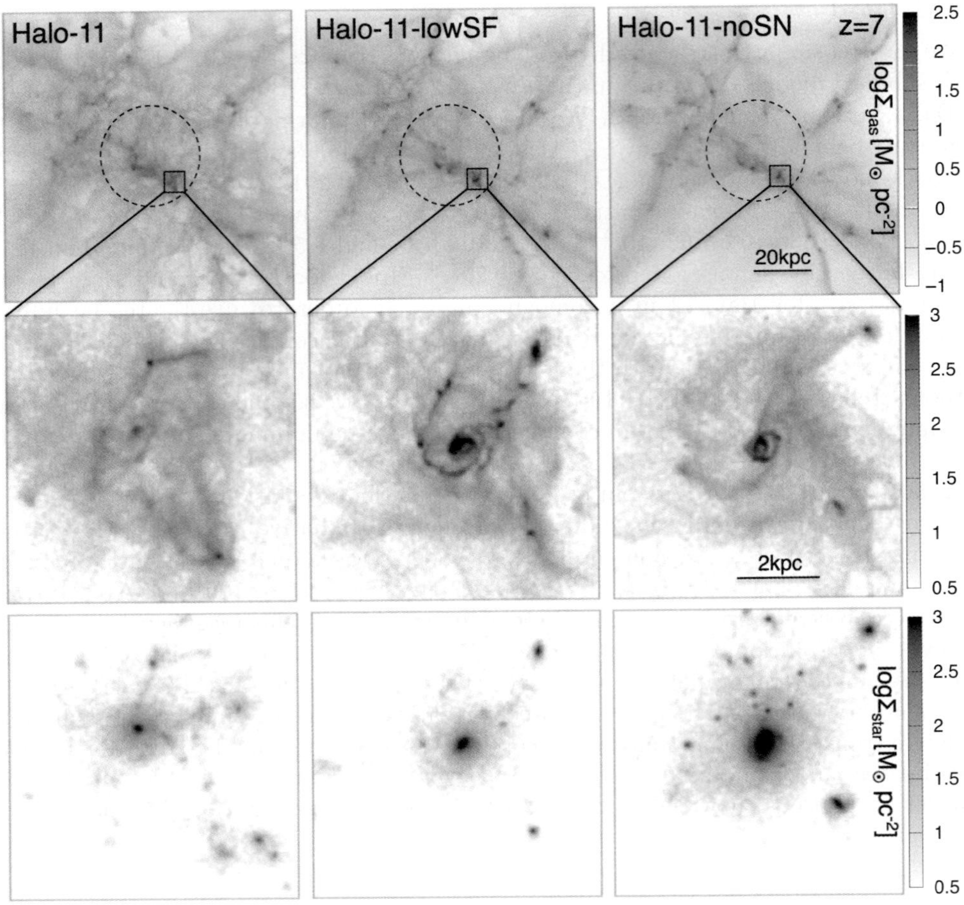

Figure 1. Column density of gas in Halo-11, Halo-11-lowSF, and Halo-11-noSN in units of $[M_\odot\,pc^{-2}]$ at $z = 7$ (Figure 2 in Yajima *et al.* 2017). The top panels show the filamentary gas structure on a scale of physical 100 kpc, and the dashed circles show the virial radius of the most massive galaxy ≈ 18 kpc in the simulation boxes. The middle panels show the zoom-in view of the most massive galaxy in the halo with a scale of physical 7 kpc. The lower panels represent the stellar surface density with the same logarithmic scale as the middle panels.

gas against SN feedback, which allows for more continuous SF at $z \lesssim 10$. We find that Halo-11 and Halo-12 show similar SFRs as the observed LAEs at $z \sim 7$ and the SMG at $z = 7.5$ discovered by ALMA (Watson *et al.* 2015).

The escape fraction of UV continuum is shown in the second panel of Figure 2. At higher redshift $z > 10$, most of UV photons can escape because of low dust contents in galaxies. As the galaxy evolution proceeds, dusty gas covers star-forming regions and absorbs UV photons. The third and fourth panels show the fluxes at infrared and UV wavelengths in the rest frame. At lower redshifts, galaxies become bright at the infrared due to the dust absorption of UV photons. Our simulations show that ALMA would be able to detect the dust thermal emission from galaxies with the halo mass of $\gtrsim 10^{11}\ M_\odot$ at $z \lesssim 9$. UV fluxes of galaxies increase with SFR as redshift decreases. We find that the bright wavelengths of galaxies rapidly change between UV and infrared due to the fluctuations of SFR and the escape fraction.

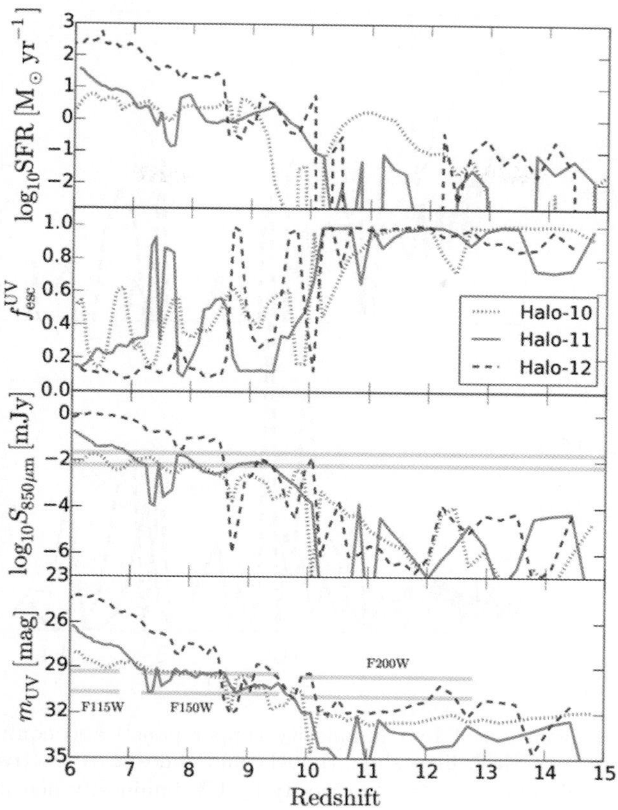

Figure 2. Redshift evolution of SFR (top panel), UV escape fraction (second panel), sub-mm flux (third panel) and emergent UV flux (bottom panel) are shown for Halo-10 (green dotted), Halo-11 (red solid), and Halo-12 (blue dashed). The gray horizontal lines in the third and forth panels show 3σ (lower line) and 10σ (higher line) detection thresholds for ALMA (full operation) and JWST with 10 hours time integration, which we obtained from the online calculators of each observatory.

Figure 3 represents the redshift evolution of Lyα luminosity ($L_{\mathrm{Ly}\alpha}$) and equivalent width (EW). We find that $L_{\mathrm{Ly}\alpha}$ fluctuates with the SF history. Halo-12 shows $L_{\mathrm{Ly}\alpha} \sim 10^{43}$ erg s^{-1} at $z \lesssim 7$, which reproduces observed bright LAEs (e.g., Finkelstein *et al.* 2013). Because of the lower SFR, $L_{\mathrm{Ly}\alpha}$ of Halo-11 is smaller than that of Halo-12 by a factor of ~ 10. At higher redshift $z \gtrsim 10$, the escape fraction of Lyα photons is almost unity because of the low dust contents in galaxies. Therefore, $L_{\mathrm{Ly}\alpha}$ depends on SFR and escape fraction of ionizing photons. At lower redshift $z < 10$, inter-stellar dust absorb Lyα photons. As SFR and the escape fraction fluctuate, EW also changes significantly. At the transition from star burst to quenched phases, EW increases rapidly. This is because galactic wind caused by supernovae allows Lyα photons to escape. At lower redshift $z \lesssim 8$, EW keeps low values of $\lesssim 40$ Å due to the large amounts of gas and dust.

Thus we show the multi-wavelength radiative properties of galaxies in the early Universe by combining cosmological hydrodynamics simulations and radiative transfer calculations. We find SFR and the radiative properties of the first galaxies rapidly change with time due to the stellar feedback. We show that massive galaxies with the halo masses of $\sim 10^{11} - 10^{12}$ M$_\odot$ reproduce observed galaxies successfully.

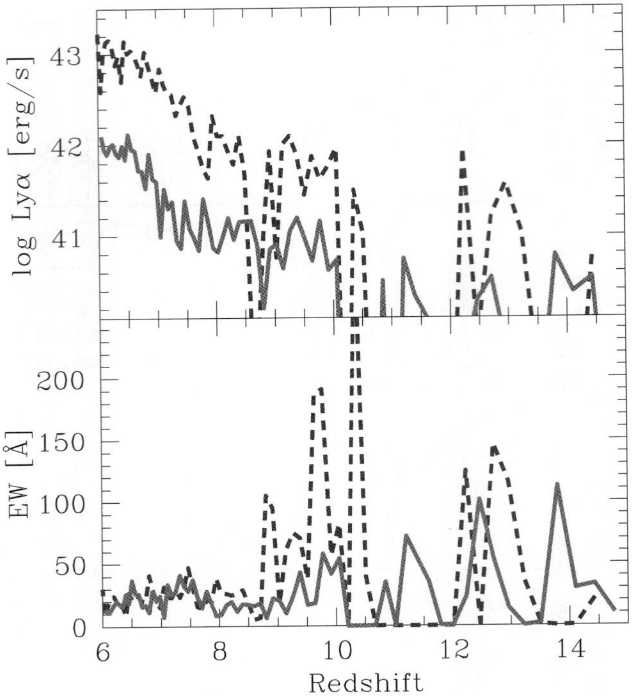

Figure 3. Redshift evolution of Lyα luminosity (upper panel) and equivalent width (lower panel). Red solid and blue dash lines show Halo-11 and Halo-12, respectively. The equivalent width is estimated by dividing the Lyα luminosity by UV luminosity density at 1300 Å in the rest frame.

References

Dalla Vecchia, C. & Schaye, J. 2012, *MNRAS*, 426, 140
Finkelstein, S. L., *et al.* 2013, *Nature*, 502, 524
Hashimoto, T. *et al.* 2018, *Nature*, 557, 392
Ouchi, M. *et al.* 2018, *PASJ*, 70, 13
Schaye, J. & Dalla Vecchia, C. 2008, *MNRAS*, 383, 1210
Springel, V. 2005, *MNRAS*, 364, 1105
Yajima, H., Choi, J. -H., & Nagamine, K. 2011, *MNRAS*, 412, 411
Yajima, H., Li, Y., Zhu, Q., & Abel, T. 2012, *MNRAS*, 424, 884
Yajima, H., Nagamine, K., Zhu, Q., Khochfar, S., & Dalla Vecchia, C. 2017, *ApJ*, 846, 30
Watson, D. *et al.* 2015, *Nature*, 519, 327

Challenges in Panchromatic Modelling with Next Generation Facilities
Proceedings IAU Symposium No. 341, 2018
M. Boquien, E. Lusso, C. Gruppioni & P. Tissera, eds.

doi:10.1017/S1743921319002175

FIRSTLIGHT: Cosmological simulations of first galaxies at cosmic dawn

Daniel Ceverino

Universitat Heidelberg, Zentrum für Astronomie, Institut fur Theoretische Astrophysik, Albert-Ueberle-Str. 2, 69120 Heidelberg, Germany

Abstract. Using the FirstLight database of 300 zoom-in cosmological simulations we provide rest-frame UV-optical spectral energy distributions of galaxies with complex star-formation histories that are coupled to the non-uniform gas accretion history of galactic halos during cosmic dawn. The population at any redshift is very diverse ranging from starbursts to quiescent galaxies even at a fixed stellar mass. The FirstLight simulations make predictions on the rest-frame UV-optical absolute magnitudes, colors and optical emission lines of galaxies at $z = 6 - 12$ that will be observed for the first time with JWST and the next generation of telescopes in the coming decade.

Keywords. galaxies: evolution – galaxies: formation – galaxies: high-redshift

1. Introduction

The period of cosmic dawn, the first billion years in the history of the Universe, is the final frontier for galaxy formation. Theory roughly predicts that the first galaxies form in the first gravitationally bound structures at high redshifts, $z \geqslant 12$. They quickly start to ionize their surroundings, driving the reionization of the Universe, which ends around $z \simeq 6$. However, little is known about these primeval galaxies. One of their most important properties is their spectral energy distribution (SED). In this talk, I will discuss SEDs coming from the FirstLight database of cosmological simulations of first galaxies at cosmic dawn, $z = 5 - 15$ (Ceverino, Glover, Klessen 2017). These cosmological simulations provide complex SFHs that are coupled with the non-uniform gas accretion history of galactic halos (Section 2 and Ceverino, Klessen, Glover 2018). The simulated SEDs are therefore consistent with the cosmological growth of structures at early times (Section 3). The final section discusses some results that can be found at Ceverino *et al.* (2018), such as BPT diagrams.

2. Dwarfs at Cosmic Dawn ($z = 6 - 15$)

Galaxies at cosmic dawn look very compact (Fig. 1). The stellar half-mass radius of this example is only 0.5 kpc. This is a factor 10 smaller than local galaxies with a similar mass. They are gas rich and they have typical SFR $\simeq 20\, M_\odot\, \mathrm{yr}^{-1}$, a factor 10 higher than their counterparts at $z = 0$. This drives a clumpy and turbulence gas morphology, more similar to local starbursts. We have used the FirstLight database to study the Star Formation (SF) histories of ∼300 galaxies with a stellar mass between $M_* = 10^6$ and $3 \times 10^9 M_\odot$ during cosmic dawn ($z = 5 - 15$).

The evolution of the SFR in each galaxy is complex and diverse, characterized by bursts of SF. Overall, first galaxies spend about 70% of their time undergoing SF bursts at $z > 5$. This diversity sets the mean and scatter of the SFMS at $z \simeq 5 - 13$. High gas fractions and short gas depletion times are common during the SF bursts. The typical

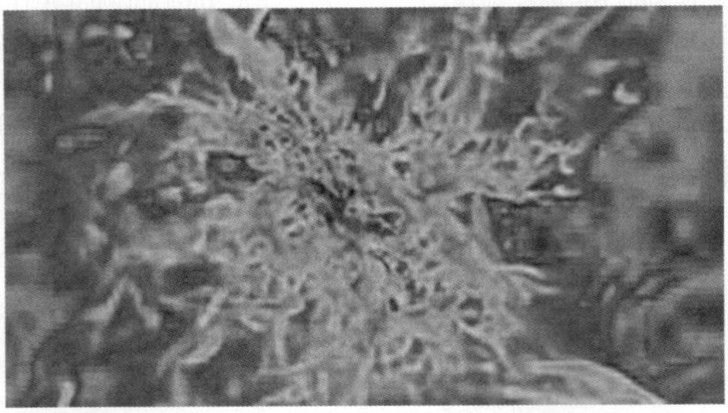

Figure 1. Gas distribution of a dwarf galaxy at $z = 6$. The horizontal size of the image is 4 kpc.

bursts at $z \simeq 6$ have a sSFR maximum of $5 - 15\,\mathrm{Gyr}^{-1}$ with a duration of ~100 Myr, one tenth of the age of the Universe. A quarter of the bursts populate a tail with very high sSFR maxima of $20 - 30\,\mathrm{Gyr}^{-1}$ and significantly shorter time-scales of $\sim 40 - 80$ Myr at all masses. The mean period of time between consecutive bursts is ~200 Myr with a small mass dependence at $z \simeq 6$. The mean sSFR increases with redshift approximately as $sSFR \propto (1+z)^{5/2}$ at all masses, as predicted by ΛCDM models. This is consistent with existing observations at $z \leq 8$. The typical sSFR height of a SF burst also increases with redshift, but it is always a factor 2 ± 0.5 higher than the mean sSFR at that redshift. This implies typical sSFR maxima of $sSFR_{\max} = 20 - 30\,\mathrm{Gyr}^{-1}$ at $z = 9 - 10$. The tail of the distribution reaches $sSFR_{\max} \simeq 60\,\mathrm{Gyr}^{-1}$ at these high redshifts. This evolution is driven by shorter time-scales at higher redshifts, proportional to the age of the Universe. These galaxies are the most efficient star formers in the history of the Universe.

3. Example SEDs

In this talk we focus on the intrinsic SEDs coming from the stellar populations and their surrounding, unresolved HII regions. They provide templates that could be used for a better understanding of the underlying stellar and gas properties in galaxies at cosmic dawn. Therefore, we ignore dust attenuation or any other radiative transfer effect from intervening gas. These effects will be consider in future works. We compute the SEDs of the simulated galaxies using publicly available tables from the Binary Population and Spectral Synthesis (BPASS) model (Eldrige *et al.* 2017) including nebular emission (Xiao *et al.* 2018).

Figure 2 shows two examples which represent the same galaxy with a stellar mass of $M_* \simeq 10^8 M_\odot$ at $z \simeq 6$, at two different phases of evolution. The left panel depicts the galaxy at the peak of a burst of star formation. The starburst brings the galaxy a factor 6 times above the star forming main sequence at that redshift (SSFR=30 Gyr^{-1}). Its luminosity at 1500 $\mathring{A}$ is correspondingly high, $M_{1500} \simeq -19$. The nebular emission dominates the SED at all wavelengths. Lines, such as Lyα, Hα, OIII or SiIII, are particularly prominent. For example, the equivalent width (EW) of Hα+NII exceeds 1000 $\mathring{A}$. OIII+Hβ is the next prominent feature with EW$\simeq 700\mathring{A}$. These lines contaminate the continuum in the B and R bands. The nebular continuum is also particularly important in the U band and it is able to completely remove the mass -sensitive Balmer break at 4000 $\mathring{A}$ The stellar continuum is also very steep, consistent with a young stellar population with extremely blue colors, $V - I = -0.2$ and $U - V = -0.4$. These are the typical features of an extreme emission-line galaxy at cosmic dawn. The SED of the same galaxy 200 Myr after the starburst looks very different (right panel of Figure 2). The starburst

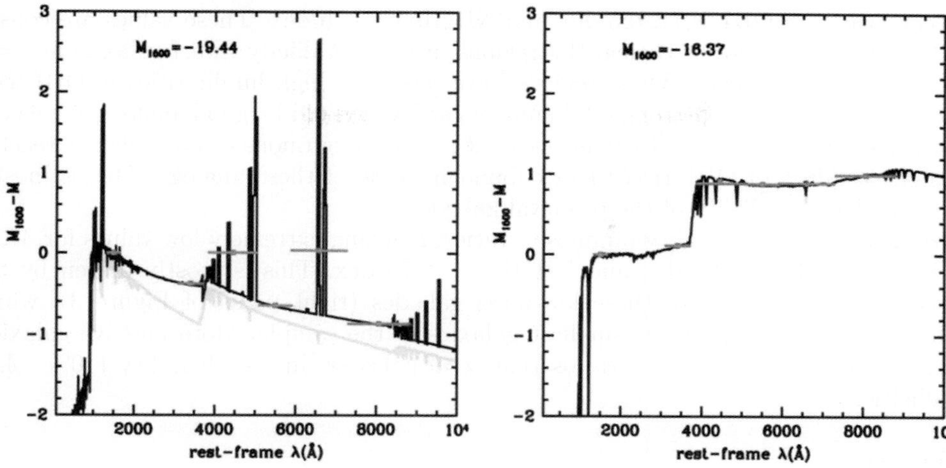

Figure 2. rest-frame SEDs of the same galaxy at the peak of a star formation burst at $z \simeq 6$ (left) and during the subsequent quiescent phase 200 Myr later (right). Cyan lines represent stellar light and black lines include nebular emission (lines+continuum). Red circles with bars represent six photometric bands. The monochromatic luminosity at each wavelength in erg/s/Hz is normalized to the absolute magnitude at 1500 $\AA$ (top label).

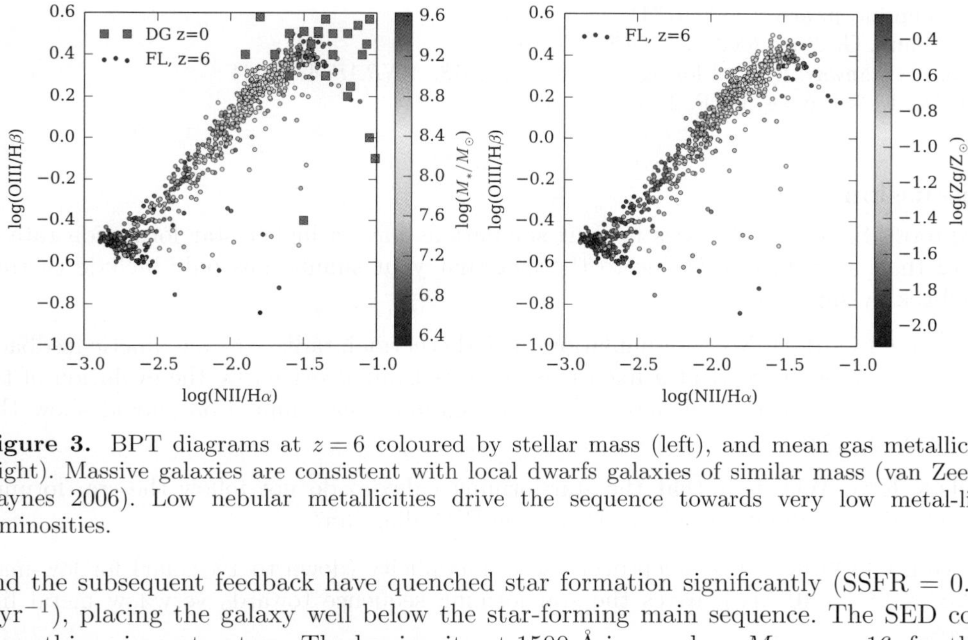

Figure 3. BPT diagrams at $z = 6$ coloured by stellar mass (left), and mean gas metallicity (right). Massive galaxies are consistent with local dwarfs galaxies of similar mass (van Zee & Haynes 2006). Low nebular metallicities drive the sequence towards very low metal-line luminosities.

and the subsequent feedback have quenched star formation significantly (SSFR = 0.02 Gyr^{-1}), placing the galaxy well below the star-forming main sequence. The SED confirms this quiescent nature. The luminosity at 1500 Å is very low, $M_{1500} \simeq -16$, for that stellar mass. There is no significant nebular emission. Instead, there is a strong Balmer break with significantly red rest-frame colors, $V - I = 0.2$ and $U - V = 0.8$, typical of a mature population.

4. BPT Diagrams

The nebular emission lines give some clues about the properties of the HII regions in galaxies using line luminosity ratios like OIII(5007 $\AA$)/Hβ and NII(6584 $\AA$)/Hα. In Figure 3 we make predictions of BPT diagrams. The FirstLight galaxies form a clear sequence at $z = 6$. The most massive galaxies, $M_* \simeq 1 - 2 \times 10^9$ M$_\odot$ occupy the tip of the sequence with OIII/H$\beta \simeq 0.5$ dex and NII/H$\alpha \simeq -1.5$ dex. These values are consistent with the line ratios of typical HII regions in low-metallicity dwarf galaxies at $z = 0$ (van Zee & Haynes 2006). These regions have relatively high luminosities in OIII with respect to Hβ. Therefore, strong OIII-emitters at low-z could be good analogs of galaxies responsible for reionization (Fletcher *et al.* 2018), although none of our galaxies reaches the highest values of OIII/H$\beta \simeq 1$ dex found in some of these analogs. This is mostly due to the low-metallicity of the primeval galaxies.

Low-mass galaxies have low luminosity ratios, reaching extremely low values for $M_* \simeq 10^6$ $M_\odot$: OIII/H$\beta \simeq -0.5$ dex and NII/H$\alpha \simeq -2.75$ dex. This is mostly driven by the low nebular metallicities of these low-mass galaxies (right panel of Figure 3), which reach $\log(Z_g/Z_\odot) = -2$ for the smallest galaxies in the sample. More massive galaxies, $M_* \simeq 10^8$ $M_\odot$, still have lower ratios than z=0 galaxies due to their low (10% solar) metallicities.

References

Ceverino, D., Glover, S. C. O., & Klessen, R. S. 2017, *MNRAS*, 470, 2791

Ceverino, D., Klessen, R. S., & Glover, S. C. O. 2018, *MNRAS*, 480, 4842

Ceverino, D., Klessen, R., Glover, S. *et al.* 2018, arXiv:1810.09754

Eldridge, J. J., Stanway, E. R., Xiao, L., McClelland, L. A. S., Taylor, G., Ng, M., Greis, S. M. L., Bray, J. C., *et al.* 2017, Publ. Astron. Soc. Australia, 34, e058

Fletcher, T. J., Robertson, B. E., Nakajima, K., Ellis, R. S., Stark, D. P., Inoue, A., *et al.* 2018, preprint, p. arXiv:1806.01741

Kauffmann, G. *et al.* 2003, *MNRAS*, 346, 1055

Xiao, L., Stanway, E. R., Eldridge, J. J., *et al.* 2018, *MNRAS*, 477, 904

van Zee, L. & Haynes, M. P. 2006, *ApJ*, 636, 214

Discussion

RAPHAEL SADOUN: Galaxies in your simulations have a higher star formation rate at z$\sim$12 than at z$\sim$6. Is this due to the fact that your simulations only include thermal feedback by supernovae?

DANIEL CEVERINO: No, the simulations include thermal, radiative, and kinetic feedback. The increase of the SFR at a fixed mass with redshift is driven by the evolution of the cosmological gas accretion onto halos. Any cosmological simulation should show this evolution at z$\geq$6.

HELEN KIM: Why is it that the star-forming galaxies do not follow the star-forming sequence of Kauffmann *et al.* (2003) on the BPT diagram?

DANIEL CEVERINO: The steep decrease in metallicity (down to 1% solar) for low-mass galaxies ($M_* \sim 10^6 M_\odot$) drives the star-forming sequence towards very low metal line ratios.

VIVIANA ACQUAVIVA: You showed us that the FIRSTLIGHT simulations can successfully reproduce several observational properties such as star formation rates and line ratios. Is there something that the simulation cannot reproduce well?

DANIEL CEVERINO: The UV slope β of bright galaxies ($M_{UV} < -19$) is inconsistent with observations because dust is not included in our spectral energy distributions. This would be addressed in future follow-ups.

Challenges in Panchromatic Modelling with Next Generation Facilities
Proceedings IAU Symposium No. 341, 2018
M. Boquien, E. Lusso, C. Gruppioni & P. Tissera, eds.

doi:10.1017/S1743921319002424

The distribution and physical properties of high-redshift [OIII] emitters in a cosmological hydrodynamics simulation

Kana Moriwaki

Department of Physics, The University of Tokyo, 7-3-1 Hongo, Bunkyo, Tokyo 113-0033, Japan
email: kana.moriwaki@utap.phys.s.u-tokyo.ac.jp

Abstract. Atacama Large Millimeter/submillimeter Array (ALMA) has enabled us to detect [OIII] 88 μm line even at $z > 9$. To study the properties of high-redshift [OIII] emitters, we calculate [OIII] luminosities of galaxies in a cosmological simulation by applying a physical model of HII region and using the photoionization code CLOUDY. We find that the [OIII] 88 μm luminosity, $L_{\mathrm{OIII,88}}$, scales with SFR with slightly larger $L_{\mathrm{OIII,88}}$ than a local relation. Some [OIII] emitters have extended disk-like structure. We propose to use the ratio between [OIII] 88 μm line and [OIII] 5007 Å line, which can be detected with James Webb Space Telescope (JWST), to estimate the gas density and the metallicity in HII region of high-redshift [OIII] emitters.

Keywords. galaxies: formation, galaxies: high-redshift, galaxies: ISM

1. Introduction

It is important to study the high-redshift galaxies at the epoch of reionization. After Hubble Space Telescope has detected many high-redshift galaxy candidates using dropout techniques, the spectroscopic identifications have primarily been done with hydrogen Lyα line. However, Lyα emission becomes increasingly weak at $z > 7$ due to the attenuation by the neutral hydrogen in the intergalactic medium before the completion of the reionization (e.g. Konno *et al.* 2014).

Thanks to the ALMA, we can now use far-infrared (FIR) lines instead of Lyα. While [CII] 158 μm line is the brightest line in local star-forming galaxies, [OIII] 88 μm line is brighter than [CII] 158 μm in local dwarf galaxies (Lebouteiller *et al.* 2012; Cormier *et al.* 2015). Motivated by this, Inoue *et al.* (2014) suggested to use [OIII] 88 μm line for the spectroscopic redshift identification of distant galaxies. Recently, [OIII] 88 μm lines have been detected in several galaxies at $z > 7$ (Inoue *et al.* 2016; Carniani *et al.* 2017; Laporte *et al.* 2017; Hashimoto *et al.* 2018a; Hashimoto *et al.* 2018b; Tamura *et al.* 2018).

Line emission can also be used to study the properties of galaxies. For example, rotation signature can be seen with line-of-sight velocity gradient (e.g. Smit *et al.* 2018). Also, emission line diagnostics using either optical or FIR lines to estimate the property of the interstellar medium are well studied (e.g. Nagao *et al.* 2011). Since ALMA and JWST provide us access to both FIR and optical lines, diagnostics using lines in both wavelength should also be considered.

2. Methods

To study the typical properties of high-redshift [OIII] emitters, we use a cosmological hydrodynamics simulation of Shimizu *et al.* (2016) using GADGET-3. This simulation has

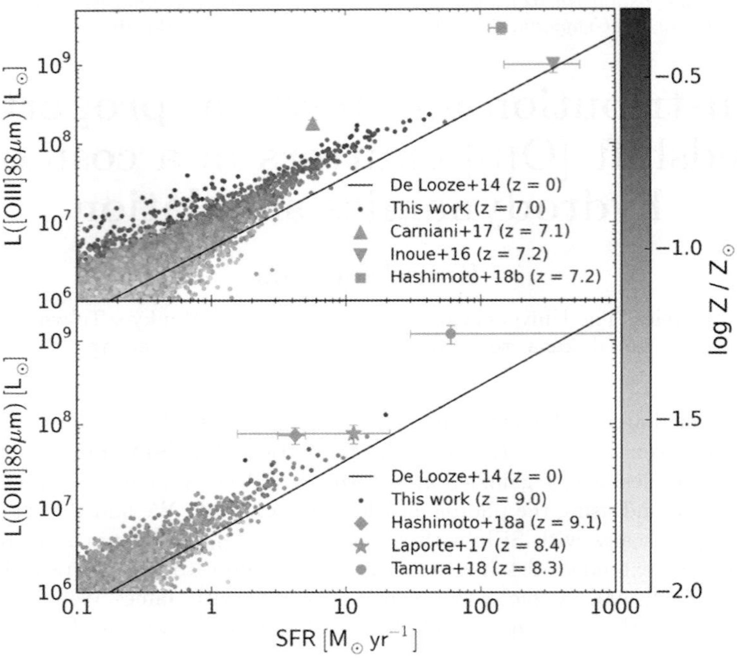

Figure 1. [OIII]-SFR relation of simulation galaxies (Moriwaki *et al.* 2018) and observed galaxies (Inoue *et al.* 2016; Carniani *et al.* 2017; Laporte *et al.* 2017; Hashimoto *et al.* 2018a; Hashimoto *et al.* 2018b; Tamura *et al.* 2018) at $z \sim 7$ (upper) and ~ 9 (bottom). The solid line is the local relation derived by De Looze *et al.* (2014).

a boxsize of a comoving $50h^{-1}$ Mpc cube. The number of particles is 2×1280^3 for dark matter ($m_{\rm DM} = 4.44 \times 10^6 h^{-1}$ M$_\odot$) and gas ($m_{\rm gas} = 8.11 \times 10^5 h^{-1}$ M$_\odot$). See Shimizu *et al.* (2016) for more details.

The detail calculation method of line luminosities is found in Moriwaki *et al.* (2018). Briefly, we use CLOUDY (Ferland *et al.* 2013) to generate a table of line luminosities with various sets of gas metallicity, ionization parameter, and gas density as in Inoue (2011) and Inoue *et al.* (2014). Then we calculate the line luminosity of each star particle by assuming that a spherical HII region forms around it.

3. Results

Fig. 1 shows the [OIII] 88 μm luminosity, $L_{\rm OIII,88}$, and SFR of simulation galaxies at $z = 7$ and 9. The solid line is the $L_{\rm OIII,88}$-SFR relation in local galaxies (De Looze *et al.* 2014). We find that $L_{\rm OIII,88}$ scales with SFR and the [OIII] emitters with $L_{\rm OIII,88} > 10^8$ L$_\odot$ has SFR > 10M$_\odot$ yr^{-1}. They reside in massive haloes with masses $> 10^{11}$ M$_\odot$ and the typical metallicity is ~ 0.1 Z$_\odot$. These indicate that well-established galaxies are selectively observed at this early universe. Our model predicts slightly larger $L_{\rm OIII,88}$ than those expected from the local relation due to the high ionization parameters and the metallicities.

Brightest [OIII] emitters have as large size as 1 kpc, which corresponds to ~ 0.2" at $z \sim 7$. These extended galaxies have a disk-like structure and rotation velocity ~ 50 km s^{-1}. In some [OIII] emitters, there is displacement between [OIII] emission and entire stellar distribution. Since [OIII] emission traces the young stellar population, this indicates the inside-out star formation history.

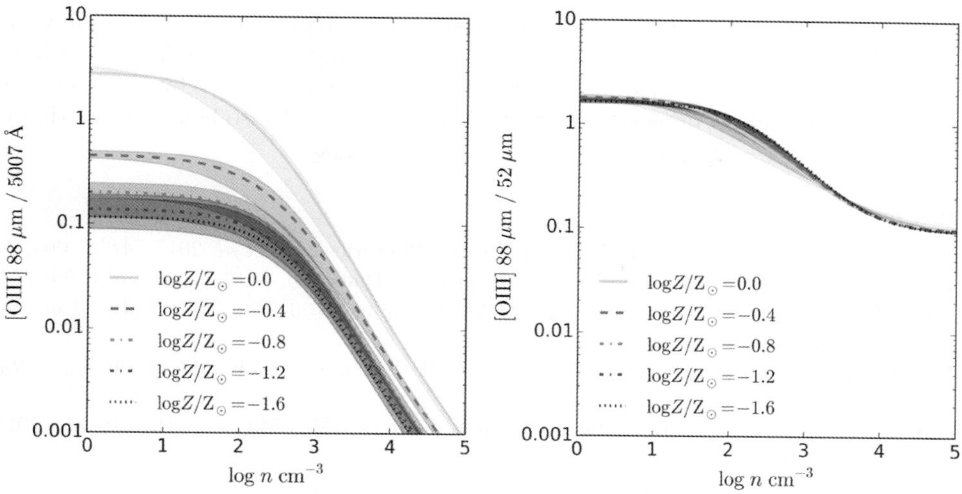

Figure 2. The line ratios of [OIII] 88 μm / 5007 Å (left) and [OIII] 88 μm / 52 μm (right) calculated with CLOUDY. Each line shows the line ratios with ionization parameter $U = 10^{-2}$ and fixed metallicities. The filled region of each color corresponds to ionization parameter range from $U = 10^{-3}$ to 10^{-1}. The input stellar continuum is calculated by assuming constant star formation duration of 10 Myr.

4. Discussion

The [OIII] 5007 Å line is roughly an order of magnitude larger than the [OIII] 88 μm line, whereas the typical detection limits of JWST is only ~ 3 times worse than that of ALMA. So, when ALMA already discovered the [OIII] 88 μm emission from a galaxy, then JWST can always detect [OIII] 5007 Å emission from that galaxy too. With the high angular resolution of ALMA and JWST, we can study the structure and the kinematics of bright [OIII] emitters.

Since [OIII] 88 μm and [OIII] 5007 Å lines have different critical densities and excitation temperatures, the ratio between them depends on the temperature and the density of the HII region. The temperature also depends on the gas metallicity because the metallicity controls the cooling efficiency. We show the [OIII] 88 μm / 5007 Å line ratio calculated with CLOUDY in the left panel of Fig. 2. The ratio decreases near the critical density of [OIII] 88 μm, ~ 500 cm^{-3}. At lower metallicity or higher ionization parameter, [OIII] 5007 Å is stronger due to high temperature and the ratio decreases. Although there is a degeneracy between parameters, properties of HII region can be extracted from this line ratio to some extent; if the ratio is greater than 0.3, then it indicates that the metallicity is larger than ~ 0.2 Z$_\odot$, or if the ratio is much lower than 0.1, then the gas density is much higher than 100 cm^{-3}. If galaxies have as high density as 1000 cm^{-3}, [OIII] 52 μm line is brighter than [OIII] 88 μm and can be detected with ALMA. As shown in the right panel of Fig. 2, [OIII] 88 μm / 52 μm ratio provide us the gas density without any degeneracy. If the gas density is fixed, the temperature or the metallicity can be more constrained from [OIII] 88 μm / 5007 Å ratio. Though we need to consider dust attenuation especially for [OIII] 5007 Å line, this diagnostics still works well for galaxies with little dust extinction. For instance, when the dust extinction A_V is 0.3, the difference between observed and intrinsic line ratios is only 0.1 dex and it is negligible.

It is possible to detect the large-scale structure by performing [OIII] line survey. Here, we consider a survey of [OIII] 5007 Å line emitters with JWST NIRCam. There are two possible ways: using grism module with F444W filter and using narrow-band F466N filter. We calculate the average number of [OIII] emitters within a field of view of $2 \times 2.2' \times 2.2'$

that can be detected with $S/N > 5$ by assuming 10^4 s integration. We find that we can detect ~ 2 [OIII] emitters at $6.8 < z < 9.0$ with grism module and ~ 0.5 at $z = 8.3$ with narrow band. We can remove the foreground Hα contaminants at $z \sim 6$ by J-dropout technique. For grism module, detecting doublet [OIII] 4959, 5007 Å also help us to distinguish these two lines.

References

Carniani, S., Maiolino, R., Pallottini, A., Vallini, L., Pentericci, L., *et al.* 2017, *A&A*, 605, A42
De Looze, I., Cormier, D., Lebouteiller, V., Madden, S., Baes, M., *et al.* 2014, *A&A*, 568, A62
Ferland, G. J., Porter, R. L., van Hoof, P. A. M., Williams, R. J. R., Abel, N. P., *et al.* 2013, *Rev. Mexicana AyA*, 49, 137
Hashimoto, T., Laporte, N., Mawatari, K., Ellis, R. S., Inoue, A. K., *et al.* 2018a, *Nature*, 557, 392
Hashimoto, T., Inoue, A. K., Mawatari, K., Tamura, Y., Matsuo, H., *et al.* 2018b, ArXiv:1806.00486
Inoue, A. K. 2011, *MNRAS*, 415, 2920
Inoue, A. K., Shimizu, I., Tamura, Y., Matsuo, H., Okamoto, T., & Yoshida, N. 2014, *ApJ*, 780, L18
Inoue, A. K., Tamura, Y., Matsuo, H., Mawatari, K., Shimizu, I., *et al.* 2016, *Science*, 352, 1559
Konno, A., Ouchi, M., Ono, Y., Shimasaku, K., Shibuya, T., *et al.* 2014, *ApJ*, 797, 16
Laporte, N., Ellis, R. S., Bauer, F. E., Quénard, D., Roberts-Borsani, G. W., *et al.* 2017, *ApJ*, 837, L21
Lebouteiller, V., Cormier, D., Madden, S. C., Galliano, F., Indebetouw, R., *et al.* 2012, *A&A*, 548, A91
Moriwaki, K., Yoshida, N., Shimizu, I., Harikane, Y., Matsuda, Y., *et al.* 2018, *MNRAS*, 481, L84
Nagao, T., Maiolino, R., Marconi, A., & Matsuhara, H. 2011 *A&A*, 526, A149
Shimizu, I., Inoue, A. K., Okamoto, T., & Yoshida, N. 2016, *MNRAS*, 461, 3563
Smit, R, Bouwens, R. J., Carniani, S. Oesch, P. A., Labbé, I., *et al.* 2018, *Nature*, 553, 178
Tamura, Y., Mawatari, K., Hashimoto, T., Inoue, A. K., Zackrisson, E., *et al.* 2018, ArXiv:1806.04132

Challenges in Panchromatic Modelling with Next Generation Facilities
Proceedings IAU Symposium No. 341, 2018
M. Boquien, E. Lusso, C. Gruppioni & P. Tissera, eds.

doi:10.1017/S1743921319002564

Pop III Supernova Feedback on the Formation of the First Galaxies

Li-Hsin Chen[1,2] and Ke-Jung Chen[1]

[1]Institute of Astronomy and Astrophysics, Academia Sinica, P.O. Box 23-141, Taipei 10617, Taiwan

[2]Institute of Astrophysics, National Taiwan University, Taipei 10617, Taiwan

Abstract. Modern cosmological simulations suggest that the hierarchical assembly of dark matter halos provided the gravitational wells that allowed the primordial gases to form stars and galaxies inside them. The first galaxies comprised of the first systems of stars gravitationally bound in dark matter halos are naturally recognized as the building blocks of early Universe. To understand the formation of the first galaxies, we use an adaptive mesh refinement (AMR) cosmological code, `Enzo` to simulate the formation and evolution of the first galaxies. We first model an isolated galaxy by considering much microphysics such as star formation, stellar feedback, and primordial gas cooling. To examine the effect of Pop III stellar feedback to the first galaxy formation, we adjust the initial temperature, density distribution and metallicity distributions by assuming different IMFs of the first stars. Our results suggest that star formation in the first galaxies is sensitive to the initial conditions of Pop III supernovae and their remnants. Our study can help to correlate the populations of the first stars and supernovae to star formation inside these first galaxies which may be soon observed by the (James Webb Space Telescope JWST).

Keywords. methods: numerical, galaxies: formation, galaxies: evolution, (ISM:) supernova remnant

1. Introduction

Recent theoretical studies suggest that the first galaxies form about several hundred million years after the Big Bang. The next generation of facilities such as *James Webb Space Telescope (JWST)* will be able to probe the Universe at redshift $\gtrsim 10$. If these first galaxies are bright enough, they are likely to be observed by such the coming big telescopes. The key to understand the formation of the first galaxies is to study the chemical and radiative feedback from the stars formed prior to them. In this work, we aim to study the influence of the first supernovae by testing different Pop III initial mass functions (IMFs) on the Pop II star formation inside the first galaxies. We mainly simulate the merging process of the Pop III supernova remnants (SNR) in an isolated galaxy. Therefore, our simulations achieve a much higher resolution than other typical cosmological simulations. The initial conditions are constructed manually to examine the effect of different assumptions of Pop III IMF.

2. Numerical Setup

We perform the simulations using the adaptive mesh refinement code `Enzo`. Our simulation includes a dark matter halo with an NFW profile, self-gravity of baryon, and chemical cooling. The fiducial setup for our halo is with a mass of 10^9 $M_\odot$ and a 1 kpc

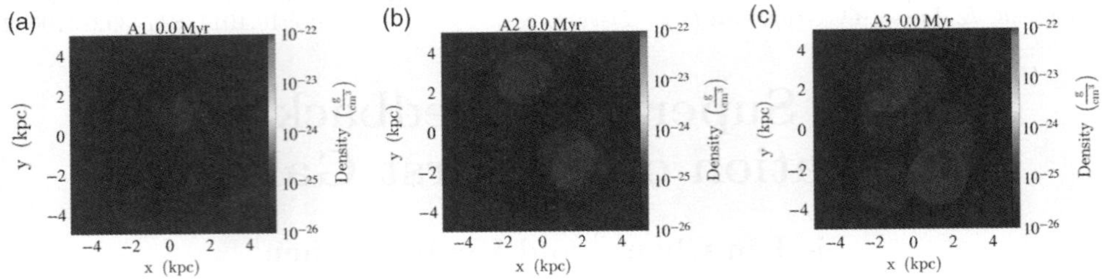

Figure 1. Density-weighted projection plots of the initial gas density distribution of three simulations. Left, middle and right panels show model A1 (Salpeter IMF), model A2 (Flat IMF), model A3 (Hirano IMF), respectively.

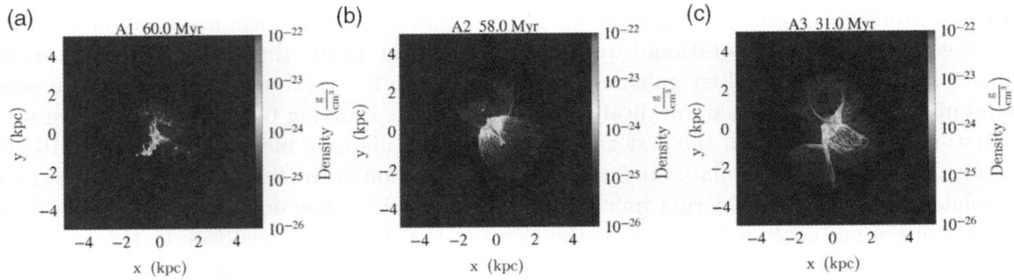

Figure 2. Density-weighted projection plots of gas density distribution at the last snapshots of the three simulations. Left, middle and right panels show model A1 (Salpeter IMF), model A2 (Flat IMF), model A3 (Hirano IMF), respectively.

core radius. We utilize the nine-species (H, H^+, He, He^+, He^{++}, e^-, H_2, H_2^+, H^-) non-equilibrium chemistry model in `Enzo` to calculate the cooling rate of the primordial gas. We calculate the metal cooling rate following the method of Glover & Jappsen (2007). For star formation and stellar feedback, we follow the model of Cen & Ostriker (1992). We adopt a minimum stellar mass of 1 $M_\odot$ with a star formation efficiency of 0.05.

The initial condition of our isolated galaxy is constructed by a spherical cloud of pristine gas with a uniform density profile. SNRs are implemented into the galaxy based on different Pop III IMFs and are added a drift velocity to the SNRs based on the free fall velocity. In this work, we approximate SNRs into spheres and adopt radial physical profiles. We consider three types of them based on the simulations from Chen *et al.* (2014, 2015); core-collapse supernovae (CCSNe), hypernovae, and pair-instability supernovae (PISNe), respectively. The remnants coming from CCSNe have a radius of 500 pc and lowest metallicity and density, the remnants coming from hypernovae have a radius of 1 kpc and intermediate metallicity and density, and the remnants coming from psne have a radius of 1.5 kpc and highest metallicity and density. We adopt three different Pop III IMFs and construct three models (A1, A2 and A3). In model A1, we adopt a Salpeter IMF Salpeter (1955) for the Pop III stars with a peak at 10 $M_\odot$. As a result, most of the Pop III stars dies as CCSN. In model A2, we adopt a Flat IMF and thus, the number of different types of SNRs are equal. In model A3, we reference from the cosmological simulation by Hirano *et al.* (2015) that most of the Pop III stars are massive and die as PISN. The initial gas density distribution of the three models is shown in Fig. 1.

3. Results

We show the gas distribution in Fig. 2, the zoomed-in gas density ditribution at star-forming sites in Fig. 3, and the stellar mass distribution in Fig. 4 at the last snapshot of the three simulations, respectively. We find that the merging process is different among

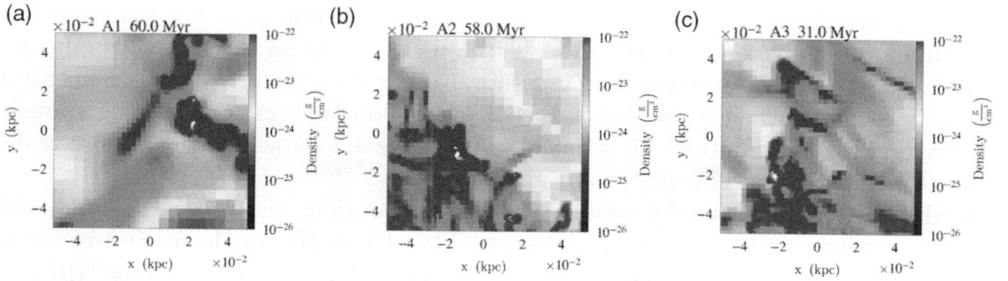

Figure 3. Zoomed-in density-weighted projection plots of gas density distribution at the last snapshots of the three simulations. Left, middle and right panels show model A1 (Salpeter IMF), model A2 (Flat IMF), model A3 (Hirano IMF), respectively. The coordinate of the center of the plot in the left panel is chosen to be (4.6, 4.2, 4.4) kpc from the lower-left corner instead of (5, 5, 5) kpc as in the other two panels.

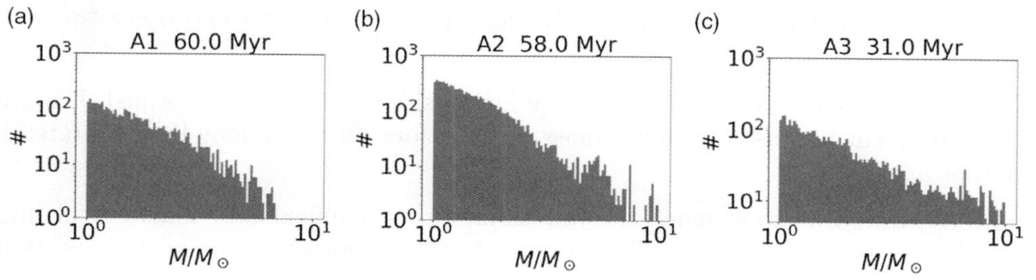

Figure 4. Stellar mass distribution at the last snapshots of three models. Left, middle and right panels show the stellar mass distribution of model A1 (Salpeter IMF), model A2 (Flat IMF), model A3 (Hirano IMF), respectively.

the three models. In model A1, we observe a bar structure appearing in a region deviated from the center. We believe this is because of the asymmetric distribution of the SNRs. When the SNRs merge and collide with each other, part of the gas is squeezed out of the central region of the halo. Therefore, the densest regions and star-forming sites are not in the central part of the halo. In Fig. 3(a), we show the star-forming regions and the white dots represent individual stars. Note that the center of the plot is shifted that it is not the center of the halo anymore. There are ∼3,600 stars at 60 Myr and the resulting stellar mass distribution in this model can be described by a power law. In model A2, the region with highest density remains in the center but we also observe a bar structure along with filaments. In Fig. 3(b), we find more star-forming sites than in model A1. There are over 8,000 stars at 58 Myr. The stellar mass distribution of this simulated galaxy can also be described by a power law with a steeper slope than A1. We find a second peak in the stellar mass distribution that we think it comes from the contribution of star formation in another star-forming site. In this model, we find more star-forming sites. In model A3, the galaxy is initially occupied by the remnants and the region with highest density locates in the center. We observe some filamentary structures in the outer regions where the SNRs collide with each other before entering the central part of the halo. The galaxy has evolved shorter than the other two galaxies but has a much earlier star-forming time. There are ∼4,800 stars at 31 Myr. The stellar mass distribution of this simulation can be described by a power law with a much flatter slope than A1 and A2.

4. Discussion

We perform a series of simulations with various initial conditions based on different assumptions of Pop III IMFs. We find that the star formation history is sensitive to the initial Pop III IMFs. The slope of the stellar mass distribution is flat when all of the first stars are massive and die as PISN. The Pop II stars in all three models inherit the metallicity information from the first supernovae and their star formation is concentrated in certain regions, because the thermal feedback of Pop II stars is not strong enough to halt the rapid star formation. To resolve the issue, we are implementing a more realistic feedback with radiative transfer to our simulations.

Besides the uniform density profile, we plan to consider the pristine clouds with a gradient density profile to examine the possibility of Pop III star formation in the first galaxies. We will follow the method in Wise *et al.* (2012) to consider Pop III SF and PopII SF separately by adopting the metallicity criterion. If the gas metallicity is $> 10^{-6}$ $Z_{\odot}$, PopII SF is used, and if the metallicity is $< 10^{-6}$ $Z_{\odot}$, Pop III SF is used. We will also investigate the dark matter halos of first galaxies by considering different halo mass and core radius. Our ultimate goal is to develop observational diagnostics for JWST and provide useful insight into the IMF of the first stars.

AKIO INOUE: Do you have any idea to distinguish the three scenarios observationally in the future?

LI-HSIN CHEN: We will be calculating the luminosity of the stars we formed. If we find it being different because the stellar mass distributions differ, we may be able to tell the scenarios apart.

ANDREA MERLONI: How much do you think the dynamical and kinematical initial conditions of your simulation influence the statistical properties of subsequent stellar generations?

LI-HSIN CHEN: I think the dynamics and kinematics have more impact on the star formation site distribution. Unless the dynamics of the turbulence is so strong to change the mixing of the metals. I believe the stellar mass is more related to the cooling, which is affected by the metal content.

Acknowledgements

LHC and KC are supported by the Ministry of Education, Taiwan, R.O.C. under Grant no. MOST 107-2922-I-002-432. and MOST 107-2112-M-001 -044 -MY3. All simulations and analysis were performed on the supercomputer Edison of the National Energy Research Scientific Computing Center (NERSC), a U.S. Department of Energy Office of Science User Facility operated under Contract No. DE-AC02-05CH11231 and the TIARA Cluster at the Academia Sinica Institute of Astronomy and Astrophysics (ASIAA). Computations described in this work were performed using the publicly-available `Enzo` code (http://enzo-project.org), which is the product of a collaborative effort of many independent scientists from numerous institutions around the world. Their commitment to open science has helped make this work possible. The analysis and plots were done with yt (Turk *et al.* 2011).

References

Cen, R. & Ostriker, J. P. 1992, *ApJ*, 399(2), 113
Chen, K.-J., Heger, A., Woosley, S., Almgren, A., & Whalen, D. J. 2014, *ApJ*, 792, 44
Chen, K.-J., Bromm, V., Heger, A., Jeon, M., & Woosley, S. 2015, *ApJ*, 802, 13
Hirano, S., Hosokawa, T., Yoshida, N., Omukai, K., & Yorke, H. W. 2015, *MNRAS*, 448(1), 568
Glover, S. C. O. & Jappsen, A.-K. 2007, *ApJ*, 666(1), 1
Salpeter, E. E. 1955, *ApJS*, 121, 1
Turk, Matthew J., Smith, Britton D., Oishi, Jeffrey S., Skory, Stephen, Skillman, Samuel W., Abel, Tom & Norman, Michael L., *et al.* 2011, *ApJ*, 192, 9
Wise, J. H., Turk, M. J., Norman, M. L., & Abel, T. 2012, *ApJ*, 745(1), 50

Challenges in Panchromatic Modelling with Next Generation Facilities
Proceedings IAU Symposium No. 341, 2018
M. Boquien, E. Lusso, C. Gruppioni & P. Tissera, eds.

doi:10.1017/S1743921319002229

Chemical enrichment of Pop III supernovae in the first galaxies

Ke-Jung Chen

Institute of Astronomy and Astrophysics, Academia Sinica, Taipei 10617, Taiwan

Abstract. Understanding the formation of the first galaxies (FGs) is one of the most important topics in modern cosmology. In this proceeding, we briefly summarize the results of chemical enrichment from the Pop III supernovae during the assembly of the FGs. This early chemical enrichment plays an important role in triggering the Pop II star formation. Generally speaking, there are two major enrichment channels, inside-out (internal) and outside-in (external). Our results suggest that the external channel of chemical enrichment only works if the Pop III stars are very massive stars of 200–260 $M_{\odot}$, which produce strong enough radiative feedback and supernova to derive the external metal mixing down to the center of the nearby halo.

Keywords. The first stars, the first galaxies, supernovae, and early universe

---

Discussions

The formation of Pop III stars transformed the simple early Universe into one of many complexes. One of the significant Pop III stellar feedback is by producing supernovae (SNe), which are extremely powerful explosions injecting a tremendous amount of energy into the interstellar medium. More critically, these SNe bring the first heavy elements to a primordial Universe only containing hydrogen and helium left from the Big Bang, and completely reshape the later stekkar and galaxy formation, see examples by Greif *et al.* (2010); Bromm & Yoshida (2011); Wise *et al.* (2012); Jeon *et al.* (2012). We show a schematic picture of how the dark matter, primordial gases, and feedback from the Pop III stars jointly formed into the FGs in Fig. 1.

One single SN can rapidly pollute the pristine intergalactic medium (IGM) of a size of 1 kpc with $10^{-4} - 10^{-3} Z_{\odot}$ as shown in Fig. 2. However, how these heavy elements mixing with the primordial gas to form the Pop II stars inside the FGs is still unclear. Previous studies by Smith *et al.* (2009, 2015); Chiaki *et al.* (2013) suggested external and internal channels for Pop III SNe to dispense the metals to the Pop II star-forming cloud. For an internal channel, the ejected metal eventually re-collapses to its host halo and triggers the consequent star formation in the FGs. For an external channel, the metal pollutes a nearby halo and mixes up the pristine gas to drive the later star formation. If these scenarios did happen in the cosmic dawn, one might ask which channel might be dominating and can it explain the origins of metal-poor stars?

We discuss the results of an external enrichment scenario based on Chen *et al.* (2017), that investigated how the first supernovae regulated the star formation in the nearby halos. Before the incoming metal, the pre-enriched minihalo is first photo-evaporation by the Pop III radiative feedback and the inputs of ionizing photons of hydrogen/helium are from the one-dimensional stellar evolution model. During the photo-heating process, the radiation pressure compresses the spherical halo into a bullet shape, and removes much gas out of the halo, that facilitates the following enrichment process. The massive Pop III star explodes as an SN after its short lifetime of several million years. SNe blow metal to

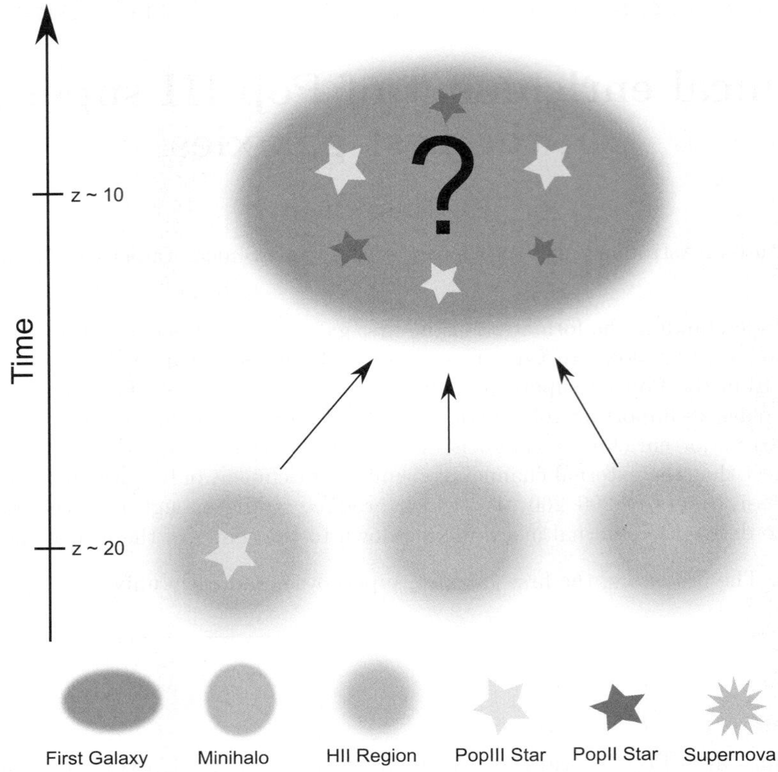

Figure 1. Formation of the FG. The FG contains a masses of about $10^8 M_\odot$ and forms at $z \sim 10$. Its assembly is determined by the Pop III stellar feedback, that builds up an extensive H II region of several kpc and forges the first heavy elements beyond the hydrogen and helium made, to be dispensed to the intergalactic medium through supernova explosions. The chemical enrichment would trigger the Pop II star formation in the FG.

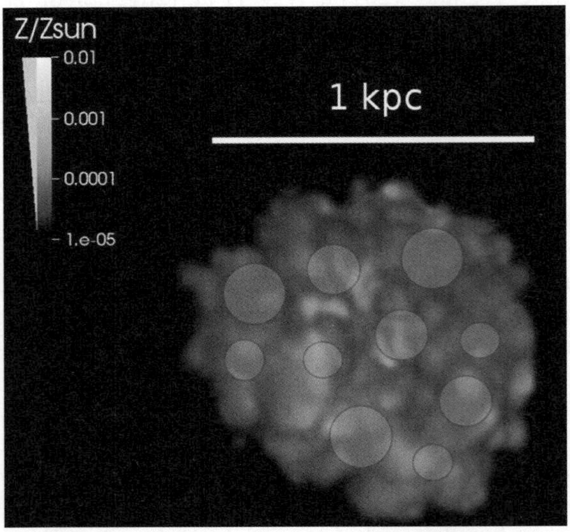

Figure 2. An extent metal region created by a Pop III SN. A Pop III SN can produce a big metal bubble of a size about 1 kpc with a metallicy of $10^{-3} - 10^{-4} Z_\odot$. There are several minihalos bathed inside this bubble. (reproduced from Chen *et al.* (2015))

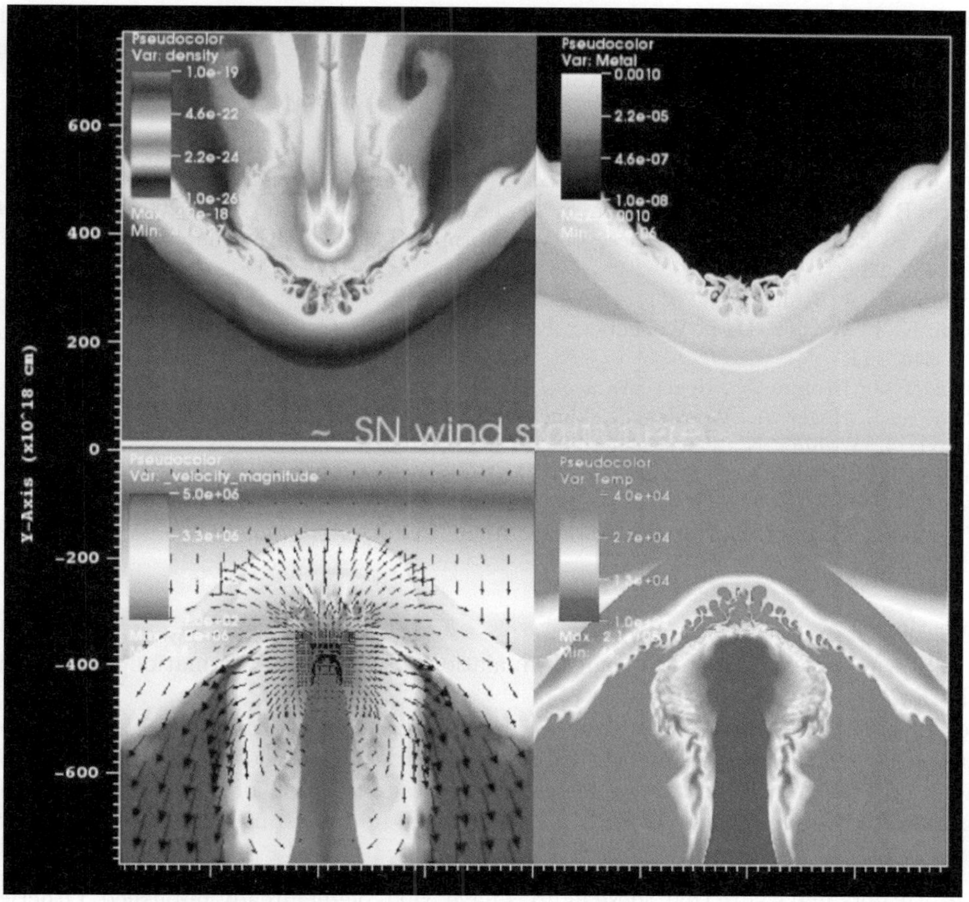

Figure 3. The external enrichment of SN metal. Panels show the physical quantities of; density, metallicity, velocity, temperature when the SN ejecta swipes through a nearby minihalo. The densest region indicates the center of minihalo and metal mixing only occurs around at outskirts without reaching to its center.

collide into the photo-heated halo and drive mixing at the outer shell of the halo. From five out of six models in Chen *et al.* (2017), most of the halos seem to be able to prevent from the enrichment of SN metals. If it were true for the real situation, the Pop III stars might have still formed in these nearby halos. Some mixing of metal appears at an outer skirt of the shell, even if the metal cooling acts effectively. However, the low gas density prevents the Pop II stars forming in this mixing region as shown in Fig. 3.

Based on the highly-resolved simulations of Chen *et al.* (2017), we conclude the external enrichment only work for very massive Pop III stars of 200 - 250 $M_{\odot}$, which have strong radiative feedback and die as powerful pair-instability supernovae (Chen *et al.* (2014)) to break out the dense shell of the nearby halo and mix metal deep into the center to initialize the Pop II star formation. Therefore, the external enrichment channel can be promising only if the Pop III IMF is very top-heavy.

Acknowledgements

We thank Stan Woosley, Alexander Heger, and Volker Bromm for many useful discussions. This research is supported by an EACOA Fellowship, and by the Ministry of Education, Taiwan, R.O.C. under Grant no. MOST 107-2112-M-001 -044 -MY3.

Numerical simulations are supported by the National Energy Research Scientific Computing Center (NERSC), and the TIARA Cluster at the Academia Sinica Institute of Astronomy and Astrophysics (ASIAA).

References

Almgren, A. S., Beckner, V. E., Bell, J. B., *et al.* 2010, *ApJ*, 715, 1221

Smith, B. D., Turk, M. J., Sigurdsson, S., O'Shea, B. W., & Norman, M. L. 2009, *ApJ*, 691, 441

Smith, B. D., Wise, J. H., O'Shea, B. W., Norman, M. L., & Khochfar, S. 2015, *MNRAS*, 452, 2822

Bromm, V., & Yoshida, N. 2011, *ARAA*, 49, 373

Chen, K.-J., Whalen, D. J., Wollenberg, K. M. J., Glover, S. C. O., & Klessen, R. S. 2017, *ApJ*, 844, 111

Chen, K.-J., Bromm, V., Heger, A., Jeon, M., & Woosley, S. 2015, *ApJ*, 802, 13

Chen, K.-J., Heger, A., Woosley, S., Almgren, A., & Whalen, D. J. 2014, *ApJ*, 792, 44

Chiaki, G., Yoshida, N., & Kitayama, T. 2013, *ApJ*, 762, 50

Greif, T. H., Glover, S. C. O., Bromm, V., & Klessen, R. S. 2010, *ApJ*, 716, 510

Jeon, M., Pawlik, A. H., Greif, T. H., *et al.* 2012, *ApJ*, 754, 34

Wise, J. H., Turk, M. J., Norman, M. L., & Abel, T. 2012, *ApJ*, 745, 50

Questions:

1. How reliable can the cosmological simulations predict binary formation in population III stars?

Recent simulations of the first star formation have achieved a resolution of sub-solar mass scale and included many detailed micro-physics such as cooling, magnetic field, and radiation transfer. Their results suggested the fragmentation of star-forming disk and formation of binary stars. At least above 50% of the first stars may have formed into binaries.

2. Authors make sure that these figure s have NOT being already published. Otherwise you need to submit a permission from the publisher.

Figures in this manuscript have not be published before.

Challenges in Panchromatic Modelling with Next Generation Facilities
Proceedings IAU Symposium No. 341, 2018
M. Boquien, E. Lusso, C. Gruppioni & P. Tissera, eds.

doi:10.1017/S1743921319003004

Radiative properties of the first galaxies: Rapid transition from blue to red

Shohei Arata[1], Hidenobu Yajima[2], Kentaro Nagamine[1,3,4], Yuexing Li[5] and Sadegh Khochfar[6]

[1]Theoretical Astrophysics, Department of Earth and Space Science, Graduate School of Science, Osaka University, Toyonaka, Osaka 560-0043, Japan
email: arata@astro-osaka.jp

[2]Center for Computational Sciences University of Tsukuba, Ibaraki 305-8577, Japan

[3]Department of Physics & Astronomy, University of Nevada, Las Vegas, 4505 S. Maryland Pkwy, Las Vegas, NV 89154-4002, USA

[4]Kavli IPMU (WPI), The University of Tokyo, 5-1-5 Kashiwanoha, Kashiwa, Chiba, 277-8583, Japan

[5]Department of Astronomy & Astrophysics, The Pennsylvania State University, 525 Davey Lab, University Park, PA 16802, USA

[6]SUPA, Institute for Astronomy, University of Edinburgh, Royal Observatory, Edinburgh, EH9 3HJ, UK

Abstract. Combining cosmological hydrodynamic simulations and radiative transfer (RT) calculations, we present predictions of multi-wavelength radiative properties of the first galaxies at $z \sim 6-15$. We find that intermittent star formation due to supernova (SN) feedback causes the escape fraction of UV photons to fluctuate rapidly, which then produces the observed diversity of SEDs for high-z galaxies. The simulated galaxies make rapid transition between UV-bright and IR-bright phase, and our RT calculations suggest that dust temperatures in the first galaxies are higher than $z < 3$ galaxies with $\sim 60\,\mathrm{K}$.

Keywords. galaxies: formation, galaxies: high-redshift, methods: numerical, radiative transfer

1. Introduction

Recent observations have successfully detected galaxies in the early Universe, the so-called *first galaxies*, in optical and near-infrared wavelengths (e.g. Bouwens *et al.* 2015), or in sub-millimeter wavelengths (e.g. Hashimoto *et al.* 2018). This variety of radiative properties probably reflects different physical conditions in the first galaxies, such as the escape fraction of ionizing and UV photons. Here we investigate the radiative output of first galaxies by applying RT calculations on zoom-in hydrodynamic simulations, as reported in Arata *et al.* (2018).

2. Method & Result

We use zoom-in cosmological hydrodynamic simulations from Yajima *et al.* (2017), which focused on massive galaxies and satellites with halo masses $1.6 \times 10^{11}\ h^{-1}\,\mathrm{M}_\odot$ (Halo-11) and $0.7 \times 10^{12}\ h^{-1}\,\mathrm{M}_\odot$ (Halo-12) at $z = 6$. We also use the multi-wavelength RT code ART^2 (Li *et al.* 2008; Yajima *et al.* 2012), which computes the RT on an adaptive mesh refinement grid. The minimum spatial resolution is less than 10 pc at $z \sim 10$.

Figure 1 shows the redshift evolution of physical properties of Halo-11. At the end of their lifetime, massive stars explode as SNe and disrupt central star-forming gas clouds.

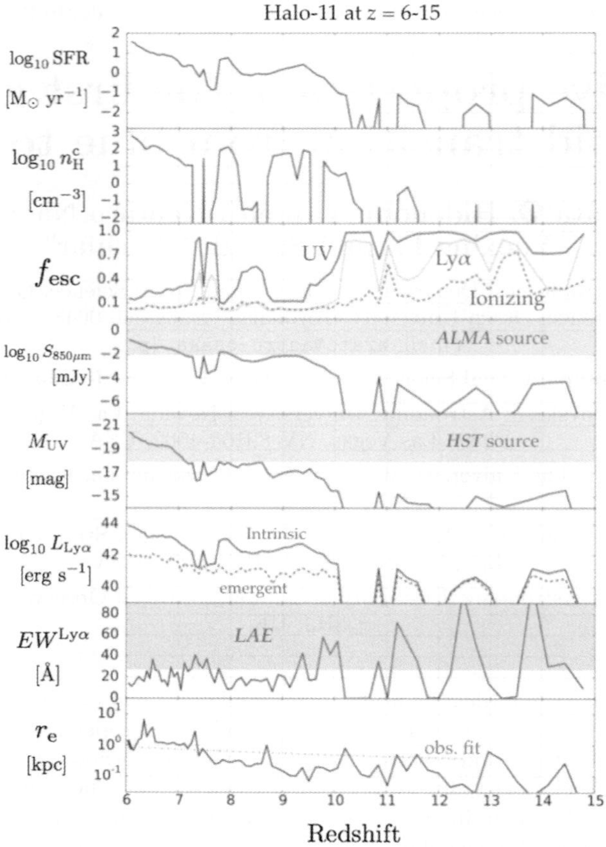

Figure 1. From top to bottom panels: Redshift evolution of SFR; central gas density; escape fraction of UV (solid), Lyα (dotted) and ionizing photons (dashed); observed-frame sub-millimeter flux; UV absolute magnitude; Lyα luminosity; Lyα equivalent width; and UV half-light radius of Halo-11 at $z = 6 - 15$. In the fourth (fifth) panel, shaded area represents the sensitivity of 10-hr observation by ALMA (HST). In the second bottom panel, the shaded region represents $EW^{\rm Ly\alpha} > 30$ Å, which is often used as a criterion to define LAEs. In the bottom panel, the dashed line represents the observational fit by Kawamata *et al.* (2018), who used bright galaxies with $-21 < M_{\rm UV} < -20.5$ mag. The size of simulated galaxy increases similarly to virial radius ($\propto M_{\rm h}^{1/3}(1+z)^{-1}$) with decreasing redshift.

This results in intermittent star formation (see also Yajima *et al.* 2017). The UV photons escape more efficiently when the gas density becomes lower in galactic center. Whereas in the starbursting phase, high density gas clumps absorb $\sim 80\,\%$ of UV photons and re-emit the energy in infrared (IR) photons, thus the sub-millimeter flux increases. We find that the peak of SED shifts rapidly between UV and IR wavelengths on a time-scale of ~ 100 Myrs. We predict the detectability of high-z galaxies with the Atacama Large Millimeter Array (ALMA). For a sensitivity limit of 0.1 mJy (0.01 mJy) at 850 μm, the detection probability of galaxies in halos $M_{\rm h} \gtrsim 10^{11}\,{\rm M}_\odot$ ($M_{\rm h} \gtrsim 10^{10.5}\,{\rm M}_\odot$) at $z \lesssim 7$ ($z \lesssim 9$) exceeds 50 %.

The Lyα luminosity and equivalent width (EW) also change with intermittent star formation. In starburst phase at $z > 10$, the maximum luminosity reaches $\sim 10^{41}$ erg s^{-1} and $EW^{\rm Ly\alpha} \gtrsim 30$ Å, thus such galaxies may be detected by JWST as a dust-free Lyα emitter. At $z < 10$, $\sim 90\,\%$ of Lyα photons are absorbed by dust, resulting in almost constant luminosity of $\sim 10^{41} - 10^{42}$ erg s^{-1} and low EW even in starburst phase. The galactic size increases with decreasing redshift, and becomes ~ 1 physical kpc at $z \sim 6$, which is consistent with observations (Kawamata *et al.* 2018). This compactness of high-z galaxies leads to stronger UV absorption by dust, resulting in high dust temperature.

References

Bouwens, R. J., Illingworth, G. D., Oesch, P. A., Trenti, M., Labbé, I., Bradley, L., Carollo, M., van Dokkum, P. G., *et al.* 2015, *ApJ*, 803, 34

Hashimoto, T., Laporte, N., Mawatari, K., Ellis, R. S., Inoue, A. K., Zackrisson, E., Roberts-Borsani, G., Zheng, W., *et al.* 2018, *Nature*, 557, 392

Arata, S., Yajima, H., Nagamine, K., Li, Y., & Khochfar, S. 2018, arXiv:1810.07621

Yajima, H., Nagamine, K., Zhu, Q., Khochfar, S., & Dalla Vecchia, C. 2017, *ApJ*, 846, 30

Li, Y., Hopkins, P. F., Hernquist, L., Finkbeiner, D. P., Cox, T. J., Springel, V., Jiang, L., Fan, X., & Yoshida, N. 2008, *ApJ*, 678, 41

Yajima, H., Li, Y., Zhu, Q., & Abel, T. 2012, *MNRAS*, 424, 884

Kawamata, R., Ishigaki, M., Shimasaku, K., Oguri, M., Ouchi, M., & Tanigawa, S. 2018, *ApJ*, 855, 4

Challenges in Panchromatic Modelling with Next Generation Facilities
Proceedings IAU Symposium No. 341, 2018
M. Boquien, E. Lusso, C. Gruppioni & P. Tissera, eds.

doi:10.1017/S1743921319001364

Helium rich stars produce the UV upturn

R. De Propris[1], S. Ali[2], M. N. Bremer[2] and S. Phillipps[2]

[1] FINCA, University of Turku, 20014, Finland
email: rodepr@utu.fi

[2]H. H. Wills Physics Laboratory, University of Bristol, United Kingdom
emails: s.ali, m.bremer, s. phillipps @bristol.ac.uk

Abstract. We measure the evolution of the UV upturn color for galaxies on the red sequence in clusters at $0 < z < 0.7$ and to luminosity levels $L \sim L^*$. We show that the UV upturn color does not change until at least $z = 0.55$ but becomes significantly redder at $z = 0.7$. This is the first detection of evolution in the UV upturn. Our observations are inconsistent with all models proposed for its origin except the presence of a population of helium enriched stars, with helium abundances above 42 % and formed at $z > 4$.

Keywords. ultraviolet: galaxies, galaxies: stellar content, galaxies: elliptical and lenticular, cD, stars: horizontal-branch

1. Introduction

The Ultraviolet Upturn is an unexpected rise in the spectra of early-type galaxies at $\lambda < 2500$ Å, where no flux is expected to be produced by their old, metal-rich and quiescent stellar populations (e.g., see reviews by Yi 2008, Yi 2010). It is generally agreed that hot horizontal branch stars are the most likely sources of the UV flux (Greggio & Renzini 1990) but they are not expected to occur in significant numbers given the stellar populations that predominate in early-type galaxies. Possible origins include: very metal poor stars (Park & Lee 1997), metal-rich stars with extra mass loss (Yi *et al.* 1998), binaries (Han *et al.* 2007) and helium-rich stars (Chung *et al.* 2017). Each of these has significant implications for the dependence of the UV upturn in galaxy properties and especially its evolution with redshift. Observations so far have been limited to very bright galaxies in distant clusters and show little significant evolution in the FUV-V upturn color (e.g., Boissier *et al.* 2018).

2. Observations

We have used a set of archival data from the Hubble Space Telescope to measure the evolution of the UV upturn in red sequence (therefore quiescent and cluster members) galaxies in a series of clusters at $z = 0.2$ (Abell 1689), $z = 0.31$ (Abell 2744), $z = 0.55$ (MACS 0717+37 and MACS 1149+22) and $z = 0.68$ (SDSS 1004+41). These clusters have deep optical-IR images to select a sample of red sequence early-type cluster members and deep images in the UV (HST filters F225W or F275W) to probe the rest-frame ~ 1550 Å bandpass (slight mismatches with the local FUV band are of course unavoidable). Unlike previous studies, our samples of galaxies reach to the level of normal ($\sim L^*$) galaxies and not just the most massive systems.

Figure 1 shows the main results of our investigations, that can be found in a series of papers by Ali *et al.* 2018a,c,c. We find no evolution to $z = 0.55$, an observation that in itself rules out metal poor or metal rich stars. However, in our highest redshift cluster we find no galaxy that can be detected with our data, except the brightest cluster

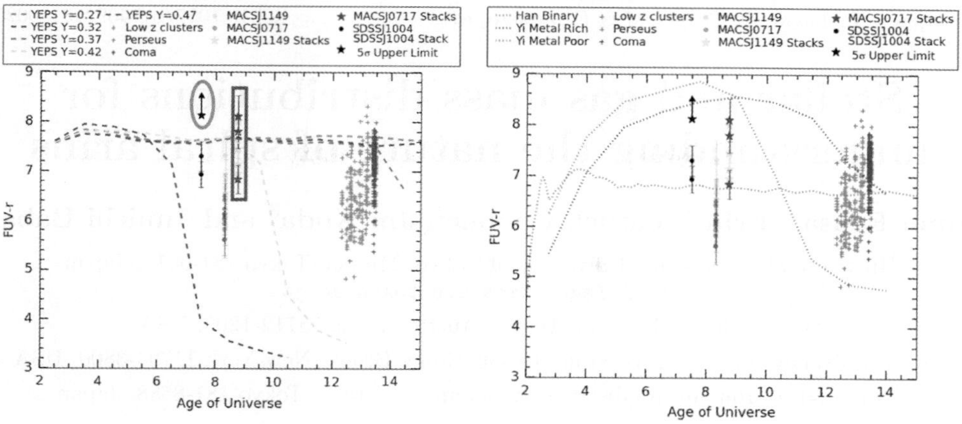

Figure 1. Left: Evolution of the UV upturn color and YEPS models from Chung *et al.* (2017). We observe a significant reddening in the UV upturn color at $z > 0.6$. Right: Comparison with other models for the evolution of the UV upturn. No model fits the data.

galaxy (which has usually bluer UV colors). We derive a significant upper limit to the stacked FUV-V color for red sequence galaxies in this cluster and demonstrate that it is significantly redder than similar stacks for galaxies in the two $z = 0.55$ clusters.

3. Implications

As shown in Figure 1 the evolution observed in the UV upturn is only consistent with a stellar population that is enriched in helium, at $Y > 0.42$. Given the timescale for the evolution of the UV upturn sources, these stars were formed at least by $z = 4$ and possibly earlier. We rule out all other proposed explanations as they do not fit the data across the entire redshift range.

This suggests the existence of a new nucleosynthesis channel in early stellar populations, as a similar phenomenon must account for the second parameter effect in local globular clusters (Bastian & Lardo 2018).

References

Ali, S., Bremer, M., Phillipps, S., & De Propris, R. 2018a, *MNRAS*, 476, 1010
Ali, S., Bremer, M., Phillipps, S., & De Propris, R. 2018b, *MNRAS*, 478, 541
Ali, S., Bremer, M., Phillipps, S., & De Propris, R. 2018c, *MNRAS*, 480, 2236
Bastian, N. & Lardo, C. 2018, *ARA&A*, 56, 83
Boissier, S., Cucciati, O., Boselli, A., Mei, S., & Ferrarese, L. 2018, *A&A*, 611, A42
Chung, C., Yook, S.-J., & Lee, Y.-W. 2017, *ApJ*, 842, A91
Greggio, L. & Renzini, A. 1990, *ApJ*, 364, 35
Han, Z., Podsiadlowski P., & Lynas-Gray, A. E. 2007, *MNRAS*, 380, 1098
Park, J.-H. & Lee, Y.-W. 1997, *ApJ*, 476, 28
Yi, S. K., Demarque, P., & Oemler, A. 1998, *ApJ*, 492, 480
Yi, S. K. 2008, *Stellar Populations - Planning for the Next Decade, IAU Symposium*, 262, 147
Yi, S. K. 2010, *Hot Subdwarf Stars and Related Objects, ASP Conference Series*, 392, 3

Challenges in Panchromatic Modelling with
Next Generation Facilities
Proceedings IAU Symposium No. 341, 2018
M. Boquien, E. Lusso, C. Gruppioni & P. Tissera, eds.

doi:10.1017/S1743921319002503

Stellar and gas mass distributions for understanding the nature of spiral arms

Fumi Egusa[1], Erin Mentuch Cooper[2], Jin Koda[3] and Junichi Baba[4]

[1]Institute of Astronomy, University of Tokyo, Mitaka, Tokyo 181-0015, Japan
email: fegusa@ioa.s.u-tokyo.ac.jp

[2]University of Texas at Austin, Austin, Texas 78712-1205, USA

[3]State University of New York Stony Brook, Stony Brook, New York 11794-3800, USA

[4]National Astronomical Observatory of Japan, Mitaka, Tokyo 181-8588, Japan

Abstract. Numerical simulations of disk galaxies with steady (long-lived) and dynamic (short-lived) spiral arms suggest that offsets between stellar and gas spiral arms depend on their nature or lifetime (Baba *et al.* 2015). Based on this theoretical study, we investigated gas-star offsets in the nearby grand-design spiral galaxy M51, and found that its two spiral arms exhibit different offset dependences against radius. One arm is consistent with a steady arm, while the other is consistent with a dynamic arm. We deduce that this difference is likely due to a tidal interaction with the companion galaxy (Egusa *et al.* 2017). For this study, a stellar mass distribution with a high accuracy at a high spatial resolution is essential, which has come to be available by applying recent SED fitting techniques to multi-wavelength images. We are now working to extend this study to other nearby spiral galaxies.

Keywords. galaxies: individual (M51 or NGC 5194, M74 or NGC 0628), galaxies: spiral, galaxies: structure

1. Introduction

In the local universe, more than half of galaxies are categorized as spiral galaxies (e.g. Delgado-Serrano *et al.* 2010). However, we do not have a conclusive answer to the very fundamental question: *"What is the nature of spiral arms?"* Table 1 lists theoretical models for spiral arms and their predictions for spiral arm properties, but these models have not been fully confirmed by observations yet.

Baba *et al.* (2015) performed numerical simulations of disk galaxies with steady (long-lived) and dynamic (short-lived) spiral arms, and found that the location of gas spiral arms (relative to stellar arms) is different between the two models (their Figure 2). For the steady models, the gas arm is found downstream to the stellar arm at inner disk, and then gradually moves upstream with increasing radius. This trend is consistent with the classical view of galactic shocks (Fujimoto 1968; Roberts 1969), and does not change depending on parameters (e.g. pitch angle, arm strength) in the steady models. On the other hand, the gas arm in the dynamic model almost coincides with the stellar arm. These results suggest that offsets between stellar and gas spiral arms (which can be observationally measured) should tell whether that spiral arm is steady or dynamic.

2. Result: a case study with M51

Based on the simulation results above, we measure spatial offsets between stellar and gas spiral arms in the nearby grand-design spiral galaxy M51 (NGC 5194). Here we

Table 1. Theoretical models for spiral arms

Model	Lifetime	Predicted spiral feature	References[1]
Density wave	>1 Gyr	2-armed?	Lin and Shu (1964)
Tidal interaction	~ 1 Gyr	2-armed	Pettitt *et al.* (2016)
Bar driven	~ 0.2 Gyr?	2-armed	Baba (2015)
Swing amplification	~ 0.1 Gyr[2]	multi-armed or flocculent?	Fujii *et al.* (2011)

Notes:
[1]Not a complete list. See Dobbs and Baba (2014) for a comprehensive review.
[2]The lifetime of each structure is this short, but spiral structures are recurrent.

describe a brief overview of our study, while the full detail is presented in Egusa *et al.* (2017).

Data. The stellar mass distribution is obtained by fitting SEDs to multi-wavelength images from optical to NIR (Mentuch Cooper *et al.* 2012). The gas mass distribution is obtained by combining an atomic gas map (from HI; Walter *et al.* 2008) and molecular gas map (from CO; Koda *et al.* 2009). Both stellar and gas mass data are smoothed to a common angular resolution of $6''$ (FWHM), and regridded to a pixel size of $2''$.

Gas-star offset versus radius. At a radial range of $30''$–$220''$, we measure locations of two spiral arms (called arm1 and arm2) at an interval of $4''$ for both stellar and gas arms and calculate their offsets. We find that offsets for only the inner part ($r < 150''$) of arm2 decrease with radius, while those for the others (inner arm1 and outer two arms) appear independent on radius (see Fig. 7 of Egusa *et al.* 2017). We conclude that only inner arm2 is consistent with the steady spiral arm, while the others are not.

The gas-star offsets in M51 suggest that the nature of spiral arms can be different even within a galaxy. In this particular case, this difference could be due to the interaction with the companion galaxy (Pettitt *et al.* 2018).

3. Work in progress

We are now working to extend this study for other nearby spiral galaxies. The very first step is to apply the SED fitting code `CIGALE` (Boquien *et al.* 2019) for multi-wavelength images including GALEX UV, SINGS optical, SDSS, 2MASS, and Spitzer NIR images of M74 (NGC 0628). This is another well-studied nearby grand-design spiral galaxy, but lacks apparent perturbers (e.g. bar, companion) which clearly contrasts to M51.

References

Baba, J. 2015, *MNRAS* 454, 2954
Baba, J., Morokuma-Matsui, K., & Egusa, F. 2015, *PASJ* 67, L4
Boquien, M., Burgarella, D., Roehlly, Y., Buat, V., Ciesla, L., *et al.* 2019, *A&A* 622, A103
Delgado-Serrano, R., Hammer, F., Yang, Y. B., Puech, M., *et al.* 2010, *A&A* 509, A78
Dobbs, C. & Baba, J. 2014, *PASA* 31, 35
Egusa, F., Mentuch Cooper, E., Koda, J., & Baba, J. 2017, *MNRAS* 465, 460
Fujii, M. S., Baba, J., Saitoh, T. R., Makino, J., Kokubo, E., *et al.* 2011, *ApJ* 730, 109
Fujimoto, M. 1968, in *Non-stable Phenomena in Galaxies*, Vol. 29 of *IAU Symposium*, p. 453
Koda, J., Scoville, N., Sawada, T., La Vigne, M. A., Vogel, S. N., *et al.* 2009, *ApJL* 700, L132
Lin, C. C. & Shu, F. H. 1964, *ApJ* 140, 646
Mentuch Cooper, E., Wilson, C. D., Foyle, K., Bendo, G., Koda, J., *et al.* 2012, *ApJ* 755, 165
Pettitt, A. R., Egusa, F., Dobbs, C. L., Tasker, E. J., *et al.* 2018, *MNRAS* 480, 3356
Pettitt, A. R., Tasker, E. J., & Wadsley, J. W. 2016, *MNRAS* 458, 3990
Roberts, W. W. 1969, *ApJ* 158, 123
Walter, F., Brinks, E., de Blok, W. J. G., Bigiel, F., *et al.* 2008, *AJ* 136, 2563

Challenges in Panchromatic Modelling with Next Generation Facilities
Proceedings IAU Symposium No. 341, 2018
M. Boquien, E. Lusso, C. Gruppioni & P. Tissera, eds.

doi:10.1017/S1743921319001352

Panchromatic modeling of the extremely luminous dust-obscured quasars at the cosmic noon

Lulu Fan[1], Yunkun Han[2] and Kirsten K. Knudsen[3]

[1]Shandong Key Laboratory of Optical Astronomy and Solar-Terrestrial Environment, Institute of Space Sciences, Shandong University, Weihai, Shandong 264209, China
email: llfan@sdu.edu.cn

[2]Yunnan Observatories, Chinese Academy of Sciences, Kunming, 650011, China

[3]Department of Earth and Space Sciences, Chalmers University of Technology, SE-439 92, Sweden

Abstract. The massive galaxies and their central supermassive black holes (SMBHs) co-evolution scenario proposes that a gas-rich major merger can trigger the central starburst and feeding the SMBH accretion, and then star formation is eventually quenched by quasar feedback. In this evolutionary sequence, dust-obscured quasars may represent the critical transition phase between starburst and unobscured quasars. Modeling the panchromatic emission of these hidden monsters provides a unique way to explore their physical properties and therefore the co-evolution between SMBHs and their hosts. However, most of modelling methods are not suitable for the extremely luminous systems with obscured Active Galactic Nucleus (AGN) emission. Here we present two case studies of panchromatic modeling of the extremely luminous dust-obscured quasars at the cosmic noon.

Keywords. galaxies: evolution, galaxies: high-redshift, galaxies: active,quasars: general, infrared: galaxies.

1. Introduction

Recently, a new population of luminous, dust-obscured galaxies (Wu *et al.* 2012) has been successfully discovered with a so-called *W1W2*-dropout color-selected method (Eisenhardt *et al.* 2012). Several works have suggested that those dust-obscured quasars may represent a key transition phase during the evolution of massive galaxies, linking starbursts and luminous unobscured quasars (Wu *et al.* 2012, Díaz-Santos *et al.* 2016, Fan *et al.* 2016a, Fan *et al.* 2016b, Fan *et al.* 2017, Fan *et al.* 2018b). Modeling the panchromatic emission of these hidden monsters provides a unique way to explore their physical properties and therefore the co-evolution between SMBHs and their hosts. Here we present two case studies of panchromatic modeling of the extremely luminous dust-obscured quasars at the cosmic noon.

2. Panchromatic modeling

Sample: Our sources studied here are selected from the *WISE* All-Sky Source catalog. The basic idea of sample selection is to search for more heavily obscured galaxies at high redshift ($z > 1.5$), whose W1 (3.4 μm) and W2 (4.6 μm), sampling the rest-frame near-infrared obscuration, are faint or undetected by *WISE*, but whose W3 (12 μm) and W4 (22 μm) emission, tracing the hot dust heated by starbursts and/or AGN, are well

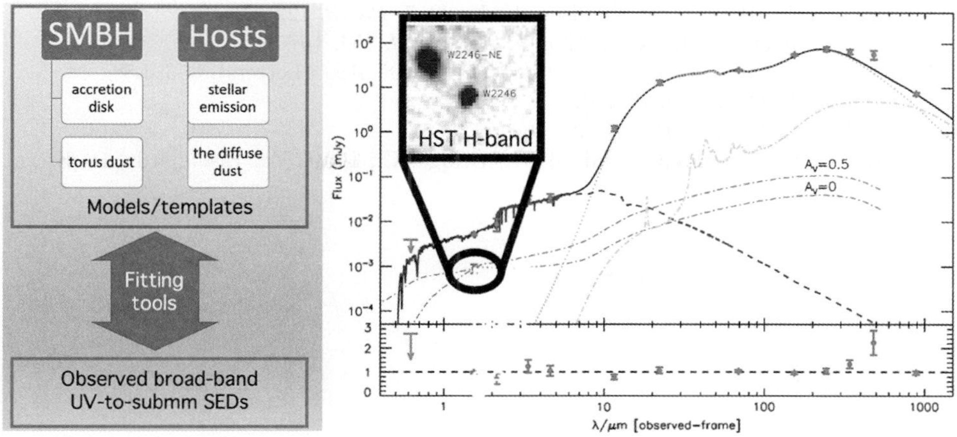

Figure 1. *left:* A schematic description of our basic idea for panchromatic modelling of the extremely dust-obscured quasars at $z \sim 3$. *Right:* An example of UV-to-FIR SED fitting of W2246-0526 at $z = 4.6$ with SED3FIT. The contribution of AGN accretion disk to the H-band has been constrained by the HST F160W image decomposition.

detected with SNR>5. Most of them are extremely luminous dust-obscured quasars at $1 < z < 4$ suggested by UV/optical spectra and X-ray data (Wu *et al.* 2012).

IR SED decomposition (Fan *et al.* 2016a): We select 22 submm-detected dust-obscured quasars with spectroscopic redshift. Their observed IR SEDs have been constructed by combining *WISE*, *Herschel* PACS and SPIRE, *JCMT* SCUBA-2 850 μm data and other available millimeter observations. We use a Bayesian SED analysis approach (Han & Han 2012, Han & Han 2014) to decompose the observed IR SEDs into two components: torus and cold dust. We can compare the different models by using the Bayes factor. One-dimensional and two-dimensional marginalized posterior probability distributions of the model parameters and the derived quantities can be obtained.

An example of UV-to-FIR SED fitting (Fan *et al.* 2018a) We construct the rest-frame UV/optical-to-far-IR SED of W2246-0526, a *WISE*-selected, hyperluminous dust-obscured galaxy at $z = 4.6$. We use the three-component SED-fitting code SEDFIT by (Berta *et al.* 2013). In the right panel of Figure 1, we present the best-fit model SED (solid line), which provides a rather good description of the rest-frame UV/optical-to-far-IR SED. The high spatial resolution HST H-band image of W2246-0526 has been decomposed with a Sérsic + PSF model. The flux of the decomposed PSF component is about six times weaker than the Sérsic component. We assume that the PSF component comes from the scattered AGN emission. AGN contamination has the negligible effect on the stellar mass measurement of W2246-0526.

References

Berta, S., Lutz, D., Santini, P., Wuyts, S., Rosario, D., Brisbin, D., Cooray, A., Franceschini, A., *et al.* 2013, *A&A*, 551, 100

Díaz-Santos, T., Assef, R. J., Blain, A. W., Tsai, C.-W., Aravena, M., Eisenhardt, P., Wu, J., Stern, D., & Bridge, C. 2016, *ApJ* (Letters), 816, L6

Eisenhardt, P. R. M., Wu, J., Tsai, C.-W., Assef, R., Benford, D., Blain, A., Bridge, C., & Condon, J. J. 2012, *ApJ*, 755, 173

Fan, L., Han, Y., Fang, G., Gao, Y., Zhang, D., Jiang, X., Wu, Q., Yang, J., *et al.* 2016, *ApJ* (Letters), 822, L32

Fan, L., Han, Y., Nikutta, R., Drouart, G., & Knudsen, K. K. 2016, *ApJ*, 823, 107

Fan, L., Jones, S. F., Han, Y., & Knudsen, K. K. 2017, *PASP*, 129, 124101
Fan, L., Gao, Y., Knudsen, K. K., & Shu, X. 2018, *ApJ*, 854, 157
Fan, L., Knudsen, K. K., Fogasy, J., & Drouart, G. 2018, *ApJ* (Letters), 856, L5
Han, Y., & Han, Z. 2012, *ApJ*, 749, 123
Han, Y., & Han, Z. 2014, *ApJS*, 215, 2
Wu, J., Tsai, C.-W., Sayers, J., Benford, D., Bridge, C., Blain, A., Eisenhardt, P., Stern, D. *et al.* 2012, *ApJ*, 756, 96

Challenges in Panchromatic Modelling with Next Generation Facilities
Proceedings IAU Symposium No. 341, 2018
M. Boquien, E. Lusso, C. Gruppioni & P. Tissera, eds.

doi:10.1017/S1743921319001376

The new fundamental plane dictating galaxy cluster evolution

Yutaka Fujita[1], Keiichi Umetsu[2], Elena Rasia[3], Massimo Meneghetti[4,5], Megan Donahue[6], Elinor Medezinski[7], Nobuhiro Okabe[8], Marc Postman[9] and Stefano Ettori[4,5]

[1]Department of Earth and Space Science, Graduate School of Science, Osaka University, Toyonaka, Osaka 560-0043, Japan
email: fujita@astro-osaka.jp

[2]Institute of Astronomy and Astrophysics, Academia Sinica, P.O. Box 23-141, Taipei 10617, Taiwan

[3]INAF, Osservatorio Astronomico di Trieste, via Tiepolo 11, I-34131, Trieste, Italy

[4]Osservatorio di Astrofisica e Scienza dello Spazio, via Pietro Gobetti 93/3, 40129 Bologna, Italy

[5]Sezione di Bologna, viale Berti Pichat 6/2, I-40127 Bologna, Italy

[6]Physics and Astronomy Department, Michigan State University, East Lansing, MI 48824, USA

[7]Department of Astrophysical Sciences, 4 Ivy Lane, Princeton, NJ 08544, USA

[8]Department of Physical Science, Hiroshima University, 1-3-1 Kagamiyama, Higashi-Hiroshima, Hiroshima 739-8526, Japan

[9]Space Telescope Science Institute, 3700 San Martin Drive, Baltimore, MD 21208, USA

Abstract. In this study, we show that the characteristic radius r_s, mass M_s, and the X-ray temperature, T_X, of galaxy clusters form a thin plane in the space of $(\log r_s, \log M_s, \log T_X)$. This tight correlation indicates that the cluster structure including the temperature is affected by the formation time of individual clusters. Numerical simulations show that clusters move along the fundamental plane as they evolve. The plane and the cluster evolution within the plane can be explained by a similarity solution of structure formation. The angle of the plane shows that clusters have not achieved "virial equilibrium". The details of this study are written in Fujita *et al.* (2018a,b).

Keywords. galaxies: clusters: general — large-scale structure of universe

1. Fundamental plane

The structure of galaxy clusters can be represented by the NFW profile (Navarro *et al.* 1997):

$$\rho_{\rm DM}(r) = \frac{\delta_c \rho_c}{(r/r_s)(1+r/r_s)^2} \,. \tag{1.1}$$

The characteristic density r_s and mass M_s, which is the mass inside r_s, include information of cluster formation history (e.g. Wechsler *et al.* 2002; Zhao *et al.* 2003). For example, older clusters are more concentrated. Since the X-ray temperature reflects cluster structure, it should also be affected by cluster formation history. Thus, we studied the relation among r_s, M_s, and T_X for the CLASH cluster sample (Donahue *et al.* 2014; Umetsu *et al.* 2016). We found that clusters form a plane (Fig. 1a), which means that

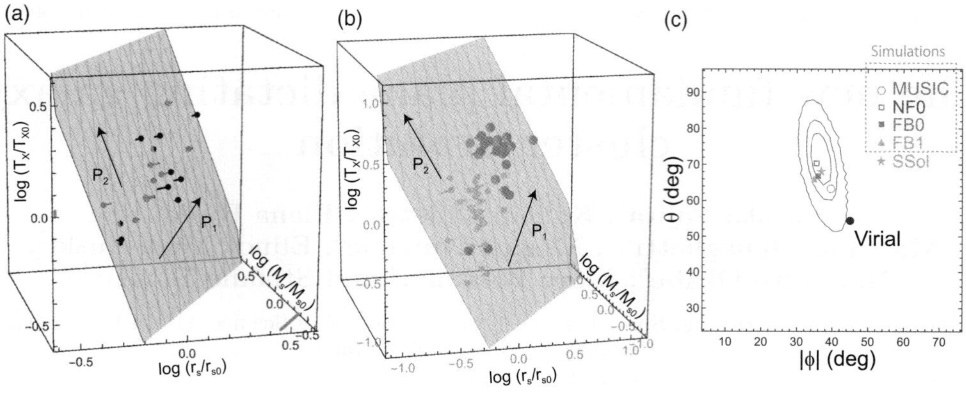

Figure 1. (a) CLASH clusters distributed on the fundamental plane. (b) Distribution of simulated clusters (blue: $z=0$, green: $z=1$). (c) Direction of the plane normal. While CLASH results (contours) are inconsistent with virial equilibrium (black dot), they are consistent with simulation results and the similarity solution (SSol). Figures are reconstructed from Fujita *et al.* (2018a).

T_X is actually determined by the formation history. Numerical simulations confirmed the plane and showed that clusters evolve along the plane in the direction of P_1 (Fig. 1b). We also found that the direction of the plane normal is not consistent with virial equilibrium ($T_X \propto M_s/r_s$; Fig. 1c).

2. Discussion

In Fujita *et al.* (2018a), we showed that the properties of the plane can be explained by a similarity solution by Bertschinger (1985). The solution shows that clusters are not in virial equilibrium because of matter accretion. It predicts a relation, $T_X \propto M_s^{3/2}/r_s^2$, that is consistent with observations and simulations (SSol in Fig. 1c). Moreover, it can also explain the direction of the cluster motion on the plane (P_1 in Fig. 1b).

The reason why clusters form a plane is the following. In the universe, smaller (less massive) density perturbations tend to have larger amplitudes (first dimension). However, there is a dispersion about the relation. For example, there is a range of mass scale for a given amplitude. This introduces another dimension and makes the planar distribution.

The well-known mass-temperature relation of clusters ($M_{200} \propto T_X^{3/2}$) can also be explained by the fundamental plane and the mass dependence of halo concentration without assuming that clusters are in virial equilibrium Fujita *et al.* (2018b).

References

Bertschinger, E. 1985, *ApJ*, 58, 39

Donahue, M., Voit, G. M., Mahdavi, A., Umetsu, K., Ettori, S., Merten, J., Postman, M., Hoffer, A. *et al.* 2014, *ApJ*, 794, 136

Fujita, Y., Umetsu, K., Rasia, E., Meneghetti, M., Donahue, M., Medezinski, E., Okabe, N., & Postman, M. 2018, *ApJ*, 857, 118

Fujita, Y., Umetsu, K., Ettori, S., Rasia, E., Okabe, N., & Meneghetti, M. 2018, *ApJ*, 863, 37

Navarro, J. F., Frenk, C. S., & White, S. D. M. 1997, *ApJ*, 490, 493

Umetsu, K., Zitrin, A., Gruen, D., Merten, J., Donahue, M., & Postman, M. 2016, *ApJ*, 821, 116

Wechsler, R. H., Bullock, J. S., Primack, J. R. Kravtsov, A. V., & Dekel, A. 2002, *ApJ*, 568, 52

Zhao, D. H. Mo, H. J. Jing, Y. P., & Börner, G. 2003, *MNRAS*, 339, 12

Challenges in Panchromatic Modelling with Next Generation Facilities
Proceedings IAU Symposium No. 341, 2018
M. Boquien, E. Lusso, C. Gruppioni & P. Tissera, eds.

doi:10.1017/S1743921319002199

Radial resolved galaxy disks models based on semi-analytic models of galaxy formation†

Jian Fu

Key Laboratory for Research in Galaxies and Cosmology, Shanghai Astronomical Observatory, CAS, 80 Nandan Road., Shanghai, 200030, China
email: fujian@shao.ac.cn

Abstract. We show our work on the L-Galaxies semi-analytic models of galaxy formation, which includes the radial resolved distribution of star, gas, SFR and metallicity on each galaxy disk. The newest version of the codes include the H_2-to-HI gas transition prescriptions and the chemical enrichment of various elements. Our revised model can give results on cold gas components, radial metallicity gradients and scaling relations, which can fit the recent observations.

Keywords. galaxies: formation, galaxies: ISM, ISM: atoms, galaxies: abundances

1. Model Introduction

Our work are based on the newest version of the Munich group L-Galaxies semi-analytic models of galaxy formation (Henriques *et al.* 2020), which combines the previous work on the prescriptions of radial profiles, atomic-to-molecular gas components (Fu *et al.* 2013) and chemical enrichment models (Yates *et al.* 2014) in AGB star, SN-Ia and SN-II for many different chemical elements. The models now run on Millennium and Millennium II haloes to study disk and elliptical galaxies in different mass ranges. Aquarius haloes are adopted to study the dwarf satellite galaxies in the local group. On the other hand, ELUCID (Wang *et al.* 2014) haloes give opportunities to compare the model results directly to the local universe, which are based on the reconstruction of the initial density field according to local observations from SDSS.

2. Results

The following are the main results from our current models:
(i) The new models can fit the HI mass function from ALFALFA 100 to very low mass end, which gives a success for ELUCID simulations (see Fig. 1).
(ii) Our radial resolved galaxy disks can give the results of chemical gradient of various elements, which are consistent with recent IFU results (see Fig. 2).
(iii) We construct mock catalogues for 21cm HI gas survey for future SKA and FAST (see Fig. 3), which offers opportunities to test the models and ΛCDM cosmology.
(iv) Current single phase cooling prescription predicts too high X-ray luminosity compared to the X-ray observations (e.g Chandra), which implies more realistic cooling prescriptions (e.g multiple phase cooling) and hot gas radial distribution prescriptions are needed in the models.

† Supported by Natural Science Foundation of China (No. U1531123) and the Youth innovation Promotion Association, Chinese Academy of Sciences.

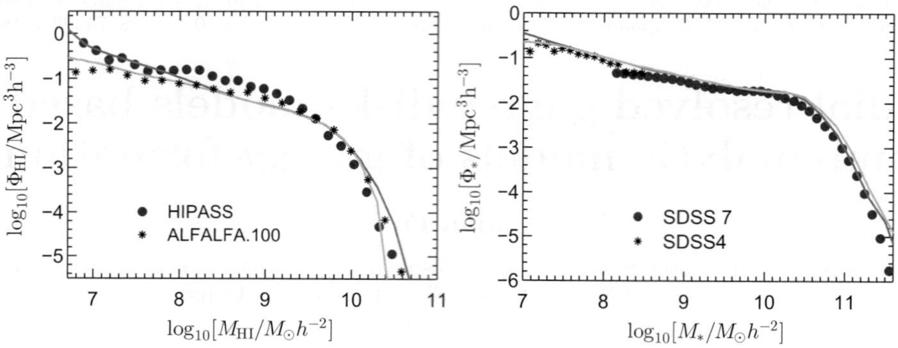

Figure 1. Model results of HI mass functions and stellar mass functions at z=0. Green curves are based on ELUCID haloes and red curves are based on Millennium haloes. Black dots and blue dots are from various observations.

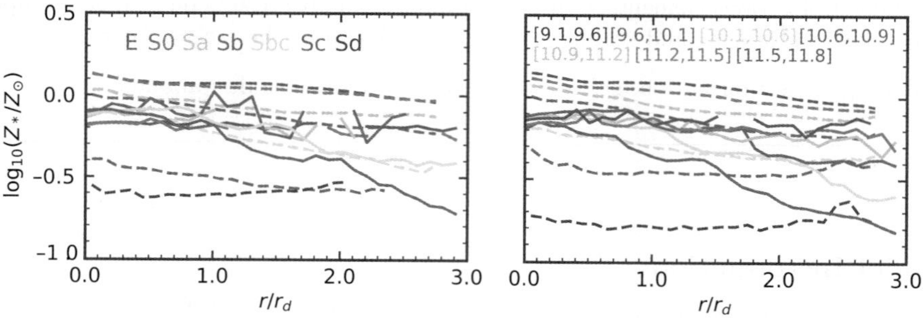

Figure 2. The model results of chemical radial gradient (dashed curves) vs. observational results from CALIFA (solid curves). In the left panel, galaxy samples are divided into different galaxy types. In the right panel, the samples are divided into different stellar mass bins.

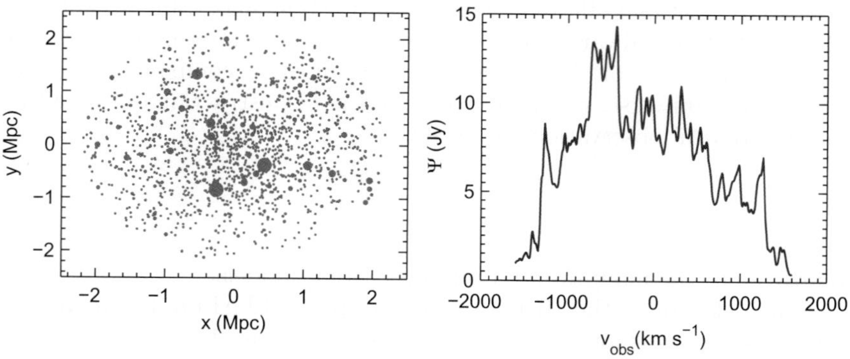

Figure 3. The synthetic observations of a Virgo sized galaxy cluster in the model samples. The left panel shows the HI gas distribution, in which the size represents the mass of HI gas. The right panel shows the 21cm flux profiles, in which the velocity resolution is 5 km/s.

References

Fu, J., Kauffmann, G., Huang, M.-L., *et al.* 2013, *MNRAS*, 434, 1531

Wang, H., Mo, H. J., Yang, X., Jing, Y. P., & Lin, W. P. 2014, *ApJ*, 794, 94

Yates, R. M. & Kauffmann, G. 2014, *MNRAS*, 439, 3817

Henriques, B. M., Yates, R. M., Fu J., *et al.* 2020, *MNRAS*, 491, 5795H

Challenges in Panchromatic Modelling with Next Generation Facilities
Proceedings IAU Symposium No. 341, 2018
M. Boquien, E. Lusso, C. Gruppioni & P. Tissera, eds.

doi:10.1017/S1743921319002163

A comparison of star formation history between NGC 300 and M33

Xiaoyu Kang[1,2,3], **Fenghui Zhang**[1,2,3] **and Ruixiang Chang**[4]

[1]Yunnan Observatories, Chinese Academy of Sciences, 396 Yangfangwang, Guandu District, Kunming, 650216, P. R. China
email: kxyysl@ynao.ac.cn

[2]Key Laboratory for the Structure and Evolution of Celestial Objects, Chinese Academy of Sciences, 396 Yangfangwang, Guandu District, Kunming, 650216, P. R. China

[3]Center for Astronomical Mega-Science, Chinese Academy of Sciences, 20A Datun Road, Chaoyang District, Beijing, 100012, P. R. China

[4]Key Laboratory for Research in Galaxies and Cosmology, Shanghai Astronomical Observatory, CAS, 80 Nandan Road, Shanghai, 200030, China

Abstract. NGC 300 is a near-optical twin of the Local Group galaxy M33, which are benchmarks for understanding late-type spiral galaxies. They are two bulgeless and low-mass spiral galaxies in different environments. In order to explore the common properties and differences between the two nearby low-mass systems, we first use the simple chemical evolution model to explore the star formation history (SFH) of NGC300 and M33, and then compare the feasible model predicted SFH of NGC 300 with that of M33. Through comparing the SFHs between them, it can be found that the mean stellar age of NGC 300 is older than that of M33, there is a recent lack of primordial gas infall onto the disk of NGC 300, recent star formation along the disk of NGC 300 is less active than that of M33, and the local environment may play a key role in the secular evolution of a galaxy.

Keywords. galaxies: evolution, galaxies: spiral, galaxies: individual: NGC 300 and M33

1. Introduction

Motivated by the result of Guglielmo *et al.* (2015) that the average star formation history (SFH) of a galaxy depends on its stellar mass, but galaxies of a given stellar mass have different SFHs depending on their environments. NGC 300 is a near-optical twin to the Local Group galaxy M33 in Hubble type and mass, and they are two bulgeless, low-mass pure-disk galaxies. Although they are similar in appearance, M33 has a disk break at ~8 kpc, while NGC 300 has a pure exponential disk out to ~14 kpc. Furthermore, an HI bridge exists between M33 and M31, indicating that they interacted with each other in the past. Compared to M33, NGC 300 is a relatively isolated system. The aforementioned differences indicate that they may experience different evolutionary histories, therefore it is necessary to study and compare the SFH of NGC 300 with that of M33.

2. Model

Simple chemical evolution models for NGC 300 (Kang *et al.* 2016) and M33 (Kang *et al.* 2012) are constructed to build a bridge between their SFHs and their observed properties. Here we briefly introduce the main ingredients and basic assumptions of the models. Their disks are assumed to grow up gradually by continuous infall of primordial gas ($X = 0.7571, Y_{\rm p} = 0.2429, Z = 0$) from their halos. To describe the rate at which stars are forming from the cold gas, the molecular-hydrogen-correlated star formation

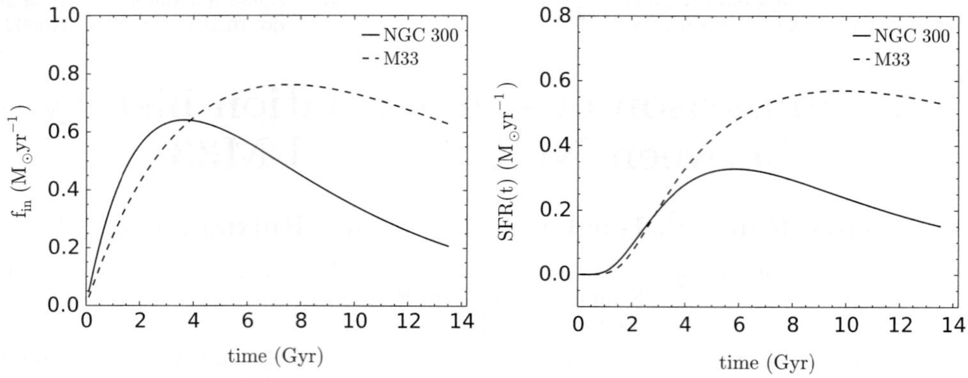

Figure 1. The feasible model predicted time evolution of gas infall rate (left) and star formation rate (right) for NGC 300 (solid) and for M33 (dashed).

law is adopted in the models (Leroy *et al.* 2008). Since NGC 300 and M33 are low-mass galaxies, the gas-outflow process is also considered in the models, and the gas outflow rate is assumed to be proportional to star formation rate (SFR). Both instantaneous recycling assumption and instantaneous mixing of the interstellar medium with stellar ejecta are also assumed in the models.

3. Result

Investigations of the feasible models for NGC 300 and M33 are in Kang *et al.* (2016) and Kang *et al.* (2012), respectively. The left panel and the right panel of Figure 1 respectively plots the feasible model predicted time evolution of gas infall rate and SFR. It can be seen from the left panel of Figure 1 that the gas infall rate of NGC 300 is gradually increasing with time and reaches its peak at about 10 Gyr ago, and then slowly drops down, while that of M33 achieves its peak around 6 Gyr ago and then gradually falls. Moreover, the gas infall rate of NGC 300 increases and falls faster than that of M33, and the present-day gas infall rate of NGC 300 is much lower than that of M33. On the other hand, the right panel of Figure 1 reveals that the SFR of NGC 300 is lower than that of M33 after $z \sim 2$, and the difference between them increases with time and reaches the maximum at the present time. Furthermore, the feasible model predicted values of present-day SFR are $0.51\,M_\odot yr^{-1}$ for M33 and $0.18\,M_\odot yr^{-1}$ for NGC 300, which are in good agreement with the observed data for them. All these indicate that, compared to M33, less gas infall and weaker star formation occurred recently in the disk of NGC 300. Our results show that, even though NGC 300 and M33 have similar stellar mass and morphology, they have experienced different SFHs due to their environmental differences (See Kang *et al.* 2016 for more details).

Acknowledgements

This work is supported by the National Natural Science Foundation of China (NSFC) grant No. 11403092, 11373053, 11573062, 11390373, 11390374, and the Youth Project of Western Light of Chinese Academy of Sciences.

References

Guglielmo, V., Poggianti, B. M., Moretti, A., *et al.* 2015, *MNRAS*, 450, 2749
Kang, X. Y., Zhang, F. H., Chang, R. X., *et al.* 2016, *A&A*, 585, A20
Kang, X. Y., Chang, R. X., Yin, J., *et al.* 2012, *MNRAS*, 426, 1455
Leroy, A. K., Walter, F., Brinks, E., *et al.* 2008, *AJ*, 136, 2782

Challenges in Panchromatic Modelling with Next Generation Facilities
Proceedings IAU Symposium No. 341, 2018
M. Boquien, E. Lusso, C. Gruppioni & P. Tissera, eds.

doi:10.1017/S1743921319002631

The environment of large equivalent width Lyα emitters at $z \sim 3$

Arisa Kida[1], **Satoshi Kikuta**[2,3], **Yuichi Matsuda**[2,3], ***et al.***

[1]Kwansei Gakuin University, 2-1 Gakuen, Sanda, Hyogo, 669-1337, Japan
email: arisa.kd@gmail.com

[2]National Astronomical Observatory of Japan, 2-21-1 Osawa, Mitaka, Tokyo, 181-8588, Japan

[3]SOKENDAI (the Graduate University for Advanced Studies), 2-21-1 Osawa, Mitaka, Tokyo 181-8588, Japan

Abstract. We present the environmental properties of a sample of 260 Lyα emitters (LAEs) with extremely large equivalent widths (EW_0 >240 Å) at $z \sim 3$. Such large EW LAEs may be very low-metal / young starbursts in void regions or HI gas / galaxies illuminated by ionizing radiation from the QSO. We undertook Subaru/HSC Lyα imaging observations around a hyper luminous QSO at $z = 2.84$. We identified 3490 LAEs including 260 large EW LAEs. We found that the large EW LAEs tend to lie at galaxy over-dense regions surrounding the QSO, suggesting that most of the large EW LAEs are illuminated by the QSO. From the radial distribution of the large EW LAEs from the QSO, we found that the fraction of large EW LAE shows two excesses in the range of 2 to 5 Mpc and 14 to 16 Mpc from the QSO. If the two peaks are reflected with the QSO activity, the QSO should have episodic lifetimes with ~10 Myr duration and ~30 Myr separations. However, we require future spectroscopic observations to investigate if the large EW LAEs far from the QSO are low-metal young starbursts.

Keywords. Lyα emitters, QSO, Environment

1. Introduction

The accretion disk created when gas or dust falls into the black hole is shining. It is a very compact and heavily time varying object. QSO lifetime is generally said to be 1 Myr to 100 Myr (Martini 2004), but we do not know what kind of activities have occurred among them. In this time, we thought about triggers of the QSO activity by investigating the QSO activity. The equivalent width (EW) of emission line shows the line strength relative to the continuum. Lyα emitters (LAEs) with extremely large EW (EW_0 >240 Å) are thought to be good analogues of the PopIII galaxies or first galaxies, because such large EW could only be produced from star formation with very low metallicity or in a short period (Schaerer 2002). Theoretical works predict that such galaxies could remain in un-evolved region (i.e., void regions) even at $z \sim 3$ (Tornatore *et al.* 2007). An alternative case is HI gas / galaxy illuminated by ionizing radiation from the QSO (Cantalupo *et al.* 2012). In this case, the large EW LAEs should distribute in the vicinity of the QSOs, which are likely to be formed at the peak of over-dense regions.

2. Data and LAE Selection

We undertook deep, wide-field Lyα imaging observations around a hyper luminous QSO at $z = 2.843$ (HS1549+1919, Trainor & Steidel 2013) by using a custom narrow-band filter, NB468, on Subaru/HSC. We obtained 2.2 hours g-band and 6.3 hours

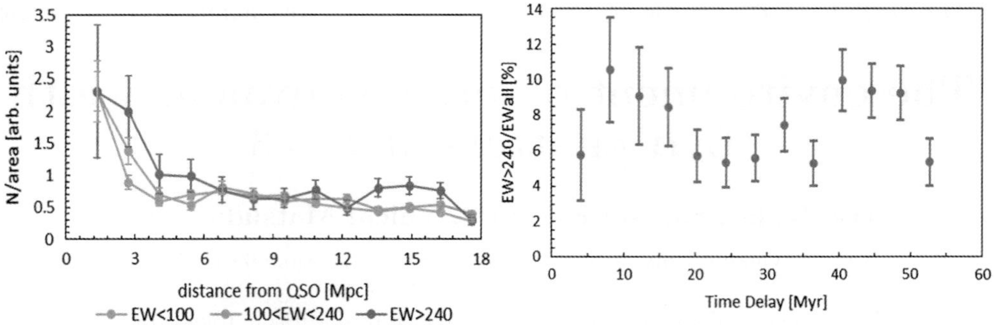

Figure 1. **(Left)** Comparison of surface densities of LAEs with $EW_0<100$ Å (green), 100 Å $<EW_0<240$ Å (orange), and $EW_0>240$ Å (blue) as a function of projected distance from the QSO. The surface densities are normalized at the shortest distance (~1 Mpc). We found that large EW LAEs show excess in the range of 2 to 5 Mpc and 14 to 16 Mpc. **(Right)** Fraction of large EW LAEs over all LAEs as a function of time delay (i.e., light travel time from the QSO). This fraction shows excesses at 10 to 20 Myr and 40 to 50 Myr, suggesting that QSO has episodic lifetimes with ~10 Myr duration and ~30 Myr separation.

NB468 data. We identified 3490 LAEs brighter than $NB468 < 26.6$mag and rest-frame EW ($EW_0 \geqslant 12$ Å) including 260 large EW LAEs ($EW_0 \geqslant 240$ Å) in this field.

3. Results and Discussion

We found that the large EW LAEs tend to lie at over dense regions of normal LAEs, suggesting that the large EW LAEs are HI gas / galaxy illuminated by ionizing radiation from the QSO rather than low metal galaxies in void regions (Figure 1, left). It seems that these two peaks indicate that there is a wave in QSO activity. Furthermore, considering the history of activities based on the distance from QSO, the fraction of large EW LAEs shows excesses at 10 to 20 Myr and 40 to 50 Myr, suggesting that the QSO has episodic lifetimes with ~10 Myr duration and ~30 Myr separations (Figure 1, right).

In this work, we investigated the QSO lifetime 11 Gyrs ago with Lyα light echo. Although we found a hint of episodic lifetime, each duration of 10 Myr is consistent with the estimates from the previous studies (1 Myr $\lesssim t_Q \lesssim$ 20 Myr, Trainor & Steidel 2013). We note that the lowest fraction seems to have a constant value around 6% in Figure 1 (right). This is possibly due to the uncertainty of the true distance from the QSO because we used the projected distance assuming the LAEs have the same redshifts. Another issue is that our large EW LAE sample far from the QSO potentially includes young low-metal galaxies in un-evolved regions, because the second excess of large EW LAE at ~15 Mpc are relatively far from the QSO. Therefore we require future spectroscopic follow-up observations to measure the true 3D distance from the QSO and to search for other emission lines for metallicity measurements.

Acknowledgement

This work was partially supported by Summer Student Program (2018) of NAOJ / SOKENDAI and KAKENHI (25287043, 17H04831 and 17KK0098).

References

Martini, P. 2004, Coevolution of Black Holes and Galaxies, 169
Schaerer, D. 2002, *A&A*, 382, 28
Tornatore, L., Ferrara, A., & Schneider, R. 2007, *MNRAS*, 382, 945
Cantalupo, S., Lilly, S. J., & Haehnelt, M. G. 2012, *MNRAS*, 425, 1992
Trainor, R. & Steidel, C. C. 2013, *ApJ*, 775, L3

Challenges in Panchromatic Modelling with Next Generation Facilities
Proceedings IAU Symposium No. 341, 2018
M. Boquien, E. Lusso, C. Gruppioni & P. Tissera, eds.

doi:10.1017/S1743921319002448

DEIMOS and MOSFIRE spectroscopy of star-forming galaxies in the AKARI NEP-Deep field

Helen K. Kim[1], Matt Malkan[1], Nagisa Oi[2], Toshinobu Takagi[3], Denis Burgarella[4], Véronique Buat[4], Samir Salim[5], Chris Pearson[6] and Hideo Matsuhara[7]

[1]Department of Physics and Astronomy, UCLA, Los Angeles, CA 90095-1547, USA
email: hkim@astro.ucla.edu

[2]Tokyo University of Science, 1-3 Kagurazaka, Shinjuku-ku, Tokyo 162-8601, Japan

[3]Japan Space Forum, 3-2-1, Kandasurugadai, Chiyoda-ku, Tokyo 101-0062, Japan

[4]Laboratoire d'Astrophysique de Marseille, Aix-Marseille Université, CNRS, F-13013 Marseille, France

[5]Department of Astronomy, Indiana University, Bloomington, IN 47404, USA

[6]Department of Physical Sciences, The Open University, Milton Keynes, MK7 6AA, UK

[7]Institute of Space and Astronautical Science, Japan Aerospace Exploration Agency, 3-1-1 Yoshinodai, Chuo-ku, Sagamihara, Kanagawa 252-5210, Japan

Abstract. Observing high resolution optical to infrared spectra is crucial to understanding how energy is generated in galaxies. We present follow-up optical Keck-II/DEIMOS and infrared Keck-I/MOSFIRE spectra of ~200 galaxies in the AKARI/North Ecliptic Pole Deep survey region at intermediate redshift. From rest-frame optical emission lines, we classify most of our objects as star-forming (53%), with the MIR selection favoring relatively massive galaxies (median log $M/M_{\odot}$~10.3). In addition, we combine our spectroscopic redshifts with UV to FIR photometry as inputs in order to model SEDs with CIGALE, and we measure the PAH 7.7 μm luminosity as an SFR indicator.

Keywords. techniques: spectroscopic, galaxies: evolution, infrared: galaxies

1. Introduction

The AKARI North Ecliptic Pole (NEP) Deep Field is a unique region with continuous 2-24 μm coverage in ~0.5 square degrees from the Infrared Camera (IRC), providing crucial mid-infrared data of the dust-obscured universe out to z~1 (Murata *et al.* 2013, Oi *et al.* 2014). We present optical to NIR spectroscopic results of ~200 galaxies in the NEP-Deep field, combining Keck-I/MOSFIRE J- and Y-band and Keck-II/DEIMOS observations. DEIMOS targets prioritized intermediate redshift (z~1) sources with AKARI/IRC photometry and Chandra X-ray sources to include possible (obscured) AGN. MOSFIRE targets prioritized strong emission line galaxies with previous optical spectroscopy and AKARI sources in the range $0.7<z<1.1$ in order to measure Hα/[NII]. In addition, we used the Y-band to target Hα/[NII] sources at $0.5<z<0.8$.

2. Source Classification

Most of our DEIMOS galaxies on the BPT classification diagram (Fig. 1, left) lie along the star-forming sequence (53%) with 30% composite, 4% Seyfert, and 13% LINERs. In comparison, the sample of AKARI NEP-Wide galaxies (Shim *et al.* 2013, median z~0.24

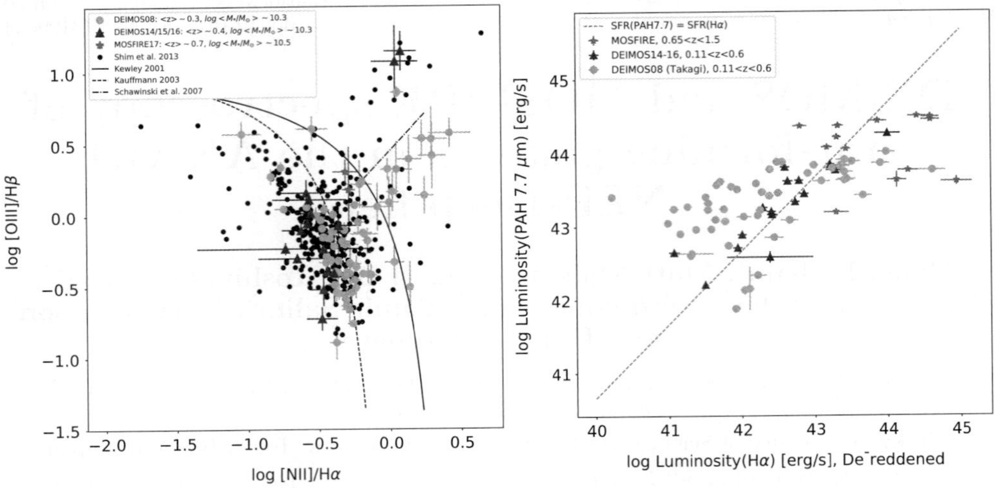

Figure 1. Left: BPT-NII source classification diagram. Right: PAH 7.7 μm luminosity vs. the dust-corrected Hα luminosity for combined DEIMOS and (higher z) MOSFIRE sample.

on BPT) contains a smaller fraction of AGN. Our IR-selected sample contains relatively massive galaxies. Although not shown here, the BPT diagram was consistent with the Mass-Excitation Diagram ([OIII]/Hβ vs. stellar mass).

3. Spectral Energy Distributions and PAH Luminosity Estimates

We combined our spectroscopic redshifts with UV to FIR photometry in order to model SEDs with CIGALE. We assumed a delayed star formation history with exponential burst, Salpeter IMF, and solar metallicity for the stellar model. Key parameters that were allowed to vary included color excess E(B-V) of the young stellar population (0–1.6), mass fraction of PAH dust (0.47–4.58), and AGN fraction (0–0.4).

In order to investigate the PAH 7.7 μm feature as a SF indicator, we compare the PAH 7.7 luminosity from CIGALE best fits to the de-reddened Hα luminosity (Fig. 1, right) assuming the Calzetti *et al.* (2000) extinction law and E(B-V) from CIGALE for our combined DEIMOS S11 (z~0.5) and MOSFIRE L15 (z~1) sample. The dashed line represents the luminosity relation such that SFR(PAH7.7)=SFR(Hα), where SFR(PAH7.7) [$M_\odot$/yr] = 4.56×10^{-9} L(PAH7.7) [$L_\odot$] (Hernán-Caballero *et al.* 2009) and SFR(Hα) [$M_\odot$/yr] = 5.49×10^{-42} L(Hα) [erg/s] (Kennicutt *et al.* 2009). Our limited MOSFIRE points at z~1 (red stars) cannot determine how the SFR(PAH7.7)/SFR(Hα) correlation evolves from low z to z>1, highlighting the need for further data.

References

Calzetti, D., Armus, L., Bohlin, R. C., Kinney, A. L., Koornneef, J., & Storchi-Bergmann, T. 2000, *ApJ*, 533, 682

Hernán-Caballero, A., Pérez-Fournon, I., Hatziminaoglou, E., Afonso-Luis, A., Rowan-Robinson, M., Rigopoulou, D., Farrah, D., Lonsdale, C. J., *et al.* 2009, *MNRAS*, 395, 1695

Kennicutt, R. C., Hao, C.-N., Calzetti, D., Moustakas, J., Dale, D. A., Bendo, G., Engelbracht, C. W., Johnson, B. D., *et al.* 2009, *ApJ*, 703, 1672

Murata, K., Matsuhara, H., Wada, T., Arimatsu, K., Oi, N., Takagi, T., Oyabu, S., Goto, T., *et al.* 2013, *A&A*, 559A, 132M

Oi, N., Matsuhara, H., Murata, K., Goto, T., Wada, T., Takagi, T., Ohyama, Y., Malkan, M., *et al.* 2014, *A&A*, 566A, 60O

Shim, H., Im, M., Ko, J., Jeon, Y., Karouzos, M., Kim, S. J., Lee, H. M., Papovich, C., *et al.* 2013, *ApJS*, 207, 37S

Challenges in Panchromatic Modelling with Next Generation Facilities
Proceedings IAU Symposium No. 341, 2018
M. Boquien, E. Lusso, C. Gruppioni & P. Tissera, eds.

doi:10.1017/S1743921319002928

Mid-infrared PAH emission from star-forming galaxies selected at 250 μm

Seong Jin Kim and AKARI NEP Collaboration

National Tsing Hua University No. 101, Section 2, Kuang-Fu Road, Hsinchu, 30013, Taiwan
email: seongini@phys.nthu.edu.tw

Abstract. The infrared (IR) galaxies detected at Herschel/SPIRE 250 μm band over the AKARI's NEP-Wide field are various types of dusty star-forming (SF) galaxies ranging from quiescent to starbursts, having mid-IR polycyclic aromatic hydrocarbon (PAH) features near 8 μm. The measurements of the 8 μm luminosity ($L_{8\mu m}$) along with the total infrared luminosity (L_{IR}) based on the physical modeling of SEDs a take unique advantage of the continuous near- to mid-IR coverage, far-IR data points, and spectroscopically determined accurate redshifts. Our sample shows shortage of 8 μm luminosity compared to the total IR luminosity. This deficit gets severe in more luminous IR galaxies, suggesting PAH molecules in these galaxies are destroyed by a strong radiation field from SF regions, or the existence of a unexpectedly large amount of cold dust in the ISM that contributes to L_{IR}.

Keywords. galaxies: evolution – infrared: galaxies

1. Introduction

The 8 μm luminosity (dominated by PAH features) of a normal SF galaxy is believed to have a relation with the overall SF activity of a whole galaxy (Elbaz *et al.* 2011; Schreiber *et al.* 2016). With expanding interests in dust contents of dusty SF galaxies, infrared (IR) wavelengths are getting more important in terms of physically motivated SED modeling and for the better understanding of dusty star formation activities. Thanks to the surveys over the north ecliptic pole (NEP) region by AKARI and Herschel (Matsuhara *et al.* 2006; Pearson *et al. in preparation*), we have a large number of mid-IR (MIR) galaxies observed at far-IR bands. The detection in the Herschel/SPIRE bands provides an important basis to select various types of IR-luminous populations of galaxies while the AKARI continuous band coverage gives more detailed information about the spectral energy distribution (SED) in the MIR. Now, we can derive more accurate 8 μm luminosity based on the continuous mid-IR coverage and carry out the consistent comparison with total infrared luminosity determined using reliable constraints on the far-IR peak.

2. Total infrared luminosity and 8μm luminosity

Our sample is composed of 250 mid-IR galaxies observed by AKARI over the NEP region (Kim *et al.* 2012), selected based on the SPIRE 250 μm detection as well as the optical spectroscopic redshift (Shim *et al.* 2013). The modeling of SED was carried out using `MAGPHYS` (da Cunha *et al.* 2008) based on the energy balance argument. We obtained physically meaningful fits to the observed photometric data from optical to sub-mm bands. This physical modeling gives important physical quantities such as star formation rate (SFR), stellar mass (M_*), and total infrared luminosity (L_{IR}). The sample galaxies analysed in this study appear to be various types, normally classified as SF/spiral, and ranges from quiescent to starburst in terms of the SFR–M_* relation. As shown in Fig. 1,

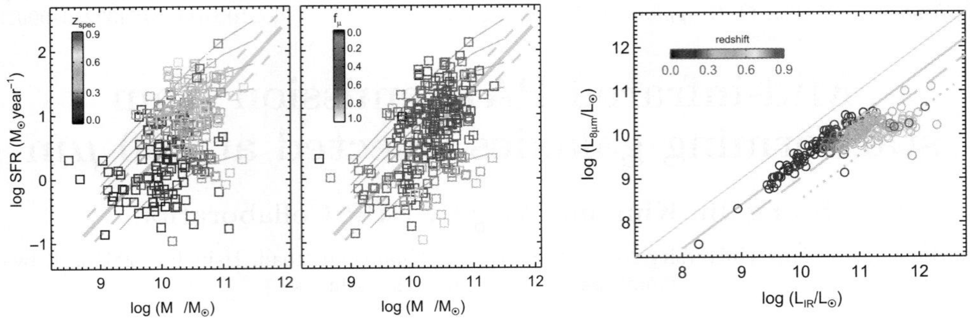

Figure 1. Distribution of our sample in the plane of SFR vs M_* with color bars showing redshift (left panel) and f_μ (middle panel). Thick gray lines in the background represent the IR galaxy main sequence (MS) and its 1-σ dispersion (Elbaz *et al.* 2011). Orange lines show the redshift evolution of MS from $z = 0$ to $z = 2$ (Schreiber *et al.* 2016). The right panel shows the distribution of $L_{8\mu m}$ as a function of L_{IR}. The long dashed line represents $L_{IR}/L_{8\mu m} = 10$. The dotted line represents $L_{IR}/L_{8\mu m} = 30$.

the wide spread of our sample seems to be due to the combination of types and the redshifts. Although it is not easy to get a clear and meaningful census on and off the main sequence (MS), we can test evolutionary properties of galaxies that have MIR PAHs. According to one of the resultant parameters, f_μ (defined by the fraction of cold dust contributing to the total IR luminosity), there seems to be an IR-luminous population which has abundant cold dust irrelevant to current SF activity. The shortage of 8 μm luminosity (dominated mostly by mid-IR PAH emission features) compared to the total IR luminosity seems mostly due to a strong radiation field from the SF region. As shown in the right panel of Fig. 1, some of our galaxies ($z > 0.6$) show an unexpectedly and surprisingly severe deficit.

References

da Cunha, E., Charlot, S., & Elbas, D. 2008, *MNRAS*, 388 1595

Elbaz, D., Dickinson, M., Hwang, H. S., Diaz-Santos, T., Magdis, G., Magnelli, B., Le Borgne, D., Galliano, F., *et al.* 2011, *A&A*, 533, A119

Kim, S. J., Lee, H. M., Matsuhara, H., Wada, T., Oyabu, S., Im, M., Jeon, Y., Kang, E., Ko, J., Lee, M. G., Takagi, T., *et al.* 2012, *A&A*, 548, A29

Matsuhara, H., Wada, T., Matsuura, S., Nakagawa, T., Kawada, M., Ohyama, Y., Pearson, C. P., Oyabu, S., *et al.* 2006, *PASJ*, 58, 673

Schreiber, C., Elbaz, D., Pannella, M., Ciesla, L., Wang, T., Koekemoer, A., Fafelski, M., & Daddi, E. 2016, *A&A*, 589.A35

Shim, H., Im, M., Ko, J., Jeon, Y., Karouzos, M., Kim, S. J., Lee, H. M., Papovich, C., *et al.* 2013, *ApJS*, 207, 37

Challenges in Panchromatic Modelling with Next Generation Facilities
Proceedings IAU Symposium No. 341, 2018
M. Boquien, E. Lusso, C. Gruppioni & P. Tissera, eds.

doi:10.1017/S1743921319004782

First [N II] 122 μm line detection in a starburst pair at $z = 4.7$

Minju M. Lee[1,2,3], Tohru Nagao[4], Carlos De Breuck[5], *et al.*

[1]Max-Planck-Institut für Extraterrestrische Physik (MPE), Giessenbachstr., D-85748 Garching, Germnay

[2]Division of Particle and Astrophysical Science, Graduate School of Science, Nagoya University, Furo-cho, Chikusa-ku, Nagoya 464-8602, Japan

[3]National Observatory of Japan, 2-21-1 Osawa, Mitaka, Tokyo 181-0015, Japan
email: minju@mpe.mpg.de

[4]Graduate School of Science and Engineering, Ehime University, 2-5 Bunkyo-cho, Matsuyama 790-8577, Japan

[5]European Southern Observatory, Karl Schwarzschild Straße 2, 85748 Garching, Germany

Abstract. We report the first detection of [N II] 122 μm line toward a QSO-SMG pair, BRI 1202-0725, at $z = 4.7$ using ALMA. Combining with [N II] 205 μm line detection and taking the line ratio of [NII]122/[NII]205, we constrain electron densities of both galaxies. The derived electron densities are 26^{+12}_{-11} and 134^{+50}_{-39} cm^{-3} for the SMG and the QSO, respectively, which are the first measurements for galaxies at $z > 4$. The electron density of the SMG is comparable to the Galactic plane and the average of local spiral galaxies, while the value for the QSO is comparable to local starbursts and optical-line based measurements for star-forming galaxies at $z \sim 2$–3. Considering the similar star-formation rates (SFRs) of $\approx$1000 $M_\odot$ yr^{-1} for both galaxies, our results suggest a large scatter of electron densities at fixed SFR and caution against using optical lines for dusty starbursts. The details of this report are presented in Lee *et al.* 2019 (submitted).

Keywords. galaxies: evolution – galaxies: high-redshift – galaxies: ISM – galaxies: starburst – submillimeter – ISM: evolution

1. Introduction

Fine structure line transitions at far-infrared (FIR) and sub-millimeter wavelengths are a powerful tool to probe dusty star-forming galaxies that form stars at a rate of $\approx$ 1000 M$_\odot$ yr^{-1} or even above, because they are less affected by dust extinction compared optical and ultra-violet emission lines. Two FIR fine-structure lines of [N II] (at 122 μm and 205 μm) originate from fully ionized gas with the ionization potential of 14.5 eV and the combination of these lines can be used as a probe of electron density.

BRI 1202-0725 at $z = 4.7$ is a pair system of two starburst galaxies (a QSO and a SMG) that remains the archetype of gas-rich mergers. Extremely high FIR luminosities of the QSO and the SMG (Omont *et al.* 1996; Iono *et al.* 2006; Yun *et al.* 2000, $\sim$10^{13} $L_\odot$) imply vigorous star formation rate of $\approx$ 1000 $M_\odot$ yr^{-1}. This unique system is an ideal target for probing ISM conditions of dusty starbursts at high-z with rich ancillary data sets. Here, we report the first detection of [N II] 122 μm with ALMA observations.

2. ALMA detection of [N II] 122 μm line from the QSO and the SMG

We detected [N II] 122 μm line from the QSO and the SMG with $> 4\sigma$. This is the first detection for $z > 4$ galaxies so far. Combining with [N II] 205 μm line detection for both galaxies (Pavesi *et al.* 2016) and taking the [N II] line ratio, we derived the electron

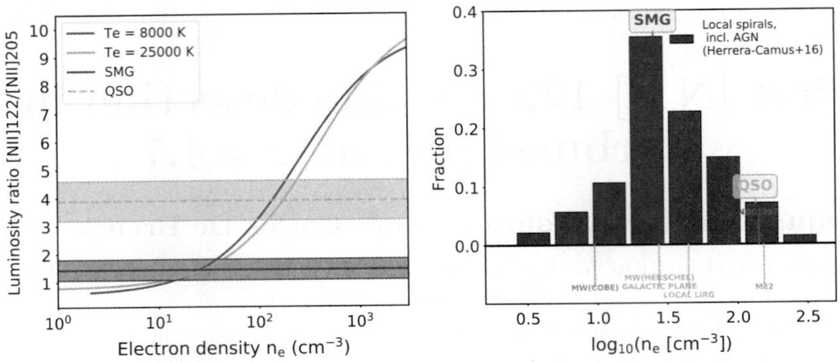

Figure 1. Left : The [NII] line luminosity ratio as a function of electron density. The two solid curves are for different electron temperatures (T_e = 8000 K and 25000 K). The line ratios for the SMG and the QSO are shown as purple solid and dark orange dashed lines, respectively. Right : The histogram for the distribution of electron density, based on the observed line ratio of local spirals from Herrera-Camus *et al.* (2016) to compare with the BR1202-0725 system. The remaining measurements are from Bennett *et al.* (1994) (MW:COBE), Goldsmith *et al.* (2015) (MW : galactic plane), Petuchowski *et al.* (1994) (M82), Beirão *et al.* (2012) (NGC1097), Díaz-Santos *et al.* (2017) (local LIRGs) (figures modified from the original in Lee *et al.* 2019, submitted)

densities for both galaxies. We used the PYNEB package (Luridiana *et al.* 2015) to perform the calculation. The derived electron densities are 26^{+12}_{-11} and 134^{+50}_{-39} cm^{-3} for the SMG and the QSO, respectively. Considering the similar star-formation rates of these galaxies, our results suggest a large scatter of electron densities at fixed SFR in global scale. The electron density of the SMG is comparable to the Galactic plane and local spirals whereas the QSO has a similar value to local starburst galaxies (Figure 1). The enhanced value for the QSO is also consistent with the measurements of high-z star-forming galaxies at $z = 2$–3 (e.g., Sanders *et al.* 2016) which are obtained from rest-frame optical lines, and with the lower limit obtained from the stacking analyses of lensed starburst galaxies at $z = 1$–3.6 using the same [NII] tracers (Zhang *et al.* 2018). Provided the varying degree of electron densities at similar SFR, our results demonstrate the power of using the FIR [N II] lines in future systematic surveys for probing electron densities of dusty starburst galaxies. The details are presented in Lee *et al.* 2019 (submitted).

References

Beirão, P., Armus, L., Helou, G., *et al.* 2012, ApJ, 751, 144
Bennett, C. L., Fixsen, D. J., Hinshaw, G., *et al.* 1994, ApJ, 434, 587
Díaz-Santos, T., Armus, L., Charmandaris, V., *et al.* 2017, ApJ, 846, 32
Goldsmith, P. F., Yıldız, U. A., Langer, W. D., & Pineda, J. L. 2015, ApJ, 814, 133
Herrera-Camus, R., Bolatto, A., Smith, J. D., *et al.* 2016, ApJ, 826, 175
Iono, D., Yun, M. S., Elvis, M., *et al.* 2006, ApJL, 645, L97
Luridiana, V., Morisset, C., & Shaw, R. A. 2015, A&A, 573, A42
Omont, A., Petitjean, P., Guilloteau, S., *et al.* 1996, Nature, 382, 428
Pavesi, R., Riechers, D. A., Capak, P. L., *et al.* 2016, ApJ, 832, 151
Petuchowski, S. J., Bennett, C. L., Haas, M. R., *et al.* 1994, ApJL, 427, L17
Sanders, R. L., Shapley, A. E., Kriek, M., *et al.* 2016, ApJ, 816, 23
Yun, M. S., Carilli, C. L., Kawabe, R., *et al.* 2000, ApJ, 528, 171
Zhang, Z.-Y., Ivison, R. J., George, R. D., *et al.* 2018, MNRAS, 481, 59

Challenges in Panchromatic Modelling with Next Generation Facilities
Proceedings IAU Symposium No. 341, 2018
M. Boquien, E. Lusso, C. Gruppioni & P. Tissera, eds.

doi:10.1017/S1743921319002606

Panchromatic Analysis for Nature of HIgh-z galaxies Tool (PANHIT)

Ken Mawatari[1,2], Akio K. Inoue[2], Satoshi Yamanaka[2], Takuya Hashimoto[2,3] and Yoichi Tamura[4]

[1]Institute for Cosmic Ray Research, University of Tokyo, 5-1-5 Kashiwanoha, Kashiwa, Chiba 277-8582, Japan
email: mawatari@icrr.u-tokyo.ac.jp

[2]Department of Environmental Science and Technology, Fuculty of Design Technology, Osaka Sangyo University, 3-1-1, Nakagaito, Daito, Osaka, 574-8530, Japan

[3]National Astronomical Observatory of Japan, Osawa 2-21-1, Mitaka, Tokyo 181-8588, Japan

[4]Division of Particle and Astrophysical Science, Graduate School of Science, Nagoya University, Nagoya 464-8602, Japan

Abstract. We have developed a new SED fitting tool specialized for frontier redshift galaxies. It is a common case for high-z galaxies that the available data are restricted to rich optical to near-infrared photometry and few far-infrared (FIR) data deep enough to detect the faint object (e.g., *HST*/WFC3 + *Spitzer*/IRAC + ALMA). In such situation, one cannot perform a complicated modeling of dust emission in FIR regime. We then adopt simple treatment for the dust emission using empirical LIRG templates. Instead, we adopt a sophisticated and physically motivated modeling for stellar and nebular emission parts in rest-frame UV-to-optical regime. Our new code fits not only broad band photometry but also spectral emission line flux. There is an option to fit observed SED with two templates with different physical properties. Our new code, PANHIT, is now in public, and was already applied to some high-z frontier galaxies.

Keywords. methods: data analysis, galaxies: high-redshift, early universe

---

1. Spectral model templates & Fitting algorithm

Our galaxy spectral models called as "SND" consist of the stellar continuum models (S: Bruzual & Charlot 2003), the nebular emission models (N: Inoue 2011), and the empirical dust emission templates (D: Rieke *et al.* 2009). Example of the three components are shown in Figure 1. Stellar continuum spectra are generated with a wide variety of age (0.1 Myr $\lesssim$ age $\lesssim$ 10 Gyr), metallicity ($0.0001 \leqslant Z \leqslant 0.05$), and star formation history (Instantaneous burst, constant SFR, exponentially declining, and rising SFHs). User defined SFHs can also be adopted. The nebular continuum and emission line fluxes are calculated as functions of the ionizing photon production rate and metallicity of the stellar component (see Mawatari *et al.* 2016 for more detail). The escape fraction of ionizing photons is a free parameter. We follow an analytic model (Inoue *et al.* 2014) for the intergalactic medium attenuation. The same dust attenuation, A_V, is applied to both the nebular line flux and the stellar continuum, where users can select an attenuation law among the Calzetti *et al.* (2000) law, Small Magellanic Cloud law (SMC law; Prevot *et al.* 1984), and Milky Way law (MW law; Seaton 1979). The attenuated energy is re-radiated in the infrared wavelengths. The dust emission is described by the empirical templates of luminous infrared galaxies (LIRG) as a function of the dust emitting infrared luminosity ($L_{\rm IR}$), where we assume $L_{\rm IR}$ is equal to the luminosity attenuated by dust.

In the template fitting, we followed the χ^2 minimization algorithm for data including flux density upper limits (Sawicki 2012). Not only broad band photometry but also

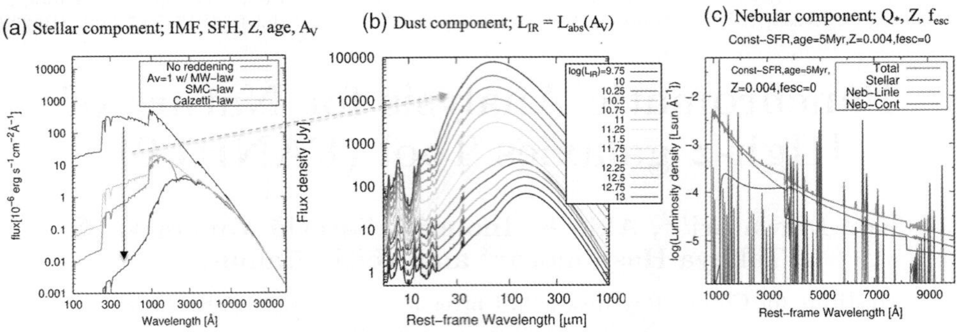

Figure 1. Our spectral models in the three components: (A) Stellar continuum with three dust attenuation laws, (B) Dust emission spectra that come from the empirical LIRG templates, and (C) Nebular continuum and line emissions.

spectral line flux can be fit. Fitting with a composite of two SND templates with different properties is available as an option, where fraction of two components' SFRs at their birth ages is parameterized (f_{SFR}). A Monte-Carlo (MC) technique is adopted to evaluate significance of the fitting solutions and uncertainties associated with the fitting parameters. The SND templates and fitting codes are available from our website: http://www.icrr.u-tokyo.ac.jp/~mawatari/PANHIT/PANHIT.html.

2. Applications

Hashimoto *et al.* (2018) applied PANHIT to the most distant spectroscopically confirmed galaxy at $z = 9.1$. Combining young and old components is required to explain the observed [O III]88μm line flux and *Spitzer*/IRAC red color, from which they suggest that this galaxy started the star-formation at $z \sim 15$. PANHIT also has been used in Hashimoto *et al.* (2019), Tamura *et al.* (2019), Mawatari *et al.* (2020), and Sunaga *et al.* (proceeding of IAU 341 symposium). These studies show that our new SED analysis tool is very useful to investigate detailed physical properties of frontier redshift galaxies.

References

Bruzual, G. & Charlot, S. 2003, *MNRAS*, 344, 1000

Inoue, A. K. 2011, *MNRAS*, 415, 2920

Rieke, G. H., Alonso-Herrero, A., Weiner, B. J., Pérez-González, P. G., Blaylock, M., Donley, J. L., & Marcillac, D. 2009, *ApJ*, 692, 556

Mawatari, K., Yamada, T., Fazio, G. G., Huang, J.-S., & Ashby, M. L. N. 2016, *PASJ*, 68, 46

Inoue, A. K., Shimizu, I., Iwata, I., & Tanaka, M. 2014, *MNRAS*, 442, 1805

Calzetti, D., Armus, L., Bohlin, R. C., Kinney, A. L., Koornneef, J., & Storchi-Bergmann, T. 2000, *ApJ*, 533, 682

Prevot, M. L., Lequeux, J., Maurice, E., Prevot, L., & Rocca-Volmerange, B. 1984, *A&A*, 132, 389

Seaton, M. J. 1979, *MNRAS*, 187, 73P

Sawicki, M. 2012, *PASP*, 124, 1208

Hashimoto, T., Laporte, N., Mawatari, K., Ellis, R. S., Inoue, A. K., Zackrisson, E., Roberts-Borsani, G., Zheng, W., *et al.* 2018, *Nature*, 557, 392

Hashimoto, T., Inoue, A. K., Mawatari, K., Tamura, Y., Matsuo, H., Furusawa, H., Harikane, Y., Shibuya, T., *et al.* 2019, *PASJ*, 71, 71

Tamura, Y., Mawatari, K., Hashimoto, T., Inoue, A. K., Zackrisson, E., Christensen, L., Binggeli, C., Matsuda, Y., *et al.* 2019, *ApJ*, 874, 27

Mawatari, K., Inoue, A. K., Hashimoto, T., Silverman, J., Kajisawa, M., Yamanaka, S., Yamada, T., Davidzon, I., *et al.* 2020, accepted for publication in ApJ (arXiv:1912.10954)

Challenges in Panchromatic Modelling with Next Generation Facilities
Proceedings IAU Symposium No. 341, 2018
M. Boquien, E. Lusso, C. Gruppioni & P. Tissera, eds.

doi:10.1017/S174392131900259X

First galaxy SED: Contribution from pre-main-sequence stars

Hiroto Mitani[1], **Naoki Yoshida**[1,2], **Kazuyuki Omukai**[3] **and Takashi Hosokawa**[4]

[1]Department of Physics, The University of Tokyo, 7-3-1 Hongo, Bunkyo, Tokyo 113-0033, Japan
email: `mitani@utap.phys.s.u-tokyo.ac.jp`

[2]Kavli Institute for the Physics and Mathematics of the Universe (WPI), UT Institutes for Advanced Study, The University of Tokyo, Kashiwa, Chiba 277-8583, Japan

[3]Astronomical Institute, Tohoku University, Sendai 980-8578, Japan

[4]Department of Physics, Kyoto University, Kyoto 606-8502, Japan

Abstract. We calculate the spectral energy distribution of the first galaxies which contain pre-main-sequence stars by using the stellar evolution code Modules for Experiments in Stellar Astrophysics, the spectra model BT-Settl, and the stellar population synthesis code PEGASE. We calculate the galaxy spectral energy distribution for Salpeter Initial Mass Function. We find that very young first galaxies are bright also in mid-infrared, and the contribution of pre-main-sequence stars can be significant over 0.1 Myr after a star-formation episode.

Keywords. galaxies: high-redshift, stars: pre–main-sequence, stars: formation

1. Introduction

Reionization is thought to be triggered by the first stars (at z∼20) with the contribution of first galaxies (z∼10) in the early Universe. Revealing the nature of these first objects is of great interest in astronomy. It is expected that the first objects will be observed by the next generation telescopes such as NASA's James Webb Space Telescope (JWST). In the conventional calculations of spectra of the first galaxies, pre-main-sequence (PMS) stars are not included (Bromm & Kudritzji (2001); Tumlinson, Giroux & Shull (2001); Schaerer (2002); Zackrisson *et al.* (2001)). In nearby galaxies, this is usually assumed because PMSs are hidden by dust grains. However, the light from PMS stars in the first galaxies where dust grains do not exist, can be directly observed. For very young ($\sim 10^5$ yr) first galaxies, which contain PMSs, contribution from PMS stars can be important in spectra of galaxies.

2. Results, Discussion

We calculate the spectral energy distribution (SED) from PMSs as well as from main-sequence (MS) stars. In order to calculate the contribution from PMSs, we calculate a number of PMS evolutions by using Modules for Experiments in Stellar Astrophysics (MESA, Paxton *et al.* 2011). The right panel of Fig. 1 shows the evolutionary tracks of PMSs. The mass range is from $0.1 M_\odot$ to $100\ M_\odot$. We obtained properties for each PMSs from MESA then used to calculate each star spectra by using the BT-Settl prescription, (Allard & Freytag 2012). Summing up the PMS spectra for an assumed initial mass function (IMF), we can then derive the PMSs contribution.

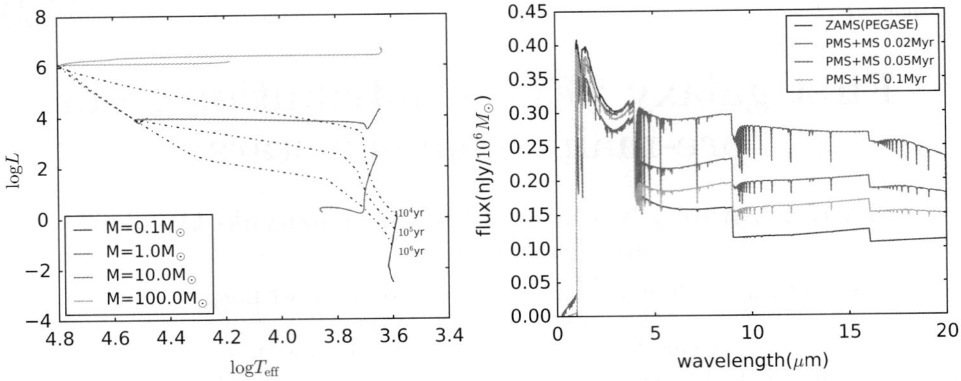

Figure 1. *left panel*: PMSs evolution in the HR diagram. Evolutionary track (solid curve) and isochrones (dotted curve) are also displayed. *right panel*: SED with PMSs at redshift $z = 10$. Observed flux (in nJy per $10^6 M_\odot$ stellar mass) vs. observed wavelength (in μm). The Blue curve represents the SED of the galaxy which consists of ZAMS and other curves represent SED with young PMSs. We assume the Salpeter IMF and instantaneous star burst galaxy at redshift $z = 10$. PMS contribution in mid-infrared ($\sim 10\mu$m) decreases with time.

In the calculation of the MS component, we use the stellar population synthesis code PEGASE (Fioc & Rocca-Volmerange (1997)). In the MS component, we also include the contribution of nebular continuum emission (Osterbrock & Ferland (2006)). We set electron temperature to be 20000K and we use zero age main-sequence (ZAMS) instantaneous star burst galaxy SEDs.

We combine the results of two components (i.e. the MS component and the PMS component) so that the mass distribution becomes Salpeter IMF, then we get the galaxy SED with PMSs contribution. The right panel of Fig. 1 shows the young galaxy SEDs with PMSs. The difference between the blue line (ZAMS) and the green line (0.02 Myr) is due to PMSs. The PMS contribution is important in mid-infrared wavelength because PMSs are low temperature and high luminosity. We find that PMSs can be significant red sources in the very young high-redshift galaxies. In the nearby galaxies, red components represent old stars, but in the distant galaxies, we may overestimate the galaxy ages and the stellar mass without considering PMSs in the SED fitting. The SED of the very young galaxies dramatically change in about 0.1 Myr. The flux of the first galaxy which have a stellar mass of $10^6 M_\odot$ and a Salpeter IMF at redshift $z = 10$ is about 0.3 nJy. It is difficult to detect such small and distant galaxies even if we use JWST, because the flux limit for the NIRCam imaging is about 10 nJy. However, if there is such a galaxy at redshift $z = 5$ and the light is magnified 10 times by gravitational lensing, we can detect not only the MS contribution but also the PMS contribution whose flux is about 10 nJy by using JWST. We thus conclude that the effect of the PMS contribution is indeed significant in the SED modeling of distant and young galaxies.

References

Allard, F., Homeier, D., & Freytag, B. 2012, *RSPTA*, 370, 2765

Bromm, V. & Kudritzki, R. P. 2001, *ApJ*, 552, 464

Fioc, M. & Rocca-Volmerange, B. 1997, *A&A*, 500, 507

Tumlinson, J., Giroux, M. L., & Shull, J. M. 2001, *ApJ*, 550, L1

Paxton, B., Bildsten, L., Dotter, A., Herwig, F., Lesaffre, P., & Timmes, F. 2011, *ApJS*, 192, 3

Osterbrock, D. E. & Ferland, G. J. 2006, *Astrophysics of gaseous nebulae and active galactic nuclei*

Schaerer, D. 2002, *A&A*, 382, 28

Zackrisson, E., Bergvall, N., Olofsson, K., & Siebert, A. 2001, *A&A*, 375, 814

Challenges in Panchromatic Modelling with Next Generation Facilities
Proceedings IAU Symposium No. 341, 2018
M. Boquien, E. Lusso, C. Gruppioni & P. Tissera, eds.

doi:10.1017/S1743921319002941

Direct collapse to SMBH seeds in cosmological halos with radiation transfer

Kentaro Nagamine[1,2,3], Isaac Shlosman[1,4] and Yang Luo[5]

[1]Theoretical Astrophysics, Department of Earth and Space Science, Graduate School of Science, Osaka University, Toyonaka, Osaka 560-0043, Japan
email: kn@astro-osaka.jp

[2]Department of Physics & Astronomy, University of Nevada, Las Vegas, 4505 S. Maryland Pkwy, Las Vegas, NV 89154-4002, USA

[3]Kavli IPMU (WPI), The University of Tokyo, 5-1-5 Kashiwanoha, Kashiwa, Chiba, 277-8583, Japan

[4]Department of Physics & Astronomy, University of Kentucky, Lexington, KY 40506-0055, USA

[5]Department of Astronomy and Jiujiang Research Institute, Xiamen University, Xiamen, Fujian 361005, People's Republic of China

Abstract. We present results of our zoom-in cosmological hydrodynamic simulations of direct collapse (DC) to supermassive black hole (SMBH) seeds with radiative transfer (RT). The DC has been modeled in dark matter halos of $\sim 10^8\,M_\odot$, using adaptive mesh refinement (AMR) code Enzo. For the first time, the baryonic collapse has been followed down to 10^{-7} pc ($\sim$0.01 AU) with on-the-fly RT and the flux-limited diffusion (FLD) approximation. We find a complex behavior involving accretion flow and associated outflows driven by the radiation force. The resulting gas dynamics around the central density peak differs profoundly from that in previous works which adopted adiabatic approximation in the core. The core forms with a photosphere at $\sim$1 AU, and its growth starts to saturate at $\sim 100\,M_\odot$. The unrelaxed core radiates intermittently near the Eddington luminosity, correlated with strong anisotropic outflows.

Keywords. hydrodynamics, black hole physics, radiative transfer, accretion, methods: numerical

---

1. Introduction

The DC appears as a leading scenario for the formation of SMBH seeds at high redshift. SMBHs with masses of $M_{\rm BH} \sim 10^9\,M_\odot$ have been discovered at $z > 6$. However, to grow such objects from Pop III star remnants ($\sim 10-100\,M_\odot$) within several 100 Myr from $z \sim 20$ to 7, it requires either continuous Eddington accretion or episodic super-Eddington accretion. DC allows formation of more massive seeds with $10^4 - 10^6\,M_\odot$ at $z \sim 10-20$, giving a head start for SMBH formation. So far, attempts to simulate DC have assumed an adiabatic equation-of-state in the optically-thick regime, when radiation is generated within the flow. In this case, the collapse is terminated and a supermassive star formation is subsequently assumed, however, its formation has never been demonstrated self-consistently in ab initio radiation hydrodynamic simulations.

2. Method and Results

We use Enzo-2.4 AMR code with an RT module (Reynolds *et al.* 2009) heavily modified by our group. For initial conditions, we use MUSIC code (Hahn & Abel 2011) on a comoving box of $1\,h^{-1}$Mpc with 128^3 DM particles. In this pathfinder run, we identify a

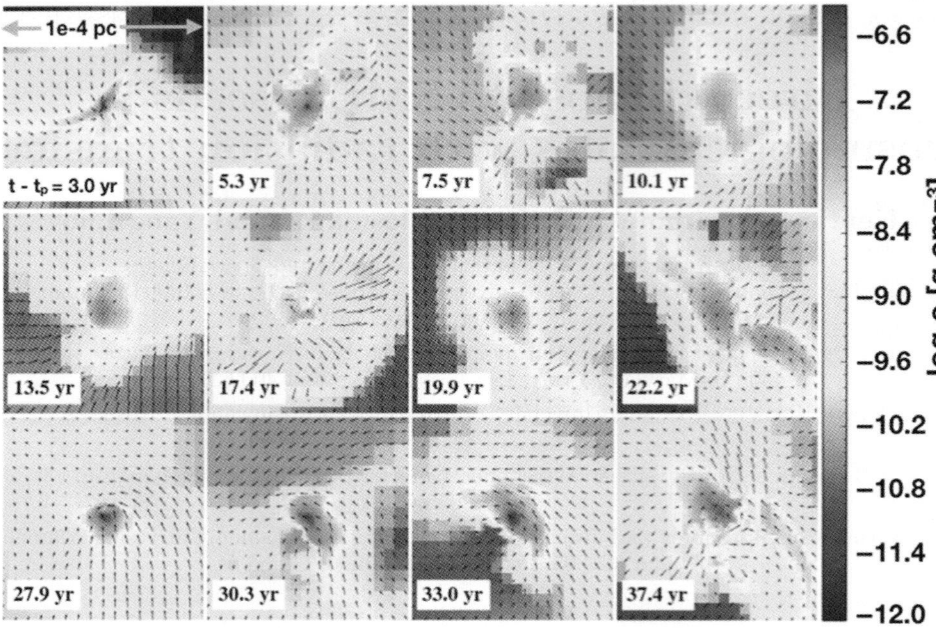

Figure 1. The FLD collapse. Evolution of the innermost density slice (color map) with superposed velocity vectors. The size of each panel is 10^{-4} pc, and the indicated time is measured from the moment when the photosphere formed. Recurrent anisotropic outflows are driven by radiation and thermal pressure gradients and form expanding shells, mixing with the accreting flow.

zoom region of $0.18\,h^{-1}$Mpc with 512^3 initial grid, and use AMR to simulate the collapse down to 10^{-7} pc with maximum refinement of 32–36 levels. The primordial opacity has been adopted from Mayer & Duschl (2005). Cooling by metals and dust are ignored for $z > 10$. Details of comparison between GADGET-3 and Enzo runs are given in Luo *et al.* (2016) and of the long-term evolution of SMBH seeds using sink particle technique in Shlosman *et al.* (2016). Collapse with the FLD in isolated halos is discussed in Luo *et al.* (2018) and in the cosmological context in Ardaneh *et al.* (2018). Runs with an adiabatic equation-of-state have been performed for comparison.

The FLD RT models demonstrate formation of the core with penetrating accretion streams, driving internal turbulence. Thermalization of the kinetic energy under the photosphere leads to the formation of hot bubbles, and drives convection. The density stratification in the core is weak, and the photosphere is far from being spherical. Strong anisotropic outflows mixes with the accretion flow. Figure 1 displays the evolution of gas density distribution in the innermost collapsing flow with superposed velocity vectors (Ardaneh *et al.* 2018). Expanding shells and hot bubbles form due to the radiation pressure generated by the core, which radiates intermittently near the Eddington limit with photospheric luminosities of $\sim 10^{38} - 10^{39}\,\mathrm{erg\,s^{-1}}$.

These features differ substantially from previous works which applied the adiabatic approximation in the optically-thick core. We confirm that the runs with adiabatic approximation lead to the formation of a central disk with spiral arms. Fragments are formed along the spiral shocks due to the Kelvin-Helmholtz shear instability and not the Jeans instability. Whereas with the FLD, the accreting gas is far less dominated by angular momentum, at least initially. The photosphere forms at a few AU, within which the gas becomes ionized and expands. At the end of the run, the central gas density has reached $10^{-6}\,\mathrm{g\,cm^{-3}}$ and the temperature increased to $\sim 10^{4.5}$ K. The mass accretion rate fluctuates between $0.1–1\,M_\odot\,yr^{-1}$. Powerful recurrent outflows compete with the accretion flow and can lead to dissolution of the core or saturate its growth.

In the future we will attempt to run the RT simulation for a longer time to examine the long-term evolution of DC cores, and extend our work to Lyα emission and its impact on gas dynamics.

References

Ardaneh, K., Luo, Y., Shlosman, I., Nagamine, K., *et al.* 2018, *MNRAS*, 479, 2277
Hahn, O. & Abel, T. 2011, *MNRAS*, 415, 2101
Mayer, M. & Duschl, W. J. 2005, *MNRAS*, 358, 614
Luo, Y., Nagamine, K., & Shlosman, I. 2016, *MNRAS*, 459, 3217
Luo, Y., Ardaneh, K., Shlosman, I., Nagamine, K., *et al.* 2018, *MNRAS*, 476, 3523
Reynolds, D. R., *et al.* 2009, *Journal of Computational Physics*, 228, 6833
Shlosman, I., Choi, J.-H., Begelman, M. C., & Nagamine, K. 2016, *MNRAS*, 456, 500

Challenges in Panchromatic Modelling with Next Generation Facilities
Proceedings IAU Symposium No. 341, 2018
M. Boquien, E. Lusso, C. Gruppioni & P. Tissera, eds.

doi:10.1017/S1743921319002400

Optical properties of infrared-bright dust-obscured galaxies viewed with Subaru Hyper Suprime-Cam

A. Noboriguchi[1], T. Nagao[2], Y. Toba[3], M. Niida[1], M. Kajisawa[1], M. Onoue[4], Y. Matsuoka[2], T. Yamashita[2], Y. Chang[5], T. Kawaguchi[6], Y. Komiyama[7], K. Nobuhara[1], Y. Terashima[1] and Y. Ueda[3]

[1]Graduate School of Science and Engineering, Ehime University, Bunkyo-cho, Matsuyama, Ehime 790-8577, Japan
email: noboriguchi@cosmos.phys.sci.ehime-u.ac.jp

[2]Research Center for Space and Cosmic Evolution, Ehime University, Bunkyo-cho, Matsuyama, Ehime 790-8577, Japan

[3]Department of Astronomy, Kyoto University, Kitashirakawa-Oiwake-cho, Kyoto 606-8502, Japan

[4]Max-Planck-Institut für Astronomie, Königstuhl 17, D-69117 Heidelberg, Germany

[5]Academia Sinica Institute of Astronomy and Astrophysics, PO Box 23-141, Taipei 10617, Taiwan

[6]Department of Economics, Onomichi City University, Onomichi, Hiroshima 722-8506, Japan

[7]National Astronomical Observatory of Japan, 2-21-1 Osawa, Mitaka, Tokyo 181-8588, Japan

Abstract. Optical properties of infrared-bright (IR-bright) dust-obscured galaxies (DOGs) are reported. DOGs are faint in optical but very bright in mid-IR, which are powered by active star formation (SF) or active galactic nucleus (AGN), or both. The DOGs is a candidate population that are evolving from a gas-rich merger to a quasar. By combining three catalogs of optical (Subaru Hyper Suprime-Cam), near-IR (VIKING), and mid-IR (ALLWISE), we have discovered 571 IR-bright DOGs. Using their spectral energy distributions, we classified the selected DOGs into the SF-dominated DOGs and the AGN-dominated DOGs. We found that the SF-dominated DOGs show a redder optical color than the AGN-dominated DOGs. Interestingly, some DOGs shows extremely blue color in optical (blue-excess DOGs: bluDOGs). A possible origin for this blue excess is either the leaked AGN light or stellar UV light from nuclear starbursts. The BluDOGs may be in the transition phase from obscured AGNs to unobscured AGNs.

Keywords. galaxies: active, galaxies: starburst, infrared: galaxies, methods: statistical

---

1. Introduction

Dust-obscured galaxies (DOGs), whose color between optical and mid-IR is very red ($(i-[22])_{\rm AB} \geqslant 7.0$; Toba *et al.* 2015), are believed to be a candidate population that are evolving from gas-rich mergers to quasars. The origin of their red color is faint optical emission reduced by dust and bright mid-IR emission powered by active star formation (SF) or active galactic nucleus (AGN), or both (Dey *et al.* 2008). DOGs are classified into SF-dominated DOGs and AGN-dominated DOGs by using the spectral energy distribution (SED). The SED of SF-dominated DOGs shows a bumpy feature at rest-wavelength 1.6 μm (hereafter bump DOGs), while the SED of AGN-dominated DOGs shows a power-law SED between optical and mid-IR (hereafter PL DOGs). However, optical properties of IR-bright DOGs are still unclear, since their number density is very

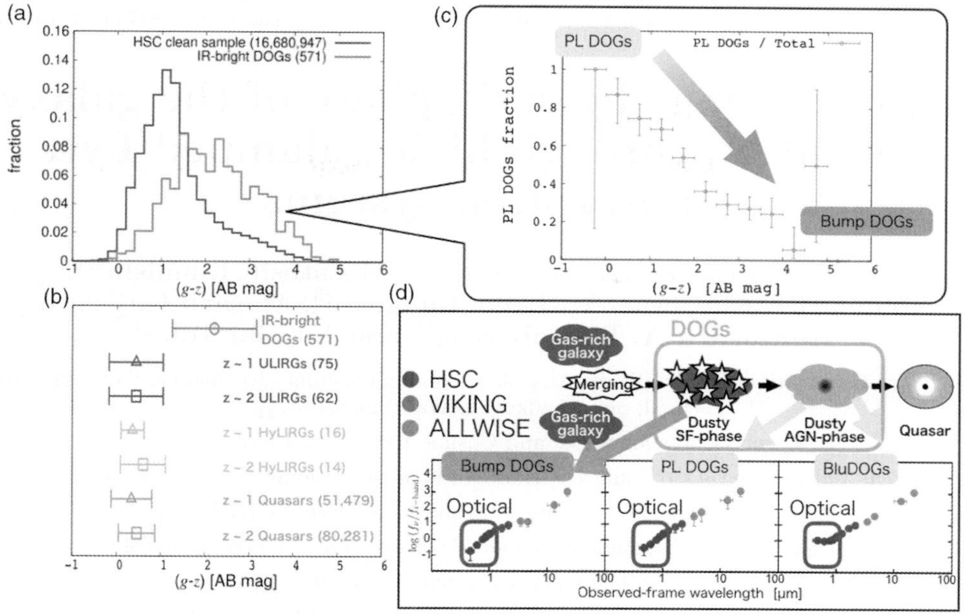

Figure 1. [A]: The $(g-z)_{\rm AB}$ color distribution of IR-bright DOGs (red) and the entire HSC sample (blue). [B]: The $(g-z)_{\rm AB}$ color means and standard deviations of IR-bright DOGs (red), ULIRGs (green), HyLIRGs (yellow), and quasars (purple). [C]: The PL fraction as a function of the $(g-z)_{\rm AB}$ color. [D]: The gas-rich major merger scenario (upper illustration; e.g., Dey *et al.* 2008), and the averaged SED of bump DOGs (lower left panel), PL DOGs (lower center panel), and BluDOGs (lower right panel).

low ($\log\phi = -6.59 \pm 0.11$ [Mpc^{-3}]: Toba *et al.* 2015), and their optical fluxes are very faint (∼23 mag or fainter: Toba *et al.* 2015). In this work (presented in Noboriguchi *et al.* 2019), we investigate optical properties of IR-bright DOGs using the Hyper Suprime-Cam (HSC) wide-field data obtained through the Subaru Strategic Program survey (Aihara *et al.* 2018).

2. Sample selection, results, and discussion

By combining three catalogs of optical (HSC), near-IR (VIKING), and mid-IR (ALLWISE), we discovered 571 IR-bright DOGs. The $(g-z)_{\rm AB}$ color of IR-bright DOGs is redder and broader than that of the entire HSC sample (as a typical object sample), dusty galaxies (e.g., ULIRG and HyLIRG) and quasars (Figure 1 [A] and [B]). By comparing the $(g-z)_{\rm AB}$ distribution of bump DOGs with PL DOGs, we found that the bump DOGs occupy the redder part while the PL DOGs occupy the bluer part in the optical color distribution of DOGs (Figure 1 [C]). Some of our DOGs show extremely blue color as also reported for dusty galaxies by Assef *et al.* (2016), and we call them blue-excess DOGs (BluDOGs). The possibilities for the origin of the blue excess are 1) leaked AGN light and 2) radiation of massive stars in the starburst region. By focusing on their optical SED, BluDOGs are suggested to be in the evolving phase between obscured phase (i.e., DOGs) and unobscured phase (i.e., quasar) as shown Figure 1 [D].

References

Aihara, H., Arimoto, N., Armstrong, R., *et al.* 2018, *PASJ*, 70, S4
Assef, R. J., Walton, D. J., Brightman, M., *et al.* 2016, *ApJ*, 819, 111
Dey, A., Soifer, B. T., Desai, V., *et al.* 2008, *ApJ*, 677, 943
Noboriguchi, A., Nagao, T., Toba, Y., *et al.* 2019, *ApJ*, 876, 132
Toba, Y., Nagao, T., Strauss, M. A., *et al.* 2015, *PASJ*, 67, 86

Challenges in Panchromatic Modelling with
Next Generation Facilities
Proceedings IAU Symposium No. 341, 2018
M. Boquien, E. Lusso, C. Gruppioni & P. Tissera, eds.

doi:10.1017/S1743921319001388

Investigating the early phase of the galaxy evolution through high-z damped Lyα absorption systems

Kazuyuki Ogura[1], Tohru Nagao[2], Masatoshi Imanishi[3,4], Nobunari Kashikawa[5], Yoshiaki Taniguchi[6], Masaru Kajisawa[7], Masakazu A. R. Kobayashi[8] and Yoshiki Toba[9]

[1]Faculty of Education, Bunkyo University, 3337 Minamiogishima, Koshigaya, Saitama, Japan
email: ogurakz@koshigaya.bunkyo.ac.jp

[2]Research Center for Space and Cosmic Evolution, Ehime University

[3]Optical and Infrared Astronomy Division, National Astronomical Observatory of Japan

[4]Department of Astronomy, School of Science, SOKENDAI

[5]Department of Astronomy, School of Science, The University of Tokyo

[6]The Open University of Japan

[7]Graduate School of Science and Engineering, Ehime University

[8]Faculty of Natural Sciences, National Institute of Technology, Kure college

[9]Department of Astronomy, Kyoto University

Abstract. We present the physical properties of Lyα emitters (LAEs) in a "DLA-concentrated regions" where there are 3 or more DLA within (50 Mpc)^3 cubic box. We observed LAEs in a DLA-concentrated region at $z = 2.3$, the J1230+34 field, with Subaru/Suprime-Cam. In the 50 Mpc scale, we found no deferences in properties of LAEs such as Lyα luminosity function in the DLA-concentrated region compared to other fields at similar redshift. On the other hand, we found a ~10 Mpc scale LAE overdensity around a strong DLA with $N_{\rm HI} = 10^{21.08}$ cm^{-2}.

Keywords. quasars: absorption lines, galaxies: emission lines, galaxies: formation, large-scale structure of universe

---

1. Investing the properties of galaxies in "DLA-concentrated regions"

The damped Lyα absorption (DLA) system is a class of quasar absorption systems with HI column density of $N_{\rm HI} \geqslant 2 \times 10^{20}$ (e.g., Wolfe *et al.* 2005). Since DLAs dominate the neutral gas content of the Universe (e.g., Noterdaeme *et al.* 2009), they are interesting systems as rich gas reservoirs for the star-formation. However the connection of DLAs with star-forming galaxies because identifying optical counterparts of DLAs is difficult (e.g., Krogager *et al.* 2017).

For investigating the relationship between DLAs and galaxies, we define a "DLA-concentrated region" as a region where 3 or more DLAs exist within a (50 Mpc)^3 box (Ogura *et al.* 2017). This definition is based on the size of high-z protoclusters (e.g., Overzier (2016); Franck & McGaugh 2016). DLA-concentrated regions are interesting because (1) we can effectively search for optical counterparts of DLAs there, (2) they are good laboratories to study the connection of DLAs with galaxies, and (3) the high-density environment is a key component in the context of the λCDM cosmology. We found six DLA-concentrated regions at $2.255 \leqslant z \leqslant 2.330$ (the Lyα line at this redshift can be

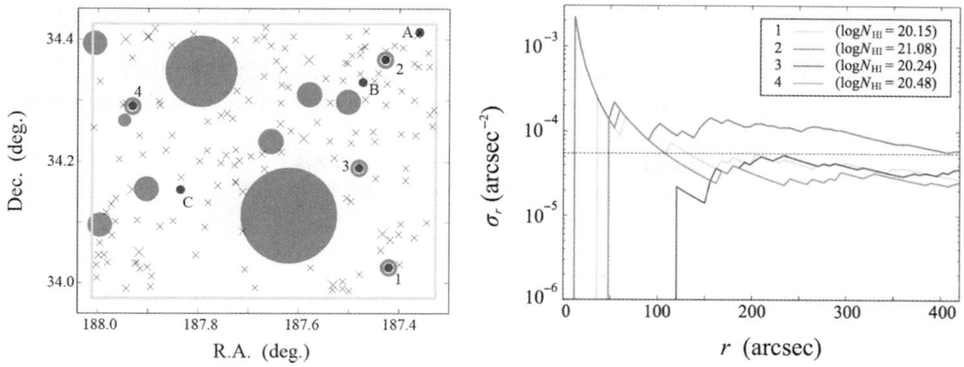

Figure 1. [**Left:**] Spatial distribution of LAEs (black crosses) and background quasars (blue filled circles) in the J1230+34 field. The blue circles with surrounding red circles show the position of DLAs at $z \sim 2.3$. LAEs shown by large red crosses have $EW_0 \geqslant 200$ Å. The green box shows the field of view of Suprime-Cam and gray filled circles show masked areas. [**Right:**] The surface number density of LAEs around each DLA in the J1230+34 field. The vertical axis indicates the surface number density of LAEs within a circular region at r. The horizontal axis indicates the radius of the circular region from the position of each DLA. The dotted horizontal black line indicates the average of the LAE density calculated for the entire J1230+34 field.

covered by the NB400 filter of Subaru/Suprime-Cam) from the the DLA catalog based on the BOSS (Noterdaeme *et al.* 2012). Among six fields we found, the J1230+34 field with 4 DLAs at $z \sim 2.3$ is the most interesting target because there are more absorbers in the J1230+34 field than the other five fields. We investigate the properties of Lyα emitters (LAEs) in the J1230+34 field through the narrow-band imaging observations with Suprime-Cam using the NB400 filter ($\lambda_{\rm eff} = 4003$ Å and $FWHM = 92$ Å).

2. Result and Discussion

We detected 149 LAEs in the J1230+34 field. The left panel of Fig. 1 shows the spatial distribution of LAEs, DLAs and background quasars. Based on this LAE sample, we derived the Lyα luminosity function (LF) and study the frequency distribution of the rest-frame equivalent width (EW_0) of the Lyα emission. In the result, we found no difference in the Lyα LF and the EW_0 distribution compared to those in other fields at similar redshift. We then investigate the surface number density of LAEs around each DLA (the right panel of Fig. 1). LAEs around DLA 2, which is the most gas-rich system in the J1230+34 field, show a density excess over $\sim$400$''$ ($\sim$11 Mpc) relative to the average density of LAEs in the entire target field. This may correspond to the gas-rich young protocluster firstly traced by the DLA. Contrary, we found no LAE overdensity around other DLAs. Due to the resonant nature of the Lyα photon, LAEs may not be a good tracer of galaxies in the gas-rich environment such as regions around DLAs. Further observations without relying on the Lyα emission (e.g., Hα emitters LBGs, and so on) is interesting to investigate detailed structures in the DLA-concentrated regions. DLA-concentrated regions are interesting target for studying the early phase of the galaxy evolutions in the future large surveys with Subaru/PFS, TMT, EELT, GMT and so on.

References

Franck, J. R. & McGaugh, S. S. 2016, *ApJ*, 817, 158

Krogager, J.-K., Møller, P., Fynbo, J. P. U., & Noterdaeme, P. 2017, *A&A*, 469, 2959

Noterdaeme, P., Petitjean, P., Carithers, W. C., Pâris, I., Font-Ribera, A., Bailey, S., Aubourg, E., Bizyaev, D., *et al.* 2012, *A&A*, 547, L1

Noterdaeme, P., Petitjean, P., Ledoux, C., & Srianand, R. 2009, *A&A*, 505, 1087

Ogura, K., Nagao, T., Imanishi, M., Kashikawa, N., Taniguchi, Y., Kajisawa, M., Kobayasi, M. A. R., Toba, Y., Nobuhara, K., *et al.* 2017, *PASJ*, 69, 51

Overzier, R. A. 2016, *A&AR*, 24, 14

Wolfe, A. M., Gawiser, E., & Prochaska, J. X. 2005, *ARA&A*, 43, 861

Challenges in Panchromatic Modelling with Next Generation Facilities
Proceedings IAU Symposium No. 341, 2018
M. Boquien, E. Lusso, C. Gruppioni & P. Tissera, eds.

doi:10.1017/S174392131900187X

Strong FeII emission in NLS1s: An unsolved mystery

Swayamtrupta Panda[1,2], **Katarzyna Małek**[3,4], **Marzena Śniegowska**[2] **and Bożena Czerny**[1]

[1]Center for Theoretical Physics — PAS, Al. Lotników 32/46, Warsaw, Poland 02—668
[2]Nicolaus Copernicus Astronomical Center, ul. Bartycka 18, Warsaw, Poland 00—716
[3]National Center for Nuclear Research, ul. Hoża 69, Warsaw, Poland 00—681
[4]Aix Marseille Univ. CNRS, CNES, LAM Marseille 13388, France

Abstract. In Panda *et al.* 2018a, we constructed a refined sample from the original Shen *et al.* (2011) QSO catalog. Based on our hypothesis — the main driver of the Quasar Main Sequence is the maximum of the accretion disk temperature (T_{BBB}) defined by the Big Blue Bump on the Spectral Energy Distribution (Panda *et al.* 2017; Panda *et al.* 2018b). We select the four extreme sources that have $R_{FeII} \geqslant 4.0$ and use CIGALE (Boquien *et al.* 2018) to fit their multi—band photometric data. We also perform detailed spectral fitting including the Fe II pseudo—continuum (based on Śniegowska *et al.* 2018) to estimate and compare the value of R_{FeII}. We show the dependence of FeII strength on changing metallicity.

Keywords. galaxies: active, (galaxies:) quasars: emission lines, accretion disks, radiative transfer, techniques: photometric

---

1. CIGALE SED analysis

We use CIGALE to fit multi—wavelength photometric data (5GHz to 1344 Å) using: (a) SFH with a delayed + exponential burst; (b) SSP: Bruzual & Charlot 2003; (c) nebular emission; (d) Calzetti *et al.* 2000 dust attenuation; (e) Draine *et al.* 2014 dust emission; and (f) Fritz *et al.* 2006 AGN model. CIGALE's SED analyses shows good agreement with the T_{BBB} derived from the observations (see Table 1†).

2. Photoionisation Predictions On Metallicity

We test the behaviour of these "strong FeII" emitting NLS1s (see Fig. 1 right panel) through photoionisation code CLOUDY (Ferland *et al.* 2017) to predict the line luminosities for FeII (integrated) and Hβ using SEDs from observation. In Panda *et al.* 2018b, we found that these BLR clouds have high density ($n_H = 10^{12}\,cm^{-3}$) and characteristic FeII emission can be modelled using high column density ($N_H = 10^{24}\,cm^{-2}$) and a low value of (micro)turbulence ($v_{turb} = 10\,km/s$). FeII strength (R_{FeII}) estimates from photoionisation are further confirmed incorporating a high value of metallicity. We performed a grid simulation that gave us a linear dependence of the FeII strength on the metallicity:

$$\langle \log (R_{FeII}) \rangle = -(0.325 \pm 0.0261) + (0.519 \pm 0.008) \left\langle \log \left(\frac{Z}{Z_{\odot}} \right) \right\rangle \ddagger \qquad (2.1)$$

† the table and the model fits for the other 3 sources can be found here
‡ average values based on the 4 sources

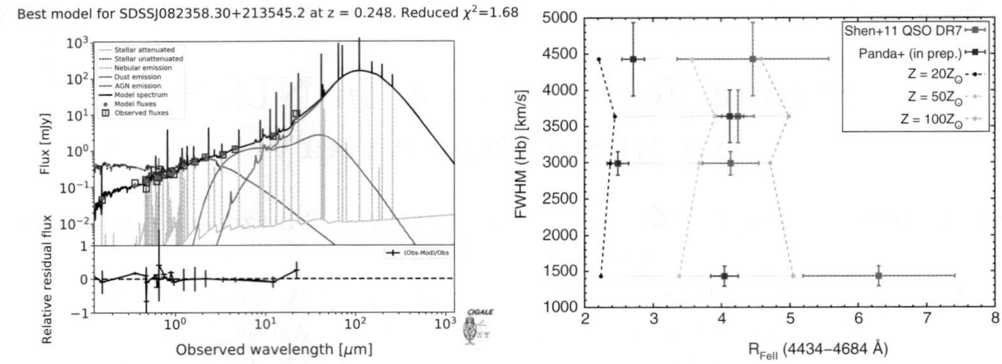

Figure 1. (a) CIGALE model fitting for SDSSJ082358.30+213545.20 — the orange line represents the stellar (attenuated) emission, the blue dashed line is for the stellar (unattenuated), the nebular, dust, AGN emission are shown in light green, red and dark green respectively. The overall model spectrum is shown in black.(b) The plot shows a comparison between values estimated by an automatic template fitting procedure (red squares) versus our semi—automatic procedure (blue squares). A subset of the photoionisation simulation results is shown in the background for varying metallicities.

The predicted average metallicities are in the range of 20—55 $Z_{\odot}$. Such high values of metallicities have been obtained for some high—redshift quasars with similar emission line features.

References

Boquien, M., Burgarella, D., Roehlly, Y., Buat, V., Ciesla, L., Corre, D., Inoue, A. K. & Salas, H. 2018. CIGALE: a python Code Investigating GALaxy Emission. *arXiv*:1811.03094

Bruzual, G. & Charlot, S. 2003. Stellar population synthesis at the resolution of 2003. *MNRAS* 344(4), 1000–1028

Calzetti, D., Armus, L., Bohlin, R. C., Kinney, A. L., Koornneef, J., & Storchi-Bergmann, T. 2000. The Dust Content and Opacity of Actively Star-forming Galaxies. *ApJ* 533(2), 682–695

Draine, B. T., Aniano, G., Krause, O., Groves, B., Sandstrom, K., Braun, R., Leroy, A., Klaas, U., *et al.* 2014. Andromeda's Dust. *ApJ* 780(2), 172–190

Ferland, G. J., Chatzikos, M., Guzmán, F., Lykins, M. L., van Hoof, P. A. M., Williams, R. J. R., Abel, N. P., *et al.* 2017. The 2017 Release Cloudy. *RMxAA* 53, 385–438

Fritz, J., Franceschini, A., & Hatziminaoglou, E. 2006. Revisiting the infrared spectra of active galactic nuclei with a new torus emission model. *MNRAS* 366(3), 767–786

Panda S., Czerny, B., & Wildy, C. 2017. The physical driver of the optical Eigenvector 1 in Quasar Main Sequence. *FrASS* 4, 33

Panda, S., Małek, K., Śniegowska, M., & Czerny, B. 2018a. Using vo tools to investigate Quasar Spectra (UNIQS). In: M. Berton, (Ed.), 2018. Revisiting narrow-line Seyfert 1 galaxies and their place in the Universe. Padova (Italy): Proceedings of Science. 50

Panda, S., Czerny, B.,, Adhikari, T. P., Hryniewicz, K., Wildy, C., Kuraszkiewicz, J., & Śniegowska, M. 2018b. Modeling of the Quasar Main Sequence in the Optical Plane. *ApJ* 866(2), 115–128

Shen, Y., Richards, G. T., Strauss, M. A., Hall, P. B., Schneider, D. P., Snedden, S., Bizyaev, D., Brewington, H., *et al.* 2011. A Catalog of Quasar Properties from Sloan Digital Sky Survey Data Release 7. *ApJS* 194(2), 45–66

Śniegowska, M., Czerny, B., You, B., Panda, S., Wang, J.-M., Hryniewicz, K., & Wildy, C. 2018. Properties of active galaxies at the extreme of Eigenvector 1. *A&A* 613, A38

Challenges in Panchromatic Modelling with Next Generation Facilities
Proceedings IAU Symposium No. 341, 2018
M. Boquien, E. Lusso, C. Gruppioni & P. Tissera, eds.

doi:10.1017/S1743921319004721

Modeling galaxy evolution at high-redshift in highly overdense and normal regions

Raphael Sadoun[1,2], Emilio Romano-Díaz[3], Isaac Shlosman[1,4] and Zheng Zheng[2]

[1]Department of Earth and Space Science, Graduate School of Science, Osaka University, Osaka 560-0043, Japan
email: sadoun@astro-osaka.jp

[2]Department of Physics & Astronomy, University of Utah, Salt Lake City, UT 84112-0830, USA

[3]Argelander Institut fuer Astronomie, Auf dem Haegel 71, D-53121 Bonn, Germany

[4]Department of Physics & Astronomy, University of Kentucky, Lexington, KY 40506-0055, USA

Abstract. We present results from high-resolution, zoom-in cosmological simulations to study the effect of feedback from galactic outflows on the physical and Lyα properties of high-redshift galaxies in highly overdense and normal environments at $z \gtrsim 6$. The Lyα properties have been obtained by post-processing the simulations with a Monte-Carlo radiative transfer (RT) code. Our results demonstrate that galactic outflows play an important role in regulating the growth of massive galaxies in overdense regions as well as the temperature and metallicity of the intergalactic medium. In particular, we find that galactic outflows are necessary to reproduce the observed Lyα luminosity functions as well as the apparent Lyα luminosity, line width and equivalent width distributions of luminous Lyα emitters at $z \sim 6$.

Keywords. cosmology: theory, galaxies: formation, galaxies: evolution, galaxies: high-redshift, methods: numerical, radiative transfer

1. Introduction

Massive high-z galaxies are found preferentially in rare, overdense regions of the universe. In dense environments, galaxy evolution is expected to proceed more rapidly than in average fields, with elevated gas accretion and star formation rates (SFRs) that can power strong galactic outflows (Romano-Díaz *et al.* 2011, 2014). Understanding the dependence of galaxy formation on environment is thus crucial for interpreting observations at high-z. An increasingly large number of high-z galaxies have been detected as Lyα Emitters (LAEs, Ouchi *et al.* 2018). Because Lyα photons interact strongly with HI, LAEs constitute a promising probe of the ionization state of the intergalactic medium (IGM) in the late stages of the reionization era at $z \gtrsim 6$. However, due to the resonant scattering nature of the interaction, extracting constraints on reionization from apparent Lyα properties of LAEs is challenging and requires detailed modeling of the Lyα line and its transfer through the IGM (Dijkstra 2014; Sadoun *et al.* 2017).

We have performed high-resolution, cosmological zoom-in simulations to study galaxy evolution in highly overdense and normal regions at $z \sim 6 - 12$ (Sadoun *et al.* 2016). Various galactic feedback models were implemented in order to investigate the effects of galactic outflows on the physical and Lyα properties of galaxies. We have obtained the apparent Lyα properties of galaxies by post-processing the output of the simulations with a 3-dimensional Monte-Carlo Lyα radiative transfer code (Sadoun *et al.* 2018).

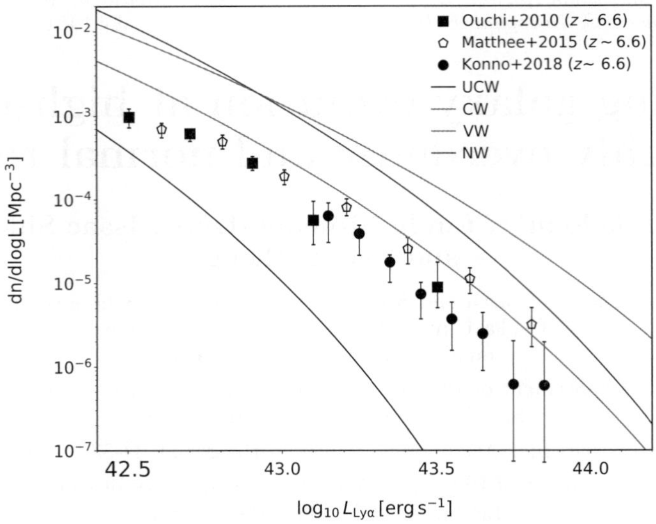

Figure 1. Comparison between the Lyα luminosity functions (LFs) in the unconstrained (UCW) and constrained (CW, VW and NW) simulation runs with observational data. The Lyα LFs are obtained by convolving the $L_{\rm Ly\alpha}$-$M_{\rm h}$ relation predicted by our radiative transfer calculations with the theoretical halo MF.

2. Numerical methods

We use the hydro+N-body code PGADGET-3, with both constrained (CR) and unconstrained (UCR) initial conditions. The CR runs contain the seed of a DM halo of $10^{12}\,h^{-1}\,{\rm M}_\odot$ that should collapse by $z \sim 6$. The UCR runs represent the parent realization of its constrained counterpart. The simulations include cooling, SF, stellar and SNe feedback. The CR suite has been run with three different prescriptions for galactic winds: constant velocity wind (CW), variable velocity wind (VW), and no-wind (NW) models. The UCR have been run with CW and VW models.

Both the ionization state of the gas and the Lyα radiative transfer (RT) were calculated *a posteriori* on a grid constructed from interpolating the SPH data obtained from the simulation outputs at $z \sim 6.6$. The former was computed with self-shielding of dense gas taken into account, assuming photoionization by a uniform UV background and by the local ionizing flux produced by galaxies in the simulation. The Lyα RT was performed using the Monte-Carlo radiative transfer code from Zheng & Miralda-Escudé (2002).

3. Results

We find that the Lyα properties of massive galaxies in the overdense region match well recent observations of luminous LAEs at $z \sim 6$–7, in terms of apparent Lyα luminosity ($L_{\rm Ly\alpha}$), line width and Lyα equivalent width distributions. Without winds, these galaxies appear less Lyα bright due to differences in the line profile emerging from galaxies (which depends on the SFR and outflow velocity) and, in the distributions of neutral gas in the IGM. The $L_{\rm Ly\alpha}$-halo mass ($M_{\rm h}$) relation appears sensitive to the environment and the wind model, in contrast to $L_{\rm Ly\alpha}$-stellar mass or SFR relations. The Lyα LFs constructed from convolving the $L_{\rm Ly\alpha}$-$M_{\rm h}$ relation with the theoretical halo mass function (MF) reproduce reasonable well the observed one at $z \sim 6.6$ (Fig. 1). Our results are consistent with a patchy reionization scenario in which observed luminous LAEs at $z \gtrsim 6$ are found preferentially in ionized, overdense regions.

References

Dijkstra, M. 2014, *PASA*, 31, e040
Ouchi, M., Harikane, Y., Shibuya, T., *et al.* 2018, *PASJ*, 70, S13
Romano-Díaz, E., Choi, J.-H., Shlosman, I., & Trenti, M. 2011, *ApJ* (Letters), 738, L19
Romano-Díaz, E., Shlosman, I., Choi, J.-H., & Sadoun, R. 2014, *ApJ* (Letters), 790, L32
Sadoun, R., Shlosman, I., Choi, J.-H., & Romano-Díaz, E. 2016, *ApJ*, 829, 71
Sadoun, R., Zheng, Z., & Miralda-Escudé, J. 2017, *ApJ*, 839, 44
Sadoun, R., Romano-Diaz, E., Shlosman, I., & Zheng, Z. 2018, *arXiv*:1810.12957
Zheng, Z. & Miralda-Escudé, J. 2002, *ApJ*, 578, 33

Challenges in Panchromatic Modelling with
Next Generation Facilities
Proceedings IAU Symposium No. 341, 2018
M. Boquien, E. Lusso, C. Gruppioni & P. Tissera, eds.

doi:10.1017/S1743921319002990

Spectral study of scattered light by interstellar dust grains

Kei Sano

Kwansei Gakuin University, 2-1, Gakuen, Sanda, Hyogo
email: `sano0410@kwansei.ac.jp`

Abstract. Interstellar dust is traced by not only thermal emission but also scattered light. The scattered light spectrum observed from ultraviolet (UV) to near-infrared (IR) is useful to constrain some dust properties, such as size distribution, albedo, and composition. Milky Way Galaxy is a unique environment to observe the diffuse scattered light because we can extract it by removing the contribution of starlight. We have observed the UV to near-IR scattered light with space instruments, including Diffuse Infrared Background Experiment (DIRBE), Hubble Space Telescope (HST), and Multi-purpose Infra-Red Imaging System (MIRIS). The scattered light spectrum is marginally consistent with prediction from a recent dust model including carbonaceous and silicate grains with polycyclic aromatic hydrocarbon (PAH). Based on the MIRIS observation of a diffuse cloud, we compare the scattered light color with the dust model with or without grains larger than 1 micrometer. The result shows that the color is consistent with the model without the large grains, which is consistent with recent simulations of dust growth in low-density regions. However, some observations have shown the spectral excess at $\sim$0.6 micrometer wavelength, suggesting the presence of extended red emission (ERE) which cannot be explained by the conventional dust model.

Keywords. scattering, dust, extinction, infrared: ISM

1. Introduction

Properties of interstellar dust grains can be studied by spectral observations of extinction curve, thermal emission, and scattered light by starlight. From ultraviolet (UV) to near-infrared (IR), the scattered light has been observed toward different regions with various instruments (e.g., Ienaka *et al.* 2013; Arai *et al.* 2015; Sano *et al.* 2015; Kawara *et al.* 2017).

Spectral shape of scattered light is expected from models of interstellar dust and interstellar radiation field. Brandt & Draine (2012) provide a model of scattered light spectrum by assuming the dust models of Weingartner & Draine (2001) and Zubko *et al.* (2004). Both models consist of sub-micrometer-sized carbonaceous and silicate grains. The observed spectrum of scattered light is marginally consistent with the prediction from these models.

Several observational studies suggest a flat extinction curve in the mid-IR, which cannot be explained by the dust models including grains less than 1 μm (e.g., Indebetouw *et al.* 2005; Nishiyama *et al.* 2009). Wang *et al.* (2015) explain this problem by adding some micrometer-sized carbonaceous grains to the dust model of Weingartner & Draine (2001). A more precise observation of scattered light is expected to constrain grain distribution of interstellar dust.

2. Overview

We present new observations of scattered light with Multi-purpose Infra-Red Imaging System (MIRIS) at 1.1 and 1.6 μm by utilizing its large field of view of $3.67° \times 3.67°$ (Han *et al.* 2014). Observing a diffuse region in the edge of the molecular cloud MBM32, we found a precise linear correlation between near-IR scattered light and thermal dust emission at 100 μm (Schlegel *et al.* 1998).

On the basis of the dust models of Weingartner & Draine (2001) and Wang *et al.* (2015), we predict the spectrum of scattered light. The result shows that the observed near-IR color of scattered light prefers the model of Weingartner & Draine (2001), suggesting little contribution of very large grains of >1 μm in the region. Combined with the implication of flat extinction curve, grain size may depend on the density of the regions because the present observation of scattered light is toward an optically thin region.

At the wavelength of $\sim$0.6 μm, we find an excess contribution in addition to the scattered light, which is the so-called extended red emission (ERE). The ERE might originate from interaction between far-UV photons and nm-sized grains (e.g., Li & Draine 2002), though not all the observations suggest the presence of ERE.

References

Arai, T., Matsuura, S., Bock, J., Cooray, A., Kim, M. G., Lanz, A., Lee, D. H., Sano, K., *et al.* 2015, *ApJ*, 806, 69

Brandt, T. D. & Draine, B. T. 2012, *ApJ*, 744, 129

Han, W., Lee, D. H., Jeong, W. S., Park, Y., Moon, B., Park, S. J., Pyo, J., Kim, I. J., *et al.* 2014, *PASP*, 126, 853

Ienaka, N., Kawara, K., Matsuoka, Y., Sameshima, H., Oyabu, S., Tsujimoto, T., & Peterson, B. A. 2013, *ApJ*, 767, 80

Indebedouw, R., Mathis, J. S., Babler, B. L., Meade, M. R., Watson, C., Whitney, B. A., Wolff, M. J., Wolfire, M. G., *et al.* 2005, *ApJ*, 619, 931

Kawara, K., Matsuoka, Y., Sano, K., Brandt, T. D., Sameshima, H., Tsumura, K., Oyabu, Y., & Ienaka, N. 2017, *PASJ*, 69, 31

Li, A. & Draine, B. T. 2002, *ApJ*, 564, 803

Nishiyama, S., Tamura, M., Hatano, H., Kato, D., Tanabe, T., Sugitani, K., & Nagata, T. 2009, *ApJ*, 696, 1407

Sano, K., Kawara, K., Matsuura, S., Kataza, H., Arai, T., & Matsuoka, Y. 2015, *ApJ*, 811, 77

Schlegel, D. J., Finkbeiner, D. P., & Davis, M. 1998, *ApJ*, 500, 525

Wang, S., Li, A., & Jiang, B. W. 2015, *ApJ*, 811, 38

Weingartner, J. C. & Draine, B. T. 2001, *ApJ*, 548, 296

Zubko, V., Dwek, E., & Arendt, R. G. 2004, *ApJ*, 152, 211

Challenges in Panchromatic Modelling with Next Generation Facilities
Proceedings IAU Symposium No. 341, 2018
M. Boquien, E. Lusso, C. Gruppioni & P. Tissera, eds.

doi:10.1017/S1743921319001406

Caught in the web: A tale of filament galaxies

Ankit Singh, Smriti Mahajan and Devika Shobhana

Indian Institute of Science Education and Research, Mohali, Punjab, India
email: ankitsingh@iisermohali.ac.in

Abstract. The Coma supercluster is one of the largest, nearby ($\sim 100h^{-1}$Mpc) gravitationally bound structures known in the universe. It comprises two large clusters of galaxies and several galaxy groups intersected by a complex network of filaments, providing the perfect laboratory for studying the evolution of galaxies in a range of 'continuous' environments. We characterised the different components of the environment to study the properties of galaxies in the optical and ultraviolet (UV) wavebands. Our analysis shows that galaxies experience accelerated evolution as they approach the spine of the filament, suggesting that the intermediate-density environment prevalent in the filaments can accelerate the evolution of galaxies.

Keywords. galaxies, galaxy-environment, star-formation

1. Introduction

Observations suggest that most of the galaxies reside in large-scale filaments. Yet the intermediate-density environment prevalent in the filaments is poorly studied due to their large angular size on the sky. In Mahajan *et al.* (2018) we explore the impact of the filament environment on the properties of galaxies in the Coma supercluster, and compare it to the other components of the environment which are clusters and groups and, voids.

2. Method

We use the optical spectroscopic and photometric data from the Sloan digital sky survey (SDSS) data release 12 and ultraviolet data from the GALaxy evolution EXplorer (GALEX) mission, to characterise different environments in the ~500 sq. deg. region of the Coma supercluster. We employed the Hierarchical Density-Based Spatial Clustering of Applications with Noise (HDBSCAN) algorithm (McInnes *et al.* (2017)) for identifying groups, while the filaments were characterised using the Discrete Persistent Structures Extractor (DisPerSE; Sousbie *et al.* (2011)). Galaxies belonging to neither of the above environments were classified as void galaxies (Figure 1).

3. Results

Figure 2(a) shows the distribution of galaxies in different environments in the colour-magnitude plane. The spectroscopically-selected star-forming galaxies (EW(Hα) $\geqslant 2$Å) can be well distinguished from their passively-evolving counterparts in all environments. Our main finding is displayed in Figure 2(b) which shows the median trend in the broad-band colours $g-r$ and $FUV-NUV$ and the EW(Hα) for galaxies as a function of their distance from the spine of the filaments. We statistically confirm that the galaxies become redder and less star-forming as they approach the spine of the filaments.

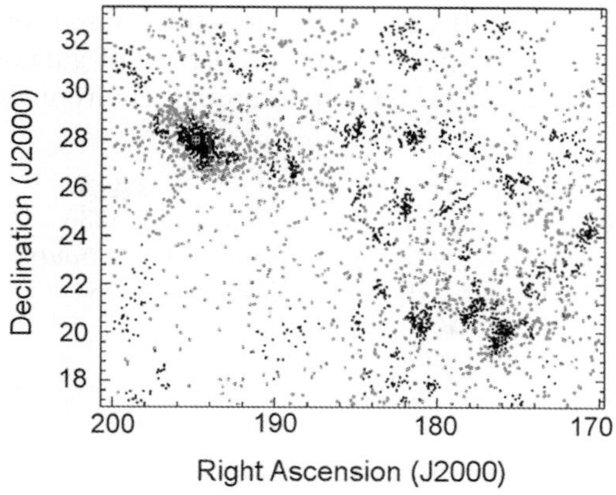

Figure 1. Environments in the Coma supercluster. The grey open circles, red points, and black points represent the galaxies in the voids, filaments and clusters or groups galaxies, respectively.

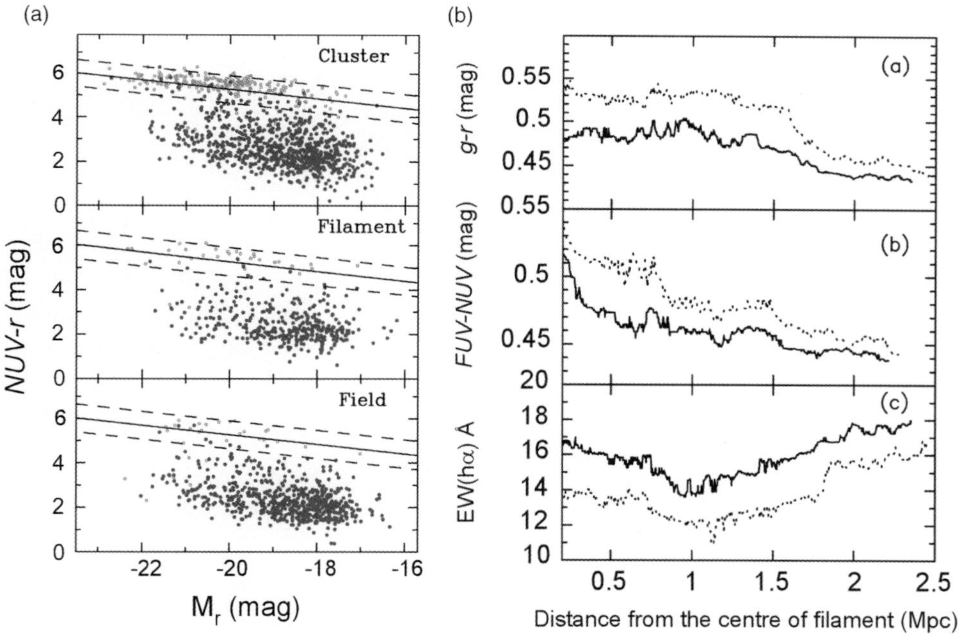

Figure 2. (a) The distribution of galaxies in the colour-magnitude plane. The red points represent passively-evolving galaxies (EW(Hα) ≤ 2Å), while their star-forming counterparts are represented by blue points. (b) The $g-r$ and $FUV-NUV$ colours, and the median EW(Hα) of all (dotted line) and non-AGN (solid line) galaxies are shown as a function of their distance from the spine of nearest filament.

4. Epilogue

The availability of multi-wavelength data from large all-sky surveys provides a great opportunity to study the large-scale environment of galaxies at $z \sim 0$. It remains to be seen whether the properties of the cosmic-web as found in the Coma supercluster duplicate themselves in other nearby superclusters, and at other redshifts. However, our finding that the intermediate density environment prevalent in the filaments influences the galaxies at least out to a radius of 1 Mpc, lays the ground for exploring the properties of the cosmic-web and its impact on the properties of galaxies in simulations, and encourage deeper observations of galaxies spanning all environments.

References

Mahajan, S., Singh, A., & Shobhana, D. 2018, *MNRAS*, 478, 4336
McInnes, L., Healy, J., & Astels, S. 2017, *The Journal of Open Source Software*, 2, 205
Sousbie, T., Pichon, C., & Kawahara, H. 2011, *MNRAS*, 414, 384

Challenges in Panchromatic Modelling with Next Generation Facilities
Proceedings IAU Symposium No. 341, 2018
M. Boquien, E. Lusso, C. Gruppioni & P. Tissera, eds.

doi:10.1017/S1743921319002126

Galactic outflows in star-forming galaxies at $z \sim 6$ studied with deep UV spectra and ALMA emission line

Yuma Sugahara[1,2], Masami Ouchi[1,3], Yuichi Harikane[1,2], Nicolas Bouché[4], Peter D. Mitchell[5] and Jérémy Blaizot[4]

[1]Institute for Cosmic Ray Research, The University of Tokyo, 5-1-5 Kashiwanoha, Kashiwa, Chiba 277-8582, Japan
email: sugayu@icrr.u-tokyo.ac.jp

[2]Department of Physics, Graduate School of Science, The University of Tokyo, 7-3-1 Hongo, Bunkyo, Tokyo, 113-0033, Japan

[3]Kavli Institute for the Physics and Mathematics of the Universe (WPI), The University of Tokyo, 5-1-5 Kashiwanoha, Kashiwa, Chiba 277-8583, Japan

[4]Univ Lyon, Univ Lyon1, Ens de Lyon, CNRS, Centre de Recherche Astrophysique de Lyon UMR5574, F-69230 Saint-Genis-Laval, France

[5]Leiden Observatory, Leiden University, PO Box 9513, NL-2300 RA Leiden, the Netherlands

Abstract. We present a velocity of galactic outflows in star-forming galaxies at the highest redshift, $z \sim 6$, so far studied with metal absorption lines. Absorption-line studies of galactic outflows need well-determined redshifts, but there are few strong emission lines in the observed-frame optical spectra of galaxies at high redshifts. In this work, we use the systemic redshifts determined by the ALMA [CII]158 μm emission lines. The sample consists of seven Lyman break galaxies at $5.1 < z < 5.7$ whose Keck/DEIMOS and ALMA data are available in the archive. The outflow maximum velocity ($v_{\rm max}$) is estimated by a fitting of line profiles to metal absorption lines in a composite spectrum. We find that $v_{\rm max}$ monotonically increases from $z \sim 0$ to 6 and that $v_{\rm max}$ tightly correlates with the halo circular velocity estimated from the stellar mass.

Keywords. galaxies: evolution, galaxies: high-redshift, galaxies: kinematics and dynamics

1. Introduction

Galactic outflows in star-forming galaxies are thought to play an important role in regulating the galaxy and IGM evolution (Somerville & Davé 2015). Recently, deep optical spectra in the observed frame enable us to probe gas kinematics in high-redshift galaxies using metal absorption lines. Sugahara *et al.* (2017) show that the outflow velocity increases from $z \sim 0$ to 2 in galaxies with similar stellar mass and star formation rate (SFR) ranges. An important parameter to measure the outflow velocity is the precise systemic redshifts. This makes it difficult to study the outflows at $z > 2$ because there are few strong emission lines in the rest-frame FUV spectral bands. Here, we present the outflow velocity of galaxies at $z \sim 5$–6 measured with absorption lines in deep rest-frame FUV spectra, using the systemic redshifts determined by ALMA [CII] 158 μm observations.

2. Redshift Evolution of Outflow Maximum Velocity

The sample consists of seven Lyman break galaxies at $z = 5$–6 observed by DEIMOS on the Keck telescope (PI: P. Capak). Their systemic redshifts have been already determined by ALMA [CII] 158 μm observations (Capak *et al.* 2015). We stack the DEIMOS spectra

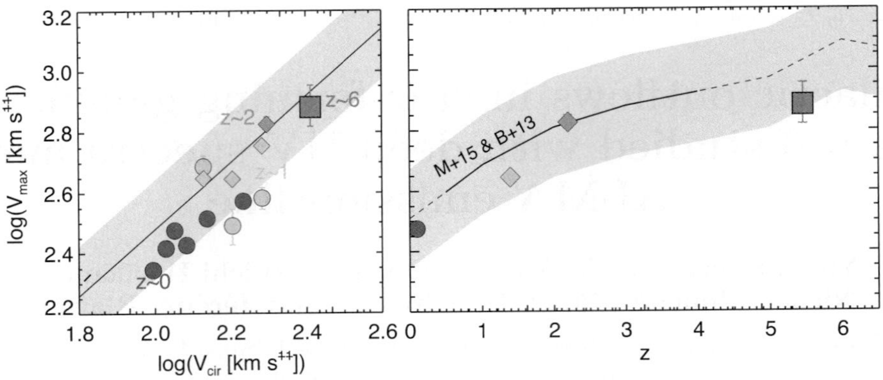

Figure 1. Left: $v_{\rm max}$ as a function of $v_{\rm cir}$. The red square is a result of this work. The data points at $z \sim 0$ (blue), $z \sim 1$ (cyan), and $z \sim 2$ (green) are presented by Sugahara *et al.* (2017). The solid line is a relation given by Muratov *et al.* (2015). Right: redshift evolution of $v_{\rm max}$ at $\log M_* \sim 10^{10.3}\ M_\odot$. The symbols are the same as in the left panel. The line indicates a relation based on Muratov *et al.* (2015) and Behroozi *et al.* (2013). The 25th and 75th percentile of the relations are indicated by the shaded region.

to obtain a composite rest-frame FUV spectrum based on the systemic redshifts, fitting a line profile to the Si IIλ1260 and C IIλ1335 absorption lines, simultaneously. From the best-fit parameters, the outflow maximum velocity $v_{\rm max}$ is defined as a velocity at which the flux density becomes 90% of the continuum from the bottom of the absorption line.

The measured value is $v_{\rm max} = 760^{+140}_{-100}$ km s^{-1}. We compare $v_{\rm max}$ at $z = 5$–6 with the velocities at $z \sim 0$–2 presented in Sugahara *et al.* (2017). Note that the value at $z \sim 2$ is re-calculated in the manner of this work. The $v_{\rm max}$ value at $z = 5$–6 is higher than the velocities at $z \sim 0$ and 1 and comparable to the one at $z \sim 2$. In the left panel of Figure 1, $\log v_{\rm max}$ is compared with the halo circular velocity $\log v_{\rm cir}$ that is converted from the stellar mass using the stellar-to-halo mass ratio given by Behroozi *et al.* (2013). Over $z \sim 0$–6, $v_{\rm max}$ has a tight correlation with $v_{\rm cir}$. In addition, interestingly, our measurements are in good agreement with the prediction of $v_{\rm max} \propto v_{\rm cir}^{1.1}$ about the hot outflows by the FIRE simulation (Muratov *et al.* 2015). This suggests that $v_{\rm max}$ physically connects with $v_{\rm cir}$ and that the cold outflows are associated with the hot outflows. In the right panel of Figure 1, $v_{\rm max}$ is plotted as a function of the redshift at the fixed stellar mass of the $\log M_* \sim 10^{10.3}$ M$_\odot$. The figure shows a monotonic increase in $v_{\rm max}$ with increasing redshift, including a large increase from $z \sim 0$ to 2 and a small increase from $z \sim 2$ to 6. This evolution is explained by a combination of the linear $v_{\rm max}$–$v_{\rm cir}$ relation and the redshift evolution of $v_{\rm cir}$. Because the halo mass does not significantly change around $\log M_* \sim 10^{10.3}$ M$_\odot$ (Behroozi *et al.* 2013), $v_{\rm cir}$ changes along with the halo radius $r_{\rm h}$. Thus, the $v_{\rm max}$ evolution is expressed as $v_{\rm max} \propto v_{\rm cir} \propto r_{\rm h}^{-0.5} \propto (1+z)^{0.5}$.

References

Behroozi, P. S., Wechsler, R. H., & Conroy, C. 2013, *ApJ*, 770, 57

Capak, P. L., Carilli, C., Jones, G., Casey, C. M., Riechers, D., Sheth, K., Carollo, C. M., Ilbert, O., *et al.* 2015, *Nature*, 522, 455

Muratov, A. L., Kereš, D., Faucher-Giguère, C.-A., Hopkins, P. F., Quataert, E., & Murray, N. 2015, *MNRAS*, 454, 2691

Somerville, R. S. & Davé, R. 2015, *ARAA*, 53, 51

Sugahara, Y., Ouchi, M., Lin, L., Martin, C. L., Ono, Y., Harikane, Y., Shibuya, T., & Yan, R., *et al.* 2017, *ApJ*, 850, 51

Challenges in Panchromatic Modelling with Next Generation Facilities
Proceedings IAU Symposium No. 341, 2018
M. Boquien, E. Lusso, C. Gruppioni & P. Tissera, eds.

doi:10.1017/S174392131900245X

ALMA detection of the [OIII] 88 μm line in a highly-magnified Lyman break galaxy at z = 6.1

Kaho Sunaga[1], Yoichi Tamura[1], Minju Lee[1,4], Ken Mawatari[2], Akio K. Inoue[3], Takuya Hashimoto[3,4], Hiroshi Matsuo[4] and Akio Taniguchi[1]

[1]Department of Physics, Nagoya University, Japan

[2]Institute for Cosmic Ray Research, The University of Tokyo, Japan

[3]Department of Environmental Science and Technology, Faculty of Design Technology, Osaka Sangyo University, Japan

[4]National Astronomical Observatory of Japan, Japan

Abstract. We present a 4.7σ detection of the [OIII] 88 μm line in a gravitationally-lensed Lyman break galaxy, RXC J2248-ID3, using the Atacama Large Millimeter/submillimeter Array (ALMA). We did not detect [CII] 158 μm and rest-frame 90 μm dust continuum emission, suggesting that the bulk of the interstellar medium (ISM) is ionized. Our two-component SED model combining the previous Hubble Space Telescope (HST) data and new photometry obtained from Very Large Telescope (VLT), Spitzer and ALMA suggests the presence of young (~ 2 Myr) and mature (~ 600 Myr) stellar components with the metallicity of $Z = 0.2 Z_{\odot}$. Our findings are in contrast with previous results claiming a very young, metal-poor stellar component.

Keywords. galaxies: high-redshift

---

1. Introduction

The [OIII] 88 μm far-infrared (FIR) fine structure line is expected to be bright enough to be detected in star-forming galaxies in the reionizaion era (Inoue *et al.* 2014). This line is a powerful probe to characterize the metallicity and the massive star formation activity. Furthermore, since the UV-to-optical nebular lines have a wide variety of ionization potentials and/or critical densities, the [OIII] 88 μm and the UV-to-optical lines will allow us to put a constraint on the physical properties of ISM even in $z > 6$ galaxies.

RXC J2248-ID3 (ID3) is a gravitationally-lensed (the lens magnification is 5.3) Lyman break galaxy at $z_{\rm Ly\alpha} = 6.110$ (Monna *et al.* 2014, Balestra *et al.* 2013). Monna *et al.* 2014 found that this object is a young (1.5 Myr), metal-poor ($Z = 0.005 Z_{\odot}$) and low-mass ($M \simeq 2.1 \times 10^8 M_{\odot}$) star-forming galaxy by spectral energy distribution (SED) modeling in the rest-frame UV. In order to constrain more physical properties of ID3, UV-to-FIR SED is needed.

2. ALMA detection of [OIII] 88 μm and SED modeling

We searched for [OIII] 88 μm in ID3 using archival ALMA band 8 data (Cycle 2, PI: S. Madden) and detected it at $z_{\rm [OIII]} = 6.1051 \pm 0.0006$ at the 4.7σ significant level, with a line flux and luminosity of $F_{\rm [OIII]} = 0.587 \pm 0.074$ Jy km/s, $L_{\rm [OIII]} = 2.0 \times 10^8\ L_{\odot}$, respectively. The 90 μm dust continuum flux was not detected with a 3σ upper limit of $S_{90\,\mu{\rm m}} < 0.61$ mJy. We do not detect the [CII] 158 μm emission line (with a 3σ upper limit of $F_{\rm [CII]} < 0.17$ Jy km/s) suggesting a depletion of the neutral component in the ISM.

Table 1. Best fit parameters.

parameters	best fit
Av [mag]	$0.00^{+0.02}_{-0.00}$
Metallicity [$Z_\odot$]	$0.2^{+0.04}_{-0.01}$
Escape fraction	$0.6^{+0.2}_{-0.1}$
Age (young) [Myr]	$2.2^{+0.2}_{-0.5}$
Age (old) [Myr]	641^{+86}_{-17}
SFH τ^{-1}(young)	0.10
SFH τ^{-1}(old)	10
Stellar mass [10^9 $M_\odot$]	$5.6^{+3.5}_{-2.6}$

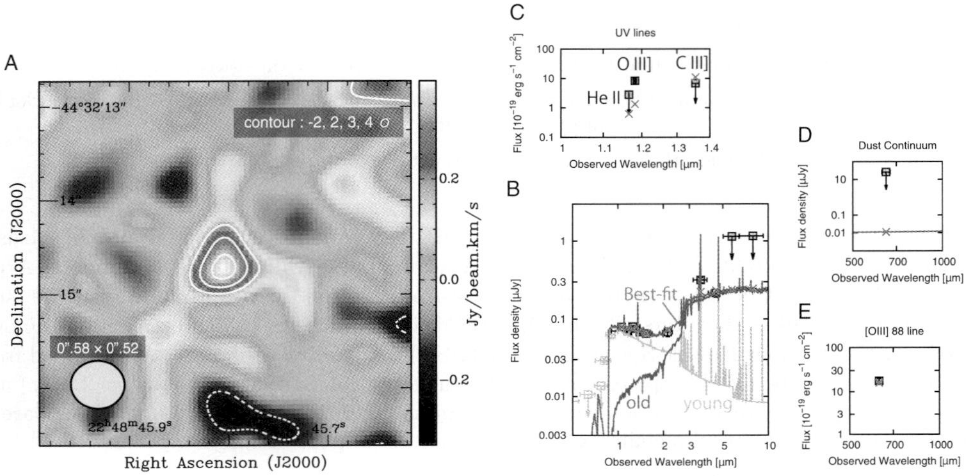

Figure 1. (A) [OIII] 88 μm integrated intensity map of ID3 obtained with ALMA. The ellipse at the bottom-left corner shows the synthesized beam size. The contours represent $(-2, 2, 3, 4) \times \sigma$. The panels from (B) to (E) show the optical-IR photometric data, UV lines, dust continuum, and [OIII] 88 μm line, respectively. Red lines show the best fit SED, while the orange and purple lines show the best fit young and old stellar component, respectively. The filled squares are observed line fluxes and continuum flux densities, while open squares are observed photometry that was not used for the SED fits. The crosses are the model predictions.

Then we perform stellar population synthesis modeling of ID3, where nebular emission and reprocessed dust emission are accounted for (PANHIT†). We use the [OIII] 88 μm flux and 90 μm upper limit in addition to the optical-to-near infrared broadband photometry obtained with HST, VLT/Hawk-I, and Spitzer/IRAC. We also account for UV line fluxes and upper limits, such as HeII λ1640, OIII] $\lambda\lambda$1661, 1666 and CIII] $\lambda\lambda$1907, 1909 (Mainali *et al.* 2017). We employ two stellar components with constant star-formation history (*cf.* Table 1), because a single component with any reasonable parameters cannot reproduce the observed SED. The best fit SED and parameters are shown in Figure 1 (B)-(E) and Table 1, respectively. The best-fitting model reproduces the broadband photometry and the [OIII] 88 μm line. However the UV lines are not reproduced very well, suggesting that the ISM has different properties from what is assumed in the model which is described by Mawatari *et al.* 2018. Photoionization modeling of the nebular emission lines will offer a unique opportunity to unveil the physical properties of ionized ISM.

† http://www.icrr.u-tokyo.ac.jp/mawatari/PANHIT/PANHIT.html

References

Balestra, I., Vanzella, E., Rosati, P., *et al.* 2013, *A&A*, 559, L9
Inoue, A. K., Shimizu, I., Tamura, Y., *et al.* 2014, *ApJ*, 780, L18
Mainali, R., Kollmeier, J. A., Stark, D. P., *et al.* 2017, *ApJ*, 836, L14
Mawatari, K., *et al.* 2018, Proc. IAU Symposium No. 341, in this volume
Monna, A., Seitz, S., Greisel, N., *et al.* 2014, *MNRAS*, 438, 1417

Challenges in Panchromatic Modelling with
Next Generation Facilities
Proceedings IAU Symposium No. 341, 2018
M. Boquien, E. Lusso, C. Gruppioni & P. Tissera, eds.

doi:10.1017/S1743921319003107

Dust evolution in galaxies at $z > 7$

Tsutomu T. Takeuchi[1], Ryosuke S. Asano[1], Sayaka Nagasaki[1], Takaya Nozawa[2], Yoichi Tamura[1], Ken Mawatari[3] and Akio K. Inoue[4]

[1]Division of Particle and Astrophysical Science, Nagoya University, Furo-cho, Chikusa-ku, Nagoya 464-8602, Japan
(e-mail: takeuchi.tsutomu@g.mbox.nagoya-u.ac.jp)

[2]National Astronomical Observatory of Japan

[3]Institute for Cosmic Ray Research, the University of Tokyo

[4]Osaka Sangyo University

Abstract. Recently huge amount of dust $M_{\rm dust} \simeq 10^{6-7}\,M_\odot$ in galaxies at $z = 7$–8 has been discovered by ALMA observations. The suggested timescale of the dust production was a few–several $\times\, 10^8$ yr, while the stellar mass was several $\times\, 10^9\, M_\odot$. This amount of dust cannot be easily explained only by a supply from supernovae if we consider the dust destruction by reverse shocks. We propose that these values can be consistently explained if we take into account the grain growth in the interstellar medium (ISM). This scenario successfully reproduces the evolution of the dust mass, as well as the SFR, and stellar mass simultaneously. We conclude that even at such an early epoch of the Universe, the dust grain growth in the ISM plays a significant role in galaxies.

Keywords. galaxies: evolution, galaxies: formation, dust, stars: formation

1. Introduction

Now the importance of dust in galaxy evolution is widely accepted. A large fraction of the star formation (SF) activity is significantly hidden by dust (up to $\geq 90\,\%$) during most of the history of the Universe (e.g., Takeuchi *et al.* 2005; Burgarella *et al.* 2013). Dust also plays an important role in the SF process itself, since it works as a catalyst of molecular formation which is needed to form stars in molecular clouds (e.g., Gould & Salpeter 1963). Then, a fundamental question arises: when dust started to form, and became important in the cosmic history? Recently, some authors discovered galaxies containing a huge dust mass at $z \sim 7$–8 (Watson *et al.* 2015; Laporte *et al.* 2017). The SFR of these galaxies was found to be only moderate ($\sim 10 - 20\, M_\odot {\rm yr}^{-1}$). This means that a galaxy of moderate SFR could already form such amount of dust at such an early epoch of the Universe. A natural explanation is desired if we develop a general theory of dust production in the early Universe. We propose a plausible explanation to solve this problem. We use $h = 0.7$, $\Omega_{\rm M} = 0.3$ and $\Omega_\Lambda = 0.7$ for the cosmological parameters throughout this article.

2. Asano Model

We have developed a theoretical framework of dust evolution in galaxies that is consistent with chemical evolution (Asano *et al.* 2013a,b, 2014; Nozawa *et al.* 2015: hereafter referred to as Asano model). Asano model takes into account the dust production in SNe and asymptotic giant branch (AGB) stars, grain destruction of dust by SN blast waves, grain growth in the ISM, and coagulation and shattering of grains. Among them, the dust supply from SNe and AGBs (stellar origin) and the grain growth play the central role to increase the net amount of dust. The strongly nonlinear behavior of dust mass is due to the onset of the grain growth Asano *et al.* (2013a,b). Asano *et al.* (2013a) discovered that the timing of the onset of grain growth is determined by metallicity, not the SF history. Asano model succeeded in reproducing the strong nonlinearity of the relation between metallicity and dust-to-gas mass ratio of local galaxies observed by *Herschel* (Rémy-Ruyer *et al.* 2014).

3. Results and Discussion

We calculated the dust amount based on Asano model using the Salpeter IMF with a mass range of $[0.1\,M_\odot$–$100\,M_\odot]$, and the SF timescale $\tau_{\rm SF} = 0.5$ Gyr and the initial gas mass $M_{\rm gas}(0) = 10^{10}\,M_\odot$. These values were not tuned to fit the data, but rather to demonstrate how naturally the data can be reproduced by the model. Other parameters are the same as those used in Asano *et al.* (2013b).

We first calculated the SFR and stellar mass evolution. We assumed a closed box and the Schmidt law with index $n = 1$. The SFR is approximately constant, and reproduces the observed SFR $\sim 20\,M_\odot\,{\rm yr}^{-1}$ in galaxies at $z \sim 8$. The stellar mass increases up to $\sim 10^8$ yr and converged to the initial gas mass because almost all the gas has turned into stars at this age of $\sim 10^8$ yr. If we use the cosmic age as a substitute of their ages, we obtain several $\times\, 10^9\,M_\odot$. These model predictions are consistent with observations (Watson *et al.* 2015; Laporte *et al.* 2017).

Now we focus on the main subject of the observation, the dust mass of these galaxies. The most striking feature of the dust mass evolution is the sudden jump at around $\sim 3 \times 10^8$ yr. This is due to the onset of the dust grain growth in the ISM, turned on when the metallicity surpasses the critical metallicity $Z_{\rm cr}$ (Asano *et al.* 2013a,b). This feature was not recognized in previous works, where a linear relation was assumed (Guiderdoni *et al.* 1998). We stress that the relation between the metal and dust, and consequently that between the gas and dust is strongly nonlinear. This is in conflict with what very widely assumed and accepted. We also compare the prediction of Asano model for the dust-to-gas mass ratio (D/G ratio) with observations. Since the direct observation of the total gas mass is not available yet, as presented in Watson *et al.* (2015), we use the inversion of the Kennicutt–Schmidt (KS) law (e.g., Kennicutt & Evans 2012). Comparing the model predictions with observed values, we reach the conclusion that the grain growth in the ISM is indeed necessary to account for the huge amount of dust discovered in such a high redshift.

References

Asano, R. S., Takeuchi, T. T., Hirashita, H., & Inoue, A. K. 2013, *EPS*, 65, 213
Asano, R. S., Takeuchi, T. T., Hirashita, H., & Nozawa, T. 2013, *MNRAS*, 432, 637
Asano, R. S., Takeuchi, T. T., Hirashita, H., & Nozawa, T. 2014, *MNRAS*, 440, 134
Burgarella, D., Buat, V., Gruppioni, C., *et al.* 2013, *A&A*, 554, A70
Gould, R. J. & Salpeter, E. E. 1963, *ApJ*, 138, 393
Guiderdoni, B., Hivon, E., Bouchet, F. R., & Maffei, B. 1998, *MNRAS*, 295, 877
Kennicutt, R. C. & Evans, N. J. 2012, *ARA&A*, 50, 531
Laporte, N., Ellis, R. S., Boone, F., *et al.* 2017, *ApJ*, 837, L21
Nozawa, T., Kozasa, T., & Habe, A. 2006, *ApJ*, 648, 435
Nozawa, T., Kozasa, T., Habe, A., *et al.* 2007, *ApJ*, 666, 955
Nozawa, T., Asano, R. S., Hirashita, H., & Takeuchi, T. T. 2015, *MNRAS*, 447, L16
Rémy-Ruyer, A., Madden, S. C., Galliano, F., *et al.* 2014, *A&A*, 563, A31
Takeuchi, T. T., Buat, V., & Burgarella, D. 2005, *A&A*, 440, L17
Watson, D., Christensen, L., Knudsen, K. K., *et al.* 2015, *Nature*, 519, 327

Challenges in Panchromatic Modelling with
Next Generation Facilities
Proceedings IAU Symposium No. 341, 2018
M. Boquien, E. Lusso, C. Gruppioni & P. Tissera, eds.

doi:10.1017/S1743921319002382

X-ray spectral model from clumpy torus and its application to Circinus galaxy

Atsushi Tanimoto

Kyoto University

Abstract. We construct an X-ray spectral model for the clumpy torus in an active galactic nucleus (AGN), utilizing the Monte Carlo simulation for Astrophysics and Cosmology framework (MONACO: Odaka *et al.* 2016). The geometry of the torus is the same as that in Nenkova *et al.* (2008), which assumes a power law distribution of clumps in the radial direction and a normal distribution in the elevation direction. We apply our model to the broadband X-ray spectrum of the Circinus galaxy observed with *XMM-Newton*, *Suzaku*, and *NuSTAR*. Our model can well reproduce the observed X-ray spectrum, yielding a hydrogen column density along the line-of-sight $N_{\rm H}^{\rm LOS} = 4.86^{+0.07}_{-0.04} \times 10^{24}$ cm^{-2} and a torus angular width $\sigma = 14.7^{+0.44}_{-0.39}$ degree.

Keywords. galaxies: active nuclei, galaxies: X-ray

1. Introduction

In the unified model framework, the obscuring matter in an active galactic nuclei, i.e. accreting supermassive black hole (SMBH) Antonucci 1993; Urry & Padovani 1995; Ramos Almeida & Ricci 2017), consists of gas and dust distributed in a toroidal shape. This "torus" is likely to play an important role in AGN feeding because the torus is a mass reservoir linking the host galaxy and the SMBH. Thus, elucidating the torus structure is essential to understand the mechanisms of the co-evolution between the host galaxy and SMBH (Kormendy & Ho 2013). Nevertheless, many of basic properties of AGN tori still remain unclear.

Many observations indicate that AGN tori must be composed of dusty clumps rather than of smooth gas (Krolik & Begelman 1988). In the infrared band, Nenkova *et al.* (2008) constructed spectral model from clumpy tori. They assumed a power law distribution of clumps in the radial direction and normal distribution in the elevation direction. This CLUMPY model has been successfully applied to the infrared spectra of nearby AGNs (Ramos Almeida & Ricci 2017).

X-ray observations are a powerful tool to investigate the torus structure. This is because X-rays can trace all matter including gas and dust at various physical conditions. It is only recently that X-ray spectral models for the clumpy torus have been developed (Liu & Li 2014; Furui *et al.* 2016. We construct a new X-ray clumpy torus model designated as "XCLUMPY", by adopting the same geometry of clump distribution as that of the CLUMPY model in the infrared band (Nenkova *et al.* 2008). This enables us to directly compare the results obtained from the infrared and X-ray and to which constrain the spatial distribution of dust and that of all matter, respectively.

2. Results and Discussion

We apply the XCLUMPY model to the broadband X-ray spectrum of the Circinus galaxy. This is because the Circinus galaxy is one of the closest (4.2 Mpc) obscured AGNs and is an ideal target for investigating the torus structure. *Suzaku* observed the Circinus

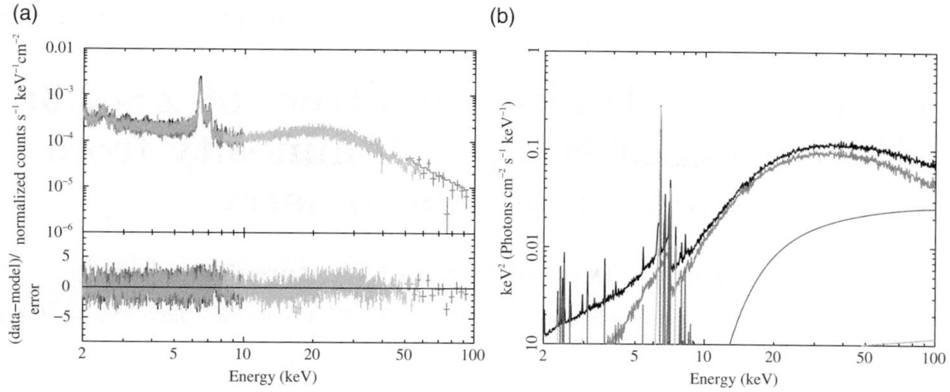

Figure 1. Left: folded X-ray spectra fitted with the XCLUMPY model. Right: best-fitting model. Left: Black crosses: *Suzaku*/BIXIS. Red crosses: *Suzaku*/FIXIS. Green crosses: *Suzaku*/PIN. Blue crosses: *Suzaku*/GSO. Pink crosses: *XMM-Newton*/EPN. Orange crosses: *XMM-Newton*/MOS. Light blue crosses: *NuSTAR*/FPMs. Right: Black line: total. Purple line: direct component. Blue line: reflection component from the torus. Light blue lines: emission line from the torus. Orange line: contamination from the CGX1. Red lines: contamination from the CGX2.

galaxy in 2006 July. Simultaneous observations with *XMM-Newton* and *NuSTAR* were performed in 2013 February. We collected X-ray data from the archive.

Our model can well reproduce the broadband X-ray spectra observed with the three satellites. Figure 1 shows the (a) folded X-ray spectra and (b) best-fitting model. We obtain the hydrogen column density along the line-of-sight $N_{\rm H} = 4.86^{+0.07}_{-0.04} \times 10^{24}$ cm^{-2} and torus angular width $\sigma = 14.7^{+0.44}_{-0.39}$ degree. Our measurement of the hydrogen column density is consistent with the one obtained by Arévalo *et al.* (2014). They analyzed the simultaneous *XMM-Newton* and *NuSTAR* observation data with the MYTorus model (Murphy & Yaqoob 2009).

We also compare the torus parameters obtained from the X-ray with those from infrared spectroscopy. The torus angular width obtained from the infrared data ($\sigma = 65^{+2}_{-5}$ degree: Ichikawa *et al.* 2015) is much larger than the X-ray results ($\sigma = 14.7^{+0.44}_{-0.39}$ degree). This can be explained by the presence of dusty polar outflow observed in the infrared interfere-metric observations (Stalevski *et al.* 2017), which makes the distribution of dust wider than that of gas.

References

Antonucci, R. 1993, Annual Review of Astronomy and Astrophysics, 31, 473
Arévalo, P., Bauer, F. E., Puccetti, S., *et al.* 2014, ApJ, 791, 81
Furui, S., Fukazawa, Y., Odaka, H., *et al.* 2016, ApJ, 818, 164
Ichikawa, K., Packham, C., Ramos Almeida, C., *et al.* 2015, ApJ, 803, 57
Kormendy, J. & Ho, L. C. 2013, Annual Review of Astronomy and Astrophysics, 51, 511
Krolik, J. H. & Begelman, M. C. 1988, ApJ, 329, 702
Liu, Y. & Li, X. 2014, ApJ, 787, 52
Murphy, K. D. & Yaqoob, T. 2009, MNRAS, 397, 1549
Nenkova, M., Sirocky, M. M., Ivezić, Ž., & Elitzur, M. 2008, ApJ, 685, 147
Odaka, H., Yoneda, H., Takahashi, T., & Fabian, A. 2016, MNRAS, 462, 2366
Ramos Almeida, C. & Ricci, C. 2017, Nature Astronomy, 1, 679
Stalevski, M., Asmus, D., & Tristram, K. R. W. 2017, MNRAS, 472, 3854
Urry, C. M. & Padovani, P. 1995, Publications of the Astronomical Society of the Pacific, 107, 803

Challenges in Panchromatic Modelling with Next Generation Facilities
Proceedings IAU Symposium No. 341, 2018
M. Boquien, E. Lusso, C. Gruppioni & P. Tissera, eds.

doi:10.1017/S1743921319002783

Propeties of H alpha emitters at z ∼ 2.3: Derivation of H alpha luminosity from multi-band photometry

Yasunori Terao[1], Lee Spitler[2] and Kentaro Motohara[1]

[1]Institute of Astronomy, Graduate School of Science, The University of Tokyo, 2-21-1, Osawa, Mitaka, Tokyo 181-0015, Japan

[2]Department of Physics and Astronomy, Macquarie University, Sydney, NSW 2109, Australia

Abstract. The measurement of Hα luminosity for large numbers of galaxies is important to investigate recent star formation history of galaxies. With SED fitting that includes emission line templates, we extract individual galaxy Hα luminosities from broad-band photometry. We compare Hα luminosity function with the result of a narrow-band survey, HiZELS, and find there are more luminous galaxies in Hα than previously reported. As a result, our derived star formation rate density at $z \sim 2.3$ turns out to be 2.2 times higher than previous studies. Most of the offset in the results can be explained by missing Hα in the HiZELS photometric aperture and different methods for dust extinction correction.

Keywords. galaxies: evolution, high-redshift, luminosity function, photometry

1. Introduction

Hα emission line is one of the most important star formation rate (SFR) indicators. Since it originates from hydrogen ionized by most massive stars with shortest lifetime, Hα is sensitive to short time variation of star formation activities in galaxies. Therefore it enables us to investigate more recent star formation history (SFH) than other SFR indicators. However, spectroscopic measurement of Hα is observationally expensive. In this proceeding, we derive Hα luminosities of galaxies at $2.1 < z < 2.5$ from broad-band (BB) photometry. An advantage of the use of BB filters is wider redshift coverage than narrow-band (NB) filters, which achieves larger survey volume at fixed field of view. Moreover, our method can provide information on other emission lines and/or other redshft ranges using archive data.

2. Method

We use ZFOURGE catalog (Straatman *et al.* 2016) which contains photometry covering from 0.3 to 8 μm and Spitzer/MIPS 24 μm data. We select galaxies in $2.1 < z < 2.5$ based on photometric redshift, where Hα emission lines from galaxies fall into the ZFOURGE K_s band. Our sample consists of 2005 galaxies in COSMOS, UDS, and CDFS.

To extract Hα flux from BB photometry, we first derive stellar continuum flux from SED fitting by FAST (Kriek *et al.* 2009). Since observed fluxes are sums of stellar continuum and nebula emission lines, we use SED templates including emission lines to obtain reliable continuum fluxes. Then total emission line fluxes are derived from observed flux excesses, which are the differences between observed K_s fluxes and continuum fluxes from SED fitting. Finally we obtain Hα fluxes assuming its fraction in total emission lines.

We correct Hα luminosities for dust extinction following the procedure of Nordon *et al.* (2013). First UV attenuation of each galaxy is derived from a ratio between IR

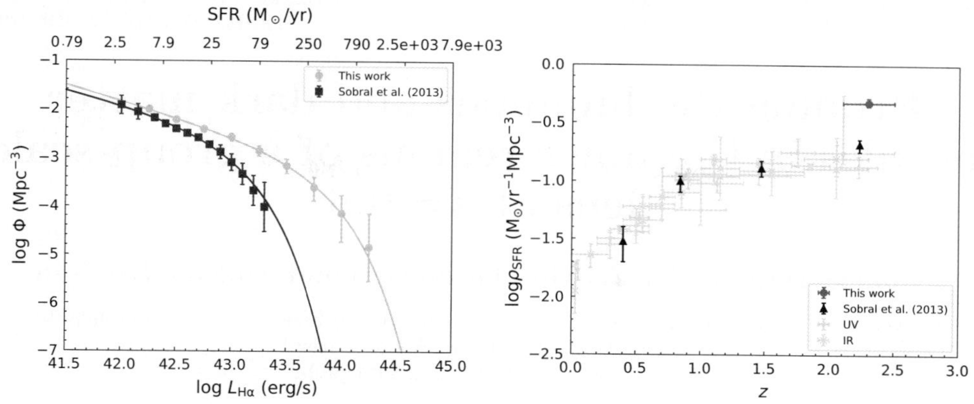

Figure 1. Left : Hα luminosity function corrected for dust extinction. Our results and that of Sobral *et al.* (2013) are shown in orange circles and blue squares, respectively. Each curve shows best-fit Schechter function. Right : SFRD as a function of redshift. Red circles and black triangles show ours and result from Sobral *et al.* (2013) measured with Hα. Results with UV or IR from previous studies summarized in Madau & Dickinson (2014) are shown in gray.

and UV SFRs. Then it is converted to Hα attenuation assuming the extinction law of Calzetti *et al.* (2000). For galaxies not detected at 24 μm, we assume no dust extinction.

3. Results & Discussions

The left panel of Figure 1 shows Hα luminosity functions corrected for dust extinction. Our luminosity function is located above the result from HiZELS (Sobral *et al.* 2013), which suggests there are more Hα luminous galaxies than previously reported. As a result, star formation rate density (SFRD) at $z \sim 2.3$ is 2.2 times higher than previous studies as shown in the right panel of Figure 1.

Most of the discrepancies between our results and Sobral *et al.* (2013) can be explained by missing Hα flux. As more massive galaxies tend to have more extended Ha profiles (Nelson *et al.* 2016), our simulation with the fixed (2") aperture and seeing of HiZELS data suggests that Sobral *et al.* (2013) may have missed at most 40% of Hα flux outside of their adopted aperture, while fluxes in ZFOURGE catalog are corrected to total fluxes using PSF models. In addition, Sobral *et al.* (2013) assumes Hα attenuation of 1 mag for all galaxies. Since more luminous galaxies are more attenuated, they may have underestimated intrinsic Hα luminosity for luminous galaxies.

References

Brammer, G. B., van Dokkum, P. G., & Coppi, P. 2008, *ApJ*, 686, 1503

Calzetti, D., Armus, L., Bohlin, R. C., Kinney, A. L., Koornneef, J., & Storchi-Bergmann, T. 2000, *ApJ*, 533, 682

Kriek, M., van Dokkum, P. G., Labbé, I., Franx, M., Illingworth, G. D, Marchesini, D., & Quadri, R. F. 2009, *ApJ*, 700, 221

Madau, P. & Dickinson, M. 2014, *ARAA*, 52, 415

Nelson, E. J., van Dokkum, P. G., Förster Schreiber, N. M., Franx, M., Brammer, G. B., Momcheva, I. G., Wuyts, S., Whitaker, K. E., *et al.* 2016, *ApJ*, 828, 27

Nordon, R., Lutz, D., Saintonge, A., Berta, S., Wuyts, S., Förster Schreiber, N. M., Genzel, R., Magnelli, B., *et al.* 2013, *ApJ*, 762, 125

Sobral, D., Smail, I., Best, P. N., Geach, J. E., Matsuda, Y., Stott, J. P., Cirasuolo, M., & Kurk, J. 2013, *MNRAS*, 428, 1128

Straatman, C. M. S., Spitler, L. R., Quadri, R. F., Labbé, I., Glazebrook, K., Persson, S. E., Papovich, C., Tran, K. H., *et al.* 2016, *ApJ*, 830, 51

Challenges in Panchromatic Modelling with Next Generation Facilities
Proceedings IAU Symposium No. 341, 2018
M. Boquien, E. Lusso, C. Gruppioni & P. Tissera, eds.

doi:10.1017/S1743921319002916

Probing the luminous and dark matter profiles in the inner regions of a group-scale lens at z = 0.6

Mônica Tergolina, Cristina Furlanetto and Marina Trevisan

Universidade Federal do Rio Grande do Sul, Instituto de Física Av. Bento Gonçalves, 9500, 91501-970, Porto Alegre, Brazil
email: monica.tergolina@ufrgs.br

Abstract. Studying the density profiles of galaxy groups offers an important insight on how large-scale structure in the Universe formed and evolved, since galaxy groups bridge the gap between individual galaxies and galaxy clusters. We aim to probe the total density profile of the galaxy group that is gravitational lensing HELMS18, a submillimeter galaxy at $z = 2.39$ from the Herschel's HerMES Large Mode Survey (HELMS), by combining strong gravitational lensing with kinematics of the centrally-located galaxies and kinematics of the group members. We will use high-resolution data of HELMS18 obtained with the Atacama Large Millimeter/submillimeter Array (ALMA) and multi-object spectroscopic data of the group members from Gemini-GMOS. Our final goal is to combine these observations to probe the stellar and dark matter density profiles and to build a complete description of this galaxy group.

Keywords. gravitational lensing: strong, galaxies: groups, galaxies: general, galaxies: high-redshift

1. Introduction

Characterizing the mass distribution of galaxy groups and clusters is one of the challenges of modern Cosmology. Detailed studies of their internal mass distributions can be used to test and constrain the model of structure formation and evolution (Bartelmann *et al.* 2013). Of particular interest is the radial density profile of dark matter, which is sensitive to baryonic processes, such as AGN and stellar feedback. As these processes vary with mass and formation history, it is important to examine the baryonic and dark matter distributions across the full mass range, from individual galaxies to galaxy clusters. It is now widely recognised that gravitational lensing is the most suitable tool for measuring the distribution of mass (dark+baryonic) for this wide range of systems. Modelling the strong lensing effect allow us to measure the total mass within the system of multiple images, defined by the Einstein radius. Combining strong lensing technique with other observations, such as weak lensing and stellar kinematics of the centrally-located galaxies, allow us to measure the density profile at different radii. While some studies suggest that baryons and dark matter follow a nearly isothermal density profile in early-type galaxies (ETGs), other studies found that the inner dark matter density profile in galaxy clusters is shallower than the Navarro-Frank-White slope, suggesting a connection between the inner halo and the baryons in the central galaxies (Sand *et al.* 2008; Newman *et al.* 2013). Such techniques have been mainly applied to galaxy-scale and cluster-scale lenses, so the main goal of this project is to bridge the gaps in the measurements of the density profiles of small and large scale structures.

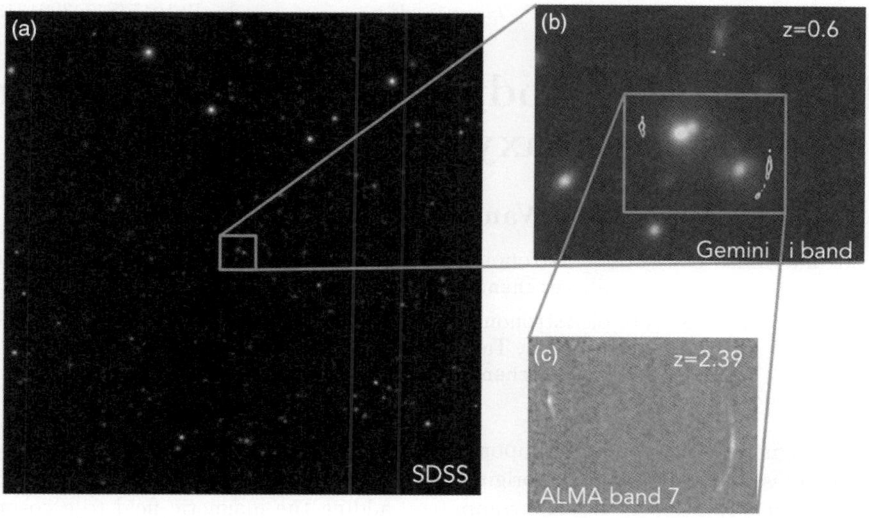

Figure 1. SDSS colour-composite image of the galaxy group (*a*). Superposition of Gemini and ALMA data (white contours) of the inner region of the group (*b*). ALMA band 7 data of HELMS18 (*c*).

2. Data and Methods

In this work we use ALMA band 7 data, which show two gravitational arcs (Figure 1c). According to Amvrosiadis *et al.* (2018), HELMS18 system has an Einstein radius of $\sim 6.5''$. Optical images from Sloan Digital Sky Survey (SDSS) show that the lensing group has two central galaxies, an ETG and a quasar candidate (Figure 1a). Spectroscopic data from SDSS (for the ETG) indicate that the group is at $z \sim 0.6$.

Currently we are modelling the ALMA data using the semi-linear inversion method outlined in Dye *et al.* (2018), which works directly in the interferometric uv-plane on the visibility data. With the modelling we aim to: *i*) measure properties of the lens; *ii*) measure the slope of the density profile; *iii*) reconstruct the image of the lensed galaxy; and *iv*) investigate properties of the lensed galaxy.

As part of this project we proposed to conduct Gemini Multi-Object Spectroscopy observations to target both the candidate group members and the two brightest central galaxies. Our proposal was accepted and observations are ongoing.

Gemini data will allow us to investigate stellar kinematics of the central galaxies, determine the group members and estimate the luminous density profile and total stellar mass. Combining strong lensing with stellar kinematics of the centrally-located galaxies and kinematics of the group members, we will measure the mass distribution at different radii.

References

Bartelmann, M., Limousin, M., Meneghetti, M., Schmidt, R., *et al.* 2013, *Space Science Reviews*, 177, 3

Sand, D. J., Treu, T., Ellis, R. S., Smith, G. P., Kneib, J.-P., *et al.* 2008, *ApJ*, 674, 711

Newman, J. A., Cooper, M. C., Davis, M., Faber, S. M., Coil, A. L., Guhathakurta, P., Koo, D. C., Phillips, A. C., *et al.* 2013, *ApJS*, 208, 5

Amvrosiadis, A., Eales, S. A., Negrello, M., Marchetti, L., Smith, M. W. L., Bourne, N., Clements, D. L., De Zotti, G., *et al.* 2018, *MNRAS*, 475, 4939

Dye, S., Furlanetto, C., Dunne, L., Eales, S. A., Negrello, M., Nayyeri, H., van der Werf, P. P., Serjeant, S., *et al.* 2018, *MNRAS*, 476, 4383

Challenges in Panchromatic Modelling with Next Generation Facilities
Proceedings IAU Symposium No. 341, 2018
M. Boquien, E. Lusso, C. Gruppioni & P. Tissera, eds.

doi:10.1017/S174392131900231X

Magneto-Hydrodynamic simulations on galaxy modeling

Wei-Chen Wang[1] **and Ke-Jung Chen**[2]

[1]Academia Sinica Institute of Astronomy and Astrophysics, National Taiwan University
email: weichenwang.network@gmail.com

[2]Academia Sinica Institute of Astronomy and Astrophysics, ASMAB R1420, No.1, Sec.4, Roosevelt Rd, Taipei 10617, Taiwan, R.O.C.
email: kjchen@asiaa.sinica.edu.tw

Abstract. Magnetic field plays an important role in star formation and galaxy evolution. Previous studies discussed about the origin of magnetic field and its effect to the environment. With the recent advancement of supercomputers, adding the magnetic field to a cosmological hydrodynamic simulations only become feasible. In this proceeding, we present the results of high-resolution magneto-hydrodynamic simulation with GIZMO and compare our simulation result with the previous literature and the observations.

Keywords. galaxies: evolution, galaxies: magnetic fields, methods: n-body simulations

---

1. Introduction

The existence of large-scale magnetic field has been supported by observations in recent years. Several authors have reported the findings of magnetic field of μG order of magnitude in nearby galaxies. For example, Fletcher *et al.* (2010) reported the mean field strength of 21 galaxies observed in the last decade is about 17 μG, and Adebahr *et al.* (2013) estimated the magnetic field strength to be 98 μG in the core region of M82. Beck *et al.* (2016) gave a summary table on the observations of magnetic field in various galaxy types.

The origin of magnetic field, however, is still under debate. Durrer and Neronov *et al.* (2013) reviewed the origin and the observational method of primordial magnetic field. If the magnetic field were primordial, how it amplified from extremely small $\sim 10^{-14}$ to μG level is one of the key questions in modern astrophysics. Since gas motion is directly influenced by the magnetic field thus it should affect the dynamics of galaxies. The associated star formation and supernova explosions inside galaxies are subject to change under the presence of magnetic field. Such studies can deepen our understanding of galaxy evolution. Therefore, studying the evolution of magnetic field and its impact to surroundings is of great importance.

The magnetic field amplification processes and its sustaining mechanism have already been discussed in many literatures. However, it is not until the recent advancement of supercomputer can high-resolution MHD simulations be conducted. The Auriga project (Pakmor *et al.* 2017) found that magnetic field is exponentially amplified in the center of Milky Way-like galaxies from z=7 to z=3 and linearly amplified at larger radii between z=2 and z=1. By comparing with thermal motions and turbulent kinetic energy, they showed that magnetic fields dominate galaxy evolution in latter epoch but can be neglected in high redshift universe for Milky Way-like galaxies. One shall note that even under these circumstances, magnetic fields might still play a role at earlier epoch when

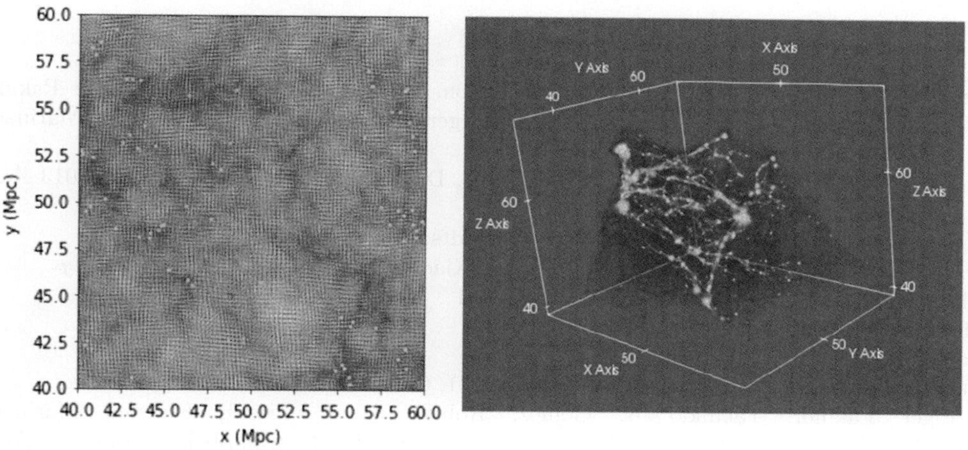

Figure 1. Left: Yellow and red dots are the high density region identified by DBSCAN algorithm. Right: Large scale structure at z=0 (without magnetic field).

anisotropic transport processes are included. The IllustrisTNG project (Pillepich *et al.* 2018) showed that magnetic field have an effect on large halos and suppressed the total star formation at low redshift.

Based on the recent simulations, we conduct a high resolution cosmological zoom-in simulation to achieve high spatial resolution while maintaining cosmological initial condition at the same time. We aim to answer open questions such as how does magnetic field effects the evolution of galaxies after anisotropic transport processes are imposed.

2. Method

In order to achieve high spatial resolution, we generated our cosmological initial condition with MUSIC (Hahn and Abel *et al.* 2011), which adopts the cosmological parameters from the WMAP data. We use a simulation box size of 100 $\mathrm{Mpc\,h^{-1}}$ at z=100 with 10^7 gas and dark matter particles each, having 10^8 solar mass per particle. However, this resolution is insufficient to resolve to the star formation process. After the universe evolves for a short period with GIZMO simulations (Hopkins *et al.* 2015), we zoom-in into a high density region with machine learning clustering algorithm DBSCAN (Ester *et al.* 1996) and apply the splitting particle method (Kitsionas and Whitworth *et al.* 2002) to increase the resolution (Figure 1). This smaller box size with split particles becomes our new initial condition for next period of simulation. By conducting this select-and-split process for ten times along with the redshift evolves from z=100 to z=15, we achieve the resolution of unit solar mass within 100 $\mathrm{kpc\,h^{-1}}$ box size. With this high resolution cosmological initial condition at z=15, we impose magnetic field of 10^{-14} G into our ongoing simulation with various processes included. We will examine our results with previous papers and compared with the observations.

Acknowledgements

This research is supported by the Ministry of Education, Taiwan, R.O.C. under Grant no. MOST 107-2112-M-001 -044 -MY3. Numerical simulations are supported by the National Energy Research Scientific Computing Center (NERSC), and the TIARA Cluster at the Academia Sinica Institute of Astronomy and Astrophysics (ASIAA).

References

A. Fletcher, 2010 *ASPC*, 438, 197

Annalisa Pillepich, Volker Springel, Dylan Nelson, Shy Gene, Jill Naiman, Rudiger Pakmor, Lars Hernquist, Paul Torrey, Mark Vogelsberger, Rainer Weinberger, & Federico Marinacci, 2018, *MNRAS*, 473, 4077

B. Adebahr, M. Krause, U. Klein, M. Weżgowie, D. J. Boman, & R.-J. Dettmar, 2013 *A&A*, 555, 23

M. Lazar, R. Schlickeiser, R. Wielebinski, S. Poedts, 2009 *APJ*, 693, 1133

Martin Ester, Hans-Peter Kriegel, Jorg Sander, Xiaowei Xu, 1996 *KDD-96 Proceedings*

Oliver Hahn, Tom Abel, 2011 *MNRAS*, 415, 2101

Philip F. Hopkins, 2015 *MNRAS*, 450, 53

Rainer Beck, 2016 *A&ARv*, 24, 4

Ruth Durrer, & Andrii Neronov, 2013 *A&ARv*, 21, 62

Rudiger Pakmor, Facundo A. Gómez, Robert J. J. Grand, Federico Marinacci, Christine M. Simpson, Volker Springel, David J. R. Campbel, Carlos S. Frenk, Thomas Guillet, Christoph Pfrommer, & Simon D. M. White, 2017 *MNRAS*, 469, 3185

S. Kitsionas, & A. P. Whitworth, 2002 *MNRAS*, 330, 129

Challenges in Panchromatic Modelling with
Next Generation Facilities
Proceedings IAU Symposium No. 341, 2018
M. Boquien, E. Lusso, C. Gruppioni & P. Tissera, eds.

doi:10.1017/S1743921319001893

Active and dust obscured star-forming galaxies at $z \sim 4$ probed with UV spectral slope beta

Satoshi Yamanaka[1] **and Toru Yamada**[2]

[1]Department of Environmental Science and Technology, Faculty of Design Technology, Osaka Sangyo University, 3-1-1, Nakagaito, Daito, Osaka 574-8530, Japan
email: s.yamanaka@est.osaka-sandai.ac.jp

[2]Institute of Space and Astronautical Science, Japan Aerospace Exploration Agency, 3-1-1, Yoshinodai, Chuo-ku, Sagamihara, Kanagawa, 252-5210, Japan
email: yamada@ir.isas.jaxa.jp

Abstract. We investigate the stellar population of star-forming galaxies at $z \sim 4$ by focusing on their slope of rest-frame ultraviolet continuum called UV spectral slope β. We analyze the sample of bright Lyman Break Galaxies (LBGs) with Subaru/$i' \leq 26.0$ in the Subaru/XMM-Newton Deep Survey field. Our detailed SED fitting analysis indicates that the LBGs with observed UV slope > -1.7, Av > 1.0, and intrinsic UV slope < -2.5 are the intrinsically active star-forming galaxies with star formation rates larger than a few $\times 10^2$ $M_\odot$ yr^{-1}. A significant fraction of the UV-selected LBGs at $z \sim 4$ is on-going active and dust obscured star-forming galaxies.

Keywords. galaxies: evolution, galaxies: formation, galaxies: high-redshift, galaxies: starburst

1. Introduction

The UV continuum spectrum of star-forming galaxies is characterized by the spectral index β in the form of $f_\lambda \propto \lambda^\beta$ (Calzetti *et al.* 1994). The UV spectral slope β is a useful probe for the physical quantities of star-forming galaxies such as the dust attenuation, the age of galaxy, the Star Formation History (SFH), and the metallicity. The UV spectral slope β is one of the most important information for star-forming galaxies.

Typical β values for given redshift and UV luminosity are individually studied in the previous literature, and it is considered that the star-forming galaxies, which are at higher redshift and/or have fainter UV Luminosity, tend to have bluer β values. In particular, because the β value varies mainly with the dust attenuation value, the "β–redshift trend" is interpreted as the dust attenuation history. Moreover recent Atacama Large Millimeter/submillimeter Array observations have revealed the dust properties of high redshift star-forming galaxies through the relation between the ratio of IR to UV (IRX) and the UV spectral slope β. In order to interpret these results, it is very important to investigate the detailed relation between the UV spectral slope β and the stellar population which is hidden by dust attenuation.

2. Data and Analysis

In this work, we select Lyman Break Galaxies (LBGs) at $z \sim 4$ in the Subaru/XMM-Newton Deep Survey (SXDS) field which is partially covered by other surveys. We make the multi-band photometry catalog which includes Subaru/*B*, *V*, *R*, *i'*, *z'*, *updated-z'*

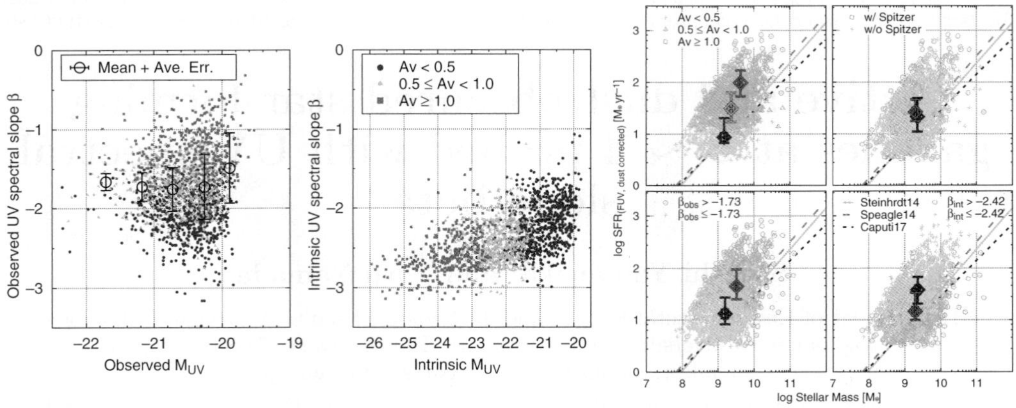

Figure 1. Left: Observed β vs. UV magnitude. Middle: Intrinsic (dust-corrected) β vs. UV magnitude. Right: Dust-corrected SFR vs. stellar mass for each sub-sample.

(Furusawa *et al.* 2008, 2016), UKIRT/*J, H, K* (Lawrence *et al.* 2007), HST/*F125W, F160W* (Grogin *et al.* 2011, Koekemoer *et al.* 2011), Spitzer/3.6μm, and 4.5μm (Ashby *et al.* 2013). By using our catalog, we calculate the observed β value, and conduct the SED fitting analysis for estimating the physical quantities. From the best-fit quantities, we estimate the *intrinsic* (dust-corrected) UV spectral slope β, the intrinsic UV magnitude, and the intrinsic Star Formation Rate (SFR).

3. Results and Discussion

The left panel of Fig. 1 shows the observed β–M_{UV} distribution, and the mean observed β value shows the weak correlation with the observed UV magnitude in the range of $-22.0 < M_{UV} < -20.0$. The middle panel of Fig. 1 shows the *intrinsic* (dust-corrected) β–M_{UV} distribution, and the intrinsic β value seems to slightly increase with the intrinsic UV magnitude. The right panel of Fig. 1 shows the distribution of the dust-corrected SFR and stellar mass for some sub-samples. The figure shows that the most intense star-forming galaxies in our sample have SFR $\gtrsim$ a few $\times\, 10^2\ M_\odot\ yr^{-1}$, and they are the objects with Av $\gtrsim 1.0$, $\beta_{\rm obs} > -1.7$, and $\beta_{\rm int} < -2.5$. Due to the dusty star-forming population, we see the flat β_{obs}–$M_{\rm UV,obs}$ distribution. Our analysis indicates that a significant fraction of $z \sim 4$ LBGs are highly dust attenuated. This population has not been recognized very well in the previous analysis but is important in understanding the early phase of galaxy formation possibly linking the typical blue LBGs and the further very red sub-mm selected galaxies.

References

Ashby, M. L. N., Willner, S. P., Fazio, G. G., Huang, J.-S., Arendt, R., Barmby, P., Barro, G., Bell, E. F., *et al.* 2013, *ApJ*, 769, 80

Calzetti, D., Kinney, A. L., & Storch-Bergmann, T. 1994, *ApJ*, 429, 582

Calzetti, D., Armus, L., Bohlin, R. C., Kinney, A. L., Koornneef, J., & Storchi-Bergmann, T. 2000, *ApJ*, 533, 682

Furusawa, H., Kosugi, G., Akiyama, M., Takata, T., Sekiguchi, K., Tanaka, I., Iwata, I., Kajisawa, M., *et al.* 2008, *ApJS*, 176, 1

Furusawa, H, Kashikawa, N, Kobayashi, M. A. R., Dunlop, J. S., Shimasaku, K., Takata, T., Sekiguchi, K., Naito, Y., *et al.* 2016, *ApJ*, 822, 46

Grogin, N. A., Kocevski, D. D., Faber, S. M., Ferguson, H. C., Koekemoer, A. M., Riess, A. G, Acquaviva, V., Alexander, D. M., *et al.* 2011, *ApJS*, 197, 35
Koekemoer, A. M., Faber, S. M., Ferguson, H. C., Grogin, N. A., Kocevski, D. D., Koo, D. C., Lai, K., Lotz, J. M., *et al.* 2011, *ApJS*, 197, 36
Lawrence, A., Warren, S. J., Almaini, O., Edge, A. C., Hambly, N. C., Jameson, R. F., Lucas, P., Casali, M., *et al.* 2007, *MNRAS*, 379, 1599

Author index

Abdurro'uf – 55
Abe, M. – 240
Abramson, L. E. – 50
Acquaviva, V. – 88
Akiyama, M. – 55
Albrecht, M. – 78
Ali, S. – 264
Arata, S. – 240, 261
Argudo-Fernández, M. – 60
Asano, R. S. – 152, 211, 312
Ashby, M. L. N. – 162

Baba, J. – 266
Baes, M. – 26, 65, 114
Beck, M. – 99
Berta, S. – 83
Béthermin, M. – 12
Bilicki, M. – 109
Binggeli, C. – 211, 226
Blaizot, J. – 307
Bonfini, P. – 162
Boquien, M. – 60, 70, 129, 201
Bouché, N. – 307
Bouwens, R. J. – 216
Brammer, G. B. – 50
Bremer, M. N. – 264
Brinchmann, J. – 235
Brown, M. J. I. – 21
Buat, V. – 39, 70, 129, 201, 279
Burgarella, D. – 39, 129, 201, 216, 279

Caminha, G. B. – 83
Capak, P. – 12
Cassata, P. – 12, 83
Ceverino, D. – 245
Chang, R. – 60, 275
Chang, Y. – 292
Chen, K.-J. – 253, 257, 320
Chen, L.-H. – 253
Chiang, Y-K. – 231
Christensen, L. – 211
Ciesla, L. – 129
Concas, A. – 83
Cooper, E. M. – 266
Corre, D. – 70, 129, 201
Cresci, G. – 83
Croton, D. – 119
Czerny, B. – 297

Daddi, E. – 44
Dai, Y. S. – 167
De Breuck, C. – 283
D'Eugenio, C. – 44
Dickinson, H. – 99
Dobbels, W. – 114
Donahue, M. – 271
Duncan, K. J. – 21

Egami,E. – 231
Egusa, F. – 266
Enia, A. – 83
Ettori, S. – 271

Fadda, D. – 83
Faisst, A. – 12
Fan, L. – 143, 268
Fanciullo, L. – 196
Fontanot, F. – 124
Fortson, L. – 99
Franceschini, A. – 83
Fritz, J. – 83
Fu, J. – 273
Fujimoto, S. – 231
Fujita, Y. – 271
Furlanetto, C. – 318

Galliano, F. – 138
Gawiser, E. – 134
Gobat, R. – 44
Goto, T. – 231

Han, Y. – 143, 268
Han, Z. – 35, 143
Hao, L. – 60
Harikane, Y. – 206, 231, 307
Hasegawa, K. – 231
Hashimoto, T. – 211, 221, 226, 285, 309
Hatsukade, B. – 211
Higuchi, R. – 231
Hirashita, H. – 216
Hopkins, P. F. – 226
Hosokawa, T. – 287

Iani, E. – 83
Imanishi, M. – 294
Inoue, A. K. – 129, 152, 211, 226, 231, 285, 309, 312
Ishikawa, S. – 231
Ito, K. – 17, 231
Iyer, K. G. – 134

Jin, S. – 44
Jones, T. – 50

Kajisawa, M. – 292, 294
Kamraj, N. – 21
Kang, X. – 275
Kashikawa, N. – 17, 231, 294
Kawaguchi, T. – 292
Kelly, P. – 50
Kemper, F. – 196
Khochfar, S. – 261
Kida, A. – 277
Kikuta, S. – 277
Kim, H. K. – 279
Kim, S. J. – 281
Kirk, M. – 21
Knudsen, K. K. – 268
Kobayashi, M. A. R. – 294
Koda, J. – 266
Kohno, K. – 211
Kojima, T. – 231
Komiyama, Y. – 231, 292
Konno, A. – 231
Kouroumpatzakis, K. – 162
Krakowski, T. – 109
Kubo, M. – 17

Landt, H. – 21
Le Fèvre, O. – 12
Lee, M. – 211, 309
Lee, M. M. – 283
Li, L. – 35
Li, Y. – 261
Liu, A. – 83
Luo, Y. – 289

Ma, X. – 226
Mahajan, S. – 304
Małek, K. – 39, 70, 109, 297
Malkan, M. – 279
Mancini, C. – 83
Matsuda, Y. – 211, 277
Matsuhara, H. – 279
Matsuo, H. – 211, 309
Matsuoka, Y. – 292
Mawatari, K. – 211, 226, 285, 309, 312
Medezinski, E. – 271
Meneghetti, M – 271
Mercurio, A. – 83
Mitani, H. – 287
Mitchell, P. D. – 307
Miyaji, T. – 172
Morishita, T. – 50
Moriwaki, K. – 249
Morselli, L. – 83
Motohara, K. – 316
Murphy, E. J. – 177

Nagamine, K. – 240, 261, 289
Nagao, T. – 283, 292, 294
Nagasaki, S. – 312
Nagata, T. – 152
Nanayakkara, T. – 235
Nersesian, A. – 65
Niida, M. – 292
Nishida, K. Y. – 152
Noboriguchi, A. – 292
Nobuhara, K. – 292
Nozawa, T. – 312

Ogura, K. – 294
Oi, N. – 279
Okabe, N. – 271
Okamoto, T. – 211
Okamura, S. – 1
Omukai, K. – 287
Ono, Y. – 231
Onoue, M. – 292
Ota, K. – 211
Ouchi, M. – 231, 307
Overzier, R. – 231

Panda, S. – 297
Pearson, C. – 279
Pearson, W. J. – 104
Pérez-González, P. G. – 83
Petrillo, C. E. – 104
Pforr, J. – 157
Phillipps, S. – 264
Pollo, A. – 109
Popesso, P. – 83
Postman, M. – 271
Propris, R. D. – 264
Puglisi, A. – 74

Rasia, E. – 271
Ricci, C. – 21
Rodighiero, G. – 83
Rodríguez-Muñoz, L. – 83
Roehlly, Y. – 39, 129
Romano-Díaz, E. – 299
Rosati, P. – 83

Sabatini, G. – 83
Sadoun, R. – 299
Salas, H. – 129
Salim, S. – 279
Sano, K. – 302
Scarlata, C. – 99
Schaerer, D. – 12
Scicluna, P. – 196
Sell, P. H. – 162
Shen, S. – 60, 147, 201
Shibuya, T. – 211, 231
Shimizu, I. – 211, 226

Shirley, R. – 39
Shlosman, I. – 289, 299
Shobhana, D. – 304
Silverman, J. – 12
Simpson, J. M. – 196
Singh, A. – 304
Sinha, M. – 119
Siudek, M. – 109
Śniegowska, M. – 297
Solarz, A. – 109
Solís-Castillo, B. – 78
Spitler, L. – 316
Srinivasan, S. – 196
Stiavelli, M. – 50
Strazzullo, V. – 44
Sugahara, Y. – 307
Sunaga, K. – 211, 309

Takagi, T. – 279
Takeuchi, T. – 109
Takeuchi, T. T. – 152, 211, 312
Tamura, Y. – 211, 285, 309, 312
Tanaka, M. – 231
Taniguchi, A. – 309
Taniguchi, Y. – 211, 294
Tanimoto, A. – 314
Terao, Y. – 316
Terashima, Y. – 292
Tergolina, M. – 318
Toba, Y. – 292, 294
Toshikawa, J. – 17, 231
Tozzi, P. – 83
Trayford, J. – 104
Trenti, M. – 50
Treu, T. – 50
Trevisan, M. – 318
Triani, D. – 119

Uchiyama, H. – 17
Ueda, Y. – 292
Umehata, H. – 211
Umetsu, K. – 271

van der Tak, F. F. S. – 104
Verstocken, S. – 65
Viaene, S. – 65
Vikaeus, A. – 187, 226
Vulcani, B. – 50

Walmsley, M. – 99
Wang, L. – 104
Wang, W.-C. – 320
Wang, X. – 50
Willner, S. P. – 162

Yajima, H. – 240, 261
Yamada, T. – 323
Yamanaka, S. – 285, 323
Yamashita, T. – 292
Yan, L. – 12
Yin, J. – 60, 147
Yoshida, N. – 211, 287
Yuan, F. – 60
Yuan, F.-T. – 201

Zackrisson, E. – 187, 226
Zackrissonm, E. – 211
Zanella, A. – 83
Zezas, A. – 162
Zhang, F. – 35, 275
Zheng, Z. – 299